Patai's guide to the

chemistry of functional groups

THE CHEMISTRY OF FUNCTIONAL GROUPS

A series of advanced treatises under the general editorship of Professor Saul Patai

The chemistry of alkenes (2 volumes)
The chemistry of the carbonyl group (2 volumes)
The chemistry of the ether linkage
The chemistry of the amino group
The chemistry of the nitro and nitroso groups (2 parts)
The chemistry of carboxylic acids and esters
The chemistry of the carbon–nitrogen double bond
The chemistry of amides
The chemistry of the cyano group
The chemistry of the hydroxyl group (2 parts)
The chemistry of the azido group
The chemistry of acyl halides
The chemistry of the carbon–halogen bond (2 parts)
The chemistry of the quinonoid compounds (2 volumes, 4 parts)
The chemistry of the thiol group (2 parts)
The chemistry of the hydrazo, azo and azoxy groups (2 parts)
The chemistry of amidines and imidates
The chemistry of cyanates and their thio derivatives (2 parts)
The chemistry of diazonium and diazo groups (2 parts)
The chemistry of the carbon–carbon triple bond (2 parts)
The chemistry of ketenes, allenes and related compounds (2 parts)
The chemistry of the sulphonium group (2 parts)
Supplement A: The chemistry of double-bonded functional groups (2 parts)
Supplement B: The chemistry of acid derivatives (2 parts)
Supplement C: The chemistry of triple-bonded functional groups (2 parts)
Supplement D: The chemistry of halides, pseudo-halides and azides (2 parts)
Supplement E: The chemistry of ethers, crown ethers, hydroxyl groups and their sulphur analogues (2 parts)
Supplement F: The chemistry of amino, nitroso and nitro compounds and their derivatives (2 parts)
The chemistry of the metal–carbon bond (4 volumes)
The chemistry of peroxides
The chemistry of organic selenium and tellurium compounds (2 volumes)
The chemistry of the cyclopropyl group (2 parts)
The chemistry of sulphones and sulphoxides
The chemistry of organic silicon compounds (2 parts)
The chemistry of enones

$$\sum_{X=A}^{Z} C{-}X$$

Patai's guide to the
chemistry of functional groups

SAUL PATAI

The Hebrew University, Jerusalem

1989

JOHN WILEY & SONS

CHICHESTER · NEW YORK · BRISBANE · TORONTO · SINGAPORE

An Interscience ® Publication

Library of Congress Cataloging-in-Publication Data:

Patai, Saul.
Patai's guide to the chemistry of functional groups/Saul Patai.
p. cm.—(The Chemistry of functional groups)
'An Interscience publication.'
Bibliography: p.
ISBN 0 471 91526 2
1. Chemistry , Organic. 2. Chemistry of functional groups.
I. Title. II. Title: Guide to the chemistry of functional groups.
III. Series.
QD251.2.P34 1989 88-20707
547—dc19 CIP

British Library Cataloguing in Publication Data:

Patai, Saul
Patai's guide to the chemistry of
functional groups.
1. Organic compounds. Functional groups.
Chemical reactions
I. Title II. Series
547
ISBN 0 471 91256 2

Phototypesetting by Thomson Press (India) Limited, New Delhi
Printed in Great Britain by Bath Press Ltd., Avon.

Contents

The Editor and some Functional Groups

Foreword

The Chemistry of Alkenes, the first volume of the 'The Chemistry of Functional Groups' series, was published in 1964. By the end of 1988, the series consisted of 40 titles bound in 59 volumes, which contained over 700 chapters and a grand total of over 43 000 rather closely printed pages.

Prices of publishing being what they are, by now very few research chemists can afford to buy the volumes of the Series for their private use and must turn to public and institutional libraries. Hence I am very grateful to Professor H. C. Brown for his suggestion of publishing a very strongly condensed version of the Series. This should present enough of the subject matter of the various chapters published in the Series to enable the reader to decide whether these chapters are relevant and useful for his or her purposes, and worthwhile perusing in full.

I hope that this volume answers this need. I will be very grateful to readers who would communicate to me any mistakes, omissions or misrepresentations relating to this volume.

Most of this book was written during a Sabbatical leave at the Department of Chemistry, University College London. I would like to thank my colleagues, Professors A. G. Davies, M. L. McGlashan and J. Ridd, for their hospitality.

Jerusalem/London SAUL PATAI

SECTION I

A. Organization of the book

Following a short explanatory section, the titles and contents of all volumes which appeared between 1964 and 1986 are presented.

The main body of the book gives a chapter-by-chapter account of all the material published in the various volumes of the Series. From each chapter, the principal parts of the Table of Contents are reproduced (including the title of the chapter, the names of the authors and a list of the main sections) with the original pagination (in parentheses) so as to give the reader an idea of the size of the chapter and the importance which is attributed to each main section in that chapter. This is followed by a brief description of the subject matter of the chapter, ending with a statement as to the total number of references in the chapter and the period which they cover.

Two more paragraphs are usually added to each summary. The first of these, marked **Complementary**, consists of a list of chapters in other volumes of the Series which deal with the same or a very similar subject matter relating to the *same functional group*. For example, eliminations involving double bonds are discussed in Vol. **1**, Chapters 2 and 3 and also Vol. **21**, Chapter 4.

The paragraph marked **Relevant** enumerates all chapters in the Series which deal with the same type of material, but relating to *different* functional groups (e.g. rearrangements of carbon—carbon double bonds are discussed in Vol. **1**, Chapter 7; while Vol. **2**, Chapter 15; Vol. **3**, Chapter 15; Vol. **4**, Chapter 10; etc., all deal with rearrangements involving other functional groups, and these are considered to be 'relevant' to Vol. **1**, Chapter 7.

Indexes

The book contains an Author Index and a Chapter/Subject Index (this will refer to the different chapters both by the name of the functional group it discusses and also by one or more key-words characterizing the chapter, e.g. the chapter on rearrangements involving amino groups will be indexed both as 'Rearrangements (amino group)' and as 'Amino group (rearrangements).'

Appendix

This contains the titles and chapters of volumes which appeared in 1987 and 1988, with some very brief explanations.

B. Symbols used

'Series' = 'The Chemistry of Functional Groups' series. Bold numbers **1**, **2**, **3** are the volume numbers in the Series, numbered in chronological order of appearance as in the list below.

In this context, each book which has separate (author and subject) indexes of its own is considered as 'a volume'. Thus **1** and **9**, **2** and **10** and also **6** and **7** are each separate volumes and numbered accordingly.

Chapter numbers (in ordinary type) are added to the volume number. Thus **5**, 6 means Chapter 6 in Volume **5**, while **7**, 4, 5, 6 specifies Chapters 4, 5 and 6 in Volume **7**. If only part of the chapter deals with the subject, the symbol(s) of the relevant sections within the chapter are added in parentheses, e.g. (II.A.3).

The following symbols, which are self-explanatory, are used as abbreviations for the various functional groups: C=C; C=O; C—O—C (ether); NH_2; COOH; COOR; NO; NO_2; C≡N; $CONH_2$; N_3; CN; OH; SH; —NH—NH—; —N=N—; —N(O)=N; C≡C.

Other less obvious symbols used are the following:

COX	acyl halides
CX	halides
Quin	quinones
Ami	amidines
X=Y	any double-bonded group, including C=C, C=O, C=N, N=N
OCN	cyanates and isocyanates
SCN	thiocyanates and isothiocyanates
N_2^+	diazonium
CN_2	diazo
C=C=O	ketenes
C=C=C	allenes
$\equiv S^+$	sulphonium
Met—C	metal—carbon
X≡Y	any triple-bonded group, including C≡C, C≡N
—O—O—	peroxide

C. Volumes and chapters of the series (1964–1986)

1. *The Chemistry of Alkenes* (Ed. S. Patai, 1964)

1. Wave mechanics and the alkene bond, C. A. Coulson and E. T. Stewart; 2. Elimination reactions in solution, H. Saunders, Jr; 3. Olefin-forming eliminations in the gas phase, A. Maccoll; 4. Alkene-forming condensation reactions, T. I. Crowell; 5. Detection and determination of the alkenes, E. J. Kuchar; 6. Alkene complexes of some transition metals, M. Cais; 7. Alkene rearrangements, K. Mackenzie; 8. Nucleophilic attacks on carbon—carbon double bonds, S. Patai and Z. Rappoport; 9. Reactions of alkenes with radicals and carbenes, J. I. G. Cadogan and M J. Perkins; 10. Allylic reactions, R. H. DeWolfe and W. G. Young; 11. Cycloaddition reactions of alkenes, R. Huisgen, R. Grashey and J. Sauer; 12. Conjugated dienes, M. Cais; 13. Cumulenes, H. Fischer; 14. Ketenes, R. N. Lacey (*See also* Vols **9, 21**).

2. *The Chemistry of the Carbonyl Group* (Ed. S. Patai, 1966)

1. General and theoretical aspects of the carbonyl group, G. Berthier and J. Serre; 2. Carbonyl-forming oxidations, C. F. Cullis and A. Fish; 3. Formation of carbonyl groups in hydrolytic reactions, P. Salomaa; 4. Formation of aldehydes and ketones from carboxylic acids and their derivatives, R. C. Fuson; 5. Formation of ketones and aldehydes by acylation, formylation and some related processes, D. P. N. Satchell and R. S. Satchell; 6. Carbonyl syntheses through organometallics, M. Cais and A. Mandelbaum; 7. Biological formation and reactions of carbonyl groups, F. Eisenberg, Jr; 8. Chemical and physical methods of analysis, J. G. Hanna; 9. Basicity of carbonyl compounds, V. A. Palm, Ü. L. Haldna and A. J. Talvik; 10. Oxidation of aldehydes by transition metals, J. Roček; 11. Reduction of carbonyl groups, O. H. Wheeler; 12. Condensations leading to double bonds, R. L. Reeves; 13. Reactions of carbonyl groups with organometallic compounds, T. Eicher; 14. Decarbonylation, W. M. Schubert and R. R. Kintner; 15. Rearrangements involving the carbonyl group, C. J. Collins and J. F. Eastham; 16. Photochemistry of ketones and aldehydes, J. N. Pitts, Jr, and J. K. S. Wan; 17. Thioketones, E. Campaigne (*See also* Vols **10, 21**).

3. *The Ether Linkage* (Ed. S. Patai, 1967)

1. Theory of the ether linkage, D. Peters; 2. Cleavage of the C—O—C bond, E. Staude and F. Patat; 3. Directive and activating effects of alkoxy and aryloxy groups in aromatic and aliphatic reactions, G. Kohnstam and D. H. L. Williams; 4. The pyrolysis of ethers, K. J. Laidler and D. J. McKenney; 5. Biological formations and reactions, K. Wallenfels and H. Diekmann; 6. Basicity and complexing ability of ethers, S. Searles, Jr, and M. Tamres; 7.

Acetals and hemiacetals, E. Schmitz and I. Eichhorn; 8. Photochemistry of the C—O—C group, D. Elad; 9. Reactions of cyclic ethers, R. J. Gritter; 10. Methods of formation of the ether linkage, H. Feuer and J. Hooz; 11. The polyethers, C. C. Price; 12. Interconversion of C—O—C, C—S—C and C—Se—C groups, V. Horak and J. Gosselck; 13. The chemistry of thioethers; differences and analogies with ethers, L. Brandsma and J. F. Arens; 14. Rearrangements of ethers, D. L. Dalrymple, T. L. Kruger and W. N. White; 15. Detection and estimation of ethers, J. S. Fritz; 16. Appendix on safety measures, N. V. Steere (*See also* Vol. **27**).

4. *The Amino Group* (Ed. S. Patai, 1968)

1. General and theoretical, R. Daudel; 2. The introduction of the amino group, M. S. Gibson; 3. Determination and characterization of amines, J. Zabicky; 4. Basicity and complex formation, J. W. Smith; 5. Directing and activating effects, G. Chuchani; 6. Substitution at an amino nitrogen, B. C. Challis and A. R. Butler; 7. Carbon—nitrogen and nitrogen—nitrogen double bond condensation reactions, P. Y. Sollenberger and R. B. Martin; 8. Cleavage of the carbon—nitrogen bond, E. H. White and D. J. Woodcock; 9. Biological formation and reactions of the amino group, B. E. C. Banks; 10. Rearrangements involving amino groups, D. V. Banthorpe; 11. Protection of the amino group, Y. Wolman; 12. Tetraaminoethylenes, D. M. Lemal (*See also* Vols **20**, **29**).

5. *The Chemistry of Carboxylic Acids and Esters* (Ed. S. Patai, 1969)

1. General and theoretical aspects of the COOH and COOR groups, M. Simonetta and S. Carra; 2. Electrochemical reactions of carboxylic acids and related processes, L. Eberson; 3. Alcoholysis, acidolysis and redistribution of esters, J. Koskikallio; 4. The formation of carboxylic acids and their derivatives from organometallic compounds, R. P. A. Sneeden; 5. Synthesis of di- and polycarboxylic acids and esters, V. F. Kucherov and L. A. Yanovskaya; 6. Acidity and hydrogen bonding of carboxyl groups, L. Eberson; 7. Introduction of COOH groups by carbonyl olefination, L. D. Bergelson and M. M. Shemyakin; 8. Rearrangement and cyclization reactions of carboxylic acids and esters, H. Kwart and K. King; 9. Substitution in the groups COOH and COOR, D. P. N. Satchell and R. S. Satchell; 10. Syntheses and uses of isotopically labelled carboxylic acids, M. Zielinski; 11. Esterification and ester hydrolysis, E. K. Euranto; 12. The decarboxylation reaction, L. W. Clark; 13. Ortho esters, E. H. Cordes; 14. Peracids and peresters, S.-O. Lawesson and G. Schroll; 15. Thiolo, thiono and dithio acids and esters, M. J. Janssen; 16. Directive and activating effects of CO_2H and CO_2R groups in aromatic and aliphatic reactions, G. Kohnstam and D. L. H. Williams; 17. Analysis of carboxylic acids and esters, T. S. Ma; 18. Biological formation and reactions of the —COOH and —COOR groups, S. Doonan (*See also* Vol. **25**).

6. *The Chemistry of the Nitro and Nitroso Groups. Part 1* (Ed. H. Feuer, 1969)

1. Theoretical aspects of the C—NO and C—NO_2 bonds, G. H. Wagnière; 2. Spectroscopy of the nitro group, C. N. R. Rao; 3. Spectroscopy of the nitroso group, C. N. R. Rao and K. R. Bhaskar; 4. The photochemistry of the nitro and nitroso groups, H. A. Morrison; 5. Methods of formation of the nitroso group and its reactions, J. H. Boyer; 6. Methods of formation of the nitro group in aliphatic and alicyclic systems, H. O. Larson; 7. Nitronic acids and esters, A. T. Nielsen; 8. Activating effects of the nitro group in nucleophilic aromatic substitutions, Th. J. de Boer and I. P. Dirkx; 9. Methods of formation of the nitramino group, its properties and reactions, G. F. Wright (*See also* Vols **7**, **29**).

7. *The Chemistry of the Nitro and Nitroso Groups, Part 2* (Ed. H. Feuer, 1970)

1. Introduction of the nitro group into aromatic systems, W. M. Weaver; 2. Directing effects of the nitro group in electrophilic and radical aromatic substitutions, T. Urbanski; 3. Activating and directing effects of the nitro group in aliphatic systems, H. H. Baer and L. Urbas; 4. Biochemistry and pharmacology of the nitro and nitroso groups, J. Venulet and R. L. VanEtten; 5. The synthesis and reactions of trinitromethyl compounds, L. A. Kaplan; 6. Polynitroaromatic addition compounds, T. N. Hall and C. F. Poranski, Jr (*See also* Vols **6**, **29**).

8. *The Chemistry of the Carbon–Nitrogen Double Bond* (Ed. S. Patai, 1969)

1. General and theoretical aspects, C. Sandorfy; 2. Methods of formation of the carbon—nitrogen double bond, S. Dayagi and Y. Degani; 3. Analysis of azomethines, D. J. Curran and S. Siggia; 4. The optical rotatory dispersion and circular dichroism of azomethines, R. Bonnett; 5. Basic and complex-forming properties, J. W. Smith; 6. Additions to the azomethine group, K. Harada; 7. Cycloaddition reactions of carbon—nitrogen double bonds, J.-P. Anselme; 8. Substitution reactions at the azomethine carbon and nitrogen atoms, R. J. Morath and G. W. Stacy; 9. *syn-anti* Isomerizations and rearrangements, C. G. McCarty; 10. Cleavage of the carbon—nitrogen double bond, A. Bruylants and E. Feytmants-de Medicis; 11. Electrochemistry of the carbon—nitrogen double bond. H. Lund; 12. Photochemistry of the carbon—nitrogen double bond, G. Wettermark; 13. Imidoyl halides, R. Bonnett; 14. Quinonediimines and related compounds, K. T. Finley and L. K. J. Tong (*See also* Vol. **21**).

9. *The Chemistry of Alkenes. Vol. 2* (Ed. J. Zabicky, 1970)

1. Nuclear magnetic resonance spectra of alkenes, V. S. Watts and J. H. Goldstein; 2. The properties of alkene carbonium ions and carbanions, H. G. Richey, Jr; 3. Alkene rearrangements, K. Mackenzie; 4. Hydrogenation of alkenes, S. Mitsui and A. Kasahara; 5. Alkene complexes of transition metals as reactive intermediates, J.-F. Biellmann, H. Hemmer and J. Levisalles; 6. Photochemistry of olefins, K. J. Crowley and P. H. Mazzocchi; 7. The mass spectrometry of the double bond, A. G. Loudon and A. Maccoll; 8. The radiolysis of olefins, G. G. Meisels; 9. Polymers containing C=C bonds, M. A. Golub; 10. Olefinic properties of cyclopropanes, M. Charton (*See also* Vols **1**, **21**).

10. *The Chemistry of the Carbonyl Group. Vol. 2* (Ed. J. Zabicky, 1970)

1. Equilibrium additions to carbonyl compounds, Y. Ogata and A. Kawasaki; 2. Oxidation of aldehydes and ketones, H. S. Verter; 3. Enolization, S. Forsen and M. Nilsson; 4. Oxocarbons and their reactions, R. West and J. Niu; 5. Mass spectrometry of carbonyl compounds, J. H. Bowie; 6. Radiation chemistry of ketones and aldehydes, G. R. Freeman (*See also* Vols **2**, **21**).

11. *The Chemistry of Amides* (Ed. J. Zabicky, 1970)

1. Molecular and electronic structure of the amide group, M. B. Robin, F. A. Bovey and H. Basch; 2. Synthesis of amides, A. L. J. Beckwith; 3. Acid–base and complexing properties of amides, R. B. Homer and C. D. Johnson; 4. Rearrangement and elimination of the amido group, J. F. Bieron and F. J. Dinan; 5. Photochemistry of the amido group, I. Rosenthal; 6. Radiation chemistry of amides, O. H. Wheeler; 7. Chemistry of imidic

compounds, O. H. Wheeler and O. Rosado; 8. The chemistry of thioamides, W. Walter and J. Voss; 9. The chemistry of the thiohydrazide group, W. Walter and K. J. Reubke; 10. The chemistry of hydrazides, H. Paulsen and D. Stoye; 11. Biological formation and reactions of the amido group, J. E. Reimann and R. U. Byerrum; 12. Directing and activating effects of the amido group, J. A. Shafer; 13. Reactions of the carboxamide group, B. C. Challis and J. A. Challis (*See also* Vol. **25**).

12. *The Chemistry of the Cyano Group* (Ed. Z. Rappoport, 1970)

1. Electronic structure in molecules, especially in cyano compounds, E. Clementi; 2. Introduction of the cyano group into the molecule, K. Friedrich and K. Wallenfels; 3. Basicity, hydrogen bonding and complex formation, J. Grundnes and P. Klaboe; 4. Detection and determination of nitriles, D. J. Curran and S. Siggia; 5. Directing and activating effects of the cyano group, W. A. Sheppard; 6. Nitrile reactivity, F. C. Schaefer; 7. Reduction of the cyano group, M. Rabinowitz; 8. Additions to the cyano group to form heterocycles, A. I. Meyers and J. C. Sircar; 9. Cyanocarbon and polycyano compounds, E. Ciganek, W. J. Linn and O. W. Webster; 10. Molecular complexes of polycyano compounds, L. R. Melby; 11. Radicals with cyano groups, H. D. Hartzler; 12. The biological function and formation of the cyano group, J. P. Ferris; 13. Syntheses and uses of isotopically labelled cyanides, L. Pichat; 14. Nitrile oxides, Ch. Grundmann; 15. Isonitriles, P. Hoffmann, D. Marquarding, H. Kliimann and I. Ugi; 16. Rearrangement reactions involving the cyano group, J. Casanova, Jr (*See also* Vol. **31**).

13. *The Chemistry of the Hydroxyl Group. Parts 1 and 2* (Ed. S. Patai, 1971)

1. Theoretical aspects of the chemistry of the hydroxyl group, R. F. W. Bader; 2. Nucleophilic attack by hydroxide and alkoxide ions, C. A. Fyfe; 3. Free radical and electrophilic hydroxylation, D. F. Sangster; 4. Formation of hydroxyl groups via oxymetallation, oxidation and reduction, I. R. L. Barker; 5. Electrochemistry of the hydroxyl group, H. Lund; 6. Detection and determination of hydroxyl groups, S. Siggia, J. G. Hanna and T. R. Stengle; 7. Acidity and inter- and intra-molecular H-bonds, C. H. Rochester; 8. Directing and activating effects, D. A. R. Happer and J. Vaughan; 9. Electrophilic attacks on the hydroxyl group, P. Salomaa, A. Kankaanperä and K. Pihlaja; 10. Oxidation and reduction of phenols, M. Lj. Mihailović and Ž. Čeković; 11. Displacement of hydroxyl groups, G. W. Brown; 12. The dehydration of alcohols, H. Knözinger; 13. Uncatalysed rearrangements involving the hydroxyl group, E. N. Marvell and W. Whalley; 14. Biological formation and reactions of the hydroxyl group, S. K. Erickson, J. Schadelin, U. Schmeling, H.-H. Schott, V. Ullrich and Hj. Staudinger; 15. Syntheses and uses of ^{18}O-labelled hydroxylic compounds, S. Oae and S. Tamagaki; 16. Photochemistry of alcohols and phenols, H.-D. Becker; 17. The radiation chemistry of the hydroxyl group, R. A. Basson; 18. Protection of the hydroxyl group, H. M. Flowers; 19. The mass spectra of hydroxyl compounds, R. G. Cooks; 20. Hydroxide–alkoxide ion equilibria and their influence on chemical reactions, J. Murto (*See also* Vol. **27**).

14. *The Chemistry of the Azido Group* (Ed. S. Patai, 1971)

1. General and theoretical aspects, A. Treinin; 2. Introduction of the azido group, M. E. C. Biffin, J. Miller and D. B. Paul; 3. Characterization and determination of organic azides, J. E. Gurst; 4. Directing and activating effects of azido groups, M. E. C. Biffin, J. Miller and D. B. Paul; 5. Decomposition of organic azides, R. A. Abramovitch and E. P. Kyba; 6. Azides as synthetic starting materials, T. Sheradsky; 7. Rearrangements involving azido

groups, D. V. Banthorpe; 8. Photochemistry of the azido group, A. Reiser and H. M. Wagner; 9. Acyl azides, W. Lwowski; 10. The chemistry of vinyl azides, G. Smolinsky and C. A. Pryde (*See also* Vol. **32**).

15. *The Chemistry of Acyl Halides* (Ed. S. Patai, 1972)

1. General and theoretical aspects, M. Simonetta and P. Beltrame; 2. Preparation of acyl halides, M. F. Ansell; 3. Detection, determination and characterization, H. Weiler-Feilchenfeld; 4. Acid–base behaviour and complex formation, D. P. N. Satchell and R. S. Satchell; 5. Directing and activating effects of COX groups, P. H. Gore; 6. Mechanisms of substitution at the COX group, A. Kivinen; 7. Reduction, O. H. Wheeler; 8. Rearrangements involving acyl halides, D. V. Banthorpe and B. V. Smith; 9. Photochemistry and radiation chemistry of carbonyl halides, U. Schmidt and H. Egger; 10. Biological reactions of carbonyl halides, S. Cohen; 11. Thiocarbonyl halides, K. T. Potts and C. Sapino; 12. Chloroformate esters and related compounds, D. N. Kevill; 13. The acyl hypohalites, D. D. Tanner and N. J. Bunce (*See also* Vol. **25**).

16. *The Chemistry of the Carbon–Halogen Bond. Parts 1 and 2* (Ed. S. Patai, 1973)

1. General and theoretical aspects of the carbon—halogen bond, G. H. Wagnière; 2. Structural chemistry of the C—X bond, J. Trotter; 3. Analysis of organic halogen compounds, J. Zabicky and S. Ehrlich-Rogozinski; 4. Mass spectrometry and the carbon—halogen bond, A. G. Loudon; 5. Hydrogen-bonding and complex-forming properties, J. W. Smith; 6. Directing, activating and deactivating effects, G. Modena and G. Scorrano; 7. Heterolytic mechanisms of substitution involving carbon—halogen bonds, P. B. D. de la Mare and B. E. Swedlund; 8. Homolytic mechanisms of substitution, E. S. Huyser; 9. Elimination reactions in solution, R. A. More O'Ferral; 10. Pyrolysis reactions involving carbon—halogen bonds, K. W. Egger and A. T. Cocks; 11. Photochemistry of the C—X group, P. G. Sammes; 12. Radiation chemistry of the carbon—halogen bond, R. E. Buhler; 13. The biochemistry of carbon—halogen compounds, S. Doonan; 14. Perchloro-, perbromo- and periodo-compounds, T. Chivers; 15. Electrochemistry of the carbon—halogen bond, J. Casanova and L. Eberson. 16. Thermochemistry of organic halides, R. Shaw; 17. Rearrangements involving halides, C. Rappe (*See also* Vol. **32**).

17. *The Chemistry of the Quinonoid Compounds. Parts 1 and 2* (Ed. S. Patai, 1974)

1. Theoretical and general aspects, G. J. Gleicher; 2. The structural chemistry of quinones, J. Bernstein, M. D. Cohen and L. Leiserowitz; 3. Synthesis, R. H. Thomas; 4. Identification and determination of quinones, St. Berger and A. Rieker; 5. Mass spectra of quinones, K.-P. Zeller; 6. Quinone complexes, R. Foster and M. I. Foreman; 7. Quinones as oxidants and dehydrogenating agents, H.-D. Becker; 8. Rearrangements of quinones, H. W. Moore and R. J. Wikholm; 9. Photochemistry of quinones, J. M. Bruce; 10. Radiation chemistry of quinones, J. H. Fendler and E. J. Fendler; 11. Fragmentation reactions of quinones, P. Hodge; 12. Syntheses and uses of isotopically labelled quinones, M. Zielinski; 13. Biological reactions of quinones, R. Bentley and I. M. Campbell; 14. Electrochemistry of quinones, J. Q. Chambers; 15. Polymeric quinones, A. S. Lindsey; 16. Non-benzenoid quinones, T. A. Turney; 17. The addition and substitution chemistry of quinones, K. T. Finley; 18. Quinone methides, R. Gompper and H.-U. Wagner.

18. *The Chemistry of the Thiol Group. Parts 1 and 2* (Ed. S. Patai, 1974)

1. General and theoretical aspects of the thiol group, I. G. Csizmadia; 2. Structural chemistry of the thiol group, I. C. Paul; 3. Thermochemistry of thiols, R. Shaw; 4. Preparation of thiols, J. L. Wardell; 5. Detection and determination of thiols, A. Fontana and C. Toniolo; 6. The mass spectra of thiols, C. Lifshitz and Z. V. Zaretskii, 7. The optical rotatory dispersion and circular dichroism of thiols, C. Toniolo and A. Fontana; 8. Acidity and hydrogen bonding, M. R. Crampton; 9. Directing and activating effects, G. Maccagnani and G. Mazzanti; 10. Photochemistry of thiols, A. R. Knight; 11. The radiation chemistry of thiols, J. E. Packer; 12. Synthetic uses of thiols, R. K. Olsen and J. O. Currie, Jr; 13. Biochemistry of the thiol group, A. L. Fluharty; 14. Protection of the thiol group, Y. Wolman; 15. Rearrangements involving thiols, T. Sheradsky; 16. Thiols as nucleophiles, M. E. Peach; 17. Oxidation of thiols, G. Capozzi and G. Modena; 18. The synthesis and uses of isotopically labelled thiols, A. Lapidot and C. S. Irving (*See also* Vol. **27**).

19. *The Chemistry of the Hydrazo, Azo and Azoxy Groups. Parts 1 and 2* (Ed. S. Patai, 1975)

1. Electronic structures of the azo, azoxy and hydrazo groups, M. B. Robin; 2. Structural chemistry, R. Allmann; 3. Thermochemistry of hydrazo, azo and azoxy groups, R. Shaw; 4. Preparative procedures, J. W. Timberlake and J. C. Stowell; 5. Mass spectra of hydrazo, azo and azoxy compounds, S. W. Tam; 6. Chiroptical properties of the azo and the azoxy chromophores, G. Snatzke and F. Snatzke; 7. Basicity, hydrogen bonding and complex formation involving hydrazo, azo and azoxy groups, P. J. Krueger; 8. Directing and activating effects, D. A. R. Happer and J. Vaughan; 9. The transition metal chemistry of azo compounds, M. I. Bruce and B. L. Goodall; 10. Radiation chemistry of hydrazo and azo compounds, A. Horowitz and L. A. Rajbenbach; 11. Formation and fragmentation of cyclic azo compounds, K. Mackenzie; 12. The electrochemistry of azoxy, azo and hydrazo compounds, F. G. Thomas and K. G. Boto; 13. Biological formation and reactions of hydrazo, azo and azoxy groups, T. Miyadera; 14. Oxidation and synthetic uses of hydrazo, azo and azoxy compounds, B. T. Newbold; 15. Reduction and synthetic uses of hydrazo, azo and azoxy compounds, B. T. Newbold; 16. Ionic reactions involving hydrazo, azo and azoxy groups, A. F. Hegarty; 17. Syntheses and uses of isotopically labelled hydrazo, azo and azoxy compounds, A. J. Dolenko and E. Buncel; 18. Rearrangements of hydrazo, azoxy and azo compounds, R. A. Cox and E. Buncel; 19. Radical reactions of azo, hydrazo and azoxy compounds, G. Koga, N. Koga and J.-P. Amselme; 20. The photochemistry of the hydrazo, azo and azoxy groups, R. J. Drewer; 21. Conformational analysis of hydrazines, Y. Shvo (*See also* Vols **21**, **23**).

20. *The Chemistry of Amidines and Imidates* (Ed. S. Patai, 1975)

1. General and theoretical aspects of amidines and imidic acid derivatives, G. Häfelinger; 2. Constitution, configurational and conformational aspects and chiroptical properties of imidic acid derivatives, G. Fodor and B. A. Phillips; 3. Detection and determination of imidic acid derivatives, W. H. Pritchard; 4. Rearrangements involving imidic acid derivatives, C. G. McCarty and L. A. Garner; 5. The electrochemistry of imidic esters and amidines, H. Lund; 6. Biological reactions and pharmaceutical uses of imidic acid derivatives, R. J. Grout; 7. Preparation and synthetic uses of amidines, J.-A. Gautier, M. Miocque and C. C. Farnoux; 8. Kinetics and mechanisms of reactions of amidines, R. H. De Wolfe; 9. Imidates including cyclic imidates, D. G. Neilson; 10. The chemistry of amidrazones, K. M. Watson and D. G. Neilson; 11. Estimation of the thermochemistry of

imidic acid derivatives, R. Shaw; 12. Complex formation, H-bonding and basicity of imidic acid derivatives, J. Sevcik and F. Grambal (*See also* Vol. **21**, Ch. 13).

21. *Supplement A: The Chemistry of Double-Bonded Functional Groups. Parts 1 and 2* (Ed. S. Patai, 1977)

1. Dipole moments, configurations and conformations of molecules containing X=Y groups, O. Exner; 2. Liquid crystals with X=Y groups, J. P. Van Meter; 3. Thermochemistry of X=Y groups, R. Shaw; 4. Mechanisms of elimination and addition reactions involving the X=Y groups, A. F. Cockerill and R. G. Harrison; 5. The electrochemistry of X=Y groups, A. J. Fry and R. G. Reed; 6. 1,3-Dipolar cycloadditions involving X=Y groups, G. Bianchi, C. De Micheli and R. Gandolfi; 7. Reactions of carbenes with X=Y groups, A. P. Marchand; 8. The formation of unsaturated groups by heterolytic fragmentation, K. B. Becker and C. A. Grob; 9. Electrophilic additions to carbon—carbon double bonds, G. H. Schmid and D. G. Garratt; 10. The olefin metathesis reaction, N. Calderon; 11. Oxidation of C=C and C=N groups, P. M. Henry and G. L. Lange; 12. Transition metal catalysed carbonylation of olefins, J. K. Stille and D. E. James; 13. Imidines and diamidides (1, 3, 5-triazapentadienes), J. A. Elvidge and N. R. Barot (*See also* Vols **1**, **2**, **8**, **9**, **10**).

22. *The Chemistry of the Cyanates and Their Thio Derivatives. Parts 1 and 2* (Ed. S. Patai, 1977)

1. Electronic structure of the cyanato and thiocyanato groups—ground state and excited states, K. Wittel, J. L. Meeks and S. P. McGlynn; 2. Structural chemistry of the cyanates and their thio derivatives, I. Hargittai and I. C. Paul; 3. Stereochemical and conformational aspects of cyanates and related groups, G. C. Corfield; 4. The optical rotatory dispersion and circular dichroism of the cyanates and related groups, C. Toniolo; 5. Detection and determination of cyanates, isocyanates, isothiocyanates and thiocyanates, D. A. Ben-Efraim; 6. Thermochemistry of cyanates, isocyanates and their thio derivatives, R. Shaw; 7. Mass spectra of cyanates, isocyanates and related compounds, K. A. Jensen and S. Schroll; 8. Hydrogen bonding and complex formation, D. Hadži and S. Milićev; 9. The electrochemistry of cyanates and related compounds, O. Hammerich and V. D. Parker; 10. Photochemistry of cyanates and related groups, R. Jahn and U. Schmidt; 11. Radiation chemistry of organic compounds containing OCN, CNO, SCN and CNS groups, A. Horowitz; 12. Kinetics and mechanisms of reactions of cyanates and related compounds, D. E. Giles; 13. Acyl and thioacyl derivatives of isocyanates, thiocyanates and isothiocyanates, O. Tsuge; 14. Syntheses and uses of isotopically labelled cyanates and related groups, A. Ceccon and U. Miotti; 15. Pyrolytic reactions of cyanates and related compounds, N. Barroeta; 16. Syntheses and preparative applications of cyanates (esters of cyanic acid), K. A. Jensen and A. Holm; 17. Syntheses and preparative applications of isocyanates, R. Richter and H. Ulrich; 18. Syntheses and preparative applications of thiocyanates, R. G. Guy; 19. Selenocyanates and related compounds, E. Bulka; 20. Biological formation and reactions of cyanates, S. Cohen and E. Oppenheimer; 21. Syntheses and reactions of isocyanide dihalides, H. Ulrich and R. Richter; 22. The chemistry of the—NCS group, L. Drobnica, P. Kristián and J. Augustín (*See also* Vol. **32**).

23. *The Chemistry of Diazonium and Diazo Groups. Parts 1 and 2* (Ed. S. Patai, 1978)

1. General and theoretical aspects of the diazonium and diazo groups, J. B. Moffat; 2. Diazonium–diazo equilibrium, V. Štěrba; 3. Structural chemistry, S. Sorriso; 4. Thermo-

chemistry of diazo compounds and organic azides, R. Shaw; 5. Detection and determination of diazo and diazonium groups, D. A. Ben-Efraim; 6. Basicity, acidity and hydrogen bonding, J. F. McGarrity; 7. Complex formation, H. M. Niemeyer; 8. Synthetic applications of diazonium ions, D. S. Wulfman; 9. Photochemistry of the diazonium and diazo groups, W. Ando; 10. The electrochemistry of the diazo and diazonium groups, A. J. Fry; 11. The influences of the diazo and diazonium groups, E. S. Lewis; 12. Kinetics and mechanisms of reactions involving diazonium and diazo groups, A. F. Hegarty; 13. Rearrangements involving the diazo and diazonium groups, D. Whittaker; 14. Preparation of diazonium groups, K. Schank; 15. Synthesis of diazoalkanes, M. Regitz; 16. Preparation and uses of isotopically labelled diazonium and diazo compounds, P. J. Smith and K. C. Westaway; 17. Carbonyl, phosphoryl and sulphonyl diazo compounds, M. Regitz; 18. Synthetic application of diazoalkanes, diazocyclopentadienes and diazoazacyclopentadienes, D. S. Wulfman, G. Linstrumelle and C. F. Cooper (*See also* Vol. **31**).

24. *The Chemistry of the Carbon–Carbon Triple Bond. Parts 1 and 2* (Ed. S. Patai, 1978)

1. General and theoretical aspects of the acetylenic compounds, M. Simonetta and A. Gavezzotti; 2. The structural chemistry of the C≡C bond, J. L. Hencher; 3. Thermochemistry of acetylenes, R. Shaw; 4. Acidity, hydrogen bonding and complex formation, A. C. Hopkinson; 5. Detection and determination of alkynes, K. A. Connors; 6. Mass spectrometry of acetylenes, C. Lifshitz and A. Mandelbaum; 7. Applications of acetylenes in organic synthesis, P. F. Hudrlik and A. M. Hudrlik; 8. Electrophilic additions to carbon—carbon triple bonds, G. H. Schmid; 9. Propargylic metalation, J. Klein; 10. Rearrangements involving acetylenes, F. Théron, M. Verny and R. Vessière; 11. Cycloadditions and cyclizations involving triple bonds, J. Bastide and O. Henri-Rousseau; 12. Photochemistry of the C≡C bond, J. D. Coyle; 13. Synthetic acyclic polyacetylenes, W. D. Huntsman; 14. Natural acetylenes, Sir E. R. H. Jones and V. Thaller; 15. Cyclic acetylenes, M. Nakagawa; 16. Proximity interactions of acetylenes, S. Misumi and T. Kaneda; 17. The electrochemistry of the carbon—carbon triple bond, J. H. P. Utley and R. Lines; 18. The preparation of acetylenes and their protection, D. A. Ben-Efraim; 19. Nucleophilic attacks on acetylenes, J. I. Dickstein and S. I. Miller; 20. Synthesis and uses of isotopically labelled acetylenes, J. C. Lavalley and J. Saussey (*See also* Vol. **31**).

25. *Supplement B: The Chemistry of Acid Derivatives. Parts 1 and 2* (Ed. S. Patai, 1970)

1. Recent advances in the theoretical treatment of acid derivatives, I. G. Csizmadia, M. R. Peterson, C. Kozmuta and M. A. Robb; 2. Thermochemistry of acid derivatives, R. Shaw; 3. Chiroptical properties of acid derivatives, R. Håkansson; 4. Mass spectra of acid derivatives, S. W. Tam; 5. Complexes of acid anhydrides, R. Foster; 6. Hydrogen bonding in carboxylic acids and derivatives, D. Hadži and S. Detoni; 7. The synthesis of carboxylic acids and esters and their derivatives, M. A. Ogliaruso and J. F. Wolfe; 8. The chemistry of lactones and lactams, G. V. Boyd; 9. The chemistry of orthoamides of carboxylic acids and carbonic acid, W. Kantlehner; 10. Detection and determination of acid derivatives, W. H. Prichard; 11. The photochemistry of organic acids, esters, anhydrides, lactones and imides, R. S. Givens and N. Levi; 12. Radiation chemistry of acids, esters, anhydrides, lactones and lactams, O. I. Mícić and O. S. Gal; 13. The electrochemistry of carboxylic acids and derivatives; cathodic reductions, J. P. Coleman; 14. Decarbonylation reactions of acid halides and aldehydes by chlorotris(triphenylphosphine)rhodium(I), M. C. Baird; 15. Pyrolysis of acids and their derivatives, R. Taylor; 16. Transcarboxylation reactions of salts of aromatic carboxylic acids, J. Ratuský; 17. Micellar effects upon

deacylation, C. A. Bunton and L. S. Romsted; 18. The chemistry of thio acid derivatives, J. Voss; 19. The synthesis of lactones and lactams, J. F. Wolfe and M. A. Ogliaruso (*See also* Vols **5**, **11**, **15**).

26. *The Chemistry of Ketenes, Allenes and Related Compounds. Parts 1 and 2* (Ed. S. Patai, 1980)

1. Theoretical methods and their application to ketenes and allenes, C. E. Dykstra and H. F. Schaefer, III; 2. Structural chemistry, W. Runge; 3. Chirality and chiroptical properties, W. Runge; 4. The thermodynamics of allenes, ketenes and related compounds, R. L. Deming and C. A. Wulff; 5. Detection, determination and identification of allenes and ketenes, J. W. Munson; 6. The generation of neutral and ionized allenes, cumulenes and heterocumulenes by electron impact, H. Schwarz and C. Köppel; 7. The preparation of ketenes, R. S. Ward; 8. Synthetic uses of ketenes and allenes, W. T. Brady; 9. Kinetics and mechanisms (excepting cycloadditions), P. Blake; 10. Organometallic derivatives of allenes and ketenes, J.-L. Moreau; 11. Preparation and uses of isotopically labelled allenes, M. D. Schiavelli; 12. Electrochemistry of allenes and cumulenes, D. G. Peters, W. F. Carroll, Jr, D. M. La Perrière and B. C. Willett; 13. Biological formation and reactions, C. H. Robinson and D. F. Covey; 14. Ketene *O,O*-acetals, P. Brassard; 15. Rearrangements involving allenes, W. D. Huntsman; 16. Ketene thioacetals, M. Kolb; 17. Ketene imines, M. W. Barker and W. E. McHenry; 18. Carbodiimides, Y. Wolman; 19. Methyleneketenes, R. F. C. Brown and F. W. Eastwood; 20. The preparation of allenes and cumulenes, H. Hopf (*See also* Vol. **1**, Ch. 13, 14).

27. *Supplement E: The Chemistry of Ethers, Crown Ethers, Hydroxyl Groups and their Sulphur Analogues. Parts 1 and 2* (Ed. S. Patai, 1980)

1. Synthesis of crown ethers and analogues, D. A. Laidler and L. F. Stoddart; 2. Crown ethers—complexes and selectivity, F. Vögtle and E. Weber; 3. Organic transformations mediated by macrocyclic multidentate ligands, C. L. Liotta; 4. Geometry of the ether, sulphide and hydroxyl groups and structural chemistry of macrocyclic and noncyclic polyether compounds, I. Goldberg; 5. Stereodynamics of alcohols, ethers, thio ethers and related compounds, C. H. Bushweller and M. H. Gianni; 6. Chiroptical properties of alcohols, ethers, thio ethers and disulphides, G. Gottarelli and B. Samorì; 7. The mass spectra of ethers and sulphides, C. C. Van de Sande; 8. The electrochemistry of ethers, hydroxyl groups and their sulphur analogues, T. Shono; 9. Electronic structures and thermochemistry of phenols, J. Royer, G. Bertholon, R. Perrin, R. Lamartine and M. Perrin; 10. Syntheses and uses of isotopically labelled ethers and sulphides, M. Zieliński; 11. Gas-phase thermal decomposition of simple alcohols, thiols and sulphides, R. L. Failes, J. S. Shapiro and V. R. Stimson; 12. Oxidation and reduction of alcohols and ethers, P. Müller; 13. Oxidation and reduction of sulphides, E. Block; 14. Oxiranes, M. Bartók and K. L. Láng; 15. Cyclic ethers, M. Bartók; 16. Dehydration of diols, M. Bartók and A. Molnár; 17. Enol ethers—structure, synthesis and reactions, P. Fischer; 18. Oxathiacyclanes: preparation, structure and reactions, K. Philaja and P. Pasanen; 19. Allene oxides and related species, P. J. Stang; 20. Advances in the chemistry of acetals, ketals and ortho esters, R. G. Bergstrom; 21. The photochemistry of saturated alcohols, ethers and acetals, C. von Sonntag and H.-P. Schuchmann; 22. The photolysis of saturated thiols, sulphides and disulphides, C. von Sontag and H.-P. Schuchmann; 23. Radiation chemistry of alcohols and ethers, C. von Sonntag and H.-P. Schuchmann; 24. Radiation chemistry of thiols, sulphides and disulphides, C. von Sonntag and H.-P. Schuchmann (*See also* Vols **3**, **13**, **18**).

28. *The Chemistry of the Sulphonium Group. Parts 1 and 2* (Ed. C. J. M. Stirling, 1981)

1. General and theoretical aspects, M. Simonetta and A. Gavezzotti; 2. The structural chemistry of sulphonium salts, E. F. Perozzi and I. C. Paul; 3. Analysis and determination, M. R. F. Ashworth; 4. Thermochemistry of the sulphonium group, R. Shaw; 5. Photochemistry of sulphonium compounds, J. D. Coyle; 6. Electronic spectra, G. C. Barrett; 7. Electrochemistry of the sulphonium group, J. Grimshaw; 8. Isotopically labelled sulphonium salts, L. F. Blackwell; 9. Electronic effects of the sulphonium group, J. Shorter; 10. Stereochemistry and chiroptical properties of the sulphonium group, K. K. Andersen; 11. Synthesis of sulphonium salts, P. A. Lowe; 12. Reactivity of sulphonium salts, A. C. Knipe; 13. Cyclic sulphonium salts, D. C. Dittmer and B. H. Patwardhan; 14. Organosulphur cation radicals, H. J. Shine; 15. Heterosulphonium salts, S. Oae, T. Numata and T. Yoshimura; 16. Synthetic applications of sulphonium salts and sulphonium ylides, E. Block; 17. The biochemistry of sulphonium salts, G. A. Maw.

29. *Supplement F: The Chemistry of Amino, Nitroso and Nitro Compounds and Their Derivatives. Parts 1 and 2* (Ed. S. Patai, 1982)

1. Structural chemistry, S. Sorriso; 2. Stereochemistry and conformations, M. Raban and J. Greenblatt; 3. The chemistry of ionized amino, nitroso and nitro compounds in the gas phase, H. Schwarz and K. Levsen; 4. Rearrangements involving nitroso and nitro compounds, D. I. H. Williams; 5. The spectroscopy, photophysics and photochemistry of saturated amines, A. M. Halpern; 6. Photochemistry of nitro and nitroso compounds, Y. L. Chow; 7. Radiation chemistry of amines, nitro and nitroso compounds, C. L. Greenstock; 8. Electrochemistry of nitro compounds, A. J. Fry; 9. Electrochemistry of amines, W. E. Britton; 10. Radical anion reactions of nitro compounds, N. Kornblum; 11. Direct aminations, T. Sheradsky; 12. Pyrolysis of nitrites, nitrates, nitro compounds, nitroso compounds and amines, L. Batt; 13. Nitrones and nitronic acid derivatives; their structure and their roles in synthesis, E. Breuer; 14. Nitroxides, H. G. Aurich; 15. Enamines and ynamines, G. Pitacco and E. Valentin; 16. Nitro-activated carbon acids, E. S. Lewis; 17. Gas-phase basicity and acidity of amines, D. K. Bohme; 18. Special properties of di- and poly-amines, R. W. Alder and R. B. Sessions; 19. Alkyl nitrate nitrations, H. Feuer; 20. Aminals, L. Duhamel; 21. Detection and determination of nitro and nitroso compounds, Y. Tapuhi and E. Grushka; 22. Deaminations (carbon—nitrogen bond cleavages), R. J. Baumgarten and V. A. Curtis; 23. Chiroptical properties of amino, nitroso and nitro compounds, H. E. Smith; 24. Thermochemistry of nitro compounds, amines and nitroso compounds, L. Batt and G. N. Robinson; 25. Oxidation of amines, D. H. Rosenblatt and E. P. Burrows; 26. *N*-Nitrosamines and *N*-nitrosoimines, B. C. Challis and J. A. Challis; 27. The role of Meisenheimer or σ-complexes in nitroarene–base interactions, E. Buncel; 28. Uses of isotopically labelled amino, quarternary ammonium and nitro compounds, P. J. Smith and K. C. Westaway (*See also* Vols **4**, **6**, **7**).

30. *The Chemistry of the Metal–Carbon Bond, Volume 1. The Structure, Preparation, Thermochemistry and Characterization of Organometallic Compounds* (Eds F. R. Hartley and S. Patai, 1983)

1. Structure of organometallic compounds, A. D. Redhouse; 2. Thermochemistry of organometallic compounds, G. Pilcher and H. A. Skinner; 3. Synthesis of ylide complexes, L. Weber; 4. Synthesis of transition metal–carbene complexes. H. Fischer; 5. Synthesis of transition metal–carbyne complexes, U. Schubert; 6. Synthesis of transition metal alkyl and aryl complexes, R. J. Puddephatt; 7. Synthesis of olefin and acetylene complexes of the

transition metals, G. S. Lewandos; 8. Synthesis of η^3-allyl complexes, P. Powell; 9. Synthesis of η^4-butadiene and cyclobutadiene complexes, G. Marr and B. W. Rockett; 10. Synthesis of complexes of η^5-bonded ligands, G. Marr and B. W. Rockett; 11. Synthesis of complexes of η^6-, η^7-, η^8-bonded ligands, G. Marr and B. W. Rockett; 12. Synthesis of organolanthanide and organoactinide complexes, W. J. Evans; 13. Metal atoms in organometallic synthesis, M. J. McGlinchey; 14. Analysis of organometallic compounds: determination of elements and functional groups, T. R. Crompton; 15. Analysis of organometallic compounds: titration procedures, T. R. Crompton; 16. Analysis of organometallic compounds; spectroscopic methods, T. R. Crompton; 17. Analysis of organometallic compounds: polarographic techniques, T. R. Crompton; 18. Analysis of organometallic compounds: gas chromatography, T. R. Crompton; 19. Analysis of organometallic compounds: other chromatographic techniques, T. R. Crompton; 20. Infrared and Raman spectroscopy of organometallic compounds, M. J. Taylor; 21. Multinuclear magnetic resonance methods in the study of organometallic compounds, J. A. Davies; 22. Mass spectrometry of organometallic compounds, T. R. Spalding (*See also* Vols **34**, **35**).

31. *Supplement C: The Chemistry of Triple-Bonded Functional Groups. Parts 1 and 2* (Eds S. Patai and Z. Rappoport, 1983)

1. Chiroptical properties of compounds containing triple-bonded functional groups, W. Runge; 2. Thermochemistry of the cyano and isocyano groups, L. Batt; 3. Mass spectra of cyano, isocyano and diazo compounds, K.-P. Zeller; 4. Infrared spectra of cyano and isocyano groups, I. N. Juchnovski and I. G. Binev; 5. Photoelectron spectra of cyano compounds, H. Stafast and H. Bock; 6. Radiation chemistry of triple-bonded molecules, Z. B. Alfassi; 7. Electrochemistry of the cyano group, K. Yoshida; 8. The directing and activating effects of triple-bonded groups, M. Charton; 9. Biological formation and metabolic transformations of compounds containing the cyano group, J. P. Ferris; 10. Free-radical reactions involving the C≡C group, Y. Amiel; 11. Arynes, T. L. Gilchrist; 12. Six-membered didehydroheteroarenes, H. C. van der Plas and F. Roeterdink; 13. Oxidation of triple-bonded groups, L. I. Simándi; 14. Reduction of triple-bonded groups, R. O. Hutchins and M. G. K. Hutchins; 15. Dediazoniations of arenediazonium ions and related compounds, H. Zollinger; 16. Alkenediazonium compounds, K. Bott; 17. Acidity and proton transfer of cyanocarbon acids, F. Hibbert; 18. Recent developments on nitrile oxides, nitrile sulphides and nitrile selenides, G. Bianchi, R. Gandolfi and P. Grünanger; 19. Conformation of cyano and isocyano compounds, C. A. Kingsbury; 20. Recent advances in isocyanide chemistry, H. Walborsky and M. P. Periasamy; 21. Complexation of aryldiazonium ions by polyethers, R. A. Bartsch; 22. Poly(diacetylenes) and polyyne polymers containing transition-metal atoms in the main chain, W. D. Huntsman; 23. Cyclodimerization of alkynes and reactivity of aluminium halide complexes of cyclobutadienes, H. Hogeveen and D. M. Kok; 24. Structure of triple-bonded molecules, J. B. Moffat; 25. NMR spectra of acetylenes, D. G. Morris; 26. Preparation and synthetic applications of cyano compounds, A. J. Fatiadi; 27. General and theoretical properties of triple-bonded molecules, J. B. Moffat; 28. Recent advances in the synthesis of triple-bonded groups, K. Friedrich (*See also* Vols **12**, **23**).

32. *Supplement D: The Chemistry of Halides, Pseudo-Halides and Azides. Parts 1 and 2* (Eds S. Patai and Z. Rappoport, 1983)

1. Molecular mechanics and conformation, A. Meyer; 2. Diamagnetic behaviour of compounds containing carbon—halogen bonds, R. R. Gupta; 3. The mass spectra of azides and halides, J. M. Miller and T. R. B. Jones; 4. Nuclear quadrupole resonance of

carbon-bonded halogens, E. A. C. Lucken; 5. 1,2-Dehalogenations and related reactions, E. Baciocchi; 6. Electrochemical oxidation, reduction and formation of the C—X bond—direct and indirect processes, J. Y. Becker; 7. Pyrolysis of aryl azides, L. K. Dyall; 8. Vinyl, aryl and acyl azides, H. W. Moore and D. M. Goldish; 9. Recent advances in the radiation chemistry of halocarbons, A. Horowitz; 10. Organic chemistry of astatine, K. Berei and L. Vasáros; 11. Positive halogen compounds, A. Foucaud; 12. Aspects of the chemistry of halophenols and halodienones, J. M. Brittain and P. B. D. de la Mare; 13. α-Halogenated imines, N. De Kimpe and R. Verhé; 14. Fluorocarbons, B. E. Smart; 15. Xenon halide halogenations, M. Zupan; 16. The $S_{RN}1$ reaction of organic halides, R. K. Norris; 17. Reactions involving solid organic halides, E. Hadjoudis; 18. Hypervalent halogen compounds, G. F. Koser; 19. Synthesis and reactivity of α-halogenated ketones, R. Verhé and N. De Kimpe; 20. Electrophilic assistance to reactions at a C—X bond, D. N. Kevill; 21. Molecular interactions involving organic halides, J.-M. Dumas, M. Gomel and M. Guerin; 22. Formation of carbon—halogen bonds, M. Hudlicky and T. Hudlicky; 23. Alkene-forming eliminations involving the carbon—halogen bond, E. Baciocchi; 24. Structural chemistry of the carbon—halogen and carbon—pseudohalogen bonds, M. Kaftory; 25. Halonium ion, G. F. Koser; 26. Carbon—carbon bond formation involving organic halides and transition metal compounds, F. Naso and G. Marchese; 27. Dihalocyclopropanes, P. Weyerstahl; 28. Photoelectron spectra of organic halogen compounds, K. Wittel and H. Bock; 29. Recent advances in the photochemistry of the carbon—halogen bond, G. Lodder (*See also* Vols **14**, **16**, **22**).

33. *The Chemistry of Peroxides* (Ed. S. Patai, 1983)

1. General and theoretical aspects of the peroxide group, D. Cremer; 2. Stereochemical and conformational aspects of peroxy compounds, O. Exner; 3. Thermochemistry of peroxides, A. C. Baldwin; 4. Mass spectrometry of organic peroxides, H. Schwarz and H.-M. Schiebel; 5. Acidity, hydrogen bonding and complex formation, W. H. Richardson; 6. Synthesis and uses of alkyl hydroperoxides and dialkyl peroxides, R. A. Sheldon; 7. Singlet oxygen in peroxide chemistry, A. A. Frimer; 8. Free-radical mechanisms involving peroxides in solution, J. A. Howard; 9. Organic sulphur and phosphorus peroxides, R. V. Hoffman; 10. Diacyl peroxides, peroxycarboxylic acids and peroxy esters, G. Bouillon, C. Lick and K. Schank; 11. Endoperoxides, I. Saito and S. S. Nittala; 12. Structural aspects of organic peroxides, J. Z. Gougoutas; 13. Polymeric peroxides, R. Ceresa; 14. Organic reactions involving the superoxide anion, A. A. Frimer; 15. Transition-metal peroxides—synthesis and use as oxidizing agents, H. Mimoun; 16. Organic polyoxides, B. Plesničar; 17. Polar reaction mechanisms involving peroxides in solution, B. Plesničar; 18. Preparation and uses of isotopically labelled peroxides, S. Oae and K. Fujimori; 19. Ozonation of single bonds, E. Keinan and H. T. Varkony; 20. Pyrolysis of peroxides in the gas phase, L. Batt and M. T. H. Liu; 21. Photochemistry and radiation chemistry of peroxides, Y. Ogata, K. Tomizawa and K. Furuta; 22. Solid-state reactions of peroxides, M. Lazar; 23. Organometallic peroxides, P. B. Brindley; 24. Four-membered ring peroxides: 1,2-dioxetanes and α-peroxylactones, W. Adam.

34. *The Chemistry of Metal–Carbon Bond. Volume 2. The Nature and Cleavage of Metal–Carbon Bonds* (Eds F. R. Hartley and S. Patai, 1985)

1. Electrochemical cleavage of metal—carbon bonds, C. J. Pickett; 2. Heterolytic cleavage of main group metal—carbon bonds, M. H. Abraham and P. L. Grellier; 3. Homolytic cleavage of metal—carbon bonds: Groups I to V, P. J. Barker and J. N. Winter; 4. Insertions into main group metal—carbon bonds, J. L. Wardell and E. S. Paterson; 5. Insertions into transition metal—carbon bonds, J. J. Alexander; 6. Nucleophilic attack on

transition metal organometallic compounds, L. S. Hegedus; 7. Electrophilic attack on transition metal η^1-organometallic compounds, M. D. Johnson; 8. Transition metal—carbon bond cleavage through β-hydrogen elimination, R. J. Cross; 9. Oxidative addition and reductive elimination, J. K. Stille; 10. Structure and bonding of main group organometallic compounds, J. P. Oliver (*See also* Vols **30** and **35**).

35. *The Chemistry of the Metal–Carbon Bond. Volume 3. Carbon–Carbon Bond Formation Using Organometallic Compounds* (Eds F. R. Hartley and S. Patai, 1985)

1. Carbon—carbon bond formation using tin and lead organometallics, V. G. Kumar Das and C.-K. Chu; 2. Carbon—carbon bond formation using organometallic compounds of zinc, cadmium and mercury, L. Miginiac; 3. Carbon—carbon bond formation using η^3-allyl complexes, Part 1: η^3-Allylnickel complexes, G. P. Chiusoli and G. Salerno; Part 2: η^3-Allylpalladium complexes, J. Tsuji; Part 3: Other η^3-allyl transition metal complexes, F. Sato; 4. Olefin oligomerization, O.-T. Onsager and J. E. Johansen; 5. Alkyne oligomerization, M. J. Winter; 6. Transition metal carbonyls in organic synthesis, J. A. Davies and R. J. Shaver; 7. Olefin and alcohol carbonylation, G. K. Anderson and J. A. Davies; 8. Olefin hydroformylation, J. A. Davies; 9. The Fischer–Tropsch synthesis, G. Henrici-Olivé and S. Olivé; 10. Olefin carbonylation, D. M. Fenton and E. L. Moorehead (*See also* Vols **30, 34**).

36. *The Chemistry of Organic Selenium and Tellurium Compounds. Volume 1* (Eds S. Patai and Z. Rappoport, 1986)

1. Organic derivatives of sulphur, selenium and tellurium—an overview, K. A. Jensen and A. Kjaer; 2. General and theoretical aspects of organic compounds containing selenium or tellurium, R. A. Poirier and I. G. Csizmadia; 3. Structural chemistry of organic compounds containing selenium or tellurium, I. Hargittai and B. Rozsondai; 4. Thermochemistry of selenium and tellurium compounds, L. Batt; 5. Detection and determination of organic selenium and tellurium compounds, K. J. Irgolic and D. Chakraborti; 6. Nuclear magnetic resonance and electron spin resonance studies of organic selenium and tellurium compounds, N. P. Luthra and J. D. Odom; 7. Mass spectrometry of organic selenium and tellurium compounds, G. D. Sturgeon and M. L. Gross; 8. Radiation chemistry of organic selenium and tellurium compounds, R. Badiello; 9. Selenium-stabilized carbenium ions and free radicals, L. Hevesi; 10. Selenium- and tellurium-containing organic polymers, Y. Okamoto; 11. Organometallic compounds with selenium and tellurium atoms bonded to main group elements of Groups IIIa, IVa and Va, R. A. Zingaro; 12. Synthesis and uses of isotopically labelled selenium and tellurium compounds, K. Fujimori and S. Oae; 13. Selenium and tellurium heterocycles, M. Renson; 14. Tetra- and higher-valent (hypervalent) derivatives of selenium and tellurium, J. Bergman, L. Engman and J. Sidén; 15. Directing and activating effects involving selenium and tellurium, F. Fringuelli and A. Taticchi; 16. Functional groups containing selenium and tellurium in various oxidation states, N. Sonoda and A. Ogawa; 17. Stereochemistry and chiroptical properties of organic selenium and tellurium compounds, G. Snatzke; 18. Ligand properties of organic selenium and tellurium compounds, H. J. Gysling.
(Volume 2 of the same title, see p. 19).

D. Appendix: Volumes published in 1987 and in 1988

The publication process of this book took much longer than planned and expected, for various reasons. During this period (from the beginning of 1987 to the end of 1988), five more volumes appeared in the Series. So as not to delay publication, we decided to add only a list of these volumes and the chapters included in them, in the form of this Appendix. This, we hope, will enable the reader ot assimilate at least the most basic information in these chapters such as names of authors and subjects, with in most cases a very brief explanatory note where we thought the title was not self-explanatory. Hence at least these data could be included in the Author and Subject indexes and will enable the reader to locate material of interest.

The numbering of volumes and chapters in this Appendix is a continuation of that in the volume proper.

37. *The Chemistry of the Metal–Carbon Bond. Volume 4. The Use of Organometallic Compounds in Organic Synthesis* (Ed. F. R. Hartley, 1987)

Part 1. *Preparation and use of Main Group Organometallics in Organic Synthesis.*

1. Preparation and use in organic synthesis of organolithium and Group IA organometallics, J. L. Wardell. (Replacement of halogens, hydrogen or metals by alkali metals and reaction of organoalkali metal compounds.) Pp. 1–157, 1247 references.
2. Preparation and use of Grignard and Group II organometallics in organic synthesis, C. L. Raston and G. Salem. Pp. 161–306, 539 references.
3. Preparation and use of organoboranes in organic synthesis, D. S. Matteson. Pp. 309–409, 625 references.
4. Preparation and use of organoaluminium compounds in organic synthesis, P. A. Chaloner. Pp. 412–472, 281 references.
5. Preparation and use of organothallium(III) compounds in organic synthesis, S. Uemura. Pp. 473–538, 325 references.
6. Preparation and use of organosilicon compounds in organic synthesis, E. W. Colvin. Pp. 540–621, 455 references.

Part 2. *Use of Transition Metal Organometallics in Organic Synthesis*

7. Use of organoiron compounds in organic synthesis, D. Astruc. (Reactions of iron complexes with nucleophiles and with electrophiles; stabilization; iron carbonyls.) Pp. 626–731, 303 references.
8. Use of organorhodium compounds in organic synthesis, F. H. Jardine. (Isomeriz-

38. *The Chemistry of Organic Selenium and Tellurium Compounds. Volume 2* (Ed. S. Patai, 1987)

39. *The Chemistry of the Cyclopropyl Group. Parts 1 and 2* (Ed. Z. Rappoport, 1987)

(Compounds substituted by cyclopropyl groups; cyclopropanes substituted with carbon or heteroatom groups; cyclopropyl-substituted carbocations, anions and radicals.) Pp. 565–632, 345 references (cf. **10**, 10).

11. Solvolysis of cyclopropyl-substituted derivatives, E. C. Friedrich. (Mainly studies on halides, carboxylate and sulphonate esters.) Pp. 633–700, 285 references.
12. Cyclopropyl radicals, anion radicals and anions, G. Boche and H. M. Walborsky. (Stereochemistry and mechanism of reactions of the title compounds.) Pp. 701–808, 279 references.
13. Rearrangements involving the cyclopropyl group, J. Salaun. (Ring openings, enlargements and contractions.) Pp. 809–878, 298 references.
14. Radiation chemistry and hot atom chemistry of cyclopropane and its derivatives, Z. B. Alfassi. Pp. 879–914, 87 references.
15. Electrochemistry of the cyclopropyl group, J. Y. Becker. (Cathodic and anodic formations and reactions.) Pp. 915–958, 118 references.
16. Biochemistry of the cyclopropyl group, H.-W. Liu and C. T. Walsh. (Representative structures, routes of formation, degradation and enzymatic mechanisms.) Pp. 959–1025, 264 references.
17. Preparation and uses of isotopically labelled derivatives, B. K. Carpenter. (Synthetic methods, thermal rearrangements; cation, anion and radical reactions.) Pp. 1027–1082, 126 references.
18. Cyclopropanes having extra strain. J. F. Liebman and A. Greenberg. (Qualitative and quantitative effects; highly fluorinated molecules; radical cations.) Pp. 1083–1119, 146 references.
19. Bicyclo[1.1.0]butane, S. Hoz. (Structure, bonding, synthesis, reactions.) Pp. 1121–1192, 303 references.
20. [*m.n.l*]Propellanes, D. Ginsburg. (Preparations, properties and reactions.) Pp. 1193–1221, 91 references.
21. Cyclopropenes, B. Halton and M. G. Banwell. (Synthesis, properties and reactions of cyclopropenes, cyclopropenones and methylenecyclopropenes.) Pp. 1223–1330, 564 references.
22. Aminocyclopropanes, E. Vilsmaier. (Preparation, reactions and biochemical activities.) Pp. 1341–1454, 927 references.
23. The chemistry of cyclopropanones, H. H. Wasserman, D. R. Berdahl and T.-J. Lu. (Preparation, MW, UV, ORD and NMR; theoretical studies; reactions—also including cyclopropanone equivalents.) Pp. 1455–1532, 197 references.
24. Cyclopropenyl compounds, W. E. Billups and A. W. Moorehead. (Synthesis, properties and chemistry of cyclopropenyl ions, cyclopropenones and methylenecyclopropenones.) Pp. 1533–1574, 214 references.

Author index, pp. 1575–1707.
Subject index, pp. 1709–1739.
(*See also* Vol. **9**, 10 and Vol. **32**, 27.)

40. *The Chemistry of the Quinonoid Compounds. Volume 2. Parts 1 and 2* (Eds S. Patai and Z. Rappoport, 1988) (cf. Vol. **17**, Parts 1 and 2, 1974)

1. General and theoretical aspects of quinones, A. Skancke and P. N. Skancke. (Structures, symmetries and orbital topologies; physical properties of ground states; spectroscopic properties of excited states.) Pp. 1–28, 212 references (cf. **17**, 1).
2. Physical and chemical analysis of quinones, S. Berger, P. Hertl and A. Rieker. (NMR, IR, UV–visible, ESR, electrochemical, chromatography, chemical.) Pp. 29–86, 270 references (cf. **17**, 4).
3. Mass spectra of quinones, K.-P. Zeller and R. Muller. (Positive and negative ion MS.) Pp. 87–109, 43 references (cf. **17**, 5).

4. Chiroptical properties and absolute configurations of chiral quinones, H. E. Smith. Pp. 111–153, 138 references.
5. Photoelectron spectra of quinonoid compounds, L. Klasinc and S. P. McGlynn. Pp. 155–201, 121 references.
6. Photochromism and thermochromism in bianthrones and bianthrylidenes, K. A. Muszkat. Pp. 203–224, 91 references.
7. Chemiluminescence of quinonoid compounds, K.-D. Gundermann and D. Lieske. Pp. 225–240, 48 references.
8. Recent advances in the synthesis of quinonoid compounds, Y. Naruta and K. Maruyama. (Oxidation, annulation, cyclization and condensation methods.) Pp. 242–402, 498 references (cf. **17**, 3).
9. *ortho*-Quinonoid compounds, C. W. G. Fishwick and D. W. Jones. (*o*-Quinodimethanes, *o*-quinone methides, their generation and reactions.) Pp. 403–453, 212 references (cf. **17**, 18).
10. *meta*-Quinonoid compounds, J. A. Berson. (Theory and spectroscopy of *m*-quinodimethanes, *m*-quinomethanes, bis-*m*-quinomethanes and their chemistry.) Pp. 455–536, 233 references.
11. Quinones as synthones, K. T. Finley. (Heteroatom additions, reactions at the carbonyl, alkylation, cycloaddition, arylation, active methylene and substitution chemistry.) Pp. 537–717, 796 references (cf. **17**, 17).
12. Electrochemistry of quinones, J. Q. Chambers. (Studies in non-aqueous solvents and in aqueous solutions). Pp. 719–757, 325 references (cf. **17**, 14).
13. Recent advances in the photochemistry of quinones, K. Maruyama and A. Osuka. (Photoreduction, photoaddition, photosubstitution and unimolecular photoreactions.) Pp. 759–878, 413 references (cf. **17**, 9).
14. Radiation chemistry of quinonoid compounds, P. Neta. (Formation and properties of semiquinones; biological electron transport.) Pp. 879–898, 141 references (cf. **17**, 10).
15. Chemistry of quinone bis- and monoketals, J. S. Swenton. (Preparation and reactions.) Pp. 899–962, 104 references.
16. Quinhydrones and semiquinones, M. C. Depew and J. K. S. Wan. (Quinhydrones; ESR and CIDMP, physical chemistry and some chemical and biological aspects of semiquinones.) Pp. 963–1018, 298 references (cf. **17**, 2, 6).
17. Heterocyclic quinones, R. W. Middleton and J. Parrick. (Quinonoid compounds fused to a heterocycle, their preparation and properties; heterocyclic quinone methides; diazaquinones.) Pp. 1019–1066, 125 references.
18. Polymerization and polymers of quinonoid compounds, S. Iwatsuki. (Polymerization of *p*-xylene, halo-*p*-xylenes and quinodimethanes.) Pp. 1067–1111, 124 references (cf. **17**. 15).
19. Isotopically labelled quinones, M. Zieliński and M. Kanska. (Syntheses with various labels; isotopic chemical studies; biochemical syntheses and uses.) Pp. 1113–1198, 189 references (cf. **17**, 12).
20. The solid-state photochemistry of tetrahydronaphthoquinones: crystal structure–reactivity relationships, J. R. Scheffer and J. Trotter. (Crystallography, photochemical results; solid-state structure–reactivity correlations; intramolecular H-abstraction.) Pp. 1199–1230, 38 references.
21. Quinone diimines, monoimines and related compounds, E. R. Brown. (Deamination, coupling, sulphonation, etc.) Pp. 1231–1292, 115 references (cf. **8**, 14).
22. Biochemistry of quinones, H. Inouye and E. Leistner. (Biosynthesis and metabolism.) Pp. 1293–1349, 251 references (cf. **17**, 13).
23. Quinones as oxidants and dehydrogenating agents, H.-D. Becker and A. B. Turner. (Dehydrogenation of hydroaromatics; aromatic hydroxy and amino compounds and heterocyclic compounds; oxidation of benzylic and allylic compounds, ketones, enols and silyl enol ethers.) Pp. 1351–1384, 191 references (cf. **17**, 7).

24. Azulene quinones, L. T. Scott. (Theory, synthesis and properties.) Pp. 1385–1417, 52 references.
25. Extended quinones, P. Boldt. (Quinonoid CO groups in different rings; polyquinocycloalkanes, diquinoethylenes; quinones of polycyclic arenes with two to thirteen rings; annulenediones.) Pp. 1419–1499, 260 references.
26. Non-benzenoid quinones, H. N. C. Wong, T.-L. Chan and T.-Y. Luh. (Cyclic dicarbonyl compounds with even- or odd-membered rings; 1, 2- and other dicarbonyl systems.) Pp. 1501–1563, 210 references (cf. **17**, 16).

Author index, pp. 1565–1673.
Subject index, pp. 1675–1711.

41. *The Chemistry of Sulphones and Sulphoxides* (Eds S. Patai, Z. Rappoport and C. Stirling, 1988)

1. General and theoretical aspects, A. Gavezotti. (MO theory, analysis of results, valence MO treatment, steric parameters.) Pp. 1–32, 32 references.
2. Structural chemistry of gaseous sulfoxides and sulfones, I. Hargittai. Pp. 33–54, 74 references.
3. Stereochemistry, conformation and chiroptical properties of sulfoxides, K. K. Andersen. (Deals also with methods of preparation, stereochemistry of reactions at sulphoxide sulphur, etc.) Pp. 55–94, 166 references.
4. Thermochemistry of sulfoxides and sulfones, J. T. Herron. (Data and estimation; pyrolysis.) Pp. 95–106, 37 references.
5. Detection and determination of sulphones and sulphoxides, M. R. F. Ashworth. (Chemical, spectral and chromatographic methods.) Pp. 107–124, 147 references.
6. Mass spectra of sulfoxides and sulfones, K. Pihlaja. (Includes also disulphones, environmentally and pharmaceutically significant compounds and sulphones with N—S bonds.) pp. 125–164, 72 references.
7. Synthesis of open-chain sulfones, K. Schank. (One-, two- and three-component methods.) Pp. 165–232, 497 references.
8. Synthesis of sulphoxides, J. Drabowicz, P. Kielbasinski and M. Mikolajczyk. (Synthesis of achiral, racemic and optically active sulphoxides and their functionalization.) Pp. 233–378, 685 references.
9. Cyclic sulfones and sulfoxides, U. Zoller. (Three-, four-, five- and six-membered and medium-ring compounds, their structures, physical and chemical properties and synthesis.) Pp. 379–482, 354 references.
10. Electronic effects of the sulfinyl and sulfonyl groups, J. Shorter. (Substituent effects, stabilization of carbanions, etc.) Pp. 483–540, 254 references.
11. Hydrogen bonding and complexing properties of R_2SO_2 and R_2SO, N. Furukawa and H. Fujihara. Pp. 541–582, 229 references.
12. Behavior of α-sulfinyl and α-sulfonyl carbanions, S. Oae and Y. Uchida. (Stereochemistry and reactions of the title compounds.) Pp. 583–664, 448 references.
13. Rearrangements involving sulfones, S. Braverman. (Sulphinates to sulphones and vice versa; 1, 2-, 1, 3-, Ramberg–Backlund and Smiles rearrangements.) Pp. 665–716, 304 references.
14. Rearrangements involving sulfoxides, S. Braverman. (Sulphenates to sulphoxides and vice versa; 1, 2-, 1, 3- and other rearrangements.) Pp. 717–758, 250 references.
15. Synthetic uses of sulfones, K. Tanaka and A. Kaji. (Substitutions and eliminations; additions to unsaturated sulphones; addition of sulphonyl carbenes and nitrenes.) Pp. 759–822, 149 references.
16. Asymmetric synthesis using α-sulfinyl carbanions and β-unsaturated sulfoxides, G. H.

Posner. (Reactions of electrophiles with α-sulphinyl carbanions; addition of nucleophiles to β-unsaturated sulphoxides.) Pp. 823–850, 108 references.

17. Methionine sulfoxide: chemistry and biochemistry, N. Brot and H. Weissbach. Pp. 851–872, 150 references.
18. Photochemistry of sulfoxides and sulfones, I. W. J. Still. Pp. 873–888, 85 references.
19. Radiation chemistry of sulfoxides and sulfones, Z. B. Alfassi. (The primary species and the final products in the radiolysis of the title compounds.) Pp. 889–924, 84 references.
20. Reduction of sulphoxides and sulphones, J. S. Grossert. (Reduction of sulphoxides by oxygen exchange, by hydrides and by metals; reduction of S—O and C—S bonds and desulphonylations of sulphones.) Pp. 925–968, 216 references.
21. Oxidation of sulphoxides and sulphones, J. Hoyle. (Oxidations of sulphoxides to sulphones; oxidations of sulphones to sulphonic acid derivatives and to sulphate.) Pp. 969–1000, 224 references.
22. The electrochemical reactivity of sulphones and sulphoxides, J. Simonet. (Cathodic cleavage and behaviour of sulphones and sulphoxides; influence of acidity.) Pp. 1001–1046, 79 references.
23. Electron transfer reactions of sulfoxides and sulfones, M. Chanon and A. Samat. (ESR spectrometry of radical anions and cations; reactions with oxidants and reductants; donor–acceptor properties in photochemistry.) Pp. 1047–1080, 150 references.
24. Sulfinyl radicals, C. Chatgilialoglu. (Structural and chemical properties.) Pp. 1081–1088, 27 references.
25. Sulfonyl radicals, C. Chatgilialoglu. (Structural characteristics; formation; radical–radical and radical–molecule reactions.) Pp. 1089–1114, 97 references.

Author index, pp. 1115–1194.
Subject index, pp. 1195–1210.

SECTION II

The chemistry of alkenes (1964)

1, *1: Wave mechanics and the alkene bond* by *C. A. Coulson and E. T. Stewart*

I. Introduction (2); **II**. Hydrogen-like wave functions (6); **III**. Approximate wave functions for many-electron systems; **IV**. The ground state of the hydrogen molecule (32); **V**. Antisymmetrised wave functions (46); **VI**. Low-energy states of the hydrogen molecule (57); **VII**. Molecular-orbital wave functions for diatomic molecules (69); **VIII**. Molecular-orbital wave functions for simple organic molecules (88); **IX**. The π-electron hypothesis (106); **X**. The Hückel approximation (110); **XI**. Spectra (129); **XII**. References (146).

This chapter discusses wave functions for simple atoms (especially hydrogen) and presents concisely atomic orbitals, quantum numbers and the main concepts necessary for the understanding of the wave-mechanical description of the hydrogen atom and the hydrogen molecule. This leads to the MO description of other diatomic molecules, emphasizing the close resemblance between the MO and the valence bond methods. Section VIII presents the MO description of some simple organic molecules, including such which contain σ–π systems with carbon—carbon double (and triple) bonds.

The π-electron energy and π-electron wave functions are delineated in Sections IX and X and the empirical Hückel method is chosen for the description of ethylene, butadiene and higher polyenes, summarizing also the concept of delocalization energy and essential features of Pople, Pariser and Parr modification (refs 35, 36). Section XI discusses the electronic spectra of ethylene, higher polyenes and alkenes.

The authors write: 'The general purpose of this chapter is to enable the reader without previous knowledge of quantum-chemical procedures to appreciate the significance and also the limitations of wave-mechanical calculations, but not to enable him to perform such calculations himself.... The mathematical analysis required is very straightforward and well within the repertoire of the average chemist.... We have assumed nothing more than a knowledge of the differentiation and integration of simple trigonometric and exponential functions and we have explained other routine mathematical procedures wherever necessary.... More systematic study of the principles of quantum mechanics.... (is) discussed at an appropriate level of sophistication in refs 4–7.'

44 references up to 1962.
Complementary: 26, 1.
Relevant: 2, 1; **3**, 1; **4**, 1; **5**, 1; **6**, 1; **8**, 1; **11**, 1; **12**, 1; **13**, 1; **14**, 1; **15**, 1; **16**, 1; **17**, 1; **18**, 1; **19**, 1; **20**, 1; **22**, 1; **23**, 1; **24**, 1; **25**, 1; **26**, 1; **27**, 9; **28**, 1; **31**, 27; **32**, 1; **33**, 1; **36**, 2.

1, *2: Elimination reactions in solution* by *H. Saunders, Jr*

I. Introduction (149); **II**. Mechanism and stereochemistry (151); **III**. Effects of structure and conditions on reactivity (177); **IV**. References (196).

The chapter deals with the elimination of HX from two adjacent C atoms (β-elimination) in solution. It discusses all mechanisms which have been proposed (with emphasis on the *E*1 and *E*2 mechanisms) including the *E*1cB mechanism, as well as the stereochemistry of the main routes and isotope effects. The section dealing with the *E*1 mechanism describes also *E*1–*E*2 borderline mechanisms such as that proposed by Winstein ('merged mechanism,' ref. 74) and others (refs 75–77).

Other (not general) mechanisms are treated briefly, such as Wittig's α'–β (ylide) mechanism for onium salts in aprotic solvents (pp. 164–166); the α-elimination (carbene) route (pp. 166–168).

The author discusses the stereochemistry of the *E*2 reaction and its preference for *trans*-elimination (pp. 168–173), quoting several exceptions to the rule. The stereochemistry of the *E*1 reactions and of the 'minor' mechanisms are sketched on pp. 174–176.

Section III of the chapter is devoted to the treatment of the various factors which influence the reaction route and which should be properly defined so as to enable one to make legitimate comparison between different reactions. These are the structures of the reagents, of the intermediates and of the transition states; the factors which cause the applicability of the Hofmann and the Saytzeff rules; the effects of changing the (alkyl) groups and the leaving groups involved in the reaction and finally the effects of catalysts and solvents.

177 references up to 1962.
Complementary: 1, 3; **16**, a; **21**, 4; **32**, 5; **32**, 23.
Relevant: 13, 12; **21**, 8; **24**, 18; **26**, 7; **31**, 28.

1, *3: Olefin-forming eliminations in the gas phase* by *A. Maccoll*

I. General introduction (203); **II**. Homogeneous elimination (206); **III**. The catalysis of gas-phase elimination (225); **IV**. Equilibria in elimination reactions (229); **V**. Theories of gas-phase elimination reactions (232); **VI**. References (236).

The title reactions may be homogeneous, heterogeneous or may follow a chain mode. Homogeneous eliminations have been widely studied with hydrocarbons, halides, esters, xanthates, alcohols, amines and chloroformates. Tables for Arrhenius parameters are given for eliminations from hydrocarbons (Table 1); from chlorides, bromides and iodides (Tables 2–4); from esters and xanthates (Tables 10 and 14); from chloroformates (Table 15). The reaction conditions, the stoichiometry, the kinetic order and the molecularity and the mechanism of these reactions, as well as the influence of α- and β-substitution (alkyl and halogen), the product composition and the stereochemistry of these reactions is discussed in detail. Section III deals with homogeneous acid (HX) catalysis in the dehydration of alcohols (Tables 16 and 17) and with the influence of α- and β-methylation in this reaction (Table 18). Heterogeneous catalysis of alkyl halides and of alcohols is discussed subsequently.

In Section IV, isomeric equilibria in the products and dissociative equilibria in the total reaction mixture are treated, while Section V is mainly devoted to theories relating to the mechanisms of the reactions and the conclusions which may be drawn from mass spectrometric studies.

143 references up to 1963.
Complementary: 1, 2; **13**, 12; **16**, 10; **21**, 4; **21**, 8; **27**, 11; **27**, 16.
Relevant: 5, 12; **22**, 15; **25**, 15; **29**, 12; **32**, 5; **32**, 7; **32**, 23; **33**, 20.

1, *4: Alkene-forming condensation reactions* by *T. I. Crowell*

Section I treats the essential facts of the aldol, Knoevenagel and related double-bond forming condensations (Reformatsky, Perkin and Stobbe) and also the Wittig reaction and a few other condensations.

Section II discusses the kinetics of consecutive reactions (including the steady-state method) and then treats more extensively the mechanisms of the above-mentioned reactions with aliphatic and aromatic aldehydes and various active methylene compounds in the presence of various basic or acid catalysts.

Section III summarizes the effect of structural changes on the rates of these reactions and the correlations obtained with the Hammett equation. In Fig. 1, calculated Hammett plots are shown for different cases of the reactions $A \rightleftharpoons B \rightarrow C$.

108 references up to 1962.
Complementary: 2, 12.
Relevant: 4, 7; **8**, 2.

1, *5: Detection and determination of the alkenes* by *E. J. Kuchar*

The chapter discusses mainly those analytical methods which are extensively used, while others are mentioned but not treated in detail. Section II deals with chemical methods, first describing quantitative analyses based almost exclusively on additions (halogenation, hydrogentation, ozonization, etc.) to the C=C group. Bromine number data are given in Table 1. Some qualitative methods are also mentioned (pp. 286–288), while cautioning that with modern instrumentation positive identification of unknown compounds is so facile as to make qualitative identification of C=C groups in general almost superfluous.

Section III describes the quantitative and qualitative use of various forms of chromatography and IR, mass, NMR, UV–VIS and Raman spectroscopy. Tables 5 and 6 compare the known values of the composition of 20 hexene isomers with the values found using two different sets of chromatographic experimental conditions (ref. 150). Tables 7 and 8 give relative retention volumes for miscellaneous alkene mixtures.

281 references, up to 1962
Complementary: 9, 1, 7; **24**, 5, 6; **26**, 5; **31**, 3, 4, 25.
Relevant: 2, 8; **3**, 15; **4**, 3; **5**, 17; **6**, 2, 3; **8**, 3; **10**, 5; **12**, 4; **13**, 6; **14**, 3; **15**, 3; **16**, 3; **17**, 4; **18**, 5, 6; **19**, 5; **20**, 3; **22**, 5; **23**, 5; **25**, 10; **28**, 3, 6; **29**, 21; **30**, 14–22; **36**, 5.

1, *6: Alkene complexes of some transition metals* by *M. Cais*

Transition metals are able to form organic derivatives having a direct carbon-to-metal bond. Ten books or reviews have been published on this subject between 1959 and 1963 (refs 1–10).

The chapter describes metal π-complexes of open-chain and cyclic unsaturated compounds, but does not deal extensively with metallocenes (e.g. ferrocene), which have been reviewed in detail close to the publication date of the chapter (refs 1, 2, 10).

Section II treats the transition metal complexes of butadiene and several of its variously substituted derivatives, emphasizing the necessary presence of conjugated double bonds in order to obtain stable complexes. Section III deals with Fe, Ni, Rh, Ir, Co, Mn and Mo complexes of cyclopentadiene derivatives and Sections IV–VI treat the complexes of the same metals with six-, seven- and eight-membered homocyclic rings. Spectra (especially NMR) and the structures of the complexes are described, together with their methods of formation and reactions. Detailed X-ray crystallographic data for several complexes are presented.

The concluding Section VII speculates on the future use of such complexes in organic syntheses and presents two prominent cases (cyclobutadiene and pentalene) which have been predicted to be inaccessible but were synthesized by using routes involving the appropriate metal complexes.

116 references up to 1963
Complementary: **9**, 5; **21**, 12; **24**, 4; **30**, 7–11.
Relevant: **17**, 6; **19**, 7; **20**, 12; **23**, 7; **32**, 26; **34**, 5–8; **35**, 3, 6; **36**, 18.

1, *7: Alkene rearrangements* by *K. Mackenzie*

Section I discusses the *cis–trans* isomerization alkyl-, halogen- and aryl-substituted ethylenes by thermal and photochemical means, and catalysis of the reaction by acids and bases. Subsequently the *cis–trans* isomerism in synthetic and natural polyenes (e.g. carotenoids) is discussed, together with the relative stabilities of the geometrical isomers of the latter and of unsaturated fatty acids.

Section II deals with prototropic rearrangements of alkenes (involving both carbonium ions and carbanions), of alkene ethers and thioethers, of allyl amines and of olefinic acids and carbonyl compounds and of polyenes (e.g. fatty acids and vitamin A).

Section III discusses rearrangements which involve the shift of a double bond with simultaneous migration of a negatively charged atom or group (together with its bonding electrons). The mechanisms presented are the unimolecular (S_N1') and the intramolecular (S_Ni'); the effects of catalysts and of substituents (alkyl, aryl, ethynyl, vinyl) are summarized, together with anionotropy in natural polyenes.

Finally, in Section IV the Cope and Claisen rearrangements and also rearrangements accompanying decarboxylation of unsaturated acids and of *N*-alkyl-*N*-vinylsulphonamides are sketched briefly.

240 references up to 1962.
Complementary: **9**, 3; **24**, 10; **26**, 15.
Relevant: **2**, 15; **3**, 14; **4**, 10; **5**, 8; **11**, 4; **12**, 16; **13**, 13; **14**, 7; **15**, 8; **16**, 17; **17**, 8; **18**, 15; **19**, 18; **20**, 4; **29**, 4.

1, *8: Nucleophilic attacks on carbon—carbon double bonds* by *S. Patai and Z. Rappoport*

In contrast to the usually facile electrophilic attacks on C=C groups, nucleophilic attacks usually require activating factors in the alkene and reactive nucleophiles. Even so, a wide variety of reaction routes including additions, additions with rearrangements or with cyclization, substitution, fragmentation, oligomerization and isomerization are available (Scheme 1).

Section II summarizes nucleophilic additions, the scope of substrates and of nucleophiles, the kinetics and mechanism of these reactions and their stereochemistry. Next, additions of organometallic compounds (especially Grignard reagents) and of phosphines are discussed, followed by rearrangements in the intermediate anion.

Section III deals with ring formations, including alkaline epoxidation, cyclopropane and aziridine formation.

Section IV discusses vinylic substitution and the five main mechanisms proposed for it (addition–elimination, elimination–addition, substitution with rearrangement, the vinyl carbanion route and the S_N1 route). Section V describes fragmentations (in which the α- and the β-carbon atoms appear in two different fragments in the products) caused by O-, S- and N-containing reagents and arylmethylene and alkylidene transfers. The last two sections discuss cyclooligomerizations and nucleophilic *cis–trans* isomerizations with special attention to the nature of the catalysts and to reaction mechanisms and stereochemistry.

362 references up to 1963.
Complementary: 17, 17; **21**, 4; **24**, 19.
Relevant: 8, 6; **19**, 16; **21**, 9; **26**, 9.

1, *9: Reactions of alkenes with radicals and carbenes* by *J. I. G. Cadogan and M. J. Perkins*

The addition of a variety of addenda to unsymmetrical alkenes usually follows the Markownikoff rule in a slow reaction. However, in the presence of free-radical catalysts (especially peroxides or oxygen, introduced either deliberately or fortuitously) the addition takes place contrary to the Markownikoff rule and very rapidly, involving a free radical intermediate in a radical chain process (refs 5–7), in which the initiating radical need to be present only in minute catalytic amounts (Schemes 1 and 2).

Section I discusses radical additions to C=C groups, involving as the addenda polyhalogenomethanes, carboxylic acids and their derivatives, aldehydes, ketones, alcohols, thiols and compounds containing P, Ge, Si or N. The orientation in the products obtained from unsymmetric alkenes depends on the nature of both the alkene and the addend; the relative reactivity of the C=C group depends on the stability of the intermediate formed, on steric effects and on the rate-enhancing effect of electron-donating groups. The mechanism of the title reaction is discussed in detail, in addition to the stereochemistry and the occurrence of (homolytic) rearrangements.

Section II deals with additions of carbenes to C=C groups. Halo- and other substituted carbenes undergo such reactions, with the rate depending on the structure of both reactants. Subsequently, the formation and reactions of unsubstituted carbene (CH_2) and of other carbenes are summarized, including some carbenoid species the nature of which is not completely certain.

Finally, reactions of carbenes with conjugated dienes, allenes, aromatic compounds and acetylenes are presented.

295 references up to 1963.
Complementary: 21, 7; **31**, 10.
Relevant: 12, 11; **16**, 8; **17**, 17; **19**, 19; **30**, 4, 5; **33**, 8.

1, *10: Allylic reactions* by *R. H. DeWolfe and W. G. Young*

In allylic compounds $\overset{\gamma}{C}=\overset{\beta}{C}-\overset{\alpha}{C}Y$ the group Y is much more reactive than in the analogous saturated compounds, and they undergo facile substitution, isomerization and rearrangement.

Normal bimolecular (S_N2) substitutions involving different nucleophiles take place about 30–300 times faster with allylic than with the corresponding saturated compounds (Table 1). Reaction rates depend strongly on substituents present in the allylic molecule as well as on the natures of the leaving and attacking groups (Table 2).

In the 'abnormal' (S_N2') mechanism the nucleophile attacks the unsaturated γ-carbon, while the α-substituent is displaced in a concerted process (Section IIB). The unimolecular S_N1 mechanism may occur with or without rearrangement of the ionic intermediate, thus leading to two isomeric products (Section IIC). In allylic solvolyses any of the S_N1, S_N2 and S_N2' mechanisms may be operative, and the determination of the actual mechanism is usually based on indirect evidence (Section IID).

The S_Ni' (substitution, nucleophilic, internal, with rearrangement) is described in Section IIE, and this and other mechanisms of allylic isomerizations are dealt with also in Section III, involving allylic alcohols, esters and halides. Finally, concerted intramolecular isomerizations of allyl aryl and allyl vinyl ethers, allylic thiocyanates and azides are summarized (Section IIID), and Section IV discusses some special characteristics of allylic Grignard reagents, which have been reviewed in detail elsewhere (refs 233, 234).

248 references up to 1962.
Complementary: 3, 14; **13**, 13 (III); **18**, 5 (IV).
Relevant: 9, 2, 3; **26**, 15 (VIII 5).

1, *11: Cycloaddition reactions of alkenes* by *R. Huisgen, R. Grashey and J. Sauer*

Section I sets out the criteria according to which a reaction can be defined as a cycloaddition and classifies these based on the number of new bonds formed and on the size of the ring formed (again divided according to the number of ring atoms which each reactant contributes, e.g. $4 + 2 \rightarrow 6$ for Diels–Alder).

Sections II and III discuss the formation of three- and four-membered products, respectively, including cyclopropanes, aziridines, oxiranes, cyclobutanes by thermal and photochemical routes, and four-membered heterocycles containing O, S and N heteroatoms.

Section IV describes the formation of five-membered rings by $3+2 \rightarrow 5$ additions. The 1, 3-dipoles may consist of any combination of C, O and N atoms and have necessarily to bear formal charges. They are classified and enumerated in Tables 4 and 5, including 'octet stabilized 1, 3-dipoles' such as nitrile ylides, nitrilimines, nitrile oxides, diazoalkanes, azides, N_2O, azomethine ylides and imines, nitrones, nitro compounds, carbonyl ylides and oxides and ozone. Finally, the rare cyclizations of 1, 3-dipoles without octet stabilization, such as various carbenes and azenes (Table 6), are sketched. All the above reactions are illustrated with olefinic dipolarophiles (non-olefinic reactants are treated in ref. 323). With each 1, 3-dipole, its preparation, reactivity, mechanisms and stereochemistry are discussed.

In Section V the $4+2 \rightarrow 6$ (Diels–Alder) cycloadditions are presented, dealing with a wide variety of dienic components such as open-chain dienes containing either C atoms only or heteroatoms in addition; polyenes, bismethylenecycloalkanes, alicyclic dienes, fulvenes, cyclopentadienone derivatives, *o*-quinones, aromatic carbocyclic and heterocyclic compounds and with simple olefins and non-conjugated dienes. Throughout, emphasis is laid on the post-1955 literature (for pre-1955 reviews, see refs 492–499). The dienophiles described include open-chain olefinic and acetylenic, allenic and cyclic compounds and also some cyclic azo compounds and dienophiles with heteroatoms (O, S, N). Subsequently, retro-Diels–Alder processes are delineated, followed by summaries of the stereochemistry of the reaction (Section V.E), of orientation rules where applicable (V.F.) and of the kinetics, comparative reactivities and mechanisms with different components (Sections V.G and H).

44 references up to 1963.
Complementary: **8**, 7; **12**, 8; **14**, 6 (III); **17**, 17 (V); **21**, 6; **22**, 16 (V.H), 17 (III), 22 (VI); **23**, 18; **24**, 11; **31**, 18 (VI).
Relevant: **26**, 16 (IV.E), 17 (V.B), 18 (IV), 19 (IV.B); **29**, 13 (II.B.5, III.C.3), 15 (II.C.6, III.C.II); **31**, 23; **33**, 11 (III.B).

1, *12: Conjugated dienes* by *M. Cais*

I. Introduction (955); **II**. Synthesis of conjugated dienes (957); **III**. Physical properties of conjugated dienes (977); **IV**. Chemical properties of conjugated dienes (992); **V**. References (1019).

Conjugated dienes and polyenes are strikingly different from their non-conjugated isomers. A brief outline is given of the MO representation of buta-1, 3-diene, as the simplest member of the class treated. Section II discusses the most important methods of synthesis of butadiene and of some of its alkyl-, aryl-, halogeno- and cyano-substituted derivatives.

Section III describes some of the physical properties of conjugated dienes, such as boiling points and magnetic properties, and, more extensively, electronic absorption spectra of some simple and steroid dienes and polyenes (Tables 3–8), their NMR spectra and the use of the latter to distinguish between *cis* and *trans* forms.

Section IV summarizes the most important reactions of conjugated dienes (except Diels–Alder reactions, see **1**, 11), such as additions of halogens and of hydrogen, protonation, hydroboration, additions of dialkyl azodicarboxylates and of some alkyl, alkoxy and aryloxy radicals, and photochemical transformations and thermal rearrangements.

147 references up to 1962.
Complementary: **1**, 11 (V); **9**, 1 (IV), 7 (IV), 9; **27**, 16 (II.B, III.C).
Relevant: **21**, 1 (V); **30**, 9.

1, *13: Cumulenes* by *H. Fischer*

I. Introduction (1026); **II**. Cumulenes with an even number of double bonds (1030); **III**. Cumulenes with an odd number of double bonds (1088); IV. Theory of the cumulene bond (1122); **V**. Acknowledgements (1147); **VI**. References (1147).

Section II discusses the first member of the cumulene series, allene, presenting the main methods of its preparation including HX, H_2O and X_2 elimination, reduction of halopropynes and -allenes, propargylic rearrangements (Table 1–3), 1,4-additions to vinylacetylenes (Table 4) and some other minor methods. Next, the physical properties of allenes and their optical activity are described and some chemical reactions (polymerization, isomerization, H_2 and X_2 additions and other such additions such as Diels–Alder and HX additions and oxidations) are summarized. Pentatetraenes, their preparation and reactions are presented next.

Section III discusses cumulenes containing 3, 5, 7 and 9 consecutive double bonds. Three main and some minor methods of preparation of butatrienes are presented (Tables 9–12 give yields, m.p.s, λ_{max}. and references for variously substituted butatrienes, bisbutatrienes, propadienylidenequinones and bisbutatrienylarenes). Subsequently, the chemical properties and reactions of the above compounds are summarized (including *cis–trans* isomerization, hydrogenation, halogenation, isomerization to vinylacetylenes and oxidation).

All the known hexapentaenes are tetrasubstituted and have been synthesized either by reduction of hexa-2,4-diyne-1,6-diols (Tables 13 and 14) or by self-condensation of diarylpropynols (Table 15). The colours, stabilities and solubilities of different types of hexapentaenes are compared in Table 16, and their *cis–trans* isomerization, hydrogenation, halogenation and oxidation reactions are sketched.

At the time of writing, only three octaheptaenes and two decanonaenes were known, all prepared by reduction of polyyne-α,ω-diols (Table 17).

Section IV describes the thermodynamics of cumulenes, their electric and magnetic properties, their vibration–rotation phenomena and vibrational spectroscopy. Subsequently, the electronic theory of the two perpendicular π-electron systems occurring in cumulenes is presented and various MO calculations are delineated (Tables 18–23).

377 references up to 1963.
Complementary: **26**, 1 (V), 2, 3, 4, 5, 6, 8, 9 (VII–XII), 10, 11, 12, 13, 15, 20.
Relevant: **24**, 10, 19; **27**, 19; **31**, 1 (II.C.2).

1, *14: Ketenes* by *R. N. Lacey*

I. Introduction (1162); **II**. Preparation (1165); **III**. Addition reactions to give carboxylic acid derivatives (1173); **IV**. Polymerization of ketenes (1182); **V**. Addition reactions through cyclic transition states (1197); **VI**. Miscellaneous reactions (1217); **VII**. Oxidation of ketenes (1218); **VIII**. References (1220).

Ketenes hardly show carbonylic properties and almost all their reactions are those of the C=C linkage. Section II describes the preparation of ketenes from α-haloacyl halides, from ketones and acid derivatives by pyrolysis, from malonic acid derivatives, from acyl halides with tertiary amines, from diazoketones and by thermal decomposition of cyclic (mainly four-membered) compounds.

Section III describes addition reactions with hydroxyl groups, thiols, amines, amides, hydrazines, carboxylic acids, halogens and inorganic and organic halides, hydrogen cyanide, ethers and acetals. All these reactions yield carboxylic acid derivatives.

Dimerization of aldoketenes (RCH=C=O) and ketoketenes (R^1R^2C=C=O) and formations of higher polymers are treated in Section IV, followed by a description of some reactions of ketene dimers (i.e. derivatives of β-propiolactones on the one hand and of cyclobutanediones on the other).

In Section V, addition reactions of the C=C bond of ketenes with other C=C bonds, with carbonyl, azomethine and azo compounds, with N=O and C=S bonds and finally with phosphine derivatives are summarized. In the last two sections some reactions with organometallic compounds and oxidations are presented.

251 references up to 1962.
Complementary: **15**, 2 (V); **25**, 19 (II.L, III.C.2a and Tables 32, pp. 1234–1263); **26**, 1, 4–10, 13, 14, 15, 16, 17, 19.
Relevant: **15**, 10 (p. 330); **26**, 18.

The chemistry of the carbonyl group (1966)

2, *1: General and theoretical aspects of the carbonyl group* by *G. Berthier and J. Serre*

I. Introduction; **II.** Physicochemical picture of the carbonyl group (3); **III.** Theoretical approach to molecules with carbonyl bonds (39); **IV.** Acknowledgements (72); **V.** References (72).

The bonds between C and O in carbonyl groups are made up of σ and π types. This electronic structure and the geometrical position control the physicochemical properties of carbonyl compounds.

Section II.A deals with physical properties and polarity, including bond lengths (Table 1), bond energies (Table 2), dipole moments (Table 3) and ionization potentials (Table 4). Section II.A.5 considers vibrational (IR) spectra and the influence of a variety of factors (inductive, mesomeric, electronic and mass effects, ring strain, hydrogen bonding, enolization, solvation) on the carbonyl stretching frequencies (Table 5). The next section (II.B) discusses UV spectra including $n \rightarrow \pi^*$ transitions (Tables 6 and 7), and the effects on these of conjugation, of substitution and of solvents (Table 8). In continuation, the uses of ORD and CD are explained, and $\pi \rightarrow \pi^*$ transitions (Table 9) are considered.

Section III surveys the MO and VB descriptions of carbonyl bonds, both for the ground state and for some excited states. The definitions and equations used in this section are based on Chapter **1**, 1 by Coulson and Stewart and referred to in parentheses as C.S., followed by the section number in **1**, 1). Electronic structures, configurations, MOs and charge distributions are presented in Figures and Tables. Sections III.C and III.D describe electronic structures of conjugated carbonyl compounds and of radical ions, respectively.

Finally, Section III.E presents, as an example of a complete quantum-mechanical calculation, *ab initio* calculations of the ground state of the formaldehyde molecule, including tabular representations of the AO coefficients, energy data and charge distribution (Tables 19–21).

186 references up to 1965.
Complementary: 1, 1; **21**, 1.
Relevant: 3, 1; **4**, 1; **5**, 1; **6**, 1; **8**, 1; **11**, 1; **12**, 1; **13**, 1; **14**, 1; **15**, 1; **16**, 1; **17**, 1; **18**, 1; **19**, 1; **20**, 1; **22**, 1; **23**, 1; **24**, 1; **25**, 1; **26**, 1; **27**, 9; **28**, 1; **31**, 27; **32**, 1; **33**, 1; **36**, 2.

2, *2: Carbonyl-forming oxidations* by *C. F. Cullis and A. Fish*

I. Introduction (80); **II.** Oxidation of alkanes and cycloalkanes (81); **III.** Oxidation of alkenes, cycloalkenes and alkynes (100); **IV.** Oxidation of aromatic hydrocarbons (120); **V.**

Oxidation of alcohols (129); **VI**. Oxidation of carbonyl compounds to di- and tricarbonyl compounds (157); **VII**. References (162).

The chapter is limited to the treatment of those reactions where the starting material is a hydrocarbon (alkane, alkene, alkyne, arene) or an alcohol. Section II describes the oxidations of alkanes and cycloalkanes in the gaseous and liquid phases mainly by molecular oxygen, in addition to some other oxidants, in the absence and presence of homogeneous and heterogeneous catalysts.

Section III deals with the oxidation of alkenes, cycloalkenes and alkynes, again both in the gaseous and liquid phases and in the absence and presence of various homogeneous and heterogeneous catalysts, by means of molecular oxygen and of other oxidants (atomic O, H_2O_2, O_3, $KMnO_4$, etc.).

Dividing the material as above, Section IV discusses the oxidation of aromatic hydrocarbons and Section V that of alcohols. In the latter case, the facile liquid-phase oxidation can be carried out with chromates, permanganate, transition metal ions, aluminium alkoxides, Br_2, peroxides, persulphates, N_2O_4, $Pb(OAc)_4$ and some other less popular reagents. In the heterogeneous mode, over a wide variety of catalysts, dehydrogenation of alcohols to aldehydes and ketones is often rapid and complete (Section V.B.2).

The last section describes the oxidation of ketones to di- and tricarbonyl compounds by O_2, SeO_2 and nitrous acid derivatives.

536 references up to 1965.
Complementary: 10, 2; **13**, 5 (III); **21**, 11; **27**, 12;
Relevant: 21, 12; **27**, 11; **31**, 13; **33**, 19.

2, *3: Formation of carbonyl groups in hydrolytic reactions* by *P. Salomaa*

I. Introduction (177); **II**. Displacements on saturated systems (178); **III**. Unsaturated systems (192); **IV**. References (207).

Carbonyl groups are formed by replacement of two halogens or alkoxyl groups attached to the same carbon atom by two hydroxyl groups, or by hydrolytic attack on various unsaturated (alkenic, acetylenic or azomethinic) systems.

Section II discusses the hydrolysis of α, α-dihalides in aliphatic systems, and in aromatic compounds dihalogenated in the side-chain, yielding carbonyl groups. The same result is given by the hydrolysis of α-halo ethers, α-haloalkyl esters and α-halo thioethers and of acetals, ketals, acetal esters and similar compounds, and in some cases even of 1, 2-dihalo compounds after rearrangement.

Unsaturated systems (Section III) in which a halogen or similar group undergoes hydrolysis yield enols, followed by tautomerization to the corresponding carbonyl compound. C=C double bonds may be hydrolysed to the corresponding carbonyl and methylene compounds if the double bond is sufficiently activated by electron-attracting groups towards nucleophilic attack.

In C=N double bonds the doubly bonded N atom may be replaced by a doubly bonded O atom (Section III.C). The samples include oximes, semicarbazones, Schiff bases, imines, alkylidene cyanamides and the *aci* forms of nitro compounds (i.e. the Nef reaction).

Throughout the chapter many references are made to the mechanisms of these reactions.

123 references up to 1964.
Complementary: 1, 8; **3**, 7; **8**, 10; **6**, 5 (II.X.W); **24**, 7 (III.B); **21**, 20 (III); **29**, 13 (III.C.5).
Relevant: 2, 2, 4, 5, 6: **15**, 7; **21**, 12; **21**, 8 (III.D).

2, *4: Formation of aldehydes and ketones from carboxylic acids and their derivatives* by *R. C. Fuson*

I. Introduction (2.11); **II**. From acids and their salts and anhydrides by pyrolysis (212); **III**. From acid chlorides (217); **IV**. From amides (220); **V**. From nitriles (225); **VI**. From esters (227); **VII**. References (239).

Synthetic methods are available through which nearly all the various carboxylic acid derivatives (e.g. COOH, COX, $CONH_2$, CN, COOR) can be converted into aldehydes and ketones.

Section II discusses the preparation of carbonyl compounds from acids, salts and anhydrides by pyrolysis. Section III deals with the (formal) hydrogenolysis of acid chlorides to aldehydes by the methods of Rosenmund and of Reissert and by the use of metal hydrides. Similarly, tertiary acid amides can be converted into aldehydes by several methods, especially by metal hydrides (Section IV). Nitriles have to be first converted into imines, which are then hydrolysed to the corresponding aldehyde either by the Stephen method or by metal hydrides (Section V). Finally (Section VI), esters give α-hydroxy ketones by acyloin condensation, thiol esters give aldehydes by hydrogenolysis/ desulphurization and some metal hydrides give aldehydes from esters.

81 references up to 1964.
Complementary: 12, 7; **15**, 7.
Relevant: 11, 13 (VIII); **25**, 15.

2, *5: Formation of ketones and aldehydes by acylation, formylation and some related processes* by *D. P. N. Satchell and R. S. Satchell*

I. Introduction (235); **II**. Direct acylation and formylation of hydrocarbons (239); **III**. Indirect acylation and formylation of hydrocarbons (278); **IV**. Conclusions (298); **V**. References (298).

Aldehydes and ketones can be synthesized by direct or indirect formylation or acylation. In most cases these reactions are electrophilic attacks by the carbonyl carbon atom of the reagent on a uncleophilic carbon of the substrate.

Section II describes direct acylations, in which the carbonyl group is introduced into the substrate in a single step, retaining its double-bond character throughout the reaction. In the reagent RCOY, R may be H or any alkyl or aryl group and Y, the leaving group, may be NR'_2, OR′, OCOR′, Hal, OSO_3H, $OClO_3$, etc. The reaction is influenced by the nature of R and of Y, by catalysts and by the nature (reactivity) of the substrate, which also determines the position of the attack.

The substrates may be aromatic hydrocarbons (Section II.E), olefins (II.F) or unsubstituted or substituted saturated aliphatic hydrocarbons (II.G and H). In each case the scope, conditions and mechanisms of the reactions are discussed.

In indirect acylations (Section III), the carbon atom of the potential carbonyl group is introduced in the first step and the double-bonded oxygen is established in a subsequent step. Acylation by carboxylic acid amides and $POCl_3$ (Section III.A) yields aldehydes if the amide is a formamide (Vilsmeier synthesis) and ketones with all other reagents. Other routes use nitriles (III.B) and some less common reagents (III.C–E). In each case, different substrates (aromatic, olefinic, saturated) are considered, together with the scope, conditions and mechanisms.

Finally, formylation by chloroform and alkali (Reimer–Tiemann reaction, Section III.F) is reviewed.

113 references up to 1965.
Complementary: 11, 13 (IV).
Relevant: 2, 3, 4, 6; **15**, 11 (IV.D); **21**, 12; **21**, 8 (III.D).

2, *6: Carbonyl syntheses through organometallics* by *M. Cais and A. Mandelbaum*

I. Introduction (303); **II**. The use of organocadmium compounds (304); **III**. The use of organomagnesium compounds (318); **IV**. The use of organolithium compounds (325); **V**. The use of metal carbonyls (327); **VI**. References (329).

Organic derivatives of Al, Cd, Cu, Pb, Mg, Hg, Na, Sn and Zn react with acid halides and similar compounds to yield ketones:

$$R^1COX + R^2M \rightarrow R^1COR^2 + MX$$

Section II describes the reaction of organocadmium compounds with acid chlorides and anhydrides and some other functional groups and discusses experimental conditions and mechanisms.

Organomagnesium compounds are treated in Section III. Acid chlorides yield ketones in their reactions with Grignard reagents. The reaction is useful for the preparation of aromatic, alicyclic and of sterically hindered ketones. Two short sections deal with the use of organolithium reagents (IV) and of metal carbonyls (V).

53 references up to 1965.
Complementary: 21, 12.
Relevant: 2, 2–5; **21**, 8(III.D); **35**, 7–10.

2, *8: Chemical and physical methods of analysis* by *J. G. Hanna*

I. Introduction (373); **II**. Chemical methods (377); **III**. Physical methods (397); **IV**. Acknowledgements (415); **V**. References (415).

Section II.A describes the qualitative chemical methods of analysis, including formation of hydrazones, oximes, semicarbazones, hydantoins, dimedone derivatives and the use of Schiff's reagent, copper(II) salts, the iodoform reaction and some colour tests. The quantitative chemical methods (Section II.B) essentially consist of the same reactions developed to give quantitative results, and some additional quantitative oxidations and reductions.

The physical methods described are the various chromatographic techniques (Section III.A), visible and UV (III.B) and IR spectroscopy (III.D), polarography (III.E) and NMR (III.F). In each subsection, qualitative and quantitative methods are discussed.

227 references up to 1964.
Complementary: 17, 4; **26**, 5.
Relevant: 1, 5; **3**, 15; **4**, 3; **5**, 17; **8**, 3; **12**, 4; **13**, 6; **14**, 3; **15**, 3; **16**, 3; **18**, 5; **20**, 3; **22**, 5; **23**, 5; **24**, 5; **25**, 10; **28**, 3; **29**, 21; **30**, 14–22; **36**, 5.

2, *9: Basicity of carbonyl compounds* by *V. A. Palm, Ü. L. Haldna and A. J. Talvik*

I. Introduction (421); **II**. The classical acid–base concept and the quantitative measure of basicity; **III**. Modern variations in the acidity–basicity concept (424); **IV**. Basicity and acid

The first two sections outline the classical (Brønsted, Lowry, Lewis) concepts of basicity and the modern variations of them, whch consider also partial interactions between acids and bases, solvation phenomena and water activity (a_{H_2O}).

Section IV discusses acid catalysis in prototropic and other processes; equilibrium interactions between CO compounds and acids are treated further (Section V), including an extensive tabulation of pK_a and pK_h values (where $pK_h = \log a_{H_2O}$ at $H_0 = pK_a$) for 56 ketones and 17 aldehydes. Sections VI and VII deal with acid catalysis in carbonyl reactions and present a general scheme for the quantitative descriptions of these reactions.

Section VIII presents an explanation of general and specific acid catalysis. The last two sections deal with basicity in non-aqueous media and with the concept of Lewis acidity applied to non-protonic acids (e.g. MgX_2, BF_3).

104 references up to 1964.
Complementary: 10, 1 (VII).
Relevant: 3, 6; **4**, 4; **5**, 6; **8**, 5; **11**, 3; **12**, 3; **13**, 7; **15**, 4; **18**, 8; **19**, 7; **20**, 12; **23**, 6; **24**, 4; **25**, 6; **33**, 5.

2, *10: Oxidation of aldehydes by transition metals* by *J. Roček*

The subject matter has been restricted only to oxidants derived from Cr, Mn, V, Ce and Co, so as to permit a detailed discussion of both experimental findings and mechanisms.

Section II treats the CrO_3 oxidation of aromatic and aliphatic aldehydes, presenting first experimental (especially kinetic) data followed by mechanistic proposals as to the role of aldehyde hydrates and of the valency changes of the metal in the various stages of the reaction, and as to the nature of the aldehydic C—H bond cleavage involved in the loss of H from the substrate. Oxidation by permanganic acid and other manganic salts (Sections IV and V) is treated for aromatic and aliphatic aldehydes at different acidities, again discussing kinetics, activation parameters and mechanisms. Oxidation by V, Ce and Co salts is discussed in shorter sections (VI–VIII).

103 references up to 1964.
Complementary: 2, 2; **10**, 2.
Relevant: 21, 11.

2, *11: Reduction of carbonyl groups* by *O. H. Wheeler*

Aldehydes, in a reduction process involving two electrons and two protons, are converted into primary alcohols whereas ketons in the same reaction produce secondary alcohols. These are defined as *unimolecular reductions* as distinct from *bimolecular reductions*, in which addition of one electron and one proton per carbonyl molecule produces a pinacol.

H_2 gas in the presence of a metallic catalyst gives unimolecular reductions, although in special cases side reactions (especially reactions with other reactive groups in the molecules, if present) may predominate (Section II). Dissolving metals in acid, neutral or basic solutions may give, in addition to unimolecular reductions, other processes, e.g. Clemmensen reduction, $>CO \longrightarrow >CH_2$, or bimolecular reductions (Section III). Electrochemical reduction (Section IV) may yield alcohols, pinacols or hydrocarbons. The carbonyl O atom may be replaced by N by reductive animation (Section V), i.e. by catalytic hydrogenation in the presence of NH_3, or though the Leuckart reaction and its modifications.

The CO group may be reduced by an organic compound which is itself oxidized (Section VI). Examples are the Meerwein–Pondorf–Verley, Cannizzaro and Tishchenko reactions and reductions by bulky Grignard reagents. Metal hydrides reduce CO groups to COH groups or even to CH_2 groups (Section VII). Finally, Section VIII discusses miscellaneous methods, e.g. the Wolff–Kishner reduction, biochemical and photochemical and radiation-chemical methods.

In each section the mechanisms of the reactions are discussed.

185 references up to 1964.
Complementary: 13, 4 (V); **21**, 4 (IV.D).

2, *12: Condensations leading to double bonds* by *R. L. Reeves*

The chapter emphasizes principles and generalizations based on studies of these reactions. The condensation involve at least two steps: first the nucleophilic addition of the reagent to the carbonyl group and second an elimination step (Scheme, p. 568), with the overall reaction being almost always reversible and catalysed by acids and/or bases.

Section II describes formation of C=C, and discusses the kinetics, mechanism, stereochemistry, scope, conditions and various modifications of the Wittig reaction (Section II.A), the aldol condensation (II.B) and the Knoevenagel and Perkin reactions (II.C).

Section III reviews reactions leading to C=N groups, again giving primary emphasis to the mechanisms of the processes. It is suggested that reactions of CO groups with all nitrogen bases are very similar, the basic mechanism being again a two-step addition–elimination mechanism. The products discussed are oximes, semicarbazones and hydrazones (Section III.B) and Schiff bases (III.C). Factors affecting reactivity are recorded (III.B.3) and mechanistic pictures are presented for Schiff base formation in solutions of different pHs (Scheme 1, p. 613).

188 references up to 1965.
Complementary: 4, 7; **8**, 2; **10**, 1 (IX); **21**, 4 (IV).
Relevant: 1, 8.

2, *13: Reactions of carbonyl groups with organometallic compounds* by *T. Eicher*

The chapter treats only reactions of carbonyl groups with carbon—metal bonds.

Section II discusses the reactions of organoalkali metal reagents with CO groups, including 1, 2- and 1, 4- (or conjugate) additions, reductions, ion-radical transfers and the mechanisms of these. The similarities and differences between the modes of action to organoalkali metal and organomagnesium compounds are emphasized, and the concept of 'at-complexes' is explained.

Section III deals with the organic chemistry of Group II elements, mostly with organomagnesium compounds, leading to reduction, enolization and 1, 2- and 1, 4-addition of the carbonyl compounds. The structure of the Grignard reagent is depicted in detail (Section III.A.3). Further, compounds of the type RMX and R_2M, where M = Be, Zn or Cd, are presented. Reactions leading to additions and to at-complexes are outlined.

Section IV presents derivatives of Group III metals, mainly organoaluminium compounds, their participation in additions and reductions with carbonyl groups and the mechanisms of these reactions.

192 references up to 1965.
Complementary: 2, 6; **35**, 2.
Relevant: 2, 6; **35**, 2.

2, *14: Decarbonylation* by *W. M. Schubert and R. R. Kinter*

The chapter discusses reactions in which aldehydes or ketones yield carbon monoxide and also reactions in which aldehydes yield formic acid or formate ion.

In aromatic electrophilic deformylation, aromatic aldehydes substituted with electron-releasing groups undergo acid-catalysed replacement of the CHO group by H (Section II.A), or the formyl substituent may be replaced by nitro groups, halogens or other groups (II.B).

Aliphatic and aromatic aldehydes also undergo deformylation in the presence of basic catalysts (Sections III). In a reaction initiated by radicals (Sections IV), when various rearrangements may also occur, aliphatic aldehydes decarbonylate, but aromatic aldehydes undergo other reactions, e.g. intramolecular acylation.

Section V describes the thermal decarbonylation of various aliphatic and aromatic aldehydes, inhibition by propene and varying effects of nitric oxide. Open-chain aliphatic cyclic and cycloalkyl ketones and 1, 2-diketones and carbonyl bridge compounds are treated more briefly.

In Section VI the decarbonylation of saturated, unsaturated and aromatic aldehydes in the presence of solid hydrogenation catalysts is presented. The Dakin reaction, yielding dihydroxy aromatic compounds from aromatic hydroxy aldehydes (VI.B), and its mechanisms are described, together with the decarbonylation of diphenyl triketones (VI.C), of phenanthraquinones (VI.D) and 9-acylanthracenes (VI.E).

209 references up to 1965.
Complementary: 2, 16; **25**, 14.
Relevant: 10, 6; **15**, 9; **34**, 5(II); **34**, pp. 683, 736, 745.

2, *15: Rearrangements involving the carbonyl group* by *C. J. Collins and J. E. Eastham*

Rearrangements are discussed in which a ketone or aldehydic carbonyl group is either formed or destroyed during the reaction, and in which the total number of C atoms remains constant. The following acid- or base-catalyzed reactions are discussed in Section II (including aspects of yields, migratory aptitudes, mechanisms and stereochemistry): pinacol and semipinacolic rearrangements, together with those of aldehydes, ketones, α-ketols (both acid- and base-catalyzed), dienones, quinamines and of benzil (to benzilic acid). The stereochemistry of ketonization is outlined in Section II.G, followed by treatment of Favorsky-type reactions and ring–chain tautomerism.

Thermal rearrangements leading to aldehydes and ketones are treated in Section III, and oxidative rearrangements involving carbon-to-oxygen and carbon-to-carbon migrations during the decomposition of primary oxidation products in Section IV. Finally (Section V), photochemical rearrangements (usually by UV) are discussed.

226 references up to 1964.
Complementary: 2, 16; **8**, 9; **10**, 3.
Relevant: 1, 7; **3**, 14; **4**, 10; **5**, 8; **9**, 3; **11**, 4; **12**, 16; **13**, 13; **14**, 7; **15**, 8; **16**, 17; **17**, 8; **18**, 15; **19**, 18; **20**, 4; **23**, 13; **24**, 10; **26**, 15; **29**, 4.

2, *16: Photochemistry of ketones and aldehydes* by *J. N. Pitts, Jr, and J. K. S. Wan*

The chapter concentrates on generalizations regarding photochemical processes of ketones and aldehydes, while referring the reader for details and experimental aspects to reviews and books.

Section II describes the spectroscopy of the C=O group in saturated, unsaturated and aromatic ketones and aldehydes, the effects of solvents and the emission from excited states. Section III deals with primary photochemical processes of C=O groups, including decarbonylation, photocycloelimination of olefins, intramolecular rearrangements, reductions and H-abstractions, and also photocyclodimerization and C=O addition to olefins.

Section IV considers in more detail the primary photochemical processes, the nature and yields of simple aliphatic, halogen-substituted and unsaturated C=O compounds and diketones and dialdehydes. Section V deals with cyclic aldehydes and ketones, cyclic diketones and bicyclic and tricyclic ketones.

α, β-Unsaturated ketones are reviewed in Section VI, including rearrangements and additions of cyclic and heterocyclic unsaturated ketones and steroid dienones. Section VII deals exclusively with aceto- and butyrophenone and Section VIII with some relatively simple aromatic aldehydes, ketones and diketones.

Finally, Section IX discusses Hg-photosensitized reactions, some photochemical processes in the solid state (including photochromism, rearrangements and dimerizations) and the photolysis of COS.

281 references up to 1966.
Complementary: **17**, 9.
Relevant: **3**, 8; **6**, 4; **8**, 12; **9**, 6; **11**, 5; **13**, 16; **14**, 8; **15**, 9; **16**, 11; **18**, 10; **19**, 20; **22**, 10; **23**, 9; **24**, 12; **25**, 11; **27**, 21, 22; **28**, 5; **29**, 5, 6; **32**, 29; **33**, 21.

2, *17: Thioketones* by *E. Campaigne*

I. General comments (918); **II**. Preparation of thiones (919); **III**. Physical properties of the thiocarbonyl group (934); **IV**. Reactions of thiones (938); **V**. Thiocarbonyls as transient intermediates (953); **VI**. References (955).

Thioketones (thiones) are unstable, very intensly coloured (usually blue or violet) compounds. They can be prepared (Section II) by acid- or amine-catalysed addition of H_2S to carbonyl groups with subsequent elimination of H_2O. However, various side reactions (polymerization, air oxidation, etc.) complicate the procedure. Both C=O and CCl_2 groups can be converted into C=S groups by reaction with a variety of S-containing reagents. Thiones are also obtained by pyrolysis of trithianes and other S-containing compounds and by condensations of CO compounds with rhodamine and subsequent hydrolysis to yield α-thioketo acids.

Section II describes physical properties, bond distances and energies, IR, UV and visible spectra and ORD.

Section IV discusses nucleophilic and dipolar additions, complex formation, reduction, the coupling reaction $2R^1R^2C{=}S \rightarrow R^1R^2C{=}CR^1R^2$ (which distinguishes thiones from ketones), oxidations, photochemistry, the thione–enethiol tautomerism and the contribution of C=S groups to no-bond resonance.

Some cases of C=S compounds proposed as transient intermediates in alkaline cleavage of disulphides and in the Willgerodt–Kindler reaction are outlined in Section V.

135 references up to 1965.
Complementary: none.
Relevant: **3**, 13; **5**, 15; **11**, 8; **15**, 11; **25**, 18.

The chemistry of the ether linkage (1967)

3, *1: Theory of the ether linkage* by *D. Peters*

I. Introduction (1); **II**. The isolated oxygen atom (4); **III**. Bonding in simple ethers (6); **IV**. Bonding in conjugated ethers (11); **V**. Bonding in strained cyclic ethers (14); **VI**. Electronic excited states, ultraviolet spectra and ionizing potentials of ethers (14); **VII**. Chemical reactivity of ethers (16); **VIII**. Thioethers (16); **IX**. References (17).

The electron organization of the ether linkage is presented primarily in the framework of high-accuracy computational work, by analogy with the closest molecule for which such accurate wave functions are available, i.e. water.

The isolated O atom is first discussed, followed by treatment of VB and MO wave functions for bonding in simple, conjugated and (strained) cyclic ethers.

Very short sections outline excited states, UV spectra, ionization potentials and the chemical reactivity of ethers, and the analogy between ethers and thioethers.

50 references up to 1965.
Complementary: none.
Relevant: 1, 1; **2**, 1; **4**, 1; **5**, 1; **6**, 1; **7**, 1; **8**, 1; **11**, 1; **12**, 1; **13**, 1; **14**, 1; **15**, 1; **16**, 1; **17**, 1; **18**, 1; **19**, 1; **20**, 1; **22**, 1; **23**, 1; **24**, 1; **25**, 1; **26**, 1; **27**, 9; **28**, 1; **31**, 27; **32**, 1; **33**, 1.

3, *2: Cleavage of the C—O—C bond* by *E. Staude and F. Patat*

I. Introduction (22); **II**. Acid-catalysed cleavage (22); **III**. Base-catalysed cleavage (46); **IV**. Cleavage by metal halides (53); **V**. Organometallic compounds (58); **VI**. Grignard reagents (62); **VII**. Complex metal hydrides (65); **VIII**. Cleavage by alkali metals (79); **IX**. Organosilicon compounds (73); **X**. Further possibilities of ether cleavage (74); **XI**. References (75).

The kinetics and mechanisms of the acid-catalysed hydrolysis of ethers may involve either a unimolecular (*A*1) or a bimolecular (*A*2) RDS. The reaction which requires strong concentrated acids (HX, H_2SO_4, $HClO_4$), was studied in some cases also in non-aqueous media, and in others at very high pressures (up to 7000 atm).

Section II describes the mechanisms of and methods for the basic cleavage of acyclic and cyclic ethers, usually under drastic conditions of base concentrations and temperatures and in many cases also high pressures.

When heated, complexes of ethers with Lewis acids, lead to decomposition of the ether. Metal halides such as $AlCl_3$, BCl_3, BF_3, $TiCl_4$, $SnCl_4$ and $FeCl_3$ can be used and polymerizations and rearrangements occur as side reactions (Section IV).

Ether cleavage with RM compounds (EtNa, AmNa, PrNa, BuLi, etc.) usually gives β-elimination, in many cases after preliminary metalation (Section V).

Grignard reagents cleave ethers usually only at high temperatures (lower with strained cyclic ethers) and both displacements and eliminations occur (Section VI).

$LiAlH_4$, $LiBH_4$ and related hydrides attack ethers usually only in the presence of catalysis such as $CoCl_2$ or $AlCl_3$ (Section VII). Aryl and benzyl (but not alkyl) ethers react with alkali metals (Section VIII) and oxiranes with organosilicon compounds (Section IX).

219 references up to 1965.
Complementary: **27**, 14, 15, 19, 21, 23.
Relevant: 5, 3.

3, *3: Directive and activating effects of alkoxy groups in aromatic and aliphatic reactions* by *G. Kohnstam and D. L. H. Williams*

Section I gives a brief outline of the polar ($-I$, $+T$) effects of RO and ArO groups. The Hammett equation in its original and revised forms is described and mean values of the various substituent constants are given (Table 2) and compared (Section II).

The influences of RO and ArO groups on the ionization constants of acid derivatives and on the dissociation constants of O and N bases, conjugated acids, positively charged acids and some other equilibria are presented in Section III.

S_N2 processes are strongly influenced by the presence of RO and ArO groups in the nucleophile and in the substrate and so are solvolytic reactions, in addition to reactions in which neighbouring-group participation can occur, exerting a stabilizing influence on the TS (Section IV).

Section V describes the influence of RO groups on the solvolysis of carboxylic esters and of benzoyl chlorides and also on $CO–CH_2$ condensations.

Section VI deals with the influence of one or more RO and ArO groups on the mechanisms, rates and selectivity of electrophilic aromatic substitutions, tabulating isomer proportions (Table 19) and relative rates and partial rate factors (Table 20) for MeO and PhO derivatives in 16 different reactions. The influence of RO groups in the presence of other substituents is discussed (Section IV.C.2).

Finally, nucleophilic aromatic substitutions and electrophilic reactions in non-aromatic systems (alkenes, alkynes) and reactions involving radicals are described (Section VII).

176 references up to 1965.
Complementary: none.
Relevant: 4, 5: **5**, 16; **6**, 8; **7**, 2; **11**, 12; **12**, 5; **13**, 8; **14**, 4; **15**, 5; **16**, 6; **18**, 9; **23**, 11; **28**, 9; **31**, 8.

3, *4: The pyrolysis of ethers* by *K. J. Laidler and D. J. McKenney*

Contrary to previous assumption that the pyrolysis of organic compounds in general and of ethers in particular are molecular in nature, almost all such reactions can now be explained in terms of complex free radical mechanisms.

High-temperature pyrolyses of Me_2O (500–500 °C) and of Et_2O (ca 600 °C) are reviewed in detail, considering detailed mechanisms, uninhibited and inhibited processes, catalysis, kinetic parameters, pressure changes, intermediates and products.

Section IV discusses the pyrolysis of some other ethers, reacting through uncertain mechanisms, although in some cases the pathway seems to be a *cis*-elimination occurring through a cyclic activated complex.

The combination of Me radicals to yield ethane is of considerable importance in Me_2O pyrolysis. The relevant material, i.e. the low-temperature reaction $2CH_3 \rightarrow C_2H_6$, and other results obtained in the pyrolysis of acetaldehyde, of ethane and of the ethyl radical are described in Section V.

73 references up to 1965.
Complementary: 27, 14.
Relevant: 5, 12; **16**, 10; **22**, 15; **25**, 15; **32**, 7; **33**, 20.

3, *5: Biological formations and reactions* by *K. Wallenfels and H. Diekmann*

The ether linkage occurs in natural products in non-cyclic and cyclic ethers, acetals, glycosides, enol ethers and in many polymeric carbohydrates (e.g. starch, cellulose). The linkage can be formed by transmethylation by enzymatic formation of glycosidic groups in what are usually specific reactions. Biogenetic processes leading to cyclic ethers (chroman derivatives), to enol ethers and to thioethers are described (Section II).

Among the reactions of the C—O—C grouping, dealkylation, hydrolysis and transfer of *O*- and *S*-glycosides and also phosphorolysis of *O*-glycosides are described, including the specificity, structure and mechanisms involved in the enzyme action. Finally, the enzyme-catalysed mutarotation is discussed.

316 references up to 1965.
Complementary: none.
Relevant: 4, 9; **5**, 18; **7**, 7; **11**, 11; **12**, 12; **13**, 14; **15**, 10; **16**, 13; **17**, 13; **18**, 13; **19**, 13; **20**, 6; **22**, 20; **26**, 13; **28**, 17; **31**, 9.

3, *6: Basicity and complexing ability of ethers* by *S. Searles, Jr, and M. Tamres*

A wide variety of Brønsted and Lewis acids can form complexes with the O-atom of ethers, employing the unshared electron pair. The types of interactions are ion–dipole, dipole–dipole or dipole–induced dipole, 'strong complexation' occurring with acids with formation of a coordinate covalent bond, whereas weaker interactions give weaker, non-isolable complexes (Section II).

Section III classifies the complexes. Oxonium salts are formed by protonation of ethers by strong acids (HNO_3, H_2SO_4, etc.), by solvation of other (metal) cations by ethers, by alkyl halides yielding trialkyloxonium salts and by acyl halides yielding dialkylcyloxonium salts. Non-ionized complexes are formed from ethers with salts and also with organometallic compounds of alkali metals, alkaline earth metals, compounds of Group III elements (BF_3, $AlCl_3$, etc.) and compounds of heavy metals (Au, Hg, Sn, etc.). Complexes are also formed by oxides and halides of non-metals (SO_3, N_2O_4, PCl_5, etc.).

Weakly bonded complexes are also formed by halogens and by compounds in general which are able to give hydrogen bonds as shown by UV.

The strength of complexes is affected by inductive, conjugative, steric, solvation and chelation effects (Section IV).

357 references up to 1965.
Complementary: 31, 21.
Relevant: 2, 9; **4**, 4; **5**, 6; **8**, 5; **13**, 7; **15**, 4; **16**, 5; **18**, 8; **19**, 7; **20**, 12; **22**, 8; **23**, 6; **24**, 4; **25**, 5; **33**, 5.

3, *7: Acetals and hemiacetals* by *E. Schmitz and I. Eichhorn*

Hemiacetals can be isolated only in rare cases, usually when the parent carbonyl compound is strongly electron deficient or when the etheric oxygen of the hemiacetal is part of a cyclic system. They may be used as transients in various reactions, e.g. acetalization (Section II).

Acetals may be prepared in an equilibrium reaction between a CO group and an alcohol, using acid catalysis or via alkoxycarbonium ions which can be obtained by a variety of routes (from carbonium ion salts, from α-halo ethers, by transacetalization, from alkynes or by addition of alcohols to α, β-unsaturated ethers). In addition, acetals can be formed in basic media from CO compounds and from ortho esters (Section III).

The spectroscopic, NMR and MS behaviour of acetals is delineated in Section IV, while Section V deals with the use of acetals as protective groups, with their hydrolysis and other ionic reactions and finally with radical reactions such as bromination, autoxidation, sensitivity to decomposing peroxides and thermal decomposition.

181 references up to 1965.
Complementary: 10, 1; **13**, 20; **26**, 14; **27**, 20.
Relevant: 26, 16.

3, *8: Photochemistry of the C—O—C group* by *D. Elad*

Photochemical reactions of ethers, epoxides and acetals may involve both C—O and C—H cleavages, the latter especially at the reactive α-position to the ether linkage.

Section II discusses the photolysis of ethers, epoxides, aromatic ethers and esters, the photooxidation of ethers, the photochemical reactions of ethers with CO compounds and with peresters, the photochemical cyanation of ethers, photoaddition of ethers to alkenes, photochemical solvolyses and halogenations, the addition of acetals to alkenes and the conversion of acetals into esters and bromination of the former with NBS.

64 references up to 1965.
Complementary: 27, 21, 22.
Relevant: 2, 16; **6**, 4; **8**, 12; **9**, 6; **11**, 5; **13**, 16; **15**, 9; **16**, 11; **17**, 9; **18**, 10; **19**, 20; **22**, 10; **23**, 9; **24**, 12; **25**, 11; **29**, 6; **32**, 29; **33**, 21.

3, *9: Reactions of cyclic ethers* by *R. J. Gritter*

I. Introduction (374); **II**. Synthesis of epoxides (381); **III**. Reactions of epoxides (390); **IV**. Preparation of oxetanes (411); **V**. Synthesis of the larger-ring cyclic ethers (414); **VI**. Chemical reactions of cyclic ethers (419); **VII**. References (431).

The introduction outlines bond and strain energies, ORD, MS, IR and NMR data, basicity, inertness and solvent properties and polymerization. Section II describes the syntheses of epoxides by peracid oxidation of olefins, from halohydrins and using Wittig reagents, phosphorus triamide, carbenes, ylides and O_2. Attempted syntheses of chloroepoxides and of oxirenes are mentioned.

Oxidation, reduction and conversion to heterocyclic compounds of epoxides are considered, followed by descriptions of the use of special reagents (metal carbonyls, phosphines, amines, etc.). Acid- and base-catalysed cleavages, rearrangements, photochemical and free radical reactions are summarized (Section III).

Section IV deals with the preparation of oxetanes and Section V with that of larger ring cyclic ethers by both nucleophilic and oxidative ring formation.

The last section describes the cleavage of cyclic ethers (other than epoxides) and some other general and special reactions such as the participation of the tetrahydrofuranyl group in the solvolysis of brosylate esters. Finally, some free-radical reactions (replacement of hydrogen by halogens, alkylations, additions) are summarized.

424 references up to 1965.
Complementary: 27, 14, 15, 18.
Relevant: 13, 12; **26**, 14; **27**, 1–4, 14, 15.

3, *10: Methods of formation of the ether linkage* by *H. Feuer and J. Hooz*

I. Introduction (446); **II**. Nucleophilic substitutions (446); **III**. Ethers from alcohols, phenols and carbodiimides (468); **IV**. Ethers from peroxy compounds (470); **V**. Ethers from the Smiles rearrangement (474); **VI**. Ethers from diazo compounds (478); **VII**. Ethers produced by Kolbe electrolysis (488); **VIII**. References (492).

The preparation of ethers from ROH and ArOH can be carried out using alkyl or aryl halides, diphenyliodonium salts, dialkyl sulphates and arylsulphonic esters. Mixed ethers are especially easily prepared by acid-catalysed etherifications of two alcohols. The mechanism of these nucleophilic substitutions, the importance of the leaving group and the role of the solvent (Table 9) and of the alkyl halides (Tables 11–13) are described (Section II).

Carbodiimides are useful dehydrating agents (Section III). Other methods involve reactions of organometallics with dialkyl peroxides or with *tert*-butyl peroxide, R_3P compounds with dialkyl peroxides and the thermolysis of triarylmethyl peroxides (Section IV). Useful methods are also the Smiles rearrangement (Section V) and the methylation of alcohols by CH_2N_2, the preparation of α-alkoxyketones from α-diazoketones, of aryl alkyl ethers from aromatic diazonium compounds and alcohols (Section VI) and finally in the Kolbe electrolysis as a by-product of varying importance (Section VII).

183 references up to 1965.
Complementary: 3, 12.
Relevant: 1, 8; **13**, 11; **26**, 14, 16; **27**, 1, 11, 14–17.

3, *11: The polyethers* by *C. C. Price*

I. Introduction (499); **II**. Stereochemistry of polymer chains (503); **III**. Stereoselective polymerization of propylene oxide (506); **V**. Concepts of mechanism (511); **VI**. Poly(phenylene oxides) (517); **VII**. References (522).

In polyethers the ether oxygen is part of the backbone of the polymer chain. The bond strength of the C—O bond (84.0 kcal mol^{-1}) is comparable to that of C—C (83.1 kcal mol^{-1}), and the low barrier of rotation (ca 1.2 vs ca 3 kcal mol^{-1} for C—C) render the polyethers excellent structures for synthetic elastomers.

Epoxides polymerize by anionic, cationic or coordinative rearrangements. The stereochemistry of polyethers (conformations, configurations, isotactic, syndiotactic and atactic arrangements) is discussed in Section II, and the stereoselective polymerization of propylene oxide, mainly to yield an isotactic polymer, in the presence of the Fe-containing Pruitt and Baggett catalyst in Section III. Studies relating to this catalyst and other catalysts are treated in Section IV and mechanisms are proposed for the polymerization processes with different catalysts and conditions in Section V. Finally, Section VI describes the preparation and properties of poly(phenylene oxides).

45 references up to 1965.
Complementary: 27, 1–4.
Relevant: none.

3, *12: Interconversion of C—O—C, C—S—C and C—Se—C groups* by *V. Horak and J. Gosselck*

I. Introduction (526); **II**. Theoretical considerations (527); **III**. Synthetic methods (529); **IV**. Discussion of synthetic methods (543); **V**. References (548).

Ethers, sulphides and selenides can be interconverted by a number of usually general methods. The basic properties of O, S and Se atoms in their various compounds are listed in Table 1 and the similarities and differences are discussed in Section II, including basicity, complex formation and nucleophilicity.

Conversions of ethers and acetals can be carried out by thermal methods or by acid catalysis while special methods are used in the conversion of epoxides into episulphides and in that of *O*-alkyltropolones into S-containing compounds. Conversions of pyrones, pyrylium cations, alkoxymethylene ketones, acetals and glycosides to S-containing compounds usually require special methods (Section III.A). Sulphides and mercaptals (and analogous Se compounds) yield O-containing derivatives, in some cases through oxidation products (sulphoxides or sulphones) are intermediates, or by using heavy metal complexes (Hg, Ag, Pb) (Section III.B).

The synthetic methods presented in Sections II and III are discussed and mechanistic, structural, polar, energetic and reaction process features are considered in Section IV.

137 references up to 1964.
Complementary: 36, 1.
Relevant: 3, 13.

3, *13: The chemistry of thioethers; differences and analogies with ethers* by *L. Brandsma and J. F. Arens*

I. Introduction (554); **II**. Reactions involving carbanions derived from thioethers (554); **III**. Electrophilic and nucleophilic additions to the unsaturated system of alkenyl and alkynyl thioethers (566); **IV**. Reductive cleavage of thioethers (573); **V**. Cleavage of thioethers by nucleophilic substitution at sulfur and at carbon (578); **VI**. Reactions of halo and hydroxy thioethers (587); **VIII**. Miscellaneous reactions of thioethers (596); **VIII**. References (606).

The acidity of α-H atoms in alkyl and aryl thioethers is much higher than that in the corresponding ethers and in the presence of strong bases (e.g. NH_2K) stable carbanions are obtained in many cases, especially where two or three SR groups are present at the same C atom. Elimination of the α-H atom together with another group takes place easily with the formation of a new C—C bond (double or triple) and a stable molecule (H_2O, ROH, RSH, etc.). Prototropic isomerizations of unsaturated thioethers occur in the presence of catalytic amounts of bases and various condensation reactions (e.g. Dieckmann) are also facile (Section II).

Double and triple bonds in thioethers undergo both electrophilic and nucleophilic additions (Section III); thioethers undergo cleavage by alkali metals to yield thiols and also desulphurization with Raney Ni (Section IV). Thioethers are also cleaved through nucleophilic substitution at S or at the α-C atom and also by hydrogen halides, Lewis acids and acyl halides (Section V).

Hydrolytic and substitution reactions occur relatively easily with α- and β-halo thioethers and slower also with γ- and δ-halo thioethers as well as with hydroxy thioethers (Section VI).

Section VII describes some reactions of the thioethers which occur directly on the sulphur atom (formation of sulphonium compounds with halogens and with alkyl halides, formation of ylides, oxidation to sulphoxides and sulphones); halogenation of α-C atoms; reactions of CCl_2 with unsaturated thioethers; thermal decompositions, rearrangements and eliminations.

190 references up to 1965.
Complementary: 27, 6, 7, 8, 11, 13, 22, 24.
Relevant: 3, 12.

3, *14: Rearrangements of ethers* by *D. L. Dalrymple, T. L. Kruger and W. N. White*

I. Introduction (617); **II**. The Wittig rearrangement (618); **III**. Acid-catalysed ether rearrangements (628); **IV**. The Claisen rearrangement (635); **V**. Other thermal rearrangements (660); **VI**. References (663).

The Wittig rearrangement is the base-catalysed rearrangement of some ethers to isomeric alcohols. It is facile with ethers which contain a relatively acidic α-proton; its mechanism was studied extensively, including that of several side reactions leading to arenes, aldehydes and phenols. The reactions was also observed with sulphides and with acetylenic ethers (Section II).

In the presence of acid catalysts some alkyl aryl ethers rearrange to alkyl phenols. The reaction seems to involve both inter- and intramolecular pathways and gives mixtures of several products and hence is of little synthetic utility (Section III).

The Claisen rearrangement was extensively studied both synthetically and mechanistically. The reaction is intramolecular with a cyclic intermediate and, if one *ortho* position is occupied, both the other *ortho* and also the *para* position are attacked (Section IV). Simple enol ethers also undergo Claisen-like thermal rearrangements to aldehydes or ketones, which are intermolecular, however, and probably free radical in nature (Section V).

162 references up to 1966.
Complementary: **27**, 17.
Relevant: **1**, 7; **2**, 15; **4**, 10; **5**, 8; **8**, 9; **9**, 3; **11**, 4; **12**, 16; **14**, 7; **15**, 8; **16**, 17; **18**, 15; **19**, 18; **20**, 4; **23**, 13; **24**, 10; **26**, 15; **29**, 4.

3, *15: Detection and estimation of ethers* by *J. S. Fritz*

I. Introduction (669); **II**. Isolation and separation of ethers (670); **III**. Detection of ethers by classical and spectrophotometric methods (672); **IV**. Quantitative chemical methods for ethers (673); **V**. Quantitative determination of thioethers (679); **VI**. References (680).

Section II gives some data and references for the gas chromatographic separation of ethers, including salting-out and solubilization chromatography. The identification of ethers by derivatization and by IR is outlined in Section III.

Quantitative determination of alkoxy groups, oxyalkylene groups, acetals, ketals, vinyl ethers and epoxides and also some special methods are described in Section IV, while Section V deals with the quantitative determination of thioethers, especially by the oxidation of the S atom to a sulphoxide group.

33 references up to 1964.
Complementary: **27**, 6, 7.
Relevant: **1**, 5; **2**, 8; **4**, 3; **5**, 17; **8**, 3; **12**, 4; **13**, 6; **14**, 3; **15**, 3; **16**, 3; **17**, 4; **18**, 5; **20**, 3; **22**, 5; **23**, 5; **24**, 5; **26**, 5; **28**, 3; **29**, 21; **30**, 14–19; **36**, 5.

3, *16: Appendix on safety measures* by *N. V. Steere*

I. Introduction (682); **II**. Fire hazards of ethers (682); **III**. Health hazards of ethers (687); **IV**. Reactivity hazards (690); **V**. Protective measures (695); **VI**. Sources of information on hazards (698); **VIII**. References.

The extensive use of ethers as reagents and as solvents demands the consideration of hazards connected with their use.

Flash points, evaporation rates, vapour densities, limits of flammability, ignition temperatures and sources of ignition in the laboratory are considered (Section II). Health hazards due to ingestion, skin contact and inhalation are described in Section III, Table 5 quoting recommended threshold limit values for various ethers. The hazards connected with the reactions of ethers themselves with other reagents and the formation, detection, estimation, inhibition and removal of peroxides formed from ethers are discussed in Section IV. Finally, Section V recommends safety measures for ventilation, containers, labelling, storage, personal protective equipment and disposal methods.

65 references up to 1965.

The chemistry of the amino group (1968)

4, *1: General and theoretical* by *R. Daudel*

I. The nature of the chemical bond and the main methods of calculating molecular wave functions (1); **II**. Some aspects of the properties of amines in their ground state (15); **III**. Some aspects of the properties of amines in their electronically excited states (32); **IV**. References (34).

The first section recapitulates some aspects of molecular structure, the nature of the chemical bond and the main ideas which form the basis of calculating electronic wave functions.

Section II discusses electronic structures, densities, spectra, force constants, ionization energies and electric dipole moments, together with the chemical properties of aliphatic and aromatic amines in their ground state, while Section III summarizes briefly the electronic spectra and the base strength of aromatic amines in their electronically excited states.

44 references up to 1966.
Complementary: none.
Relevant: 1, 1; **2**, 1; **3**, 1; **5**, 1; **6**, 1; **8**, 1; **11**, 1; **12**, 1; **13**, 1; **14**, 1; **15**, 1; **16**, 1; **17**, 1; **18**, 1; **19**, 1; **20**, 1; **22**, 1; **23**, 1; **24**, 1; **25**, 1; **26**, 1; **27**, 9; **28**, 1; **31**, 27; **32**, 1; **33**, 1.

4, *2: The introduction of the amino group* by *M. S. Gibson*

I. Introduction (37); **II**. Introduction of amino groups by replacement of hydrogen (38); **III**. Introduction of amino groups by replacement of hydroxyl (43); **IV**. Introduction of amino groups by replacement of halogen, and related reactions (45); **V**. Preparation of amines involving the use of Grignard reagents (69); **VI**. Introduction of amino groups by addition to carbon—carbon double bonds (61); **VII**. Reductive methods for the preparation of amines (66); **VIII**. References (69).

The chapter discusses the introduction of—NR^1R^2 groups (where R^1 and R^2 may be H, R or Ar) into organic structures. Section II deals with the direct amination of aromatic compounds, with the Chichibabin reaction and with the photoamination of cyclohexene.

NH_2 groups may replace OH groups in the synthetically important Bucherer reaction (Section III); halides may be replaced by NR^1R^2 groups in the alkylation of amines, although many experimental difficulties occur, requiring more complicated methods for control of the reaction. Aryl halides substituted by electron-withdrawing groups undergo aromatic nucleophilic substitution by NH_3 and amines and even unsubstituted aromatic halides can react by the aryne mechanism (Section IV). Many amino compounds give direct amination on Grignard reagents (Section V) and amines can give both nucleophilic and electrophilic additions to C=C bonds (Section VI). Finally, almost all nitrogen-containing substituents give access to amines by reductive methods (Section VII).

88 references up to 1966.
Complementary: **1**, 8; **12**, 7; **29**, 11.
Relevant: **8**, 2; **14**, 4; **16**, 7; **19**, 15; **24**, 19; **29**, 27; **31**, 11.

4, *3: Detection, determination, and characterization of amines* by *J. Zabicky*

I. Introduction (81); **II**. Features of the group (81); **III**. Quaternary ammonium compounds (122); **IV**. Elucidation of structures (123); **V**. Dedication and acknowledgement (145); **VI**. References (145).

Section II.A describes methods based on pyrolysis, MS, nuclear activation, coordination with protonic and Lewis acids, metal ions and alkyl halides. It also deals with the determination of active H and N—H and groups with reactions of the latter with acyl, sulphonyl, nitroaryl and nitrobenzyl groups, ureas, SO_3, carbonyl compounds, ninhydrin (for α-amino acids) and HNO_2. Spectral properties (IR, Raman, NMR) are also outlined. Section II.D deals with the chemical and spectral properties of the C—N bond, II.E with hydrogen bonding and II.F with miscellaneous chemical properties of amines which can be used for detection and classification.

Section III discusses briefly quaternary ammonium compounds. Section IV outlines the elucidation of the structures of amines using NMR, MS, UV, basicity and stereochemical aspects.

610 references up to 1966.
Complementary: **29**, 5.
Relevant: **1**, 5; **2**, 8; **3**, 15; **5**, 17; **8**, 3; **12**, 4; **13**, 6; **14**, 3; **15**, 3; **16**, 3; **17**, 4; **18**, 5; **20**, 3; **22**, 5; **23**, 5; **24**, 5; **26**, 5; **28**, 3; **29**, 21; **30**, 14–19; **36**, 5.

4, *4: Basicity and complex formation* by *J. W. Smith*

I. Introduction (161); **II**. Factors determining the strengths of amines as bases (163); **III**. Aliphatic amines (170); **IV**. Aromatic amines (180); **V**. Complexes with Lewis acids and with metal ions (195); **VI**. References (200).

The lone pair of electrons on the N atom enables all amines to act as bases. The basic strength is determined by thermodynamic considerations, by the nature of the solvent used, by the temperature and by inductive, resonance and steric effects (Section II). Thermodynamic functions (Table 1), pK_a values of primary (Table 2), secondary (Table 3), tertiary (Table 4) and substituted (Tables 5 and 7 alkylamines and cycloalkylamines (Table 8) are presented (Section III).

Aniline and its *o*-, *m*- and *p*-substituted derivatives (Tables 9 and 10) and *N*-alkyl- and *N,N*-dialkylanilines (Tables 11–14) are discussed, and also some polycyclic aromatic amines (Tables 15 and 16) (Section IV).

Section V describes and tabulates data for additional products with Lewis acids (e.g. Me_3B, Table 17) and with metal ions (especially Ag^+, Table 18).

119 references up to 1965.
Complementary: none.
Relevant: **1**, 6; **3**, 6; **5**, 6; **8**, 4; **9**, 5; **11**, 3; **12**, 3; **13**, 7; **15**, 4; **16**, 5; **18**, 8; **19**, 7; **20**, 12; **22**, 8; **23**, 6; **24**, 4; **25**, 5; **29**, 16; **30**, 3–12; **31**, 17–21; **32**, 21; **33**, 5.

4, *5: Directing and activating effects* by *G. Chuchani*

Amino substituents exert both inductive and resonative effects (Section II) and can be treated well with the various forms of the Hammett equation (Table 2, Section III). The presence of amino substituents also strongly influences equilibria involving the ionization constants of aliphatic and aromatic amino carboxylic acids, aminophenols, nitrogen bases and others (Section IV).

Neighbouring group participation in haloamines and in amino esters in aliphatic S_N1 and S_N2 reactions and nucleophilic aromatic displacements is discussed (Sections V and VI). Nucleophilic addition and substitution at the carbonyl carbon of aliphatic and aromatic esters, acids and amides are treated (Section VII), and also additions of acids to C=C bonds and decarboxylation of acids, all in systems containing amino substituents (Section VIII). Section XI deals with electrophilic aromatic substitution in aromatic amines, including nitration, diazonium coupling, sulphonation and halogenation.

342 references up to 1966.
Complementary: none.
Relevant: **5**, 16; **6**, 8; **7**, 3; **11**, 12; **12**, 5; **13**, 8; **14**, 4; **15**, 5; **16**, 6; **18**, 9; **19**, 8; **28**, 9; **31**, 8; **36**, 15.

4, *6: Substitution at an amino nitrogen* by *B. C. Challis and A. R. Butler*

All (non-radical) reactions of amines result from nucleophilic attack by the lone pair of N or positively charged or neutral species. The general mechanism is a two-step process involving the formation and decomposition of a quaternary ammonium ion.

Primary and secondary amines undergo acylation with a wide variety of CO-containing compounds such as esters, acid halides and anhydrides, carboxylic acids and amides, ketenes and many other reagents. Formylations (e.g. transformylation of an amine with a more reactive formamide) present special cases (Section II).

Most amines (except sterically very hindered types) react with alkyl halides and other alkylating agents (isocyanates, olefins, acetylenes) and with aryl halides in the presence of electron-attracting groups on the ring. Reductive alkylation with aldehydes and ketones yields imines and the Leuckart and Mannich reactions are also relevant (Section III).

In the competition between N and C halogenation, the latter is usually favoured since it is non-reversible. However, hypochlorous acid, the hypochlorite ion, *N*-halosuccinimides and related compounds give *N*-haloamines (Section IV).

Nitrosation and diazotization can be carried out under widely differing reaction conditions and with various reagents, with or without acid catalysts (Section V).

Oxidation can lead to a wide variety of products with different substrates. The reactions of primary, secondary and tertiary aliphatic and aromatic amines with hydroperoxides, peroxy acids, H_2O_2, diacyl peroxides, O_3, $KMnO_4$ and some other reagents are described in Section VI.

308 references up to 1966.
Complementary: **11**, 2; **23**, 14; **29**, 11, 25, 26.
Relevant: **4**, 7; **8**, 2.

4, *7: Carbon—nitrogen and nitrogen—nitrogen double bond condensation reactions* by *P. Y. Sollenberger and R. B. Martin*

The emphasis of the chapter is on mechanism and catalysis in the title reactions (for preparative methods see refs 1 and 2).

Section II describes the formation of the C=N linkage by condensation of aldehydes and ketones with primary amines, including reversibility, stability, steric and polar effects and NMR. Metal ions (Ni^{II}, Zn^{II}, Co^{II}) catalyse the reaction. C=N groups are also obtained by condensation of primary amines with ketals and *gem*-dichloro compounds, from azides with thioketones, nitroso compounds with active methylene groups and other routes. Schiff bases and phenylhydrazones show tautomerism in addition to geometric isomerism.

The azo linkage is usually produced by diazo coupling or by oxidation of hydrazines or primary aromatic amines. Condensation routes are available from aromatic amines or hydroxylamines with nitrosarenes. Azobenzenes and azoxybenzenes exist in *cis* and (more stable) *trans* forms (Section III).

Section IV discusses the kinetics, mechanism, catalysis and structure–reactivity correlations in the condensation of carbonyl compounds with amines, considering separately effects in the two reactants and the relationships between the reactivities of carbonyl compound, nucleophile and catalyst.

Azomethine condensations are important in the activity of enzymes and are involved in the metabolism of amino acids, and much research has been performed on model systems imitating enzymatic action (Section V).

Finally, in Section VI the condensations of nitrosobenzenes with anilines and with *N*-substituted hydroxylamines are discussed.

260 references up to 1966.
Complementary: **2**, 12; **8**, 2; **19**, 4.
Relevant: none.

4, *8: Cleavage of the carbon—nitrogen bond* by *E. H. White and D. J. Woodcock*

The chemistry of C—N cleavage reactions is dominated by (1) the ability of N to form a positive tetracovalent group that both labilizes adjacent C—H bonds for attack and can also serve as a good leaving group and (2) the ability of N to combine with another N atom to form the very stable N_2 molecule in deaminations.

Section II describes the Hoffmann elimination, its mechanism and stereochemistry, amine displacements, e.g. $R\overset{+}{N}Me_3X^- \rightarrow RNMe_2 + MeX$, and both the Stevens and the Sommelet–Hauser rearrangements, both of which involve migration of alkyl groups.

Amine oxides undergo Cope elimination, forming alkenes and Meisenheimer rearrangements in which a benzyl or alkyl group migrates from the N to the O atom, and also cleavages with acylating agents giving aldehydes and acylamines (Section III).

Tertiary amines react with BrCN to give an alkyl bromide and a cyanoamine (which can hydrolyse to a secondary amine); they also undergo cleavage by HNO_2 (but not at low pH) and oxidative cleavage to give a secondary amine and a carbonyl compound using various oxidizing agents (CrO_3, MnO_2, $KMnO_4$, etc.). Amine displacements and exchanges of one amine group for another in Mannich bases are observed (Section IV).

Aziridines can be deaminated to the corresponding alkenes and benzylic and allylic amines to the corresponding hydrocarbons (Section V).

Reductive cleavages occur by treatment with H_2 and a catalyst, by dissolving metals, by electrolysis, by HNF_2, by hydroxylamine-*O*-sulphonic acid and also photochemically (Section VI).

The bulk of the chapter (Section VII) discusses the reactions in which nitrogen is the leaving group, e.g. the nitrosamide decomposition in which the amine is converted first into an amide and then into a nitrosamide, which is subsequently decomposed. Detailed mechanistic results of this reaction are presented, followed by a short treatment of the triazene method, $RNH_2 + ArN_2^+ \rightarrow RNHN{=}NAr \rightarrow RX + ArNH_2$ (Section VII.A.1.a and b).

In nitrous acid deaminations (Section VII.A.1.c) a number of concurrent mechanisms may operate and are presented with special emphasis on the role of the counter ion as to the stereochemistry and the product formation.

In some replacements of the NH_2 group in RNH_2 the leaving group is not N_2 but a cyanide (Von Braun reaction) or an amide, and in others the product may be carboxylic acids or ketones obtained by oxidation. Finally, decompositions involving diazonium salts and the Bucherer reaction are mentioned briefly.

221 references up to 1966.
Complementary: 23, 8; **29**, 22; 26.
Relevant: 8, 10; **21**, 8; **29**, 12.

4, *9: Biological formation and reactions of the amino group* by *B. E. C. Banks*

The incorporation of nitrogen in the form of amines into living organisms almost always requires organic nitrogen compounds and enzymatic processes (Fig. 1).

Sections II and III describe the properties of enzymes in general and especially their role in the formation and reactions of amino acids and proteins and, briefly, the methods used in elucidating metabolic pathways.

The formation of α-L-amino acids involves ammonia formation by N_2 fixation, nitrate reduction, nitrification and denitrification, with glutamic dehydrogenase catalysing the formation of glutamate from α-ketoglutamate and NH_3. Many pairs of α-amino and α-keto acids participate in transaminations; the mechanism of these enzyme-catalysed reactions (Fig. 6) and the origins of individual amino acids (Table 3) were studied (Section IV).

Other amino compounds (D-amino acids, L-amino acid amides, amines, purines, pyrimidines, amino sugars, vitamins B_1, B_2, B_6, H and B_3, folic acid, nicotinamide, pyrrolic compounds and plant alkaloids are discussed in Section V. The biosyntheses of these biologically important compounds *in vivo* are complicated processes involving a wide variety of usually specific enzymes and one specific process, by mechanisms which are often unclear or unknown.

The last section (VI) treats the formation of nitrogen-containing waste products. Excess amino compounds are usually first converted to NH_3 and then to the common excretion products (urea and uric acid). There is often an overlap between biosynthetic and degradative pathways, but NH_3 is mainly produced by the action of L-amino acid oxidases and by amine oxidases. Finally, the pathways leading to the formation of urea and of uric acid are considered.

137 references up to 1966.
Complementary: 11, 11.
Relevant: 2, 7; **3**, 5; **5**, 18; **12**, 12; **13**, 14; **15**, 10; **16**, 13; **17**, 13; **18**, 13; **19**, 13; **22**, 20; **24**, 14; **26**, 13; **28**, 17; **31**, 9.

4, *10: Rearrangements involving amino groups* by *D. V. Banthorpe*

I. Rearrangements accompanying deamination (586); **II**. Rearrangements of quaternary compounds to amines (612); **III**. Rearrangements with migration to an electron-deficient nitrogen (623); **IV**. Intermolecular aromatic rearrangements (634); **V**. Intramolecular aromatic rearrangements (639); **VI**. Other rearrangements (650); **VII**. References (658).

The title reactions are classified according to the nature of intermediates and the mechanisms. Deamination of primary amines with nitrosating agents is best interpreted by the assumption of a 'hot' carbonium ion intermediate. Section I discusses acyclic amines and 1, 2-amino alcohols and the stereochemistry and the migratory aptitudes in these reactions. Ring contraction and expansion in cycloalkylamines is described and the rearrangements of α-amino ketones. The problems connected with non-classical ions obtained from monocyclic and bicyclic amines are outlined, together with transannular migrations and the Pschorr reaction.

The Stevens, Sommelet and *tert*-amine oxide rearrangements produce amines from quaternary ammonium compounds (Section II). An important group of rearrangements occur by migration of an alkyl or aryl group to an electron-deficient N, and the similarities in these reactions (Beckmann, Hofman, Lossen, Curtius, Schmidt, azide and Stieglitz rearrangements) are emphasized (Section III).

The group of acid-catalysed rearrangements which involve fission followed by intermolecular recombination include the diazoaminobenzene, Hofman–Martius, Fischer–Hepp, Orton, phenylhydroxylamine and Chattaway rearrangements (Section IV). Other acid-catalysed rearrangements (nitramine, benzidine, Chapman and Smiles) are formally similar, but have been demonstrated to be intramolecular by extensive mechanistic studies (especially the benzidine rearrangement) (Section V).

Finally, Section VI presents some other rearrangements which cannot be classified in any of the above categories (phenylsulphamic acid rearrangement, acyl migration to and from nitrogen, reactions involving S_Ni displacement by NH_2 groups and some other processes, including the classical urea synthesis from NH_4CNO by Wöler).

398 references up to 1965.
Complementary: **23**, 13.
Relevant: **1**, 7; **2**, 15; **3**, 14; **5**, 8; **8**, 9; **9**, 3; **11**, 4; **12**, 16; **13**, 13; **14**, 7; **15**, 8; **16**, 17; **17**, 8; **18**, 15; **19**, 18; **20**, 4; **24**, 10; **29**, 4, 22, 26.

4, *11: Protection of the amino group* by *Y. Wolman*

Since the amino group easily undergoes a wide variety of reactions, there is a need to protect it while other parts of the molecule are subjected to chemical changes.

Among acyl-type protecting groups, the following are discussed: formyl, acetyl, benzoyl, trifluoroacetyl, acetoacetyl, diacetyl, amidine, sulphonyl, sulphenyl, phosphonyl, phlorethyl, acetylmethionyl, nitrophenoxyacetyl and chloroacetyl (Section II).

Section III describes aromatic, heterocyclic, aliphatic and alicyclic urethane-type protecting groups, and also thiourethane-type derivatives.

Section IV treats Schiff bases and derivatives of 1,3-dicarbonyl compounds and Section V benzyl, dibenzyl, triphenylmethyl, dinitrophenyl and trialkylsilyl derivatives.

Finally, protection by protonation (Section VI) and by chelate formation (Section VII) is outlined.

181 references up to 1966.
Complementary: none.
Relevant: **2**, 12; **4**, 7; **8**, 2; **29**, 22.

4, *12: Tetraaminoethylenes* by *D. M. Lemal*

Tetraaminoethylenes are powerful nucleophiles, π-bases and reducing agents. They can be prepared from perhaloethylenes with amines or from dipolar 'halves' in which a carbon atom in a molecule is bound to two aminic nitrogen atoms and two more groups (one of which is usually a proton and the other a good negative leaving group). The latter two are eliminated by thermal or catalytic routes and two ylide (or carbene) intermediates combine to give the tetraaminoethylene (Table 1, Section II).

It has been claimed, but not proved, that the title compounds can dissociate into two ylide halves. They are strongly nucleophilic: protonation of both ethylenic carbons cleaves the double bond with the formation of two formamidinium molecules in the simplest case and to more complicated molecules in others. They also attack carbonyl carbon atoms of aldehydes, acid chlorides, carbon disulphide and isothiocyanates. Tetraaminoethylenes also attack N atoms of azides and of diazo compounds, S_8 and O_2 and in all cases the product contains only one half of the attacking molecules (Section III.A and B).

The tetraaminoethylenes have very strong electron donor reactivity and both radical-cation and stable dication oxidation states are known.

Many charge-transfer (CT) complexes of peraminoethylenes are known with acceptor molecules of widely differing electron affinity, the most stable being that with 7,7,8,8-tetracyanoquinodimethane (2 mol) with tetrakis(*N*-pyrrolidino)ethylene (1 mol). Some of these CT complexes are diamagnetic and others are paramagnetic.

91 references up to 1966.
Complementary: 29, 18.
Relevant: 12, 10.

The chemistry of carboxylic acids and esters (1969)

5, *1: General and theoretical aspects of the COOH and COOR groups* by *M. Simonetta and S. Carra*

I. Introduction (2); **II**. Geometry, energy and polarity (3); **III**. Spectroscopic properties (18); **IV**. Theoretical aspects (29); **V**. References (48).

Section II describes methods and results relating to the geometry and structure of the COOH group, obtained by X-rays, electron diffraction and microwave spectroscopy. This is followed by an outline of heats of formation, ionization potentials, dipole moments and ionization and dimerization equilibria. The most important data are presented in tables for many carboxylic acids.

IR, UV and NMR spectra are dealt with in Section III, while Section IV discusses electronic structures in the ground and excited states (based on quantum mechanics), hydrogen bonds and some aspects of reactivity.

164 references up to 1966.
Complementary: 15, 1; **25**, 1.
Relevant: 1, 1; **2**, 2; **3**, 1; **4**, 1; **6**, 1; **8**, 1; **11**, 1; **12**, 1; **14**, 1; **16**, 1; **17**, 1; **23**, 1; **24**, 1; **26**, 1; **32**, 1; **33**, 1; **36**, 2.

5, *2: Electrochemical reactions of carboxylic acids and related processes* by *L. Eberson*

I. Introduction (54); **II**. Definitions (56); **III**. Experimental conditions (57); **IV**. Mechanistic considerations (63); **V**. Anodic acyloxylation and related reactions (93); **VI**. Anodic esterification, alcoholysis and hydrolysis (97); **VII**. Anodic versus lead tetraacetate oxidation of organic compounds (98); **VIII**. Bibliography (98).

Organic electrochemistry is a unique method that permits the control of the activity of the electrodes and the composition of the electrolyte (and hence the products) exactly and in a number of ways (Section I). The Kolbe electrosynthesis is defined as the anodic formation of products from both radical and carbonium ion species (Section II).

Section III discusses the factors influencing the outcome of electrochemical reactions, i.e. anode potential, current density, anode material, pH, solvent, temperature and added foreign electrolytes.

Section IV treats the reaction mechanisms involved in electrochemical processes, the electrochemical mechanism in the Kolbe reaction, the radical aspects of the chemical reaction mechanism, including radical disproportionation, attack on the solvent, dimerization, polymerization, comparison between anodically and chemically generated radicals and the stereochemistry of anodically generated radicals. This is followed by a discussion of the carbonium ion aspects, since carbonium ions may be formed from radicals by a second electron transfer to the anode. Here, fatty acids, substituted aliphatic acids, alicyclic and bicyclic, dicarboxylic and hydroxy acids are considered, anodic and non-anodic carbonium ions are compared and the factors which determine whether the radical or the carbonium ion path is favoured are surveyed. Sections V–VII outline anodic acyloxylation, cyanation, esterification, alcoholysis and hydrolysis and compare the anodic with the $Pb(OAc)_4$ oxidation.

132 references up to 1966.
Complementary: **25**, 13.
Relevant: **8**, 11; **13**, 5; **16**, 15; **17**, 14; **19**, 12; **20**, 5; **21**, 5; **22**, 9; **23**, 10; **24**, 17; **26**, 12; **27**, 8; **28**, 7; **29**, 8, 9; **31**, 6; **32**, 6; **34**, 1.

5, *3: Alchoholysis, acidolysis and redistribution of esters* by *J. Koskikallio*

The three title reactions are the following:

$$R^1COOR^2 + R^3OH = R^1COOR^3 + R^2OH \text{ (alcoholysis)}$$

$$R^1COOR^2 + R^4COOH = R^4COOR + R^1COOH \text{ (acidolysis)}$$

$$R^1COOR^2 + R^4COOR^3 = R^1COOR^3 + R^4COOR^2 \text{ (redistribution)}$$

The first reaction has been investigated extensively, and eight different reaction mechanisms are possible, acid- (A) and base- (B) catalysed, unimolecular or bimolecular, each of which can involve alkyl—oxygen or acyl—oxygen fission. There are labelled $A_{Al}1$, $A_{Al}2$, $B_{Al}1$, $B_{Al}2$, $A_{Ac}1$, $A_{Ac}2$, $B_{Ac}1$ and $B_{Ac}2$.

Section II discusses the equilibria in all three reactions, the effects of electrolytes, solvents and substituents and the calculation of equilibrium constants.

Section III treats the B_{Ac} mechanism, the effects of substituents in the acyl and alkoxy moieties, reactivities and catalysis. Subsequently, surveys of the B_{Al} mechanism (Section IV) and the uncatalysed B_{Ac} (V) and B_{Al} (VI) mechanisms and of the acid-catalysed A_{Ac} (VII) and A_{Al} (VIII) mechanisms are presented.

The synthetic uses of alcoholysis for the preparation of single esters and of esters of polyhydric alcohols with monocarboxylic acids (e.g. glycerides) and of polyhydric alcohols with dicarboxylic acids (phthalates, polycarbonates, high polymers) are surveyed (Section IX), and also the analytical uses of alcoholysis (Section X). Section XI reviews acidolysis by alkyl—oxygen and by acyl—oxygen fission, including both inter- and intramolecular reactions and preparative uses.

172 references up to 1966.
Complementary: none.
Relevant: **5**, 11; **25**, 7, 16.

5, *4: The formation of carboxylic acids and their derivatives from organometallic compounds* by *R. P. A. Sneeden*

Very little is known about the mechanism of the carbonation of organometallic compounds. Hence the chapter emphasizes the preparative uses of the reaction, its conditions, scope and steric course.

Section II discusses reactions of RM (M = Li, Na, K) of R_2M^{II} (M^{II} = Mg, Ca, Ba, Sr) and of R_3Al compounds, and Section III the carbonation of compounds with alkali, alkaline earth and transition metals, including the Kolbe–Schmidt, Henkel and Friedel–Crafts reactions.

Organometallic syntheses of α,β-olefinic acids are similarly considered in Section IV, non-conjugated olefinic acids in Section V, α,β-acetylenic acids in Section VI and non-conjugated acetylenic acids in Section VII. Finally, Section VIII, treats the direct carboxylation and carbothioxylation of alkali metal derivatives of molecules containing activated H, carboxylation with methylmagnesium carbonate, ethylsodium carbonate and with oxalyl chloride and radiation-induced attack of CO_2 on hydrocarbons.

189 references up to 1966.
Complementary: **5**, 5 (Section II).
Relevant: **2**, 6; **5**, 7; **21**, 12; **25**, 7; **34**, 4, 5; **35**, 1–3.

5, *5: Synthesis of di- and polycarboxylic acids and esters* by *V. K. Kucherov and L. A. Yanovskaya*

The interest in the preparation of di- and polycarboxylic acids and esters is based on their considerable technical and synthetic application.

Carboxylation can be achieved by CO, by CO_2, by the Kolbe–Schmidt reaction and also by intermolecular rearrangement of benzenecarboxylic acids (Section II). Another route is dimerization, which can be free-radical or oxidative and in rarer cases Wurtz-type, reductive or additive (Section III).

Base-catalysed condensations include (Section IV) the ester, Knoevenagel and oxalic ester condensations, malonic syntheses, organometallic and electrolytic condensations and Reformatsky, Wittig, Diels–Alder and Friedel–Crafts methods.

Michael (Section V) and oxidative and hydrolytic methods (Section VI) are also available. Many routes are based on existing acid derivatives, e.g. ester hydrolysis, esterification, alcoholysis, salt alkylation, hydrolysis and alcoholysis of anhydrides, acyl chlorides, amides, imides, lactones and nitriles.

Monoesters of di- and polycarboxylic acids can be prepared by partial esterification or partial hydrolysis, reactions between acids and esters, alcoholysis of anhydrides and lactones and some other special methods (Section IX).

271 references up to 1966.
Complementary: 5, 3, 4, 7; **25**, 7.
Relevant: 25, 16.

5, *6: Acidity and hydrogen bonding of carboxyl groups* by *L. Eberson*

I. Introduction (212); **II**. Definitions and standardization procedures (213); **III**. Symmetry factors (219); **IV**. Conformational aspects (222); **V**. Thermodynamic functions (225); **VI**. Kinetics of carboxylic acid ionization (236); **VII**. Solvent effects (236); **VIII**. Structural effects (255); **IX**. References (288).

The aim of the chapter is to examine ionization equilibria of carboxylic acids, considering a variety of physical and chemical factors and especially inter- and intramolecular hydrogen bonds. Section II defines ionization constants, compares the title reaction with other ionization processes, considers ionization constants in mixed (aqueous–organic) solvents and the use of relative acidity constants (K_r, in the equation which represents the basis of all comparisons between acidities and emphasizes the quantitative treatment of substituent effects: eqn. 8, p. 218).

Section III considers symmetry factors which sometimes lead to small but significant corrections in K_r.

The discussion of conformational aspects (Section IV) includes the rotations around the C—COOH and CO—OH bonds. Section V deals with the calculation of thermodynamic functions in ionization equilibria, and discusses them (Table 3 and Figs 1 and 2).

The kinetics of the ionization of carboxylic acids are surveyed (Section VI), together with solvent effects including classification, electrostatic theories, solvent–solute interactions, dispersion and isotope effects (Section VII).

The last section treats in detail conformational, steric, polar and conjugation effects and also intramolecular hydrogen bonding and acidity in dicarboxylic (Tables 16–18) and hydroxy carboxylic acids (Tables 19 and 20).

217 references up to 1966.
Complementary: 13, 7; **18**, 8; **25**, 6.
Relevant: 2, 9; **3**, 6; **4**, 4; **8**, 5; **11**, 3; **12**, 3; **15**, 4; **16**, 5; **19**, 7; **20**, 12; **22**, 8; **23**, 6; **24**, 4; **33**, 5.

5, *7: Introduction of COOH groups by carbonyl olefination* by *L. D. Bergelson and M. M. Shemyakin*

I. The carbonyl olefination reaction (296); **II**. Synthesis of carboxylic acids and their derivatives (314); **III**. References (456).

The chain elongation method is which phosphorus ylides or their P=O activated analogues react with carbonyl compounds to yield olefins occurs under mild conditions, does not induce isomerization or rearrangement and is often stereospecific:

$$\gtrdot\overset{+}{P}-\overset{-}{C}R_2 + O{=}CR_2' \rightarrow R_2C{=}CR_2'$$

The most frequently used reagents are triphenylalkylene phosphoranes ($Ph_3\overset{+}{P}-\overset{-}{C}HR$, Wittig reagents).

Section I describes the scope of the reaction, the preparation and properties of P-ylides and, the mechanism and the stereochemistry of carbonyl olefination, including the steric effect of solvation and of inorganic salts.

Carbonyl olefination can be used to prepare carboxylic acids and derivatives either when the carbonyl compound being olefinated contains a potential carboxyl function (ester, amide or nitrile) or when aldehydo or keto carboxylic acid derivatives are olefinated.

The method is available for the preparation of α, β-, β, γ- and other unsaturated acids. Since the carboxylic groups may also be attached to that moiety of the P-ylide which appears in the product, a wide variety of different types of acids can be synthesized, including polyenic, divinylmethanic, divinylethanic and branched-chain acids and amides and nitriles (Section II.A.D).

Section II.E discusses some other methods which are also based on reactions of P-ylides, but which are not carbonyl olefinations, but still useful for the synthesis of carboxylic acids and derivatives. Among these are reactions of $Ph_3P{=}CHCO_2R$ compounds (α-alkoxycarbonylalkylidenetriphenylphosphoranes) with acid chlorides, with α-bromoketones and with epoxides, in each case yielding acid derivatives (especially esters), and the addition of non-stabilized P-ylides to C=C bonds.

180 references up to 1966.
Complementary: **1**, 4; **2**, 12; **21**, 12.
Relevant: **5**, 4, 5; **25**, 7.

5, *8: Rearrangement and cyclization reactions of carboxylic acids and esters* by *H. Kwart and K. King*

Owing to the stability of the carboxyl and alkoxycarbonyl (ester) groups, these are often formed under mild conditions, whereas their destruction by rearrangement or otherwise requires drastic methods.

Section I discusses Dieckmann and acyloin condensations in both of which an ester group is converted into a cyclic carbonyl group, cyclodehydration of unsaturated acids to cyclopentanones or cyclohexanones (or in some cases to lactones) and the Fries and related rearrangements of phenolic esters to *o*- and *p*-hydroxyketones.

Section II describes the rearrangement of aromatic carboxylic acids (the Raecke process), e.g. isomerization of potassium phthalate to terephthalate and the rearrangement of α-hydroxy acids to α-diketones and related reactions.

The Varrentrapp reaction occurs with unsaturated acids and involves the migration of the double bond to the α, β-position and subsequent cleavage of the α, β-bond (Section III).

Section IV treats carbonyl-forming decarboxylations of alkaline earth and other metal carboxylates to the corresponding ketones [e.g. $(CH_3COO)_2Ca \rightarrow (CH_3)_2CO$] and discusses the mechanism of this reaction although, notwithstanding the large amount of experimental data, very few definite conclusions can be reached.

104 references up to 1966.
Complementary: **2**, 4, 15.
Relevant: **5**, 12; **25**, 15, 16.

5, *9: Substitution in the groups COOH and COOR* by *D. P. N. Satchell and R. S. Satchell*

The chapter deals with substitutions in which the OH group of an acid or the OR group of an ester is substituted by an attacking group, i.e. $R'COOH(R) + X \rightarrow R'COX + OH(R)$, and to a lesser extent with substitution, $R'COOH(R) + X \rightarrow R'COOX + H(R)$. Most of these are heterolytic reactions in which the new atom to be attached to the carbonyl carbon acts as a nucleophile. Acid catalysts increase the electrophilicity of the carbonyl C, whereas basic catalysts increase the nucleophilicity of the other reagent. These factors, and also catalytic efficiency, substituent effects in all parts of the two molecules involved and mechanistic details are discussed (Section I).

Describing substitution by halogen-containing reactants, the reactions of carboxylic acids and esters with hydrogen halides, alkyl and acyl halides and with inorganic halogen compounds ($SOCl_2$, PCl_5, PCl_3) are presented (Section II).

Section III treats the reactions of acids and esters with O and S nucleophiles and gives the available mechanistic details on reactions with H_2O, H_2O_2, RCOOH (to yield anhydrides), anhydrides (to yield a new anhydride), acyl halides and H_2SO_4 (to yield acyl hydrogen sulphate, RCO_2SO_3H).

The reactions of acids and esters with NH_3 and primary, secondary and tertiary amines, hydrazines and hydroxylamines are presented in Section IV. These [e.g. amide formation, $R'COOH(R) + HNR_2 \rightarrow R'CONR_2 + H_2O(ROH)$] are usually base catalysed or uncatalysed, since the presence of acids protonates the N atom and renders it less nucleophilic. Kinetics and mechanisms are discussed and especially catalysis by imidazoles, which is believed to be relevant to the action of a variety of hydrolytic enzymes.

Nucleophilic attack (usually catalysed by Lewis acids) on acids and esters leads to the production of ketones, e.g. Friedel–Crafts acylation of arenes, $RCOOR' + ArH \rightarrow RCOAr + R'OH$. The reaction may be intramolecular and lead to a cyclic ketone. Olefins and acetylenes are also reactive, but the process is complicated by side-reactions. Aliphatic hydrocarbons are usually only attacked if they contain an active methylene group or if they are first transformed into an organometallic compound, e.g. of the Grignard type: $RMgBr + R'COOR''$ $R'COR + MgBr(OR'')$. A special case is the reaction of esters with HCN in which acyl cyanides are formed: $RCOOR' + HCN \rightarrow RCOCN + R'OH$ (Section V).

148 references up to 1968.
Complementary: 2, 5.
Relevant: 4, 6; **11**, 2; **15**, 6; **25**, 7.

5, *10: Syntheses and uses of isotopically labelled carboxylic acids* by *M. Zielinski*

The chapter describes the preparation of acids and esters labelled with ^{13}C, ^{14}C, ^{18}O and ^{2}H(D) and the uses of these compounds for mechanistic (and also analytical) studies.

Carboxylic acids can be labelled by the use of ^{14}CO or ^{13}CO in carbonation, or by the use of labelled cyanides and also by application of nuclear recoil (Section II). The use of these products in physical and analytical determinations, ligand exchanges, equilibrium and pyrolytic studies, decarbonylations and oxidations is described (Section III).

Section IV is devoted to studies of isotope effects in reactions involving C, O and H isotopes, such as decarboxylations and decarbonylations.

Chemical methods and direct tritiation for the preparation of labelled esters are described in Section V. The use of these studies of alkaline, acid and enzymatic hydrolysis of ^{18}O-labelled esters, of the mechanism of the ketone–ester conversion and of some homolytic decompositions, in acetoxy exchange and gas-phase decomposition studies is reviewed in Section VI.

Section VII considers the ^{14}C kinetic isotope effect in the hydrolysis of esters catalysed by acids and by bases (Table 10) and also secondary deuterium and tritium kinetic isotope effects in the hydrolysis of carboxylic esters labelled in various positions, e.g. CD_3COOEt or $CH_2TC_6H_4COOMe$.

162 references up to 1966.
Complementary: 12, 13.
Relevant: 13, 15; **17,** 12; **18,** 18; **19,** 17; **22,** 14; **23,** 16; **24,** 20; **26,** 11; **27,** 10; **28,** 8; **29,** 28; **33,** 18; **36,** 12.

5, *11: Esterification and ester hydrolysis* by *E. K. Euranto*

Esterification and ester hydrolysis may be acid or base catalysed, involving acyl—oxygen or alkyl—oxygen fission, and in each case the rate determining step may be uni- or bimolecular. These mechanisms are designated $A_{AC}1$, $A_{AC}2$, $B_{AC}1$, $B_{AC}2$, $A_{AL}1$, $A_{AL}2$, $B_{AL}1$ and $B_{AL}2$. Of these, $B_{AC}1$ and $A_{AL}2$ have not been observed with certainty, and $B_{AC}2$ and $A_{AC}2$ are the most common.

Section II discusses the $B_{AC}2$ mechanism, the evidence for its validity, the addition–elimination route, the kinetic form, the temperature and pressure dependence, structural effects (polar, resonance, steric, configuration, conformation), isotope, solvent isotope, salt and solvent effects and treats the kinetic data presented in Tables 1–9.

Section III describes the B_{AC} reaction in neutral media, the scope, kinetics, structural and medium effects (Section III.A). This is followed by treatment of general base catalysis (III.B) and of nucleophilic catalysis (III.C), in the latter considering the leaving group and the nucleophile and the case of general base-catalysed nucleophilic catalysis. Metal ions are often catalysts (III.D); intramolecular catalysis (neighbouring group participation) is observed in the presence of COOH-, OH-, CO- and N-containing groups (III.E).

In the discussion of the A_{AC} mechanism (Section IV), which is of course identical in the acid hydrolysis of esters and the acid esterification of carboxylic acids, both uni- and bimolecular routes operate ($A_{AC}1$, $A_{AC}2$). H^+ ion catalysis, mechanism, kinetics, structural (polar, resonance and steric) effects and medium (solvent, salt and solvent isotope) effects are described. Finally, esters with electronegative substituents (Section IV.D) and the unimolecular $A_{AC}1$ mechanism (IV.E) are reviewed briefly.

Section V deals with the A_{AL} mechanism, with evidence for its existence [especially in heterolytic reactions of esters of tertiary alcohols and of alkoxymethyl esters ($RCOOCH_2OR'$)]. This is followed by a review of the $B_{AL}1$ and $B_{AL}2$ mechanisms.

The last section (IV) considers some hydrolyses which involve reactions in other parts of the molecule before the rupture of the O—acyl or O—alkyl group, and which do not belong to the original mechanistic classification. These include hydrolyses of dialkyl carbonates, alkyl haloformates and α-haloalkyl esters.

395 references up to 1967.
Complementary: 5, 3.
Relevant: 3, 2.

5, *12: The decarboxylation reaction* by *L. W. Clark*

Carboxylation and decarboxylation are parts of the carbon dioxide–oxygen cycle in Nature and also play a very important role in the synthesis of many organic compounds in the laboratory (Section I).

Decarboxylations by heterolytic fission occur with keto and hydroxy acids and their anions and with CCl_3COOH, heterocyclic acids and several polycarboxylic acids. Kinetic data and activation parameters are tabulated and homolytic fission is mentioned briefly (Section II).

Section III discusses the decarboxylation of malonic acid in detail and of other acids more briefly, and proposes a relationship between decarboxylation and other bimolecular heterolytic reactions, e.g. the hydrolysis of benzamide and its derivatives.

105 references up to 1967.
Complementary: 25, 15.
Relevant: 2, 14; **25**, 14.

5, *13: Ortho esters* by *E. H. Cordes*

The most useful method for the preparation of ortho esters $R'C(OR)_3$, is the alcoholysis of imido ester hydrochlorides, $R'C(OR){=}NH.HCl$, at room temperature or in refluxing diethyl ether. Mixed ortho esters, $RC(OR')(OR'')$, can be prepared by appropriate variation of the method. Another important method is the reaction of alkoxides with 1, 1, 1-trihalides; $RCX_3 + 3NaOR' \rightarrow RC(OR')_3$. Less important routes are from ketene acetals, from orthothioformates and by ortho ester exchange between ortho esters and alcohols, which may be driven to completion by removal of the lower boiling alcohol (Section II).

The basicity of ortho esters has been determined and the following order found: ethers > ketals > ortho esters > orthocarbonates (Section III). Ortho esters may generate alkoxy carbonium ions [e.g. $HC(OR) \rightarrow \overset{+}{C}(OR)$] which have high stabilization energies. The hydrolysis of ortho esters is discussed in detail (Section IV.B), considering the site of the C—O bond cleavage, molecularity, catalysis, linear free energy correlations (Table 4) and all aspects of the mechanism. Finally, the reactions of ortho esters with amines (IV.C), C-nucleophiles (IV.D) and some other reagents are reviewed.

152 references up to 1966.
Complementary: 25, 9.
Relevant: none.

5, *14: Peracids and peresters* by *S.-O. Lawesson and G. Schroll*

I. General comments (670); **II**. Preparation of peracids (670); **III**. Physical properties and structure of peracids (672); **IV**. Reactions (673); **V**. Preparation of peresters (681); **VI**. Chemical properties of peresters (683); **VII**. Acknowledgements (697); **VIII**. References (697).

Peracids are prepared from carboxylic acids or anhydrides by oxidation with H_2O, from acid chlorides and Na_2O_2, from diaroyl peroxides and alcoholates and by oxidation of aldehydes (Section II). Their physical properties and structure are outlined (Section III).

Their reactions include epoxidation and hydroxylation of double bonds, oxidative cleavage reactions of CO, C=C, arenes, S-compounds and oxidation of N-compounds (e.g. $ArNH_2 \rightarrow ArNO$) and of organic iodine compounds ($ArI \rightarrow ArIO_2$) (Section IV).

Section V discusses syntheses of peresters and Section IV their reactions with Grignard and other nucleophilic reagents, their uncatalysed decomposition (e.g. to acylals; the Criegee reaction) and, in more detail, the perester reaction, i.e. the Cu-salt catalysed decomposition of *tert*-butyl perbenzoate in different substrates, leading to substitution in the latter by the benzoyloxy group, e.g.

$$R_3CH + PhCO_3Bu\text{-}t \xrightarrow{Cu^+/Cu^{2+}} R_3COCOPh + t\text{-}BuOH$$

The reaction occurs with hydrocarbons, olefins, alcohols, thio alcohols, ethers, sulphides and carbonyl, nitrogen and silicon compounds.

274 references up to 1967.
Complementary: 33, 10.
Relevant: none.

5, *15: Thiolo, thiono and dithio acids and esters* by *M. J. Janssen*

I. Structure and general properties (706); **II**. Thio acids (715); **III**. Thiolo esters (724); **IV**. Kinetics and mechanism of substitution reactions at the thiocarboxyl function (739); **V**. Thiono esters (741); **VI**. Dithio acids (746); **VII**. Dithiocarboxylic esters (752); **VIII**. Acknowledgements (757); **IX**. References (757).

The main aim of the chapter is to compare the chemistry of thio and dithio acids and esters with the corresponding oxygen analogues. Structures, physical properties, nucleophilicity and chemical properties of the title compounds are discussed in Section I.

Section II describes the preparation and acidity of thio acids, substitution in the thiocarbonyl group and additions of thio acids to unsaturated compounds.

The preparation, substitution and addition reactions acylating properties and some other reactions of the thiolo esters, RC(O)SR, are discussed in Section III.

Section IV deals with the role of thiocarboxyl compounds in biological reactions and with the hydrolysis and aminolysis of thio and thiolo acids and esters, and Section V with the properties and reactions of thiono esters, RC(S)OR.

The preparation, acidity, salt formation and organic reactions of dithio acids, RC(S)SH, are treated in Section IV, comparing them with the related xanthic ($ROCS_2H$) and thiocarbamic (R_2NCS_2H) acid derivatives. Finally, in Section VII dithiocarboxylic esters are briefly reviewed.

309 references up to 1967.
Complementary: 25, 8.
Relevant: 2, 17; **11**, 8; **26**, 16.

5, 16: Directive and activating effects of CO_2H and CO_2R groups in aromatic and aliphatic reactions by *G. Kohnstam and D. L. H. Williams*

The chapter considers the effects of CO_2H, CO_2R and CO_2 groups on routes, equilibria and composition of the products.

Internal and overall polar effects, physical properties, proximity effects (steric inhibition of resonance, hydrogen bonding, solvation) and substituent constants are discussed in Section I.

Section II describes equilibria in the ionization of aliphatic, alicyclic, unsaturated and aromatic acids, N- and O-acids (ammonium, anilinium, pyridinium ions and phenols) and carbon acids.

Nucleophilic substitutions in aliphatic and aromatic compounds and their mechanisms are discussed in Section III. Reactions at the carbonyl group and the effects of substituents on $A_{AC}2$ and $B_{AC}2$ hydrolyses and the intramolecular catalysis of carboxyl derivatives are treated (Section IV).

The influence of the title groups on electrophilic aromatic substitution, as regards mechanism, partial rate factors and selectivity, is discussed with reference to benzenes substituted by a single COOH(R) group, or by COOH(R) and one additional substituent or by a side-chain containing the COOH(R) group (Section V).

Section VI describes the kinetics, mechanisms and kinetic substituent effects on products in electrophilic and nucleophilic additions to unsaturated systems containing a COOH(R) substituent. Reaction paths and products and especially neighbouring-group participations are treated. Finally, some cases of the influence of COOH(R) groups on the transetherification of XCH_2CH_2OPh (believed to occur by way of the formation of an olefin, $XCH{=}CH_2$) are treated briefly.

212 references up to 1967.
Complementary: none.
Relevant: **3**, 3; **4**, 5; **6**, 8; **7**, 2, 3; **11**, 12; **12**, 5; **13**, 8; **14**, 4; **15**, 5; **16**, 6; **18**, 9; **19**, 8; **23**, 11; **28**, 9; **29**, 16; **31**, 8; **36**, 15.

5, 17: Analysis of carboxylic acids and esters by *T. S. Ma*

The chapter deals only with methods concerned with the COOH or COOR function and not with those which are based on the remaining portion of the molecule.

Qualitative methods including acidity, precipitation by heavy metal salts, the iron(III) hydroxamate test, preparation of solid *p*-toluidides, bromophenacyl esters, benzylthiouronium salts and hydrazides for acids are treated, followed by the characterization of esters by saponification and conversion to acid hydrazides, dinitrobenzoates and some others (Section II).

Section III discusses the quantitative determination of the COOH function by neutralization, through active H, by decarboxylation or by the Karl Fischer reagent. The COOR function can be determined by saponification, spectrophotometrically, with a Grignard reagent or with $LiAlH_4$. Mixtures can be separated by gas, thin-layer, paper or column chromatography, ion exchange or distillation (Section IV). Finally, in Section V. UV, visible, NMR and mass spectrometry are considered.

152 references up to 1968.
Complementary: none.
Relevant: 2, 8; **3**, 15; **4**, 3; **5**, 17; **8**, 3; **12**, 4; **13**, 6; **14**, 3; **15**, 3; **16**, 3; **17**, 4; **18**, 5; **20**, 3; **22**, 5; **23**, 5; **24**, 5; **25**, 10; **26**, 5; **28**, 3; **29**, 21; **30**, 14–20; **36**, 5.

5, *18: Biological formation and reactions of the —COOH and —COOR groups* by *S. Doonan*

The chapter emphasises acids and esters which are involved in the main metabolic pathways of animals, plants and microorganisms and the enzyme systems involved.

Section II discusses the ways in which organisms incorporate CO_2 into their tissues. The carboxylases involved are classified. Photosynthesis is described, considering the composition and organization of the photosynthetic apparatus, the light reactions, photophosphorylation, the carbon reduction cycle and chemosynthetic bacteria. Among heterotropic CO_2 fixation processes, the formation of dicarboxylic acids from pyruvate, the propionate metabolism, the catabolism of leucine and isovalerate and the biosynthesis of purine nucleotides are treated.

The biosynthesis of carbamyl phosphate and its anabolic functions and the formation of urea from it are the subject of Section III.

Section IV describes decarboxylations of pyruvate, α-ketoglutamate, amino acids and β-keto acids.

Carboxylic acids are also formed in biological systems by oxidation of alcohols, through the formation and metabolism of 6-phosphogluconate, by the oxidation of aldehydes (involving dehydrogenases or metalloflavoproteins) and by oxidative or hydrolytic cleavage of C—C bonds yielding carboxylates.

The types and distribution, the degradation and the biosynthesis of fatty acids are presented in Section VI.

Finally, Section VII deals with the complex lipids, their structures, distribution and functions, their degradation (including triglycerides and phospholipids) and their biosynthesis (including phosphatidic acids, triglycerides and phospholipids).

Throughout the chapter the main metabolic pathways and cycles are illustrated in Figures.

299 references up to 1966.
Complementary: none.
Relevant: 2, 7; **3**, 5; **4**, 9; **7**, 4; **11**, 11; **12**, 12; **13**, 14; **15**, 10; **16**, 13; **17**, 13; **18**, 13; **19**, 13; **20**, 6; **26**, 13; **28**, 17; **31**, 9.

The chemistry of the nitro and nitroso groups. Part 1 (1969)

6, *1: Theoretical aspects of the C—NO and C—NO_2 bonds* by *G. H. Wagnière*

I. Quantum mechanical calculations of molecular properties (2); **II**. Aliphatic C—NO and C—NO_2 compounds (3); **III**. Aromatic C—NO and C—NO_2 compounds (39); **IV**. Some special topics (69); **V**. Acknowledgement (71); **VI**. References (71).

Section I is a short introduction to treatment of a molecule as a many-body problem, of the separation of the motions of electrons and nuclei and of the solution of the electronic problem by the LCAO–MO scheme and by semi-empirical calculations.

Section II discusses *ab initio* calculations on NO and the physical properties of nitrosoalkanes; electronic and magnetic properties of nitroxide radicals obtained from amines or from oximes; the dative bond in trimethylamine *N*-oxide and the electronic structure of NO_2, including the position of the unpaired electron, electron interactions and ionization potentials, the electronic properties of nitroalkanes and the structure of nitroethylene.

The description of aromatic nitro and nitroso compound *N*-oxides and nitroxides deals with the ground- and excited-state properties of these and with their electronic and magnetic properties (Section III). Vibrational spectra and chemical reactivity are treated briefly (Section IV).

232 references up to 1966.
Complementary: none.
Relevant: 1, 1; **2**, 1; **3**, 1; **4**, 1; **5**, 1; **8**, 1; **11**, 1; **12**, 1; **13**, 1; **14**, 1; **15**, 1; **16**, 1; **17**, 1; **18**, 1; **19**, 1, **20**, 1; **22**, 1; **23**, 1; **24**, 1; **25**, 1; **26**, 1; **27**, 9; **28**, 1; **31**, 27; **32**, 1; **33**, 1; **36**, 2.

6, *2: Spectroscopy of the nitro group* by *C. N. R. Rao*

I. Introduction (80); **II**. Infrared and Raman spectra (80); **III**. Electronic absorption spectra (91); **IV**. Microwave spectra (106); **V**. Nuclear magnetic resonance spectra (106); **VI**. Literature on the spectroscopic data of some assorted nitro compounds (111); **VII**. Spectroscopic studies of hydrogen bonding and keto–enol equilibria (112); **VII**. Spectra of charge-transfer complexes (116); **IX**. Electron spin resonance spectra (120); **X**. Miscellaneous spectroscopic studies (125); **XI**. Acknowledgement (125); **XII**. References (126).

Various aspects of the spectroscopy of nitro compounds are reviewed and in many cases experimental data are presented in tables. Section II describes the IR and Raman spectra of saturated, olefinic, aromatic and heterocyclic nitro compounds, considering asymmetric and symmetric stretching frequencies and some less important bands.

Electronic spectra are discussed in Section III, with Tables 8–11 presenting data on a variety of substituted nitrobenzenes. Solvent effects, acid–base equilibria and the anions of nitroalkanes are considered, followed by reactions on microwave spectra (Section IV) and on NMR (Section V).

Section VII treats hydrogen bonding and keto–enol equilibria, followed by charge-transfer complexes (Section VIII) and by ESR spectra (Section IX).

346 references up to 1966.
Complementary: 6, 3.
Relevant: 28, 6; **29**, 5; **31**, 4, 5 (and also sections in chapters on analysis, detection and determination in various volumes).

6, *3: Spectroscopy of the nitroso group* by *C. N. R. Rao and K. R. Bhaskar*

The chapter treats mainly nitroso groups in nitroso compounds (RN=O) and in nitrosamines (—N—N=O).

Section II discusses IR spectra of organic nitrites, of *C*-nitroso compounds and of nitrosamines, emphasizing the N=O stretching frequency but also mentioning other less important bands.

The UV–visible spectra of the same classes of compounds are treated in Section III, NMR spectra in Section IV, ESR in Section V and microwave spectra in Section VI. Many experimental results are compared and tabulated.

79 references up to 1966.
Complementary: 64, 2.
Relevant: 28, 6; **29**, 5; **31**, 4, 5 (and also sections in chapters on analysis, detection and determination in various volumes).

6, *4: The photochemistry of the nitro and nitroso groups* by *H. A. Morrison*

Section III deals with the photochemistry of nitro compounds, gives a convenient scheme for the representation of the reactive excited states, lists the possible primary photochemical processes (Table 1) and compares them with those of carbonyl compounds. Dissociation into free radicals, N—O bond cleavage, photocycloelimination, *cis–trans*-photoisomerization, olefin cycloaddition, nitrite formation, H-abstraction, radical anion formation and nitro photocycloaddition are discussed.

The presence of NO_2 (and other electron-attracting) groups in aromatic systems enables photoinduced nucleophilic substitutions to occur (Section IV).

The similar processes with nitroso compounds have been studied much less (Section VI). The primary photochemical processes are given in Table 2. The photodissociation processes leading to the formation of nitroxides and *N*-nitritoamines are described, together with MS data H-abstraction, photocycloelimination and halogen expulsion.

161 references, up to 1968 (cf. **IX**, Aldenda, p. 212).

Complementary: 29, 6.
Relevant: 3, 8; **8**, 12; **9**, 6; **11**, 5; **13**, 16; **14**, 8; **15**, 9; **16**, 11; **17**, 9; **18**, 10; **19**, 20; **22**, 10; **23**, 9; **24**, 12; **25**, 11; **27**, 21, 22; **28**, 5; **32**, 29; **33**, 21.

6, *5: Methods of formation of the nitroso group and its reactions* by *J. H. Boyer*

I. Introduction (216); **II**. Preparative methods (217); **III**. Structure and reactions (252); **IV**. References (289).

Section II describes the methods available for the preparation of compounds containing nitroso groups. These are (A) combination of nitric oxide with an organic radical ($R\cdot + NO \rightarrow RNO$); (B) irradiation of a mixture of a nitrosyl halide with an alkane leading to nitrosoalkanes; (C) pyrolysis and photolysis of nitrite esters in which the nitroso compound obtained corresponds to a fragment of the reactant (e.g. $R^1R^2R^3CONO \rightarrow R^1NO$); (D) oxidative nitrosation of aromatic compounds in the presence of Cu salts yielding *o*-nitrosophenols; (E) nitrosation of tertiary aromatic amines with HNO_2 yielding *p*-nitroso derivatives ($R^1R^2NC_6H_5 + HNO_2 \rightarrow p\text{-}R^1R^2NC_6H_4NO$); (F) nitrosation of secondary aromatic amines (Fischer–Hepp reaction) in which first the amino nitrogen is attacked, but rearranges to a *C*-nitroso compound; (G) nitrosation of primary aromatic amines (1-naphthylamine → 1-amino-4-nitrosonaphthalene); (H) nitrosation of phenols in the *o*- and *p*-positions; (I) nitrosation of ethers; (J) nitrosative decarboxylations; (K) nitrosation of olefins; (L) nitrosation of phenols and heterocycles with nitrite esters in the presence of alkoxides; (M) attack of NOCl on organometallics; (N) attack of hypohalous acids on oximes, yielding *gem*-halonitroso derivatives; (O) a related reaction yielding *gem*-nitroso alcohols; oxidative processes involving dioximes, hydroxylamines, amines, nitrones and Schiff bases; reduction of nitro compounds and a number of other less frequently used methods.

Section III discusses the tendency of *C*-nitroso compounds to dimerize (with disappearance of the blue or green colour characteristic of the monomers) and the factors which promote or hinder dimer formation. The C—NO compounds may isomerize to oximes, may be reduced to amines, deoxygenated to a nitrene, oxidized to a nitro compound and may react with free radicals, olefins, acetylenes, arynes, thioketones, *P*-ylides, azomethines, diazoalkanes, conjugated dienes and active-H compounds to give a variety of products including polymers, heterocycles and nitrones. Metal salts and Lewis acids yield complexes and the base-catalysed condensation with active methylene groups yields azomethines (Ehrlich–Sachs reaction). Amines yield azo compounds and similar condensations occur with hydroxylamines and hydrazines.

Finally, substitutions in ArNO compounds, cleavages of the C—NO bond and pyrolytic and photolytic disproportionations and some other reactions are mentioned briefly.

437 references up to 1966.
Complementary: 29, 4, 12.
Relevant: 29, 3, 6, 7, 24, 26.

6, *6: Methods of formation of the nitro group in aliphatic and alicyclic systems* by *H. O. Larson*

Oximes can be converted to nitro compounds: HNO first forms *gem*-dinitro compounds which in turn may be reduced to a mononitro derivative:

$$R_2C{=}NOH \xrightarrow{HNO_3} R_2C(NO)NO_2 \xrightarrow{HNO_3} R_2C(NO_2)_2 \longrightarrow R_2CHNO_2.$$

Another reaction of oximes (with HOBr) leads to the same product through a bromonitro intermediate and peroxytrifluoroacetic acid (CF_3CO_3H) oxidized oximes directly to nitro compounds (Section II).

Amines can be oxidized to nitro compounds by the use of peracids or $KMnO_4$ (Section III) and active methylene groups are nitrated by alkyl nitrates, by HNO_3 or by acetone cyanohydrin nitrate (Section IV).

C=C double bonds are attacked by acetyl nitrate, nitryl chloride, N_2O_4, N_2O_3, NOCl or HNO_3. In most cases both addition and substitution products are obtained and separation of the mixtures is difficult (Section V).

Aliphatic halogen atoms may be displaced by nitrite ions (in the form of $AgNO_3$, the Victor Meyer reaction) and the corresponding nitro compound is obtained usually in good yields from primary halides (but much worse from secondary and not at all from tertiary halides, with $AgNO_2$. Alkali metal nitrites, usually $NaNO_2$, give good results with primary and secondary alkyl iodides, bromides and sulphonate esters, but not with tertiary halides. The mechanism and stereochemistry of these reactions is discussed (Section VI).

Polynitro compounds (Section VII) are obtained from *gem*-halo nitro compounds (Ter Meer reaction) or by the Kaplan–Shechter reaction, which is the reaction of *aci*-salts of primary or secondary nitro compounds with $AgNO_2$ or other nitrites. This can also be carried out electrolytically. The mechanism and scope of the reaction are described, and also some less general methods (Section VIII).

195 references up to 1966.
Complementary: 29, 4, 19.
Relevant: 7, 1; **29**, 6, 7, 25.

6, *7: Nitronic acids and esters* by *A. T. Nielsen*

Nitronic acids (*aci*-nitro compounds, $R^1R^2C{=}NO_2H$) share a common anion with their parent nitroalkanes ($R^1R^2CHNO_2$). They are relatively weak acids, which, at least in the solid state, have *cis–trans* isomers. In neutral or basic media the following equilibria exist:

$$B + R^1R^2C{=}N\begin{smallmatrix}\nearrow OH\\ \searrow O\end{smallmatrix} \rightleftharpoons R^1R^2C{=}N\begin{smallmatrix}\cdot\cdot O\\ \cdot\cdot O\end{smallmatrix}^{-} + BH^+ \rightleftharpoons R^1R^2CHNO_2 + B$$

The stronger the acidity of the nitronic acid the faster is its tautomerization to the nitro compound. The mechanism of the tautomerization in both directions is discussed in detail, together with the factors affecting the ionization constants (Tables 1–5, Section II).

Section III describes the methods of preparation of nitronic acids, their physical properties (ionization constants in Table 5, melting points and half-lives in Table 6), UV and IR spectra and their chemical properties, including addition reactions (both nucleophilic and electrophilic) in which nitroso compounds, hydroxamic acids or nitro compounds may be formed. In the synthetically important Nef reaction, nitronic acids are converted to the corresponding carbonyl compounds ($R^1R^2C{=}NO_2H \rightarrow R^1R^2CO$). Oxidations and reductions, and also special reactions of α-halonitronic acids and of ketonitronic acids, are treated.

Section IV discusses the preparation of acyclic and cyclic nitronic acid esters (and tabulates methods, yields, m.p.s and half-lives; Tables 11 and 13), their UV, IR and NMR spectra and also their hydrolysis, oxidation, reduction and 1,3-dipolar addition (Tables 14–17).

Finally, Section V treats more briefly nitronic acid salts ($R^1R^2C{=}NOONa$), and anhydrides $\left(R^1R^2C{=}\underset{\downarrow \atop O}{N}{-}O{-}\underset{\downarrow \atop O}{N}{=}CR^1R^2\right)$, halides $\left(R^1RC{=}N\genfrac{}{}{0pt}{}{\rightarrow O}{\smallsetminus Cl}\right)$ and amides $\left(R^1R^2C{=}N\genfrac{}{}{0pt}{}{\diagup NR^3R^4}{\rightarrow O}\right)$ and Section VI deals with the analytical methods available for nitronic acids.

490 references up to 1967.
Complementary: 29, 13.
Relevant: 29, 3, 4,

6, *8: Activating effects of the nitro group in aromatic substitutions* by *Th. J. de Boer and I. P. Dirkx*

The chapter deals exclusively with nucleophilic aromatic substitutions (usually by the addition–elimination mechanism), emphasizing principles, mechanisms and the practice and theory of the topic in the period 1952–67.

Section II discusses the displacement of halogen by amines, hydroxide, alkoxy, phenoxy, other halogens, mercaptides, azides, thyocyanates, carbanions, sulphite, carboxylate and some other reagents; it deals with the activation process by NO_2 groups, with *ortho* effects and tabulates the results obtained in synthetic experiments.

Section III describes the kinetics and mechanism of the nucleophilic displacement of halogens and deals with the bond-making and -breaking processes, kinetics, isotope effects, intermediates and transition states, substituent and steric effects, reactivities of leaving and attacking groups and medium effects.

Section IV deals with nucleophilic reactions in which the displaced group is hydrogen, nitro, alkoxy, aryloxy, sulphonate, amino or ammonium while the attacking group may by OH^-, RO^-, NH_3, NHR_2, NR_3, etc.

Displacements in polycyclic nitro compounds (nitronaphthalenes, nitroanthracenes) and in heterocyclic nitro compounds, again involving a wide variety of leaving and entering groups, are treated in Section V, and photochemical displacement of alkoxy groups, halogens and nitro groups in Section VI.

The last section deals with the Smiles rearrangement, which is an intramolecular nucleophilic aromatic substitution, and with the Von Richter reaction, which is a substitution of an aromatic nitro group by KCN and subsequent hydrolysis to the corresponding acid in which the COOH group appears in a position *ortho* to the original NO_2 group, and finally with ring elargements of some polynitro arenes.

586 references up to 1967.
Complementary: **7**, 2, 3; **29**, 16, 27.
Relevant: **3**, 3; **4**, 5; **5**, 16; **11**, 12; **12**, 5; **13**, 8; **14**, 4; **15**, 5; **16**, 6; **18**, 9; **19**, 8; **23**, 11; **28**, 9; **31**, 8.

6, *9: Methods of formation of the nitramino group, its properties and reactions* by *G. F. Wright*

The structures associated with the Section II–IX are as follows: II, $RNHNO_2$; III, $RN{=}N{<}^{\rightarrow O}_{OH}$ (tautomer of I); IV, $ROC(O)NHNO_2$ (nitrourethane); V, R_2NNO_2; VI, $RON{=}NR'$; VII, $RN(NO_2)CONH_2$ (methylnitrourea); VIII, $RC{=}NOH$; and IX, $(CH_2{=}NNO_2)_3$. In the following the Roman numerals refer either to the Section or to the structures associated with them.

II can be prepared by alkaline hydrolysis of IV or, better, by oxidation of salts of diazotates. Several other syntheses are also mentioned. They are acids, nearly as strong as carboxylic acids and exist in tautomeric equilibrium with III. Section IV discusses nitramides and the related nitroureas and nitroguanidines. The simple urethane (IV, R = Et) is used for the preparation of more complex nitramides, which in turn may be converted to the analogous nitramines (II).

Compounds V are much more stable in acidic media than II and some are available by direct nitration of the corresponding secondary amines, if the latter are of low basicity; in other cases products of nitrosation can be oxidized to V. The mechanisms of these reactions are discussed.

Very little is known about VI, and the existence of the V$\rightleftharpoons$VI equilibrium is questionable.

Compounds VII can be usually prepared by direct nitration of the corresponding amides; by hydrolysis these are converted to II. Many powerful explosives belong to this class.

VIII are prepared by the reactions of oximes with nitrous acid ($R_2C{=}NOH + HNO_2 \rightarrow R_2C{=}NNO_2$) and by various other less general methods.

The most important representative of IX is the explosive cyclonite, a cyclic trimer of $CH_2{=}NNO_2$. Various industrial methods for the synthesis of this and similar compounds are described in detail.

Section X deals with Raman, IR, UV and NMR spectra, X-ray diffraction and electrical polarization.

171 references up to 1964.
Complementary: **7**, 1 (Section VI).
Relevant: **4**, 6; **29**, 26.

The chemistry of the nitro and nitroso groups. Part 2 (1970)

7, 1: *Introduction of the nitro group into aromatic systems* by *W. M. Weaver*

I. Electrophilic nitration (2); **II**. Nitrations under non-acidic conditions (16); **III**. Oxidation of amino and nitroso compounds to nitroarenes (29); **IV**. Replacement of diazonium ion with the nitro group (31); **V**. Nitrations with oxides of nitrogen (33); **VI**. Rearrangement of *N*-nitroamides (35); **VII**. Problems in orientation (37); **VIII**. Miscellaneous electrophilic nitrating reagents (41); **XI**. References (42).

Section I discusses the mechanisms, rates, isomer distribution, reagents, experimental conditions and side-reactions in electrophilic aromatic nitrations. The use of some special reagents [$C(NO_2)_4$, nitrate salts, $NO_2^+BF_4^-$] are described (Section II) in addition to oxidations of amines and nitroso compounds to nitroarenes (Section III) and the preparation of nitroarenes from amines by diazotization and subsequent displacement of the diazonium by a nitro group (Section IV).

Nitrogen oxides (N_2O_3, NO_2, N_2O_4 and N_2O_5), dissolved in H_2SO_4 or combined with Lewis acids, generate NO_2^+ ions and thus introduce nitro groups into arenes. A variation is the oxynitration [e.g. of benzene to picric acid with $HNO_3 + Hg(NO_2)_2$] (Section V).

Migration of the NO_2 group from *N*-nitroamines to the aromatic nucleus under acid catalysis is of great theoretical (but of almost no practical) interest (Section VI).

Finally, some methods for introducing NO_2 groups into arenes at normally unaccessible positions (Section VII) and some special combinations of Lewis acids with NO_2^+ carriers (Section VIII) are treated.

167 references up to 1968.
Complementary: 29, 19.
Relevant: 29, 4.

7, 2: *Directing effects of the nitro group in electrophilic and radical aromatic substitution* by *T. Urbanski*

I. Electrophilic substitution (49); **II**. Free-radical substitution (64); **III**. References.

Section I presents a historical review of the directing effects of nitro groups, including both early and modern theories. It deals with the *meta*-directing effect and also with the prevalence of *ortho* isomers in the non-*meta* portion of the reaction and with the effect of nitro groups in a side-chain. Subsequently the indirect nitration occurring through rearrangement of *N*- and *O*-nitration products is discussed (e.g. $C_6H_5NHNO_2$

$\rightarrow NO_2C_6H_4NH_2$), together with the rearrangement of polynitroarenes (e.g., 2, 3-dinitrophenol $\rightarrow$ 2, 5-dinitrophenol).

Section II deals with the directing effect of a nitro group in free-radical substitutions and with the activating effect of the nitro group in these reactions (free radical methylation, hydroxylation, arylation, etc.).

91 references up to 1965.
Complementary: 6, 8.
Relevant: 3, 3; **4**, 5; **5**, 16; **7**, 3; **11**, 12; **12**, 5; **13**, 8; **14**, 4; **15**, 5; **16**, 6; **18**, 9; **19**, 8; **23**, 11; **28**, 9; **29**, 16; **31**, 8.

7, 3: Activating and directing effects of the nitro group in aliphatic systems by *H. H. Baer and L. Urbas*

I. The Henry addition and related reactions (76); **II**. Mannich reactions; **III**. Michael additions (130); **IV**. Diels–Alder reactions (148); **V**. Some reactions of nitro alcohols and their derivatives (154); **VI**. Additions of nucleophiles to nitro olefins (178); **VII**. References (187).

The Henry addition is an aldol-type, base-catalysed reaction of nitromethanes with aliphatic aldehydes (and in some cases also with ketones): $RCHO + RCH_2NO_2 \rightarrow RCHOHCH(R)NO_2$. The very broad scope of the reaction justifies its separate treatment as distinct from other Knoevenagel-type reactions. The nitro alcohol produced, in the presence of base may form a nitronate salt, $RCH(OH)C(R){=}NO_2Na$, or it may dehydrate to a nitroolefin, $RCH{=}C(R)NO_2$, especially easily if one of the R groups is aromatic. The nitroalkane may be primary or secondary or substituted, and the aldehyde may be aliphatic, aromatic, halogenated or a hydroxyaldehyde (including sugars) (Section I).

Nitroalkanes react with *N*-hydroxymethylated secondary amines in a manner very similar to the Mannich reaction. Either one (at low temperature) or two amines may condense with the nitroalkane: (1) $C_5H_{10}NCH_2OH + CH_2RNO_2 \rightarrow C_5H_{10}NCH_2CHRNO_2$ or (2) $2C_5H_{10}NCH_2OH + CH_2RNO_2 \rightarrow O_2NCR(CH_2NC_5H_{10})_2$. The same products may be obtained by mixing an amine, formaldehyde and the nitroalkane in the appropriate reactive conditions. The Mannich-base products are able to alkylate reactive methylene or methine compounds. The scope and stereochemistry of the reaction are discussed (Section II).

In the Michael addition of reactive α-H-containing compounds to activated C=C bonds ($R_2CHX + R_2C{=}CRY \rightarrow XCR_2CR_2CRHY$), nitro groups may serve as activators in both components or in any one of them. All these possibilities are described in detail (Section III).

The use of nitro-activated alkenes in the Diels–Alder reaction is treated in Section IV.

The formation and cleavage of nitro acetals and ketals and the preparation of nitro alcohol esters (both inorganic and organic) and their conversion into nitroolefins is dealt with in Section V.

Finally, the addition of nucleophiles to nitroolefins (addition of alkoxy groups, thiols, H_2S, sulphites, sulphinic acids, NH_3 and other amines) is described (Section VI).

516 references up to 1968.
Complementary: 6, 8; **7**, 2; **29**, 16.
Relevant: 3, 3; **4**, 5; **5**, 16; **11**, 12; **12**, 5; **13**, 8; **14**, 4; **15**, 5; **16**, 6; **18**, 9; **19**, 8; **23**, 11; **28**, 9; **31**, 8.

7, 4: Biochemistry and pharmacology of the nitro and nitroso groups by *J. Venulet and L. VanEtten*

I. Introduction (202); **II.** Biological oxidation–reduction processes and oxidative phosphorylation (204); **III.** Biochemistry and pharmacology of naturally occurring compounds containing the nitro or nitroso group (212); **IV.** Biochemistry and pharmacology of synthetic compounds containing the nitro or nitroso group (227); **V.** The role of nitro and nitroso compounds in the formation of methaemoglobin (267); **IV.** Addendum and final remarks (269); **VII.** References (272).

As a general background to the subject matter of the chapter, Section II describes some biological oxidation–reduction reactions, oxidative phosphorylation and the electron transport sequence, the involvement of NO and NO_2 groups in biochemical oxidation–reduction processes and the uncoupling of oxidation and phosphorylation.

Natural products containing NO and NO_2 groups and having biochemical or pharmacological importance include antibiotics (chloramphenicol, nitroimidazole, *p*-nitrobenzylpenicillin, etc.), some relatively simple nitro compounds (1-phenyl-2-nitroethane) and a glycoside containing the azoxy group (cycasin) (Section III). Synthetic NO and NO_2 compounds are often used for their toxic, insecticide properties (parathion and analogues) which act by inhibition of acetylcholinesterase. Others are molluscicides, fungicides or herbicides and weedkillers. Many nitroso compounds are carcinogenic, including simple ones such as *N*-nitrosodialkylamines (Section IV.A).

Many NO and NO_2 derivatives are used for their therapeutic properties (Section IV.B). These include antibacterial, antiprotozoal and anthelmintic drugs and cytostatic and spasmolytic compounds. Some others are powerful sweetening agents. The toxic properties of some technically important compounds (aliphatic and aromatic nitro compounds) and their metabolism are discussed (Section IV.C).

Section V describes the importance of NO and NO_2 compounds in the oxidation of Fe^{II} in haemoglobin to Fe^{III}, thus yielding methaemoglobin, which is no longer able to combine with oxygen, and the physiological results resemble CO poisoning.

647 references up to 1969.
Complementary: 4, 9.
Relevant: 2, 7; **3**, 5; **5**, 18; **11**, 11; **12**, 12; **13**, 14; **15**, 10; **16**, 13; **17**, 13; **18**, 13; **19**, 13; **20**, 6; **22**, 20; **24**, 14; **26**, 13; **28**, 17; **31**, 9.

7, 5: The synthesis and reactions of trinitromethyl compounds by *L. A. Kaplan*

I. Introduction (290); **II.** General characteristics of the trinitromethyl group (290); **III.** Synthetic approaches to trinitromethyl compounds (293); **IV.** Characterization of the trinitromethyl group (317); **V.** Reactions of the trinitromethyl group (318); **VI.** References (325).

Section II discusses the characteristics of the (normal) tetrahedrally hybridized trinitromethyl group as in $RC(NO_2)_3$ compounds ($R = H$, alkyl, etc.) in contrast with the (planar) trigonally hybridized $C(NO_3)_3$ group in trinitromethide ion.

The synthetic routes to obtain trinitromethyl compounds in which the $^-C(NO_2)_3$ ion acts as a nucleophile include the following: Michael addition to α,β-unsaturated systems; addition to carbonyl compounds; the Mannich reaction yielding trinitromethyl amines; "three-body reactions" in which a product is obtained from $HC(NO_2)_3$ and an aldehyde in the presence of an alcohol, thiol or amide, e.g. $RCHO + HC(NO_2)_3 + R'OH \rightarrow$

$RCH(OR')C(NO_2)_3$; nucleophilic displacement at a saturated carbon, $RX + AgC(NO_2)_3 \rightarrow RC(NO_2)_3 + AgX$; the reactions of mercury trinitromethide which result in $-HgC(NO_2)_3$ substitution in aromatic substrates and a variety of additions with olefinic substrates; and the less useful method of converting mono- or dinitro derivatives by nitration to the corresponding trinitro compound. In all cases the scope, variations and mechanisms of the reactions are described (Section III).

The characterization of $C(NO_2)_3$ groups is discussed very briefly in Section IV, and Section V treats nucleophilic displacements on the nitro groups in which one or more NO_2 groups are cleaved off by reagents such as hydroperoxide, iodide, nitrite, thiosulphate or sulphite ions or other reagents. In contrast, bases attack the α-H atoms in trinitromethyl compounds:

$$CH_3C(NO_2)_3 + B^- \rightarrow C(NO_2)_2{=}CH_2 + BH + NO_2^-$$
$$C(NO_2)_2{=}CH_2 + B^- \rightarrow {}^-C(NO_2)_2CH_2B$$

where $B = CN$, OR, $CH(CO_2R)_2$, R_2NH, NH_3 etc.

124 references up to 1969.
Complementary: 29, 16.
Relevant: none.

7, *6: Polynitroaromatic addition compounds* by *T. N. Hall and C. F. Poransky, Jr*

Polynitroarenes form brightly coloured solutions or solids with bases. Equilibria are established with OH and OR ions; the structure of the products was studied chemically and by X-ray diffraction, NMR, IR, ESR and isotope exchange and by considering acidity function correlations. The problems involved in the measurement of equilibria are described, pK, spectroscopic and thermodynamic data are tabulated (Tables 3–5) and kinetic and mechanistic studies are discussed (Section II).

The formation of addition compounds with amines and NH_3 is treated similarly in Section III and with ketones containing an active α–H atom in Section IV (Tables 6–9).

Finally, in Section V compound formations are discussed and results are tabulated (Tables 10–17) for reactions with SO_3^{2-}, EtS^-, PhS^-, CN^-, N_3^- and HCO_3^- ions and with salts of mononitroalkanes, ethyl malonate and ethyl acetoacetate.

176 references up to 1969.
Complementary: 29, 27.
Relevant: 12, 10; **13**, 2; **17**, 6; **23**, 7.

The chemistry of the carbon–nitrogen double bond (1970)

8, *1: General and theoretical aspects* by *C. Sandorfy*

I. Introduction (1); **II**. Some physico-chemical data (2); **III**. Quantum chemical aspects (6); **IV**. Infrared and Raman spectra (37); **V**. Electronic spectra (46); **VI**. Acknowledgements (56); **VII**. References (56).

Comparisons are made between the subject of this chapter, the $>C{=}N-$ group and two related groups, namely $>C{=}C<$ and $>C{=}O$; for the latter two there are plentiful physico-chemical data available in the literature, but unfortunately this is not the case for the C=N group.

Section II presents (and compares) internuclear distances, overlap integrals, dipole moments and bond energies. Section III discusses the quantum chemistry of the N atom and its hybridization; it examines the π-electrons of the azomethine group by the Hückel method and performs some simple calculations on them, including overlap integral, semi-empirical methods, and considers also the σ framework and forms in which the azomethine N carries a positive or a negative charge. IR and Raman (Section IV) and electronic (Section V) spectra are described for conjugated and non-conjugated azomethines (Tables 10–15).

138 references up to 1967.
Complementary: 20, 1.
Relevant: 1, 1; **2**, 1; **3**, 1; **4**, 1; **5**, 1; **6**, 1; **11**, 1; **12**, 1; **13**, 1; **14**, 1; **15**, 1; **16**, 1; **17**, 1; **18**, 1; **19**, 1; **22**, 1; **23**, 1; **24**, 1; **25**, 1; **26**, 1; **28**, 1; **31**, 27; **32**, 1; **33**, 1; **36**, 2.

8, *2: Methods of formation of the carbon—nitrogen double bond* by *S. Dayagi and Y. Degani*

I. Introduction (63); **II**. Reactions of carbonyl groups with amino groups and related reactions (64); **III**. Nitroso–methylene condensations (83); **IV**. Formation of oximes via *C*-nitrosations (85); **V**. Diazonium salt–methylene condensations (91); **VI**. Additions to carbon—carbon double or triple bonds (95); **VII**. Formation of C=N bonds through ylids (98); **VIII**. Tautomerization of amides and thioamides and related reactions (103); **IX**. Addition reactions to nitriles, isonitriles, nitrile oxides and related compounds (108); **X**. Oxidation of and elimination from nitrogen compounds (117); **XI**. Reduction of nitro compounds (124); **XII**. Formation of azomethines by rearrangements and photochemical reactions (126); **XIII**. References (130).

The material is arranged according to mechanistic types of the C=N bond-forming reactions.

Section II discusses CO–amine condensations, such as reactions of aldehydes and ketones with NH_3, primary and secondary amines, hydroxylamines, hydrazines and similar compounds. Further, potential carbonyl synthones (hydrates, ketals, enols, etc.) are considered and also interchange of existing C=N groups with either amines of carbonyl compounds.

Nitroso–methylene condensation are described (Section III) and the formation of oximes through isomerization of primarily formed *C*-nitroso compounds:

$$RCH_2R' \rightarrow RCH(NO)R' \rightleftharpoons RCR'{=}NOH$$

obtained by nitrosation of primary, secondary, tertiary or unsaturated carbons in ionic or free-radical reactions (Section IV).

Diazonium salts yield aryl hydrazones by coupling with active methylene groups: $ArN_2^+ + CH_2XY \rightarrow ArNHN{=}CXY$ (Section V). Acetylenes and olefins react with amines and the product enamine tautomerizes to an azomethine (e.g. $CH{\equiv}CH + EtNH_2 \rightarrow [CH_2{=}CHNHEt] \rightarrow CH_3CH{=}NEt$) (Section VI).

Wittig-type condensations of iminophosphorones with aldehydes or ketones give azomethines, e.g. $Ph_3P{=}NR + R^1R^2CO \rightarrow R^1R^2C{=}NR$, and other condensations of ylids with diazo compounds, azides, isocyanates and nitroso compounds also occur similarly (Section VII).

Section VIII describes some less general routes, in which the tautomeric forms of amides and of thioamides can be fixed in a form containing the C=N group.

Nitriles can be reduced to imines; they also add alcohols to yield imidates, amines to yield amidines and Grignard reagents to yield imines. Isonitriles, nitrile oxides, fulminates, isocyanates and related compounds also give facile reactions which result in C=N-containing compounds (Section IX).

Some dehydration and oxidation reactions of amines and hydroxylamines produce C=N derivatives, as do also eliminations from suitably substituted amines (e.g. $RCH_2NR'SO_2Ar \rightarrow RCH{=}NR'$) (Section X).

The reduction of aliphatic nitro compounds may be stopped at the oxime stage (Section XI) and a large number of rearrangements (Beckmann, Hoffman, Lossen, Curtius, Stieglitz) and photochemical reactions give C=N compounds (Section XII).

531 references up to 1967.
Complementary: 2, 12; **4**, 7; **12**, 7; **26**, 17, 18; **32**, 13.
Relevant: 10, 1; **12**, 8; **20**, 7.

8, *3: Analysis of azomethines* by *D. J. Curran and S. Siggia*

Qualitative methods are usually not specific for azomethines but for the products obtained by their hydrolysis (i.e. CO compounds and amines) (Section II).

Quantitative chemical methods include acidimetric titration of the basic C=N group directly and hydrolytic methods followed by determination of the resulting carbonyl compounds by dinitrophenylhydrazine, bisulphite or hydroxylamine (Section III).

The polarography of the C=N group is discussed in Section IV and in carefully controlled conditions can be used for quantitative studies, quality control and the determination of substituent effects.

IR spectroscopy of aldimines and ketimines and the determined C=N stretching frequencies are described in Section V (Tables 2–6). Mass spectra and the principal peaks of C=N compounds are given in Section VII (Tables 7–10). NMR (Section VII) and fluorescence (Section VIII) are also treated briefly.

70 references up to 1967.
Complementary: none.
Relevant: 2, 8; **3**, 15; **4**, 3; **5**, 17; **12**, 4; **13**, 6; **14**, 3; **15**, 3; **16**, 3; **17**, 4; **18**, 5; **22**, 5; **23**, 5; **24**, 5; **25**, 10; **26**, 5; **28**, 3; **29**, 21; **30**, 14–22; **36**, 5.

8, *4: The optical rotatory dispersion and circular dichroism of azomethines* by *R. Bonnett*

I. Ultraviolet absorption of azomethines (182); **II**. Optical dissymmetry effects of unconjugated azomethines (202); **III**. Optical dissymmetry effects of conjugated and other substituted azomethines (221); **IV**. Acknowledgements (231); **V**. References (231).

Optical dissymmetry effects (i.e. ORD and CD) of C=N groups are discussed. Section I deals with unconjugated and substituted azomethines and their electronic spectra (Table 1–4). The C=N group is a weak chromophore, located in the far quartz region.

Section II explains the terms, abbreviations and conventions used. The optical dissymmetric effects may involve dissymmetric substituents at the C=N carbon or nitrogen atom or at both atoms simultaneously and, of the latter, cyclic azomethines are of especial interest (Tables 6–8).

Section III discusses optical dissymmetry effects of aryl azomethines, oximes and *N*, *N*-dimethylhydrazones (Tables 9–12).

111 references up to 1969.
Complementary: none.
Relevant: 18, 7; **19**, 6; **22**, 4; **25**, 3; **26**, 3; **27**, 6; **29**, 23; **31**, 1; **36**, 17.

8, *5: Basic and complex-forming properties* by *J. W. Smith*

I. Introduction (235); **II**. Basic strengths (236); **III**. Hydrogen-bonding properties (239); **IV**. Complexes with metals (239); **V**. References (251).

Azomethine groups, owing to the presence of a lone pair of electrons on the N atom to the electron-donating character of the double bond, are proton acceptors and Lewis bases.

Basic strengths of azomethine compounds are described in Section II and pK_a values of their conjugate acids are tabulated (Table 1). Hydrogen bonding properties are discussed in Section III and Table 2 gives equilibrium constants, free energies and enthalpy and entropy changes for the complex formation between derivatives of PhCH=NPh and *p*-nitrophenol. Metal complexes (coordination compounds) of salicylaldimine and related compounds, of 2-hydroxyacetophenonimine, of 2-aminobenzylidene imines and of glyoxime with Ni^{II}, Pd^{II}, Pt^{II}, Cu^{II}, Zn^{II}, Co^{II} and Co^{III} are treated in Section IV.

64 references up to 1968.
Complementary: 12, 3.
Relevant: 3, 6; **4**, 4; **5**, 6; **9**, 5; **11**, 3; **13**, 7; **15**, 4; **16**, 5; **17**, 6; **18**, 8; **19**, 7; **20**, 12; **22**, 8; **23**, 6; **24**, 4; **25**, 5, 6; **29**, 17; **31**, 17, 21; **33**, 5.

8, *6: Additions to the azomethine group* by *K. Harada*

I. Introduction (256); **II**. Addition reactions (256); **III**. Acknowledgements (293); **IV**. References (293).

Section II describes addition reactions to C=N bonds, involving organic and inorganic reagents such as HCN, $NaHSO_3$, RHal, RSH, H_2S, MeCOPh, $PhCH_2COOR$, PhCOCN, Hal_3CCOOH, ArCHO, acid chlorides, maleic anhydride, Grignard reagents, RLi, ArLi, Li, Na, Mg–MgI_2, RNO_2, Ph_3C=$CHCH_3$ and CO. In continuation, catalytic hydrogenation (which converts the C=N compound into the corresponding secondary amine) is discussed, together with reductions using metals with acids or bases, or using $LiAlH_4$, $NaBH_4$, Me_2NHBH_3, formic acid or electrolysis. Finally, the stereochemistry of the hydrogenation of azomethines, including the synthesis of optically active α-amino acids and other compounds, is presented in detail.

166 references up to 1967.
Complementary: 8, 7; **12**, 7, 8; **21**, 4.
Relevant: 1, 8, 11; **9**, 4; **10**, 1; **17**, 17; **21**, 6, 9; **26**, 17.

8, *7: Cycloaddition reactions of carbon—nitrogen double bonds* by *J.-P. Anselme*

I. Introduction (299); **II**. Three-membered rings (300); **III**. Four-membered rings (305); **IV**. Five-membered rings (310); **V**. Six-membered rings (318); **VI**. References (323).

Imines are much less stable than alkenes, alkynes and nitriles and hence are used much less in cycloadditions, even though these reactions are in many cases very useful.

Cycloadditions of carbenes or nitrenes to C=N bonds yield aziridines or diaziridines (Section II).

Four-membered rings are formed from imines with ketenes (yielding β-lactams) and with ketene diaminals or alternatively with ynamines (both reactions yielding azetines) (Section III).

Five-membered rings are obtained by using diazoalkanes or nitrilimines (both reactions yielding triazolines). Other five-membered rings are formed with azomethine imines, azomethine ylids or nitrile oxides (Section IV).

Dienes and C=N bonds yield six-membered heterocycles containing one N atom, whereas heterodienes containing N or S atoms yield the corresponding heterocycles with two N or one N and one S atom, etc. (Section V).

In most cases the mechanisms of the reactions are discussed.

95 references up to 1967.
Complementary: 1, 11; **8**, 6; **12**, 8; **21**, 6.
Relevant: 19, 11; **24**, 11; **31**, 23.

8, *8: Substitution reactions at the azomethine carbon and nitrogen atoms* by *R. J. Morath and G. W. Stacy*

I. Introduction (328); **II**. Substitutions at the carbon atom (328); **III**. Substitutions at the nitrogen atom (351); **IV**. References (356).

Section II discusses substitutions on the azomethine C atom. Thus the Cl atom of imidyl chlorides (RN=CR′Cl) can be displaced by nucleophilic reagents (alkoxides, phenoxides, thiophenoxides, NH_3 or amines etc.) and the same is true for hydroximic acid chlorides

($RON{=}CR^1Cl$). Similar nucleophilic substitutions occur with the alkoxy and thioalkoxy groups of alkyl imidates [$RN{=}CR^1(OR^2)$], *O*-alkylisoureas [$HN{=}C(NH_2)OR$], alkylhydroxamic acids [$HON{=}C(R)OR$] and the corresponding thio derivatives [$RN{=}C(R^1)SR^2$ and $HN{=}C(NH)SR$] when the most frequently used nucleophiles are NH, amines and hydrazines.

Carbon-bound H atoms of Schiff bases ($RCH{=}NR^1$) can be displaced by $NaNH_2$, hypochlorites or amines and aldoximes may be directly chlorinated, $RCH{=}NOH + Cl_2 \rightarrow RCCl{=}NOH + HCl$.

NHR groups in amidoximes [$ArC(NHR){=}NOH$] or in amidines [$RC(NH_2){=}NH$] are displaced by amines or hydrazines.

Hydrogen bound to the N atom of C=N groups is displaced by Grignard reagents (e.g. $R_2C{=}NH + R^1MgX \rightarrow R_2C{=}NMgX + R^1H$) or by sulphonyl chlorides [e.g. $2RC(OEt){=}NH + R^1SO_2Cl \rightarrow RC(OEt){=}NSO_2R^1$], and in the case of amidines also by acetic anhydride, PhNCO and even with alkyl iodides [$PhC(NH_2){=}NH + RI \rightarrow PhC(NH_2){=}NR$].

225 references up to 1966.
Complementary: none.
Relevant: **32**, 13.

8, *9: syn–anti Isomerizations and rearrangements* by *C. G. McCarty*

I. Introduction (364); **II.** *syn–anti* Isomerizations (364); **III.** Rearrangements (408); **IV.** References (455).

Section II presents the evidence existing for geometric isomerism in various classes of compounds containing C=N bonds and reviews their rates of isomerization. The classes treated are *N*-aryl and *N*-alkyl aldimines ($R^1CH{=}NR^2$) and ketimines ($R^1R^2C{=}NR^3$), oximes, oxime ethers, hydrazones, azines and related compounds, *N*-halimines ($RCH{=}NHal$) and imidates and related compounds. In most cases data are presented for the kinetic parameters for thermal relaxation of photoisomerized solutions, isomer ratios, rates of (uncatalysed) isomerizations and NMR and IR data (Tables 1–11). The section ends with a discussion of *syn–anti* isomerization mechanisms.

Other rearrangements are described in Section III. The strong acid-catalysed Beckmann rearrangement [$R^1R^2C{=}NOH \rightarrow R^1NHCOR^2 + R^2NHCOR^1$] and some of its recent applications are treated. Examples are given for the 'abnormal' Beckmann rearrangement or fragmentation [$R^1R^2C{=}NOH + HX \rightarrow R^1R^2C{=}NX \rightarrow R^1CN + R^2X$] of α-trisubstituted and α-disubstituted oximes, α-oximino ketones, α-aminoketoximes and some related compounds. The mechanism of the reaction is discussed in detail.

The mechanism and synthetic utility of the Chapman reaction [$Ar^1C(OAr^2){=}NAr^3 \longrightarrow Ar^1CONAr^2Ar^3$] are described, and also the related arrangements of alkyl and acyl imidates in which also O → N migrations are involved (Section III.B).

The Neber rearrangement of oxime tosylates to α-amino ketones [$R^1CH_2CR^2{=}NOTs \rightarrow R^1CH(NH_2)COR^2$] probably occurs through an azirine intermediate. This and other proposed mechanisms together with some related reactions are treated (Section III.C).

Finally (Section III.D), rearrangements involving α-hydroxy imines, hydrazones, guanidines and amidines are discussed more briefly.

314 references up to 1967.
Complementary: **20**, 4; **21**, 8.
Relevant: **1**, 7; **2**, 15; **3**, 14; **4**, 10; **9**, 3; **12**, 16; **13**, 13; **14**, 7; **15**, 8; **16**, 17; **17**, 8; **18**, 15; **19**, 18; **23**, 13; **24**, 10; **26**, 15; **29**, 4.

8, 10: *Cleavage of the carbon—nitrogen double bond* by *A. Bruylants and E. Feytmants-de Medicis*

The chapter discusses (nucleophilic) reactions in which both bonds of the C=N group are broken, particularly by hydrolysis. Section II describes the hydrolysis of Schiff bases, which involve a tetrahedral carbinolamine intermediate:

$$>C{=}NX + H_2O \rightleftharpoons >\underset{\displaystyle OH}{\underset{|}{C}}{-}NHX \rightleftharpoons >C{=}O + XNH_2$$

Schiff bases derived from aliphatic and aromatic amines are described, and data are given for the equilibria, kinetics, thermodynamic behaviour and structure–reactivity relationships (Tables 1–6).

Several mechanisms can be distinguished in the general catalysis of the C=N hydrolysis: the rate-determining step may be direct attack by OH^- on the neutral imine group or on the protonated imine group (the latter with or without general acid–base catalysis) or, in rare cases, the decomposition of the carbinolamine intermediate.

Section II treats the hydrolysis of iminolactones, including their mechanisms and detailed kinetics, and Section IV that of Δ_2-thiazolines.

83 references up to 1967.
Complementary: **4**, 8.
Relevant: **21**, 11; **29**, 22.

8, 11: *Electrochemistry of the carbon—nitrogen double bond* by *H. Lund*

The basics of the experimental conditions permitting control of the electrolysis and the factors involved (pH, electrolyte, electrode material, optimization) are given in Section II.

Section III describes the electrochemical preparation of C=N compounds by reduction of various aliphatic and aromatic nitro and nitroso compounds, by reduction of some acid derivatives (hydroxamic acid halides, etc.), by partial reduction of heterocyclic compounds (pyridazines, cinnolines, phthalazines, pyrimidines, quinoxalines, etc.) and by anodic oxidation of amines, hydrazines and hydroxylamines.

Azomethine compounds are usually reduced to the corresponding amines electrolytically, although complications and by-reactions frequently occur. Section IV discusses the reduction of Schiff bases, imidic esters, imidothio esters, amidines, isothiocyanates, hydrazine derivatives (azines, hydrazones, thiobenzoylhydrazones and diazoalkanes) and hydroxylamine derivatives (oximes, hydroxamic acids and amide oximes).

156 references up to 1968.
Complementary: **20**, 5; **21**, 5.
Relevant: **5**, 2; **13**, 5; **16**, 15; **17**, 14; **19**, 12; **22**, 9; **23**, 10; **24**, 17; **25**, 13; **26**, 12; **27**, 8; **28**, 7; **29**, 8, 9; **31**, 7; **32**, 6.

8, *12: Photochemistry of the carbon—nitrogen double bond* by *G. Wettermark*

Many C=N compounds are photochromic in the solid state and the kinetics of the fading of some such crystals were investigated (Table I, Section II). Hydroxyl-substituted Schiff bases show phototautomerism in solution (Section III) and five-membered heterocyclic compounds containing C=N can undergo photoisomerization to a different five-membered ring (e.g. pyrazole → imidazole) (Section IV). *cis–trans* Isomerization occurs photochemically in C=N compounds (oximes, azines, hydrazones) and kinetic data for the thermal relaxation at the C=N bond are given in Table 3 (Section V).

Nitrones and aromatic *N*-oxides undergo photoisomerization (the latter to amides) (Section VI). Oximes yield amides (Section VII) and Schiff bases undergo photochemical cyclization by oxidative ring closure (sometimes with incorporation of a solvent fragment). Two C=N groups may also dimerize to a four-membered diazetidine ring. A photoexcited C=N group may abstract hydrogen from hydrogen donor (e.g. $Ph_2C{=}NCH_3 + i\text{-}PrOH \xrightarrow{h\nu} Ph_2CHNHCH_3$).

122 references up to 1967.
Complementary: none.
Relevant: **2**, 16; **3**, 8; **6**, 4; **9**, 6; **11**, 5; **13**, 16; **14**, 8; **15**, 9; **16**, 11; **17**, 9; **18**, 10; **19**, 20; **22**, 10; **23**, 9; **24**, 12; **25**, 11; **27**, 21, 22; **28**, 5; **29**, 5; **32**, 29; **33**, 21.

8, *13: Imidoyl halides* by *R. Bonnett*

Imidoyl halides, RC(=NH)Hal, are *C*-halogeno derivatives of azomethines, and systematically they are derived from imidic acids, RC(=NH)OH, in the same way as acyl halides, RC(=O)Hal, are derived from carboxylic acids. Almost the whole of the chapter deals with chlorides and very few data are available on the other halides.

Section II describes the preparation of imidoyl halides by chlorination (usually with PCl_5) of secondary amides (e.g. $RCONHR^1 \rightarrow RCCl{=}NR^1$) or by halogenation of aldimines, dehydrohalogenation of amidohalides ($RCCl_2NHAr \rightarrow HCl + RCCl{=}NAr$), by addition HHal to nitriles (giving complicated mixtures of products, but probably including imidoyl halides), by addition of acyl halides to cyanamides ($Me_2NCN + ArCOCl \rightarrow Me_2NCCl{=}NAr$), and by high-temperature chlorination of tertiary amines and amides. Some other synthetic routes are also mentioned briefly.

The discussion of the physical properties includes the bonding of the halogen atom, the IR (Table 1) and NMR spectra of the compounds and their geometric isomerism (Section III).

The reactions of imidoyl halides include self-condensations leading to the formation of amidines, iminazoles or quinolines; eliminations to yield cyanides ($ArCCl{=}NR \rightarrow ArCN + RCl$) or ketene imines, ynamines or 1,3-dipolar systems (the last by α_N-elimination of a proton from a bound to the N atom ($RCCl{=}NOH \rightarrow HCl + RC{\equiv}N{-}\bar{O}$); salt formations; and reductions (usually followed by hydrolysis to yield aldehydes). Nucleophilic substitutions include reaction of the C=N group with Hal, OH, ROH, ArOH, RNH_2 and a large variety of other nucleophiles.

H atoms bound to an α-C in imidoyl halides are easily substituted by Hal: $R_2CHCCl{=}NR^1 \xrightarrow{Br_2} R_2CBrCCl{=}NR_2$ (Section IV).

Imidoyl halides often serve as precursors for heterocyclic systems and the methods involve intramolecular condensation, nucleophilic substitution followed by cyclization and 1, 3-dipolar cycloaddition (Section V).

Section VI discusses reactions in which imidoyl halides have been postulated as reactive intermediates and their presence has been proved satisfactorily, if not absolutely. Among these are the Beckmann rearrangement, the Stephen reduction, the reactions of nitriles with HHal and the von Braun degradation ($ArCON_2 \xrightarrow{PX_5} RX + ArCX{=}NR \rightarrow ArCN + RX$).

228 references up to 1968.
Complementary: none.
Relevant: **32**, 13.

8, *14: Quinonediimines and related compounds* by *K. T. Finley and L. K. J. Tong*

I. Introduction (664); **II**. The chemistry of quinone mono- and diimines (667); **III**. The chemistry of *N,N'*-quinonemono- and diimides (700); **IV**. Quinone monoximes (711); **V**. Studies of more limited scope (718); **VI**. References (726).

Section II deals with the preparation of the title compounds, their reactions with nucleophiles leading to deamination and their coupling with anilines and related compounds which give dyes and with the addition of nucleophiles usually yielding substituted diaminobenzenes. The kinetics and mechanisms of these reactions are discussed in detail and illustrated by reaction schemes, figures and tables showing the influence of various factors (structures, pH, etc.) on the reactions.

Section III discusses the preparation and reactions of the quinonemono- and diimides (which are much more stable than the corresponding imines) and the mechanism of ring additions to quinone diimides, again leading to substituted aromatic diamines (or to aminophenols with quinone monoimines).

The preparation, tautomerism and Beckmann rearrangement of quinone monoximes is described in Section IV, which also deals briefly with the nitrosation, hydrolysis, MO description and polarography of these compounds. Finally, in Section V Diels–Alder reactions, aldol condensations and some other reactions involving the title compounds are treated.

128 references up to 1969.
Complementary: none.
Relevant: none.

The chemistry of alkenes. Volume 2 (1970)

9, *1: Nuclear magnetic resonance spectra of alkenes* by *V. S. Watts and J. H. Goldstein*

I. Introduction (2); **II**. Spectra of ethylene and its simple derivatives (4); **III**. Multiply substituted ethylenes (15); **IV**. Compounds with more than one double bond (18); **V**. Special effects (22); **VI**. Applications to complex systems (31); **VII**. References (35).

After a brief introduction to the nature of chemical shifts and coupling constants, the NMR spectra of ethylene, vinyl halides, ethers, sulphides and formates and of acrolein, vinylacetylene, alkylethylenes and some derivatives containing Si, Hg, Al and Sn are described (Section II). Section III discusses the chemical shifts and coupling constants of polysubstituted ethylenes; Section IV deals with buta-1, 3-dienes and cumulenes and Section V with the effect of the medium on NMR parameters and the effect of hindered rotation about single bonds in alkenes, in cases where rotational comformers are distinguishable about a C—C bond adjacent to the double bond. Section VI deals with more complex systems, such as cyclic olefins with one or more double bonds, norbornenes and norbornadienes and also some ethylenic oximes and substituted styrenes. Various NMR data are presented in 14 Tables throughout the chapter.

127 references up to 1967.
Complementary: 30, 21; **31**, 25.
Relevant: 6, 2.V; **6**, 32.V; **13**, 6.III.B; **16**, 3.V; **17**, 4, 11; **20**, 3; **23**, 5.VI; **24**, 5.III.D; **26**, 5.III.C; **30**, 21; **31**, 25; **36**, 6.

9, *2: The properties of alkene carbonium ions and carbanions* by *H. G. Richey, Jr*

I. Introduction (40); **II**. Vinyl ions (42); **III**. Alkyl ions (56); **VI**. Non-conjugated ions (77); **V**. Acknowledgements (106); **VI**. References (107).

The chapter deals with the stability, geometry and bonding in organic cations and anions which are influenced by the proximity of a C=C function. Section II discusses the stability, geometry, formation and reactions of vinyl cations (II.B) and anions (II.C). Section III describes the same properties in allyl cations (III.B) and anions (III.C), while Section IV deals with some non-conjugated systems containing double bonds in the 5, 6-, 6, 7-, 4, 5- or 3, 4-position in cations, and also related anionic systems.

The mechanisms of the formation and of the reactions of these systems are treated and many data relating to equilibria, rates of reactions and UV spectra which are influenced by the nature and position of the double bonds relative to the charged atom are presented, compared and discussed throughout the chapter (11 Tables).

225 references up to 1968.
Complementary: none.
Relevant: **31**, 16.

9, *3: Alkene rearrangements* by *K. Mackenzie*

Section I describes thermal, catalysed and photochemical stereomutations including acid and base catalysis and *cis–trans* isomerism in polyenes. Hydrogen transfer and prototropic shifts conducted by thermal, photochemical or catalytic means both in simple alkenes and in polyenes and heterogeneous catalysis in the rearrangements of simple alkenes are presented in Section II.

Section III deals with the mechanisms of allylic rearrangements and with the anionotropy in polyenes. The Claisen and Cope rearrangements (Section IV) and, briefly, Wittig rearrangements and shifts catalysed by transition metal compounds (Section V) are treated.

The emphasis throughout the chapter is on mechanistic problems and the chapter is intended as an update of Chapter **1**, 7 by the same author.

174 references up to 1968.
Complementary: **1**, 7; **8**, 9; **9**, 5.IV; **24**, 10.
Relevant: **1**, 8; **2**, 15; **3**, 14; **4**, 10; **5**, 8; **11**, 4; **12**, 16; **13**, 13; **14**, 7; **15**, 8; **16**, 17; **17**, 8; **18**, 15; **19**, 18; **20**, 4; **23**, 13; **26**, 15; **29**, 4.

9, *4: Hydrogenation of alkenes* by *S. Mitsui and A. Kasahara*

Section I discusses mechanisms, stereochemistry and selectivity in heterogeneous catalytic hydrogenation, compares results obtained with different catalysts and substrates and considers pressure and temperature effects and product compositions under different circumstances (Tables 1–8).

Homogeneous hydrogenations are classified and discussed according to the nature of the catalysts, which include compounds and complexes of Ru^{III}, Co^{II}, Rh^{III}, Ir^{I}, Pt^{II}, metal–carbonyl catalysts and Ziegler-type catalysts (Section II).

Section III describes chemical reduction by dissolving metals in acidic, neutral or basic solutions (Zn, Na or Li in NH_3, etc.), by metal hydrides ($NaBH_4$, $LiAlH_4$) and by diimide (NH=NH), which is the active agent in the actual use of hydrazines, azodicarboxylate salts, etc.

118 references up to 1967.
Complementary: **9**, 5.V; **21**, 9; **24**, 8; **31**, 14.
Relevant: **2**, 11; **8**, 6; **12**, 7; **15**, 7; **19**, 15; **30**, 7.

9, *5: Alkene complexes of transition metals as reactive intermediates* by *J.-F. Biellmann, H. Hemmer and J. Levisalles*

Section II discusses the structure of the title compounds, their MO picture and stability and also π-allyl complexes.

The dynamic stereochemistry of the complexes, including the kinetic *trans* effect (in which ligands L on the metal M strongly accelerate the departure of another ligand X) and the *cis* migration or *cis* ligand insertion in which a doubly bonded ligand is inserted into an M—R bond, yielding a metal complex with carbenoid properties (Section III).

The mechanisms of double bond migration in olefins may occur through metal hydride addition–elimination, or through a π-allyl complex mechanism which may catalysed by iron carbonyl or palladium (Section IV).

The title complexes may be used as homogeneous catalysts for C═C hydrogenations, which are catalysed by pentacyanocobaltate, Rh^{I}, Ru^{II} and other catalysts, usually at high rates (Section V). Most reactions of transition metal complexes are nucleophilic—the attacking reagents include Hal^-, H_2O, ROH, alkenes, alkynes and dienes and the atached metals exhibit electrophilic properties. Thus the oxidation of alkenes to aldehydes ($C_2H_4 + PdCl_4^{2-} + H_2O \rightarrow CH_3CHO + 2Cl^- + 2HCl + Pd$) is facile and much studied, and its kinetics and mechanisms are presented in Section VI, together with some less important processes in which acetate, amine or cyanide is the nucleophile.

Section VII treats carbonylations in which CO becomes linked to an alkene, using a transition metal catalyst: $RCH{=}CH_2 \xrightarrow{M(CO)_n} RHXCH_2COY + RCH(COY)CH_2X$, where X and Y are usually H, Hal or OR′. This reaction may be intramolecular in a pre-formed CO–metal–alkene complex or may occur by stepwise reaction of the alkene with the metal carbonyl catalyst followed by the carbonylation of the CO present.

Ethylene and higher alkenes undergo oligomerization in the presence of Rh^{III} and other metals such as Ni, Ru and Pd. Dienes may undergo either linear or cyclic oligomerization and mixtures of alkenes and dienes behave similarly. Finally, alkenes can give cyclobutane derivatives in the presence of Rh, Fe or Co catalysts, while the reverse reaction also occurs in the presence of metal catalysts (Section VIII).

258 references up to 1969.
Complementary: 1, 6; **21**, 12; **30**, 7–9; **35**, 3–7.
Relevant: 2, 6; **19**, 9; **21**, 10; **25**, 14; **31**, 22, 23; **32**, 26; **36**, 18.

9, *6: Photochemistry of olefins* by *K. J. Crowley and P. H. Mazzocchi*

Section II reviews the fundamentals of the photochemical excited state, the absorption process, the fate of the initial excited state and the product formation on direct and sensitized irradiation.

Section III discusses photosensitized and unsensitized *cis–trans* photoisomerizations and bond migrations (where the *cis-trans* isomerization is sterically impeded). Stereochemical predictions following from the Woodward–Hoffmann rules, 1,3- and 1,5-shifts in sigmatropic isomerizations and electrocyclic reactions (buta-1, 3-diene → cyclobutene, etc.) and some more complex presumedly electrocyclic transformations, the mechanisms of which are not certain (e.g. photoisomerization of norbornadiene derivatives to quadricyclane derivatives or the interconversions of bullvalene and its valence isomers) are treated.

Section IV describes intramolecular additions, such as cyclizations of simple non-conjugated dienes or more complex conjugated dienes, α, β-unsaturated carbonyl compounds, reactions yielding cage products (e.g. cubane) and others yielding heterocyclic photoproducts (e.g. oxetanes).

Intermolecular photoaddition (Section V) may involve free-radical (usually chain) reactions between simple olefins and H_2S, HBr, ROH, O_2, formamide, R_2NHO, $BrCH_2COOH$, etc. Conjugated double bonds (e.g. α, β-unsaturated carbonyl compounds) undergo cycloaddition with simple and substituted ethylenes and in some cases also with acetylenes. The Paterno–Buchi reaction (reaction of carbonyl compounds with olefins to yield oxetanes) and its mechanism are discussed, followed by additions of olefins to arenes or irradiation.

Cyclobutane photodimers can be formed from both non-conjugated and conjugated olefines.

275 references up to 1968.
Complementary: 24, 12.
Relevant: 2, 16; **3**, 8; **8**, 12; **11**, 5; **13**, 16; **14**, 8; **15**, 9; **16**, 11; **17**, 9; **18**, 10; **19**, 20; **22**, 10; **23**, 9; **25**, 11; **27**, 21, 22; **28**, 5; **29**, 5, 6; **33**, 21; **32**, 29.

9, *7: The mass spectrometry of the double bond* by *A. G. Loudon and A. Maccoll*

Section I presents the MS of some simple alkenes (up to C_7), the fragmentation and McLafferty rearrangements that occur and their interpretation. Ionization and the variations in ionization and appearance potentials in different olefins are discussed in Section II. Section III describes the problem of the location of double bonds by MS, made difficult by migrations under electron impact, but made possible by the application of various chemical transformations applied prior to MS.

The retro-Diels–Alder reaction contributes to the MS of cyclohexene derivatives but the relevant peaks usually remain recognizable (Section IV).

Finally, Section V discusses the expulsion of alkyl radicals from cyclic olefins and gives MS data for cyclopentene, cyclohexene and cyclooctane.

44 references up to 1969.
Complementary: 17, 5; **24**, 6; **26**, 6.
Relevant: 10. 5; **13**, 19; **16**, 4; **17**, 5; **18**, 6; **19**, 5; **22**, 7; **23**, 5.VIII; **24**, 6; **25**, 4; **26**, 6; **27**, 7; **29**, 3; **30**, 22; **31**, 3; **32**, 3; **33**, 4; **36**, 7.

9, *8: The radiolysis of olefins* by *G. G. Meisels*

Irradiation of olefins leads to the formation of fragmentation products of lower MW and also to condensation and polymerization to higher MW compounds. Olefins can be ionized at energies as low as 8.5–10 eV (Table 1), using electron accelerators of various designs. The main features of the experimental set-ups and of the dosimetry (expressing the yields in terms of *G* values, i.e. the number of molecules produced or decomposed per 100 eV deposited in the system) are described (Section II).

General and historical aspects and the mechanisms of radiolysis are discussed, including the initial event leading to the excitation and ionization of the species, the classification of unimolecular events according to the energy deposited in the system and of bimolecular events according to the mchanisms operating (involving primary species, original and observed products, collision of excited states, ion–molecule reactions, neutralization, electron attachment and radical reactions). The processes are more complicated when using mixtures, but the addition of scavangers may enhance the yields of the required products (Section III). Section IV deals with some cases in which the determination of complete product distributions was attempted (Tables 7–9).

191 references up to 1968.
Complementary: none.
Relevant: **10**, 6; **11**, 6; **13**, 17; **16**, 12; **17**, 10; **18**, 11; **19**, 10; **22**, 11; **25**, 12; **27**, 23, 24; **29**, 7; **31**, 6; **32**, 9; **33**, 21; **36**, 8.

9, *9: Polymers containing C=C bonds* by *M. A. Golub*

The classes of polymers which contain recurring (not occasional) C=C bonds include homopolymers of isoprene, butadiene, chloroprene, piperylene, cyclopentadiene, vinylcyclohexene and various tetraenes. Copolymers which contain C=C are obtained from mixtures of butadiene and styrene, acrylonitrile or phenylbutadiene and other similar reactant pairs. Others are obtained from alkynes (mainly simple or substituted acetylenes), from cycloolefins undergoing ring opening (e.g. cyclopentene, yielding linear polypentenes), from conjugated trienes, by polycondensation of unsaturated esters, amides and urethanes, by reactions of pre-formed polymers [e.g. dehydrochlorination of poly(vinyl chloride)] or by radiation-induced formation of C=C bonds (Section II).

These polymers undergo most expected reactions of C=C bonds, such as halogen or HHal addition, cyclization, *cis–trans* isomerization, hydrogenation, vulcanization, oxidation, ozonization, degradation, epoxidation, thiol or carbene additions and additions of alkene or carbonyl compounds (Section III).

301 references up to 1969.
Complementary: none.
Relevant: 24, 13.

9, *10: Olefinic properties of cyclopropanes* by *M. Charton*

cyclopropane ring (566); **X**. Addition reactions of cyclopropane (569); **XI**. Reactions of the C—H bond in cyclopropane (592); **XII**. Recent developments (595); **XIII**. References (598).

The cyclopropane (cyPr) ring shows many analogous physical and chemical properties to C=C bonds. Section II discusses various models for bonding in cyPr, such as the trigonally hybridized, bent-bond and MO models, and the transformation of these three into each other.

The geometry of cyPr derivatives is described in Section III and Tables 4 and 5 and their conformations in Section IV and Table 6.

Section V deals with ionization potentials, UV, IR, Raman, NMR, NQR and EPRS spectra (Tables 7–23), Section VI with molar refractivities and dipole moments (Tables 24–27) and Section VII with heats of combustion and the conjugation between cyPr and dimethylamides and -thioamides (Table 28).

Section VIII describes the effects of cyPr as a substituent, including electrical effects (expressed as substituent constants, Tables 29 and 30), on ionization constants (Table 31), on reaction rates (Tables 32), on electrophilic aromatic substitution (Tables 33 and 34), on carbonium ion reactions (Tables 35–38) and on the stabilization of carbanions (Table 39) and of radicals (Table 40). The transmission of resonance effects by the cyPr ring is dealt with in Section IX (Tables 42 and 43).

Additions are the most characteristic reactions of the cyPr ring (Section X). These include donor–acceptor complexes, electrophilic additions of halogens, HHal (Tables 45–48) and addition of H_2O, carboxylic acids and metal salts (Tables 49–51). Benzene is alkylated by cyPr (with $AlCl_3$ catalysis, Table 52) and bicycloalkanes containing cyPr rings react with a variety of acid reagents (Table 54). Radical (Table 55) and nucleophilic (Table 56) additions and hydrogenations (Tables 57–60) are dealt with.

Section XI treats the acidity of the C—H bond in cyPr compared with other hydrocarbons (Tables 61–63) and the configurational stability of the cyPr carbanion (Table 64) and finally Section XII presents some results from the period 1967–68.

405 references up to 1968.
Complementary: See Vol. **39** (pp. 20–21 of this book)

The chemistry of the carbonyl group. Volume 2 (1970)

10, *1: Equilibrium additions to carbonyl compounds* by *Y. Ogata and A. Kawasaki*

Carbonyl groups have a great tendency to undergo reversible additions (usually acid or base catalysed) with polar molecules.

The extent of hydration and the stability of the *gem*-diol product depend on the structure of the carbonyl compound. Equilibrium is reached rapidly and can be detected by UV or, better, by NMR (Table 1). The kinetics and mechanism of the process are discussed (Section II).

H_2O_2 with aldehydes yields a hydroxy hydroperoxide, which reacts with a second molecule of aldehyde to give a dihydroxy peroxide (Section III).

$$\mathrm{RCHO + H_2O_2 \rightleftharpoons RCH(OH)OOH \xrightleftharpoons{RCOH} RCH(OH)OOCH(OH)R}$$

Equilibria and mechanisms involved in the formation of hemiacetals, hemiketals, acetals and ketals from alcohols and aldehydes or ketones, respectively, are treated in Section IV, and the cyanohydrin formation from carbonyl compounds and HCN, including asymmetric syntheses of cyanohydrins, in Section V. The formation of bisulphite or sulphite addition compounds is in principle reversible, but under appropriate conditions is used for the isolation, identification and quantitative analysis of CO compounds (Section VI, Tables 5–7).

Inorganic acids (H_2SO_4, HHal, HNO_3, etc.) usually give unstable addition compounds (Section VII), while thiols and H_2S similarly yield unstable hemithioacetals or *gem*-thiogylcols:

$$\mathrm{R^1R^2CO + RSH \rightleftharpoons R^1R^2C(OH)SR}$$

$$\mathrm{R^1R^2CO + H_2S \rightleftharpoons R^1R^2C(OH)SH}$$

and the equilibria, kinetics and mechanisms of these are treated (Section VIII).

Reactions with NH_3 and amines (RNH_2 or R_2NH) involve usually addition, followed by either intramolecular or intermolecular dehydration:

$$\mathrm{R^1R^2CO + RNH_2 \rightleftharpoons R^1R^2C(OH)NHR}$$

(a) $R^1R^2C(OH)NHR \rightleftharpoons R^1R^2C{=}NR + H_2O$

(b) $R^1R^2C(OH)NHR + RNH_2 \rightleftharpoons R^1R^2C(NHR)_2 + H_2O$

Route (a) is the formation of Schiff bases, oximes, hydrazones, etc., used in identification and determination, and route (b) is important in industrial processes (e.g. formation of urea or melamine resins). All these are discussed in detail and mechanistically (Section IX and Tables 9–11).

Section X deals briefly with formation of 1:1 complexes between halogen molecules and CO compounds.

374 references up to 1967.
Complementary: 2, 9, 12; **3**, 7; **26**, 14; **27**, 20.
Relevant: 21, 4.

10, *2: Oxidation of aldehydes and ketones* by *H. S. Verter*

The chapter classifies oxidation reactions according to the position of the oxidizing element in the Periodic Table. In each section, the scope and mechanism of the reactions are discussed. Sections II and III deal with oxidation by Tl and Pb^{IV}, mainly as Pb(OAc). Compounds of N (oxides, but especially HNO_3 and Bi^{III} and Bi^{IV}) are treated in Section IV. Section V describes oxidation by O_2 (especially free-radical autoxidation of aldehydes and ketones), oxidative ammonolysis, photochemical oxidation and then reactions with O_3, peroxides, molten NaOH, S and Se, followed by reactions with halogens, haloimines, iodoso compounds and periodate (Section VI).

Compounds of Cu^{II}, Ag^{I} complexes and Ag^{II} ions and Au^{III} (Section VII), of Hg^{II} (Section VIII), of Ce^{IV} (Section IX), of V^{V} (Section X), of chromium (CrO_3, or $Na_2Cr_2O_7$) (Section XI), of Mn (e.g. MnO_4^- and MnO_2) (Section XII) and of Fe, Co, Ni, Ru, Pd and Pt (Section XIII) have all been employed in CO oxidations.

Section XIV deals with microbiological oxidation by various bacteria, Section XV with dehydrogenation by quinones (especially chloro- and cyano-substituted quinones such as chloranil) and finally Section XVI treats briefly oxidations by H atoms under special conditions, by carbonium ions and by electrolysis.

410 references up to 1967.
Complementary: 2, 10.
Relevant: 17, 7; **21**, 11; **29**, 25.

10, *3: Enolization* by *S. Forsén and M. Nilsson*

Enolization of carbonyl compounds is the most important example of prototropic tautomerism, and influences the physical properties and reaction paths of carbonyl compounds.

Section II describes the halogen addition, acidimetric and spectroscopic (UV, IR, NMR) methods for the determination of equilibrium constants (K_T), while Section III presents, compares and discusses these data for ketones and aldehydes, together with thermodynamic considerations (Tables 1 and 2).

Section IV treats the cyclohexadienone–phenol type equilibria of mono- and polycyclic phenols, nitroso-, azo- and ketiminophenols and of heteroaromatic hydroxy derivatives, including total delocalization energies, K_T and the solvent dependence of K_T (Tables 3–7). β-Dicarbonyl compounds may enolize in different ways, including the formation of especially stable six-membered cyclic enols. K_T and thermodynamic data for the gas phase and liquid keto–enol equilibria, the direction of *cis*-enolization, *trans*-enolization of various open-chain and cyclic β-dicarbonyl compounds, tricarbonyl compounds and solvent effects are treated (Section V, Tables 8–12).

α-Diketones, enediols and oxaloacetic acid are dealt with in Section VI. Section VII discusses the current views on the mechanisms of enolization, the use of relaxation methods for studies of rapid equilibria, the enolization of cyclohexane-1, 2-dione, the racemization (via enolization) of ketones, the ketonization of a transient 1, 1-enediol and, finally, rates of bromination of dimedone and related compounds.

232 references up to 1968.
Complementary: 2, 9.
Relevant: 27, 17.

10, *4: Oxocarbons and their reactions* by *R. West and J. Niu*

Oxocarbons are defined as compounds (especially cyclic compounds) in which all or nearly all of the C atoms bear ketonic functions. The dianions of the four-, five- and six-membered rings, and the tetra-anion of the last, are aromatic.

CO and hot K or Na yield $K_6C_6O_6$ or $Na_6C_6O_6$, the hexapotassium or hexasodium salt of hexahydroxybenzene. These can be oxidized to the corresponding rhodionate salts ($K_6C_6O_6 \rightarrow K_2C_6O_6$), which may also yield croconates:

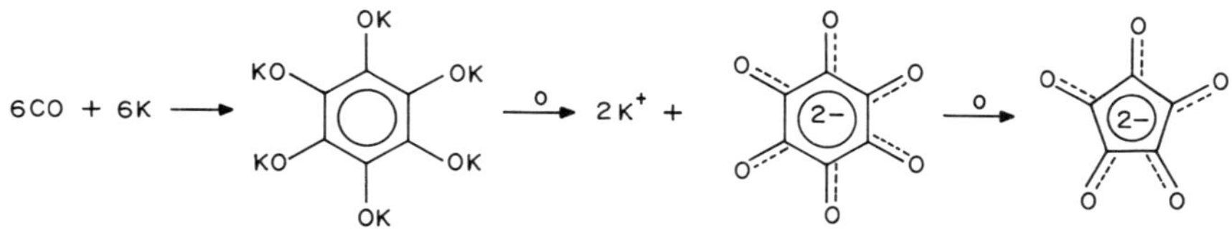

At low temperatures, the CO–K reaction yields first KOCCOK, which was identified by physical and chemical methods, and in turn at higher temperatures gives $K_6C_6O_6$. Rhodizonic and croconic acids are also obtained by oxidation of myoinositol (an isomer of hexahydroxycyclohexane), and squaric acid (dihydroxycyclobutenedione) is available by more conventional chemical methods starting from various cyclobutane derivatives (Section II). the structures and properties of oxocarbon acids are described in Section II, while Section III discusses the corresponding aromatic anions and oxocarbon radicals.

Section V deals with the oxidation and reduction products of oxocarbons and with the esterification and condensation and other reactions of squaric, croconic and rhodizonic acids. Condensations with aniline or with phenylenediamines yield complicated polycyclic products.

Rhodizonic acid and its alkali metal salts are useful analytical reagents since they form strongly coloured complexes with many metal ions and they also have biochemical applications, e.g. for the protection of yeasts from radiation damage and, as antidiabetic agents (Section VI).

109 references up to 1968.
Complementary: none.
Relevant: 17, 16.

10, *5: Mass spectrometry of carbonyl compounds* by *J. H. Bowie*

Aliphatic aldehydes fragment in the electron beam by two major processes, one being cleavage α and the other β to the C=O group. α,β-Unsaturated ketones and aldehydes, β-diketones and γ-diketones are described in Section II and in each mass spectra, major fragmentation processes, rearrangements and base peaks are given. Saturated and unsaturated monocyclic ketones and monocyclic diketones are similarly treated in Section III and bicyclic ketones (β-hydrindanones , decalones and unsaturated bicyclic ketones) in Section IV. Section V deals with saturated and unsaturated steroidal ketones and Section VI with simple aromatic aldehydes and ketones, unsaturated aromatic ketones, aromatic diketones, indanones, tetralones, quinones, tropones and some others.

Skeletal rearrangements (migrations of groups, loss of CO, etc.) are treated in Section VIII.

153 references up to 1967.
Complementary: none.
Relevant: 9, 7; **13**, 19; **16**, 4; **17**, 5; **18**, 6; **19**, 5; **22**, 7; **24**, 6; **25**, 4; **26**, 6; **27**, 7; **29**, 3; **30**, 22; **31**, 3; **32**, 3; **33**, 4; **36**, 7 (and smaller reactions in analytical chapters).

10, *6: Radiation chemistry of ketones and aldehydes* by *G. R. Freeman*

Section II reviews energy absorption processes (X- and γ-rays, electrons, charged heavy particles and neutrons and the distribution of activated species. Section III discusses reactive intermediates, including the formation and reactions (decomposition, deactivation, ion–molecule reactions, charge exchange and neutralization) of positive ions; the states and reactions (attachment to neutral molecules and reactions of the ions so formed) and neutralization of electrons; the formation, reactions (thermal decomposition, electron transfer, neutralisation, proton acception from Brønsted bases) and lifetimes of negative ions; the formation, chemical reactions, photon emission and energy transfer involving excited molecules; and the formation of neutral free radicals, their reactions and the proposed diradical and concerted mechanisms. Finally, Section IV describes in detail the radiation chemistry of acetone and of cyclopentanone (Tables 3 and 4).

134 references up to 1967.
Complementary: none.
Relevant: **9**, 8; **11**, 6; **13**, 17; **15**, 9; **16**, 12; **17**, 10; **18**, 11; **19**, 10; **22**, 11; **25**, 12; **27**, 23, 24; **29**, 7; **31**, 6; **32**, 9; **33**, 21; **36**, 8.

The chemistry of amides (1970)

11, *1: Molecular and electronic structure of the amide group* by *M. B. Robin, F. A. Bovey and H. Basch*

I. Introduction (2); **II**. Molecular structure (3); **III**. Nuclear magnetic resonance (7); **IV**. Electronic structure (40); **V**. Optical properties (53); **VI**. References (68).

Knowledge of the molecular structure of amides is based mainly on X-ray and electron diffraction. Structural parameters are given in Table 1 and Section II discusses these data. Section III deals with NMR spectra, describing *J* couplings (Table 2), chemical shifts, conformational preferences of *N*-alkyl groups and of α-C substituents, rotational barriers at the C—N bond (Table 4) and also at N—C and C′—C bonds, amide–iminol tautomerism and self-association by hydrogen bonding and by dipole–dipole association.

Section IV discusses the electronic structure, including theoretical calculations and orbital population analysis and charge densities in both the ground and excited states (especially of formamide) (Tables 5–10).

Section V treats optical properties, such as absorption, circular dichroism and photoelectron spectra.

145 references up to 1969.
Complementary: 25, 1.
Relevant: 1, 1; **2**, 1; **3**, 1; **4**, 1; **5**, 1; **6**, 1; **8**, 1; **12**, 1; **13**, 1; **14**, 1; **15**, 1; **16**, 1; **17**, 1; **18**, 1; **19**, 1; **20**, 1; **22**, 1; **23**, 1; **24**, 1; **26**, 1; **28**, 1; **31**, 27; **32**, 1; **33**, 1; **36**, 2.

11, *2: Synthesis of amides by A. L. J. Beckwith*

I. Introduction (74); **II**. Acylation of amines and amides (75); **III**. Preparation of amides from nitriles (119); **IV**. Rearrangement reactions (131); **V**. Reactions of carbamyl halides, isocyanates and isocyanides (149); **VI**. Alkylation of amides, imides and lactams (159); **VII**. Oxidation and reduction reactions (165); **VIII**. Acknowledgement (167); **IX**. References (167).

Amides, imides and lactams are obtained by *N*-acylation of NH_3, amines or amides, usually by nucleophilic substitution involving a tetrahedral intermediate. The attacking agents may be acyl halides, anhydrides or esters, carboxylic acids, aldehydes, ketones, acyl azides or ketenes. Less frequently, the exchange reaction $RNH_2 + R'CONH_2 \rightarrow R'CONHR + NH_3$ is used. Amines with thio acids, thiol esters or anhydrides undergo acylation, e.g. $RCOSH + R'NH_2 \rightarrow RCONHR' + H_2S$, and formyl derivatives are obtained from amines and CO: $RR'NH + CO \rightarrow RR'NCHO$, in the presence of various metallic catalysts.

Nitriles yield amides by hydration $RCN + H_2O \rightarrow RCONH_2$ usually in the presence of mineral acids, or with alkaline H_2O_2 and sometimes in neutral conditions in the presence of, e.g. a Zn–Ni catalyst. The Ritter reaction is an alkylative hydration of olefins in the presence of strong acids ($HCl-AlCl_3$, H_2SO_4) and HCN or nitriles: $Me_2C{=}CH_2 + MeCN \rightarrow ME_3CNHCOCH_3$. The scope and mechanism of the reaction are discussed (Section III).

Section IV describes rearrangements which yield amide products, such as the Beckmann, Schmidt, Willgerodt and Chapman reactions, together with diazotization of hydrazones and semicarbazones and rearrangements of nitrones, oxaziridines and α-aminothiol esters. In most cases the mechanism and scope of these reactions are treated.

Carbamyl halides (NH_2COHal) and their *N*-alkyl and -aryl derivatives attack arenes and yield aromatic amides: $R^1R^2NCOCl + ArH \xrightarrow{AlCl_3} ArCONR^1R^2$. The same reagents with carboxylic acids yield dialkylamides: $R^1R^2COCl + R^3COO^- \rightarrow R^1R^2NCOR^3 + CO_2$. Isocyanates undergo alkylation and arylation by nucleophilic attack on the isocyanate carbonyl carbon, involving a variety of carbanionic reagents, e.g. Grignard: $R^1NCO + R^2MgX \rightarrow R^1N(MgX)COR^2 \xrightarrow{H_2O} R^1NHCOR^2$. Similar reactions occur with carboxylate ions: $R^1NCO + R^2COO^- \rightarrow R^1NHCOCR^2 \rightarrow R\,NHCOR^1 + CO_2$ Isocyanides yield amides, e.g. by the Passerini reaction: $R^1NC + R^2R^3CO + AcOH \rightarrow R^1NHCOC(OAc)R^2R^3$ (Section V). Finally, alkylations of amides, imides and lactams (Section VI) and some oxidation and reduction reactions (Section VII) are described briefly.

600 references up to 1968.
Complementary: 25, 19.
Relevant: 4, 6; **5**, 9; **25**, 8, 9.

11, *3: Acid-base and complexing properties of amides* by *R. B. Homer and C. D. Johnson*

Amides are relatively weak bases. Their protonation, hydrogen bonding and complex formation are very important since the CONH function is the basic structural unit of peptides and proteins. Although the NH group is more basic, the significant contributions from various canonical forms to the *O*-protonated structures make the latter largely predominant. Convincing evidence for this comes from NMR, IR, UV and basicity studies. In the case of thioamides, the main site of protonation is the S atom (Section II).

Section III discusses the pK_a values of amides functioning as bases, H_A and H_0 acidity functions, protonation data, the determination of pK_a by potentiometric titration and by indicator methods and the correlation of pK_a values with structure (Figs 3–10 and Tables 1–10).

Section IV deals with hydrogen bonding, where the CONH group may act as both a proton acceptor and donor. Self-association in solutions (Table 11), amide–phenol complexes (Table 12) and associations with hydrogen acceptor, hydrogen-donor and π-electron systems (Table 13) are considered. Complexes with Lewis acids such as BCl_3, $TiCl_4$, I_2 and several metal (Cr, Co, Ni) complexes are treated (Tables 14–16, Section V).

Amides may act as weak acids and thus as proton donors, but this occurs to a significant extent only in very strongly basic media when acidity constants can be determined (Table 17, Section VI).

140 references up to 1968.
Complementary: 25, 6.
Relevant: 2, 9; **3**, 6; **4**, 4; **5**, 5; **8**, 5; **12**, 3; **13**, 7; **15**, 4; **16**, 5; **17**, 6; **18**, 8; **19**, 7; **20**, 12; **22**, 8; **23**, 6, 7; **24**, 4; **25**, 5, 6; **29**, 16, 17; **31**, 17; **33**, 5; **32**, 21.

11, *4: Rearrangement and elimination of the amido group* by *J. F. Bieron and F. J. Dinan*

The chapter does *not* discuss the Hofmann degradation, which has been reviewed repeatedly (see ref. 1).

Section I describes rearrangements in which three-membered aziridine rings are part of the amide system, such as pyrolytic rearrangements to oxazolines or to unsaturated amides:

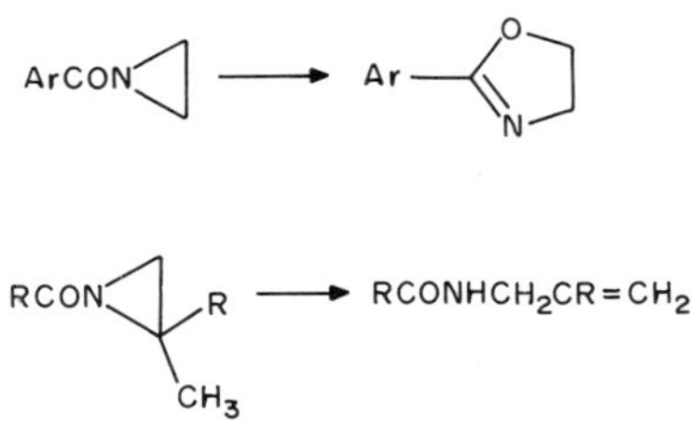

Aroylaziridines undergo both nucleophile- and electrophile-catalysed rearrangement leading in both cases to ring enlargement to aryloxazolines (as in the first reaction above). Acetanilides, under rather drastic conditions, yield quinolines.

Rearrangement of *N*-nitrosoamides may yield, with elimination of N_2, a mixture of ester, olefin and carboxylic acid and *N*-nitroamides give similar products. The Orton rearrangement of aromatic *N*-haloamides yield *o*- *p*-haloaromatic amides in a free-radical process or catalysed by HX in protic and by RCOOH in aprotic solvents. Other *N*-haloamides undergo various photolytic free-radical reactions, or may form α-lactams under basic conditions. α, *N*-dihaloamides in base yield alkylidene halides and isocyanate ions (Section II).

Acid- and base-catalysed dehydrations of amides yield nitriles by elimination of H_2O. In a similar reaction (Von Braun), *N*-alkylbenzamides with PX_5 yield benzonitrile and alkyl halide. Pyrolytic dehydration usually yields a complicated mixture of products (Section III).

Section IV deals with the Vilsmeier–Haack reaction, which is an acylation of an activated aromatic ring by a complex of a substituted amide and $POCl_3$, with subsequent elimination of the NR_2 group of the amide:

$$C_6H_5NMe_2 + R_2NCHO + POCl_3 \rightarrow p\text{-}Me_2NC_6H_4CHO$$

Throughout the chapter, mechanistic details of the reactions discussed are given.

105 references up to 1967.
Complementary: 11, 13 (IX. A, B, C).
Relevant: 1, 2, 3, 7; **2**, 15; **3**, 14; **5**, 8; **8**, 9; **9**, 3; **12**, 16; **13**, 12, 13; **14**, 7; **15**, 8; **16**, 9, 17; **17**, 8; **18**, 15; **19**, 8; **20**, 4; **21**, 4; **23**, 13; **24**, 10; **26**, 15; **29**, 4.

11, *5: Photochemistry of the amido group* by *I. Rosenthal*

Section II gives data on the main absorption bands of amides, imides and lactams.

Section III discusses the photolysis of amides and lactams, the photooxidation of amides, the photoamidation reaction (addition of $HCONH_2$ to olefins, acetylenes or aromatic systems, to yield higher amides) and the photochemical reactions of haloamides, often leading to cyclizations and rearrangements.

Detailed mechanistic schemes are given for the reactions discussed (Schemes 1–9).

79 references up to 1969.
Complementary: 25, 11.
Relevant: 2, 16; **3**, 8; **6**, 4; **8**, 12; **9**, 6; **13**, 16; **14**, 8; **15**, 9; **16**, 11; **17**, 9; **18**, 10; **19**, 20; **22**, 10; **23**, 9; **24**, 12; **27**, 21, 22; **28**, 5; **29**, 5, 6; **32**, 29; **33**, 21.

11, *6: Radiation chemistry of amides* by *O. H. Wheeler*

The chapter deals with reactions initiated by γ- and X-rays or neutrons. The last type include the formation of radioactive atoms in nuclear reactions ('hot-atom reactions'). The radicals formed are usually detected by ESR spectra.

Section II deals with solid simple amides and their *N*-alkyl and *N*, *N*-dialkyl derivatives and with some acetyl and carbobenzoxy derivatives of amino acids, all irradiated in the solid state.

Section III describes irradiations in non-aqueous and aqueous solutions of relatively simple amides ($HCONMe_2$, $HCONH_2$, $CH{=}CHCONH_2$ and acylamino acids).

Peptides and polymeric amides are dealt with in Section IV, and Tables 1 and 2 give rate data for the reaction of peptides with hydrated electrons and with hydroxyl radicals, respectively.

Section V treats hot-atom reactions in which ^{14}C is produced by irradiation of the N atoms of CONH groups.

104 references up to 1968.
Complementary: 25, 12.
Relevant: 9, 8; **9**, 6; **13**, 17; **15**, 9; **16**, 12; **17**, 10; **18**, 11; **19**, 10; **22**, 11; **27**, 23, 24; **29**, 7; **31**, 6; **32**, 9; **33**, 21; **36**, 8.

11, *7: Chemistry of imidic compounds* by *O. H. Wheeler and O. Rosado*

Imides are compounds in which the NH group is flanked by two carbonyl groups, i.e. $R'CONHCOR^2$, and in cyclic compounds such as succinimide and phthalimide.

They can be prepared (Section II) by decomposition of the diamides of dicarboxylic acids (e.g. maleic acid diamide yields maleimide) or by acylation of amides, or by intramolecular condensation of amidic acids:

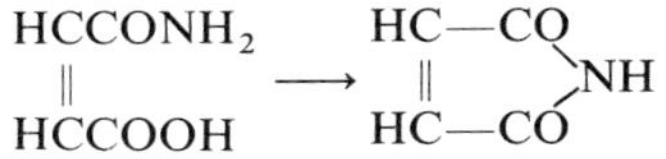

They can also be prepared from acids or acid derivatives with NH_3 or amines, from isocyanates with anhydrides and some by other less important methods.

Section III describes the dipole moments (Tables 1 and 2) and IR, UV, NMR and mass spectra of some unsubstituted and *N*-substituted imides. Section IV discusses hydrolysis to amic acids, exchange of reagents and reactions of *N*-haloimides (e.g. *N*-bromosuccinimide), extensively used in halogenations.

227 references up to 1968.
Complementary: 25, 11.
Relevant: none.

11, *8: The chemistry of thioamides* by *W. Walter and J. Voss*

I. General and theoretical aspects of the thioamide group (385); **II**. Preparation of thioamides (415); **III**. Analytical characteristics of thioamides (433); **IV**. Prototropic and chelating properties (434); **V**. Chemical reactions (439); **VI**. Spectral topics (461); **VII**. References (464).

Thioamides or thionamides, $RC(=S)NR^1R^2$, are very useful reagents, especially in the field of heterocyclic chemistry. Section I deals with their topology (Table 1) derived from X-ray, NMR, IR, UV, dipole moment, MS, polarography and EPR data (Tables 2–19) and also from quantum mechanical calculations (Table 14) and compares the —CNOH—, —CSNH— and—CSeNH groups.

Section II describes methods for the preparation of thioamides. These include thiolysis of carboxylic acid derivatives, e.g. acid- or base-catalysed addition of H_2S to nitriles: $RC\equiv N + H_2S \rightarrow RC(=S)NH_2$; thiolysis of imidoyl halojemides, amidines or imidic esters [e.g. $RC(=NH)NHR^1 + H_2S \rightarrow RC(=S)NHR^1 + RC(=S)NH_2$]; thionation of amides (i.e. conversion of an amidic CO group to CS, using P_4S_{10}; thioacylation of amines by means of thionocarboxylates [i.e. $RNH_2 + R^1(C=S)OEt \rightarrow R^1C(=S)NHR + EtOH$], dithiocarboxylic acids (RCSSH), dithiocarboxylates ($RC(=S)SR^1$], thioamides or thioketenes; addition of nucleophiles (e.g. carbonions) to isothiocyanates:

$$R_3C^-Na^+ + R^1NCS \rightarrow R_3CC(=S)NR^1Na^+ \rightarrow R_3CC(=S)NHR^1$$

the Willgerodt–Kindler reaction of alkyl aryl ketones with secondary amines and S_8:

$$ArCOCH_2Me + R_2NH + S_8 \rightarrow ArCH_2CH_2C(=S)NR_2$$

and also the cleavage of various heterocyclic compounds by thiolysis (e.g. cleavages of Δ^2-oxazolines by H_2S:

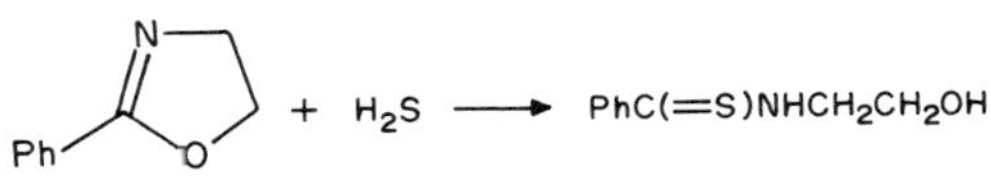

or the ammonolysis, hydrolysis or pyrolysis of various sulphur-containing heterocycles.

Section III deals with qualitative and quantitative analytical methods and Section IV with the thionamide–thiolimide equilibrium [RC(=S)NHR$\rightleftharpoons$RC(=NR)SH], with acidity and basicity of the group (Table 16) and with complex formation with I_2, phenols and various metals ions.

Section V treats nucleophilic (hydrolysis, substitution and reduction) and electrophilic (alkylation, acylation and oxidation) attacks on thioamides (Tables 17 and 18) and the formation of heterocycles in which either the S or the N atom or both are in the five- or six-membered cyclic products. Finally, reactions not involving the CSNH group [electrophilic substitution of the thioformyl proton, reactions at activated α-positions and protection of sensitive functional groups (e.g. OH)] during reactions of thioamides are treated.

Section VI mentions briefly naturally occurring thioamides, physiological activity and pharmaceutical, analytical and industrial applications.

386 references up to 1969.
Complementary: 25, 18.
Relevant: 2, 17; **3**, 13; **5**, 15; **11**, 9; **15**, 11.

11, *9: The chemistry of the thiohydrazide group* by *W. Walter and K. J. Reubke*

Section II discusses the formation of thiohydrazides [$RC(=S)NRNR^1{}_2$] by thioacylation of hydrazine derivatives with dithio acids or with dithioacid esters, thionacid esters and thioamides. Thiolysis of hydrazides to thiohydrazides can be carried out by P_2S_5 or by thioacyl hydrazones and thiohydrazides can be prepared from amidrazones or from hydrazidic halides by H_2S, e.g. $ArCH{=}NNR_2 \xrightarrow{Br_2} ArCBr{=}NNR_2 \rightarrow ArC(=S)NHNR_2$.

Section III discusses the thiohydrazide tautomerism involving 'mercapto' and zwitterionic forms and gives results of X-ray, IR, UV NMR and MS studies and some theoretical aspects of the molecular and electronic structures.

Section IV describes general features, stability, acidity and basicity (Table 1), complex formation, alkylation, addition, oxidation and cyclization of thiohydrazides, and Section V deals with some practical applications.

99 references up to 1969.
Complementary: none.
Relevant:2, 17; **3**, 13; **5**, 15; **11**, 8; **15**, 11; **25**, 18.

11, *10: The chemistry of hydrazides* by *H. Paulsen and D. Stoye*

Hydrazides are highly reactive and hence important synthones for the preparation of amines, aldehydes and heterocycles. The chapter emphasizes physicochemical and mechanistic aspects.

Section III describes X-ray analysis, NMR, dipoles, IR, UV, polarography, electrochemistry and chemiluminescence.

Section IV discusses the preparation of hydrazides by hydrolysis of nitriles to amides, followed by reaction with hydrazine; by hydrolysis of hydrazidic halides or *gem*-diphorohydrazines; by various acylation methods (with esters, lactones, acid chlorides, anhydrides, ketenes) on hydrazine and its derivatives; by dehydration of hydrazinium salts thermally or with carbodiimide; by *N*-amination of amides; by conversion of azo compounds, e.g.

$$PhN{=}NCOPh + RMgX \rightarrow PhN(R)N{=}C(OMgX)Ph \xrightarrow{H_2O} PhN(R)NHCOPh$$

by cleavage of various cyclic compounds; and finally by interconversion of hydrazides, e.g. by acyl migration: $PhCONRNH_2 \xrightarrow{\Delta} RNHNHCOPh$.

Section V describes enolization of the hydrazide group, which does not occur with monoacylhydrazines, $RCONHNH_2$, but only with diacylhydrazines, as shown by their reactions. It also deals with acidity, basicity and complex formation.

Reactions (Section VI) may involve the carbonyl group (hydrolysis, ammonolysis, reduction with $LiAlH_4$ or reactions with chlorinating agents, e.g. $R_2NNHCHO \rightarrow R_2NNHCHCl_2$).

Reactions at the hydrazine group include alkylation, acylation, nitrosation and hydrazone formation with carbonyl compounds:

$$RR'CO + H_2NNHCOR'' \rightarrow RR'C{=}NNHCOR''$$

the formation of semicarbazides with isocyanates and of thiosemicarbazides with isothiocyanates; reactions with diazonium salts or azo compounds to yield tetrazanes or tetrazenes; N—N bond cleavage by catalytic hydrogenation; several methods of oxidation yielding diimides, peptides, acid chlorides, etc., according to the conditions employed; fragmentation; rearrangements; and thermal deacylations and dehydrations.

Another route is the formation of cyclic compounds, which are usually five-membered and may contain two N atoms and one of O or S (oxadiazoles, thiadiazoles) or three N atoms (triazole derivatives); other five- and six-membered heterocycles are also available.

Section VII deals with the analysis of hydrazides, VIII with their physiological activity and IX with some industrial applications.

362 references up to 1968.
Complementary: 11, 9.
Relevant: none.

11, *11: Biological formation and reactions of the amido group* by *J. E. Reimann and R. U. Byerrum*

The chapter considers in detail the biochemical utilization of NH_3 in the formation of amides, amidine and guanidine derivatives and of carbamoyl phosphate, and the mechanisms of the reactions.

Section II discusses syntheses in which the ATP → ADP reaction is involved, including the formation of glutamine (its mechanism, stereospecificity and regulation), asparagine, glutathione, 5′-phosphoribosylglycinamide, 1-(5′-phosphoribosyl)-4-(*N*-succinocarboxamide)-5-aminoimidazole, 5′-phosphoribosylformylglycinamidine, adenylosuccinate and cytidine 5′-triphosphate. The mechanisms of these processes and their role in essential biosynthesis are described.

Section III deals with processes which involve the ATP → AMP reaction. These include syntheses of guanidine derivatives (nicotinamide adenine dinucleotide, guanosine 5′-phosphate, argininosuccinate and 5-phosphoribosylamine, leading to the formation of purine nucleotides.

Other syntheses involving glutamine include the formation of histidine, anthranilate, *p*-aminobenzoate, glucosamine 6-phosphate and γ-glutamylamides.

Section V deals with the enzymes which catalyse the hydrolysis of the amide group in α-ketoglutaromate and α-ketosuccinamate (ω-amidase) and also glutaminase and asparaginase.

The formation of carbamoyl phosphate is a major pathway for the incorporation of NH_3 into organic compounds. This formation is catalysed by carbonate kinase, carbamoyl phosphate synthetase I and II and some other minor pathways (Section VI). The reactions of carbamoyl phosphate include the biosynthesis of citrulline, carbamoyl aspartate and biotin (Section VI).

Shorter sections describe the formation of guanidines (Section VIII), *N*-acyl compounds (IX) and hydroxamates (X).

234 references up to 1968.
Complementary: 4, 9; **20**, 6.
Relevant: 2, 7; **3**, 5; **5**, 18; **7**, 4; **12**, 12; **13**, 14; **15**, 10; **16**, 13; **17**, 13; **18**, 13; **19**, 13; **22**, 20; **24**, 14; **26**, 13; **28**, 17; **31**, 9.

11, *12: Directing and activating effects of the amido group* by *J. A. Shafer*

I. Substituent effects of amido groups in aromatic compounds (685); **II**. Effects on aliphatic carbon attached to amido nitrogen (694); **III**. Effects of aliphatic carbon attached to an amido carbonyl (702); **IV**. Neighbouring group effects (704); **V**. Acknowledgement (726); **VI**. References (726).

Since amido groups are sensitive to, e.g., solvolysis, the study of their effects is difficult. RCONH substituents on arenes can donate electrons by resonance and can withdraw electrons by inductive effects. Carboxamido substituents (R_2NCO—) withdraw electrons by both effects.

Section I discusses substituent constants (Tables 1 and 2), halogenations of anilides in organic acids (Tables 3 and 4), nitration of acetanilide and benzamide, the hydroxylation of acetanilides and the *ortho*-metallation of benzamides and sulphonamides.

Section II deals with amidomethyl radicals ($\dot{C}H_2CONH_2$ and $RCONH\dot{C}H_2$); and with amidomethyl carbonium ions and carbanions.

CH_3 or CH_2 groups α to an amido carbonyl are acidic and tend to form carbanions, which in turn are effective nucleophiles (Section III).

Neighbouring amido groups facilitate solvolytic and oxidation–reduction reactions and, being potent nucleophiles, can facilitate intramolecular displacements. Thus, intramolecular displacements both base-catalysed by amido anions and also by neutral amido groups lead to a variety of acyclization reactions, e.g.

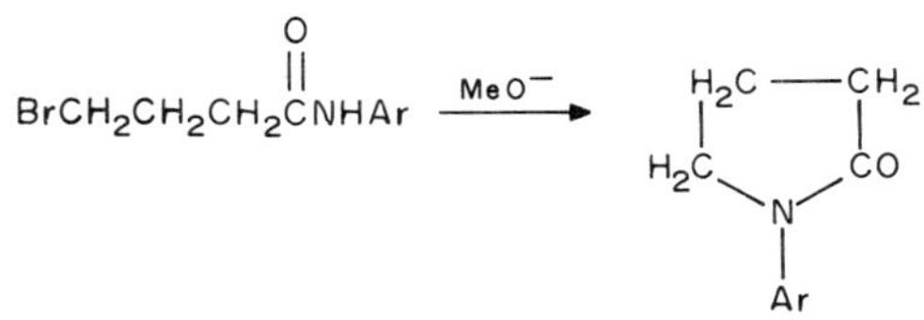

Finally, an example is given in which the amido group effects the reduction of a Co^{III} complex by Cr^{II}.

93 references up to 1968.
Complementary: none.
Relevant: 3, 3; **4**, 5; **5**, 16; **6**, 8; **7**, 2, 3; **12**, 5; **13**, 8; **15**, 5; **16**, 6; **18**, 9; **19**, 8; **23**, 11; **28**, 9; **29**, 16; **31**, 8; **36**, 15.

11, *13: Reactions of the carboxamide group* by *B. C. Challis and J. A. Challis*

I. Introduction (733); **II**. Alkylation (734); **III**. Reactions with aldehydes and ketones (754); **IV**. Acylation (759); **V**. Halogenation (775); **VI**. Nitrosation and nitration (780); **VII**. Oxidation (786); **VIII**. Reduction (795); **IX**. Inorganic acid halides (801); **X**. Hydrolysis and solvolysis (816); **XI**. Addition of organometallic reagents (847); **XII**. References (848).

Most reactions of amides occur by nucleophilic attack by the N or O atoms on electrophilic centres or, less commonly, by nucleophilic attack of a reagent on the C atom of the amide moiety. Mechanistic aspects are emphasized throughout the chapter.

Section II discusses alkylations under neutral conditions, either on N or on O, by alkyl sulphates, sulphonates, halides and diazoalkanes and also intramolecularly:

$$\mathrm{RCONR^1_2} \xrightarrow{\mathrm{R^2X}} \mathrm{RC(OR^2){=}NR^1 + RC({=}O)NR^1R^2}$$

Under basic conditions, *N*-alkylation is preferred with primary and secondary amines, but in the presence of silver salts appreciable amounts of *O*-alkylation products are obtained. *C*-alkylation occurs in the presence of concentrated strong bases.

Primary and secondary amides add to the CO groups of aldehydes or ketones. Thus with HCHO, mono- or bis-hydroxymethylation occurs:

$$\mathrm{RCONH_2 + HCHO \rightleftharpoons RCONHCH_2OH \rightleftharpoons RCON(CH_2OH)_2}$$

In strongly acidic conditions the carbinolamines undergo dehydration and further condensation reactions, whereas in strongly basic conditions attack on the amide C also occurs (Section III).

Section IV describes acylations by acyl halides (including acyl exchanges), anhydrides, amides, esters, oxalyl chloride, ketenes and isocyanates. Various side-reactions compete with the main route (e.g. dehydration to nitriles and salt formation) (Schemes 12–22 and Tables 7–9).

The usual products of halogenation by halogens and hypohalites are *N*-haloamides, although gaseous F_2 yields a mixture of *C*- and *N*-fluorinated products (Section V).

Nitrosation leads to deamination, usually to the corresponding acid ($RCONH_2 + XNO \rightarrow RCOOH + N_2 + HX$) and *N*-nitration occurs with the usual nitrating agents (Section VI).

Section VII deals with oxidation, when H abstraction from C and N competes with oxidative substitution. Autoxidation often transforms N-bound CH_2R groups to —COR or gives oxidative dealkylation, $RCONHCH_2R^1 \rightarrow RCONH_2 + R^1CHO$, but various other radical-chain reactions also occur. Similar reactions take place with H_2O_2 and diacyl peroxides and peroxydisulphates. $Pb(OAc)_4$ reacts only with primary amides, giving various rearranged products via an isocyanate intermediate.

Reduction by complex metal hydrides yields the corresponding amines, sometimes (especially with tertiary amides) complicated by reductive alkylation. $Na-NH_3$ (Birch) reduction yields an aldehyde and an amine, $RCONR'_2 \rightarrow RCHO + R'_2NH$, while electrolytic reduction gives amines without C—N bond scission (Section VIII).

Inorganic acid halides ($COCl_2$, $SOCl_2$, PCl_5, etc.) usually form imidoyl halides $[RC(Cl){=}NR'_2]^+Cl^-$, which in the case of primary amides yield nitriles. Carbonyl fluoride or SF_4 converts the amidic CO groups into CF_2, i.e. $RCONR'_2 \rightarrow RCF_2NR'_2$.

In most cases side-reactions complicate the picture and introduce products of dehydration, rearrangement and cyclization (Section IX).

Section X deals with alkaline hydrolysis to the corresponding carboxylic acid and amine, discusses ^{18}O exchange, general base and acid catalysis and mechanistic aspects. Acid hydrolysis is similarly treated, separating *N*-acyl and *N*-alkyl bond fissions and substituent effects. In continuation, general acid- and base-catalysed hydrolysis is presented, discussing its role in inter- and intramolecular reactions, and the section ends with a brief description of solvolysis reactions.

Addition of organometallic reagents to amides is mentioned (Section XI).

231 references up to 1968.
Complementary: 11, 4.
Relevant: 25, 8, 9.

The chemistry of the cyano group (1970)

12, *1: Electronic structure in molecules, especially in cyano compounds* by *E. Clementi*

I. Introduction (1); **II**. Multiconfiguration self-consistent-field theory (19); **III**. Electron population analysis (20); **IV**. Bond energy analysis (23); **V**. Correlation energy and statistical methods (27); **VI**. Analysis of the 'experimental' correlation energy (35); **VII**. Molecular extra correlation energy (43); **VIII**. Remarks of the electronic structure of the —CN group (46); **IX**. References (63).

Section I reviews the two dominant theories of the electronic structure of molecules, namely the valence-bond (VB) approximation and the molecular orbital method (MO), and the attempts to unify the two. Section II discusses the multiconfiguration self-consistent-field theory (MC SCF) and its chemical implications. Sections III and IV describe simplified versions of electron density and bond energy analyses, respectively. Problems related to the correlation energies in atoms and in molecules and their relation to the Hartree–Fock density, the use of statistical methods, data involving experimental and Hartree–Fock energies and various corrections and the molecular extra correlation energy are dealt with in Sections V, VI and VII.

Treatments of the CN group itself, orbital energies for selected RCN molecules (Table 5), charge transfer between R and CN and dipole moments are given in Section VIII.

65 references up to 1968.
Complementary: none.
Relevant: 1, 1; **2**, 1; **3**, 1; **4**, 1; **5**, 1; **6**, 1; **8**, 1; **11**, 1; **13**, 1; **14**, 1; **15**, 1; **16**, 1; **17**, 1; **18**, 1; **19**, 1; **20**, 1; **22**, 1; **23**, 1; **24**, 1; **25**, 1; **26**, 1,9; **27**, 9; **28**, 1; **31**, 27; **32**, 1; **33**, 1; **36**, 2.

12, *2: Introduction of the cyano group into the molecule* by *K. Friedrich and K. Wallenfels*

I. Introduction (67); **II**. Preparation of nitriles by addition of hydrogen cyanide (68); **III**. Preparation of nitriles by substitution (77); **IV**. Preparation of nitriles by elimination (92); **V**. Miscellaneous methods (103); **VI**. Preparation of nitriles using molecules or molecular fragments with cyano groups (105); **VII**. References (110).

Section II describes the additions of HCN to C=C, C=O and to C=N bonds with formation of the corresponding nitriles. Section III discusses substitution methods, such as the reaction of metal cyanides with organic halides or with aryl sulphonates, the conversion of alcohols, esters and ethers into nitriles in the presence of special catalysts and/or solvents, nucleophilic aromatic substitution by CN^- on activated arenes, the Sandmeyer reaction of substitution of aromatic diazonium groups by CN^- in the presence of Cu^I salts and the reaction of $(CN)_2$, CNHal and similar compounds with organometallic compounds (e.g. sodium diethylmalonate, Grignard reagents).

Elimination methods (Section IV) usually start from carbonyl compounds or their derivatives, e.g.

$$\text{RCHO} \longrightarrow \text{RCH}{=}\text{NOH} \longrightarrow \text{RCN} + \text{H}_2\text{O}$$

or from acids, their conversion into the corresponding amide and dehydration of the latter. Some less general methods are briefly treated in Section V, and methods in which the nitrile group is introduced together with another part of the final molecule (e.g. condensations of CO with active methylene compounds containing CN groups, cyanomethylation) are dealt with in Section VI.

364 references up to 1968.
Complementary: 31, 26.
Relevant: 1, 4; **2**, 12; **10**, 1; **16**, 7.

12, *3: Basicity, hydrogen bonding and complex formation* by *J. Grundnes and P. Klaboe*

I. Introduction (124); **II**. Basicity of the cyano group (125); **III**. Weak complexes (134); **IV**. Non-ionized coordination complexes (150); **V**. Donor strength of the cyano group (156); **VI**. References (159).

Section II describes methods available for the determination of the basicity of the CN group, the formation and structure of nitrilium salts and the ability of nitriles to solvate both alkali metal and heavy metal ions, which is the cause of the abnormal solubility of such salts in, e.g., acetonitrile.

Section III discusses weak complexes of nitriles with organic and inorganic molecules with a heat of formation of less than 10 kcal mol^{-1}. These are detected by UV, IR, Raman and NMR methods and include inter- and intramolecular hydrogen bonding, adsorption on solids, charge-transfer complexes, dipole–dipole hetero- and self-association and some less defined cases (with N_2O_4, by hydroquinone, arenes).

Nitriles react with Lewis acids (compounds of B, Al, Ga, heavy metals) and dinitriles give chelate complexes with Group IV halides (Section IV).

The donor strength of the cyano group and the relevant inductive, conjugative and steric effects are treated in Section V.

325 references up to 1968.
Complementary: 12, 10; **31**, 17.
Relevant: 2, 9; **3**, 6; **4**, 4; **5**, 6; **7**, 6; **8**, 5; **11**, 3; **13**, 7; **15**, 4; **16**, 5; **17**, 6; **18**, 8; **19**, 7; **20**, 12; **22**, 8; **23**, 6, 7; **24**, 4; **25**, 5, 6; **29**, 16, 17; **31**, 17; **32**, 21; **33**, 5.

12, *4: Detection and determination of nitriles* by *D. J. Curran and S. Siggia*

I. Qualitative chemical methods (168); **II**. Quantitative chemical methods (173); **III**. Separation techniques (177); **IV**. Electrochemical approaches (179); **V**. Ultraviolet absorption (190); **VI**. Infrared absorption (194); **VII**. Mass spectrometry (197); **VIII**. Nuclear magnetic resonance (201); **IX**. References (202).

Nitriles may be detected qualitatively by hydrolysis to acids, by reduction to amines, by Grignard reaction forming ketones after hydrolysis and by some less general methods (Section I).

Quantitative methods include hydrolysis to acids and titration of the latter and reduction by $LiAlH_4$ or catalytically to the corresponding amine and determination of the latter (Section II). Section III deals briefly with distillation, extraction and liquid and gas chromatography. Section IV describes electrochemical studies on the reduction of aromatic and aliphatic nitriles to amines (although aliphatic saturated nitriles are inactive under polarographic conditions).

UV (Section V), IR (IV), mass (VII), and NMR (VIII) spectroscopic methods are treated briefly.

188 references up to 1968.
Complementary: none.
Relevant: 1, 5; **2**, 8; **3**, 15; **4**, 3; **5**, 17; **8**, 3; **13**, 6; **14**, 3; **15**, 3; **16**, 3; **17**, 4; **18**, 5; **20**, 3; **22**, 5; **23**, 5; **24**, 5; **25**, 10; **26**, 5; **28**, 3; **29**, 21; **30**, 14–22; **36**, 5.

12, *5: Directing and activating effects of the cyano group* by *W. A. Sheppard*

I. Size and shape (209); **II**. Electrical character (211); **III**. Substituent effects (217); **IV**. Reactivity (226); **V**. Spectral properties (231); **VI**. Summary (235); **VII**. References (235).

The CN substituent is a powerfully electron-withdrawing, essentially rod-shaped group surrounded by a cylindrical cloud of π-electrons, and has a strong dipole with the negative end towards the N atom.

Section II discusses dipole moments and complexes, while Section III describes the electronic effects of the CN group and of the cyano carbon grops, tabulating many of the relevant p*K* data (Table 6), chemical shifts (Table 7) and acidities (Table 9).

The major effects of CN groups on chemical reactivity are treated in Section IV, including the effects of CN groups on the acidity of hydrocarbons (Table 10), the high electrophilicity of polycyanoolefins and the *meta*-directive effect of CN in cyano-substituted arenes.

Effects of CN on IR, UV and NMR spectra are dealt with in Section VI.

69 references up to 1967.
Complementary: 12, 9; **31**, 8, 17.
Relevant: 3, 13; **4**, 5; **5**, 16; **6**, 8; **7**, 2, 3; **11**, 12; **13**, 8; **14**, 4; **15**, 5; **16**, 6; **18**, 9; **19**, 8; **23**, 11; **28**, 9; **29**, 21; **36**, 5.

12, *6: Nitrile reactivity* by *F. C. Schaefer*

I. General considerations (240); **II**. Interaction of nitriles with acids (243); **III**. Nitrile–metal complexes (249); **IV**. Hydration of nitriles (256); **V**. Nucleophilic addition of hydroxy, amino and sulphydryl compounds (263); **VI**. Nucleophilic addition of carbanions (276); **VII**. Linear polymerization of nitriles (287); **VIII**. Electrophilic addition of carbonium ions (290); **IX**. Addition of free radicals to the cyano group (296); **X**. References (298).

Owing to the polarization of the C≡N bond, nucleophiles attack the C atom, while the N atom is weakly basic. Thus, halogen acids add to the nitrile group, forming imidyl halides [$RCN + HCl \rightleftharpoons RC(Cl){=}NH$], oxyacids form complexes and carboxylic acids react in various ways, yielding imides or isoimides [e.g. $RCN + R'COO^- \rightleftharpoons RC(=NH)OCOR'$] (Section II).

Section III describes complexes of nitriles with electrophilic metal halides (AlX_3, BX_3, etc.), the metal halide-catalysed electrophilic attack of the CN carbon on electron-rich arenes, e.g. (Gatterman aldehyde and Houben–Hoesch ketone synthesis) and metal carbonyl complexes.

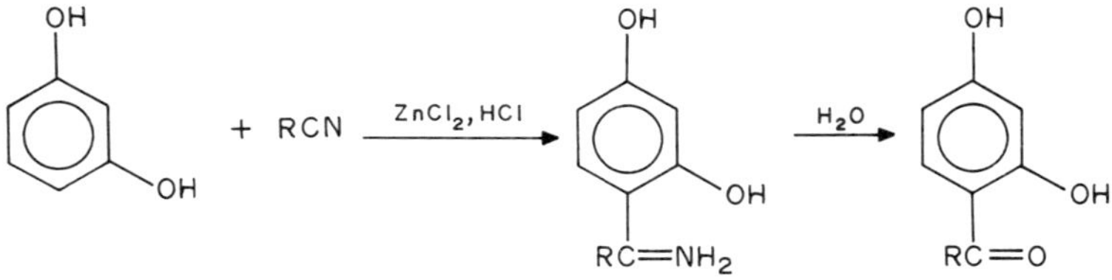

The hydration of nitriles to amides and hence to acids may be base- or acid-catalysed, and amides are also the isolable products for the action of weakly basic H_2O_2 on nitriles (Section IV).

Section V deals with the addition of alcohols and phenols, yielding imidates (Pinner synthesis) in the presence of acids:

$$RCN + R'OH + HCl \longrightarrow [RC(=\overset{+}{N}H_2)OR']Cl^-$$

or the same products by base-catalysed alcoholysis (Nef synthesis). Amidines are obtained by the reaction of NH_3 or amines with imidates, or also by direct addition of NH_3 or RNH_2 to nitriles [$RCN + NH_3 \rightleftharpoons RC(=NH)NH_2$] and also with $NaNH_2$ or KNH_2.

H_2S with nitriles yields thioamides [$RCN + H_2S \rightarrow RC(=S)NH_2$] and thiols or thiophenols give the corresponding thioimidates under basic catalysis.

Section IV deals with the Grignard reaction of nitriles, yielding imines and by subsequent hydrolysis ketones, and also with the base-catalysed dimerization and trimerization of nitriles, while Section VII treats linear polymerizations yielding recurrent $>C=N-$ units.

Olefins and alcohols which are protonated in strong acids give electrophilic attack on the nitrile N atom, and aliphatic aldehydes react with nitriles to yield *N*,*N*′-alkylidenebisamides under acid catalysis ($2RCN + R'CHO + H_2O \rightarrow RCONHCHR'NHCOR$), a reaction applied in the synthesis of industrially important polyamides.

Finally (Section IX), free-radical fluorination and some other free-radical attacks are considered briefly.

244 references up to 1968.
Complementary: none.
Relevant: 12, 7–10; **31**, 26.

12, *7: Reduction of the cyano group* by *M. Rabinowitz*

I. Introduction (307); **II**. Reduction to aldehydes (308); **III**. Reduction to aldimines (319); **IV**. Reduction to amines (319); **V**. Reduction to hydrocarbons (331); **VI**. Miscellaneous (336); **VII**. References (337).

Among the reduction products of RCN the following may be obtained under different conditions: $RCH=NH_2$, RCH_2NH_2, RCH_2NHCH_2R and RCH_3.

Section II discusses reductions (through imines) to aldehydes (Stephen synthesis) $[RCN \rightarrow RC(=NH)Cl \xrightarrow{H_2O} RCHO]$ and selective hydrogenations by Raney Ni, by metals and acids or bases, by $LiAlH_4$ or by Al compounds, all finally yielding aldehydes (Tables 1–3). The reaction can be stopped at the aldimine stage (Table 4, Section III).

Amines may be obtained by catalytic hydrogenation with Pt, Pd, Rh Raney Ni and other catalysts (Tables 5–8), by metal-hydrogen donor combinations (Na + ROH; Na–Hg + H_2O; Sn + HCl, etc.) or by $LiAlH_4$ or similar reagents (Tables 9–14).

Section V treats reductions to hydrocarbons ($RCN \rightarrow RCH_3$), which can only be carried out in special cases (Table 15), and reductive decyanation in which polycyclic arenes are formed, e.g. the corresponding anthracenes from 2-arylmethylbenzonitriles (Table 16).

118 references up to 1967.
Complementary: 31, 14, 26.
Relevant: 2, 11; **8**, 6.

12, *8: Additions to the cyano group to form heterocycles* by *A. I. Meyers and J. C. Sircar*

Cyclization reactions to form heterocycles are classified (see list of contents above) into three main types (A, B and C) and two minor (intramolecular) variations (B_i and C_i).

Section II describes cycloadditions which involve concerted reactions. Five-membered heterocycles are obtained from nitriles with 1, 3-dipoles. Thus nitrile oxides produce 1, 2, 4-oxadiazoles (Tables 1 and 2), nitrile imines produce 1, 2, 4-triazoles, diazoalkanes yield 1, 2, 3-triazoles, nitrile ylids 1, 3-diazoles and azides yield tetrazoles. Other cycloadditions are discussed with azomethine imines and ylids, with ketocarbenes, ketonitrenes and thioketenes, all producing five-membered rings. Six-membered rings are obtained with 1, 4-dipoles, e.g. butadiene with acrylonitrile gives a very poor yield of 2-vinylpyridine.

Section III deals with reactions of nitrilium salts with various C=C bonds yielding pyridine or isoquinoline derivatives; with C=O bonds or with OH groups, oxazole or oxazine rings are obtained. The analogous cyclizations with SH yield thiazoles or thiazines and N-containing compounds give imidazoles or quinazolines, e.g.

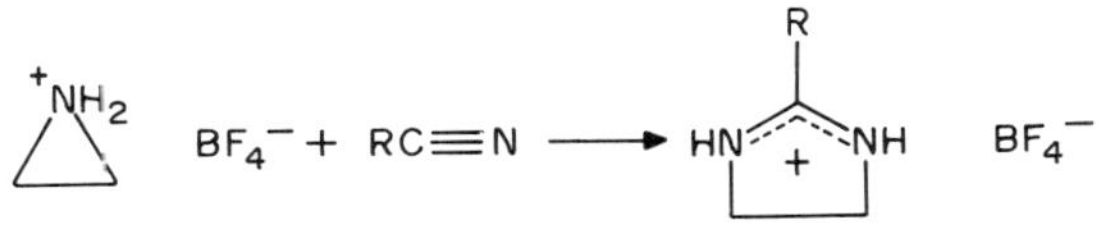

Section IV treats B_i reactions, such as the intramolecular cyclizations of olefinic nitriles, e.g.

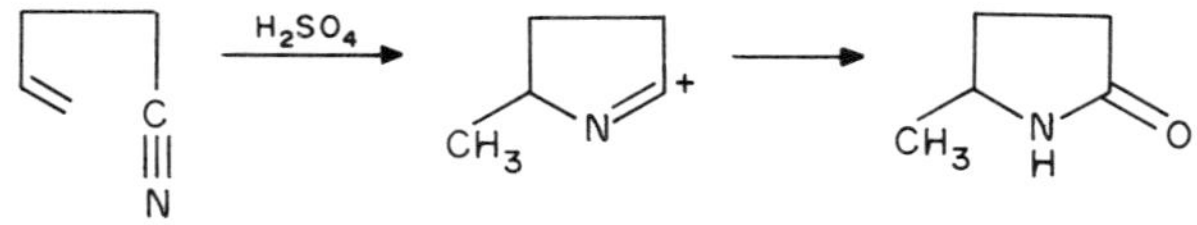

Analogously, δ-ketonitriles yield 2-pyridones; dinitriles also cyclize, e.g. $NCCHPhCH_2CHPhCN$ in the presence of HBr gives an aminodihydropyridine derivative.

Type C_i reactions (Section V) are intermolecular and nucleophilic and may involve primary addition of carbanions, amino groups, oxygen or sulphur nucleophiles, in each case followed by cyclization to a heterocycle. Analogous reactions may take place intramolecularly (Type C_i).

In all sections the mechanism and scope of the reactions are discussed.

241 references up to 1968.
Complementary: 31, 26.
Relevant: 8, 6, 7.

12, *9: Cyanocarbon and polycyano compounds* by *E. Ciganek, W. J. Linn and O. W. Webster*

The presence of several CN groups in a molecule imparts to it unusual physical and chemical properties regarding acidity, electrical conductivity, oxidizing ability, etc.

Cyanogen adds two molecules of primary amines to yield oxamidines [$2RNH_2 + (CN)_2 \rightarrow RHNC({=}NH)C({=}NH)NHR$] and similar reactions occur with other N-containing compounds, with alcohols [$2ROH + (CN)_2 \rightarrow ROC({=}NH){-}C({=}NH)OR$] and with thiols. More complicated products are formed with active methylene, Grignard and aromatic compounds, dienes, AgF_2 and SO_3 (Section II).

Section III deals with the synthesis, structure, salt formation and addition of $HC(CN)_3$ and its halo, alkyl and trifluoromethyl derivatives and with polycyanoethanes.

Most of the section on polycyanoolefins treats tetracyanoethylene (TCE), its syntheses, physical properties and chemical reactions. A wide variety of additions to the double bond of TCE are described, such as Diels–Alder and transannular additions with dienes, additions of free radicals, carbenes, H_2O_2, CH_2N_2, CN^-, phenols, anilines, Pt and Pd compounds and Ph_3P. A few reagents (N_3^-, FSO_3H, etc.) add to the CN groups of TCE with formation of both open-chain and cyclic products. Electron-rich reagents may displace one or two CN groups of TCE [$C_6H_5NMe_2 + (NC)_2C{=}C(CN)_2 \rightarrow Me_2CC_6H_4C(CN){=}C(CN)_2$, i.e. tricyanovinylation]. Some reagents cause fragmentation of the central double bond of TCE, and others (metals, cyanides, iodides, R_3N) give the tetracyanoethylene anion radical. In continuation, tricyanoethylenes substituted with alkyl, aryl or halogen are described, followed by dicyanoethylenes containing other electronegative substituents (halogens, fluroalkyl, sulphonyl, carboxyl groups) or containing electron-donating substituents, as in dicyanoketene acetals and thioacetals and diaminomaleo- and diaminofumaronitrile. Finally, hexacyanobutadiene is treated.

Section V describes the synthesis, physical properties (Tables 2 and 4), salt and complex formation, addition and cycloaddition reactions and polymerizations of cyanoacetylenes. Section VI deals with polycyanobenzenes and Section VII with tetracyanoquinodimethanes, their synthesis, properties and reactions. Section VIII treats (ring-substituted) polycyanoquinones, including 2, 3-dichloro-5, 6-dicyano-*p*-benzoquinone (DDQ), which forms CT complexes with many donor molecules and is also a very useful dehydrogenating agent.

Azacyanocarbons [$HN(CN)_2$, $N(CN)_3$, N_3CN, $(NC)_2C{=}C(R)N_3$], NCN=NCN and cyanodiazo compounds ($NCCHN_2$) are discussed in Section VIII.

Section IX deals with polycyano-substituted heterocycles. Among these, polycyanooxiranes are obtained by epoxidation of the corresponding olefins (Table 9) and they are easily cleaved by simultaneous scission of both a C—O and the C—C bond by amines, sulphides and thiocarbonyl compounds, e.g.

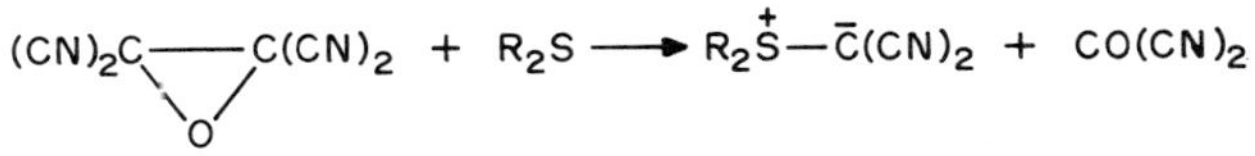

Other reactions lead to ring opening with cleavage of the C—C bond, e.g.

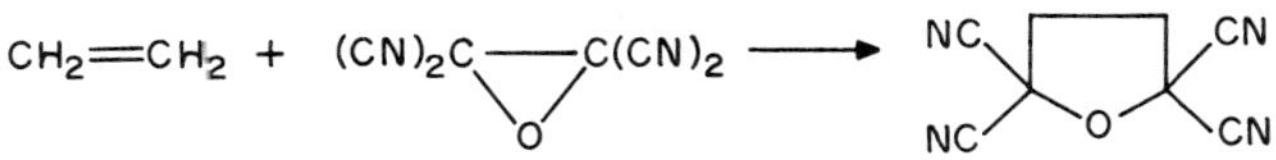

or analogous additions to aromatic compounds. Other reactions (e.g. with acid halides or anhydrides) lead to C—O bond cleavage; electron-rich olefins attack the O atom in a trans-epoxidation reaction:

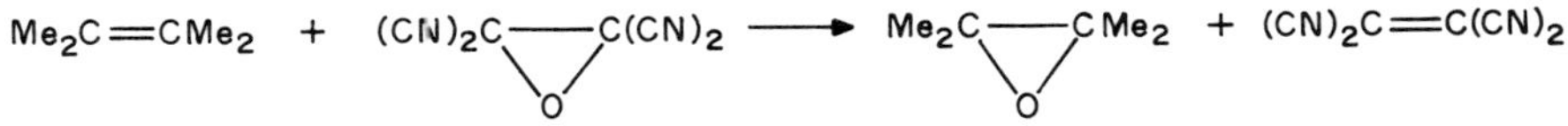

Subsequently tetracyanofuran and some polycyano pyrroles, pyrazoles, triazoles, pyridines, triazines, dithiins and thiophenes are described.

Carbonyl cyanide [$CO(CN)_2$] and some of its trifluoromethyl derivatives are dealt with in Section X.

The final section (XI) treats the anions and salts of cyanocarbon acids, derived from malononitrile [$HC(CN)_3$], polycyanoethanes, -propanes, -butanes, -cyclopentadienes and analogous compounds. The formation, reactions and physical properties of these anions are reviewed (Tables 13–17), including pK_a values of the cyanocarbon acids (Table 18).

825 references up to 1969.
Complementary: 12, 10; **13**, 17.
Relevant: 31, 26.

12, *10: Molecular complexes of polycyano compounds* by *L. R. Melby*

I. Introduction (639); **II.** Theoretical aspects and implications (640); **III.** Acceptor strengths of polycyano π-acids (643); **IV.** Electronic absorption spectroscopy (649); **V.** Ion-radical formation (656); **VI.** Electrical properties of polycyano complexes (662); **VII.** Acknowledgements (667); **VIII.** References (667).

Polycyano compounds such as tetracyanoethylene (TCNE) and tetracyanoquinodimethane (TCNQ) are strong π-bases and form stable complexes. Some theoretical aspects of the interaction between electron donors and acceptors are treated in Section II.

Section III discusses the acceptor strength of polycyano π-acids, based on MO calculations, polarography (Table 1) and electron affinities (Table 2).

The absorption maxima of various donor–acceptor complexes are described in Section IV (Tables 3–5), together with equilibrium constants (Table 6) and thermodynamic considerations.

Section V deals with anion radicals derived from the complexes of TCNE, DDQ and TCNQ in solutions and in the solid state.

Finally, Section VI reviews the electrical properties of the complexes (Tables 9 and 10).

78 references up to 1967.
Complementary: 12, 3; **31**, 17.
Relevant: 1, 6; **3**, 6; **4**, 4; **7**, 6; **8**, 5; **9**, 5; **15**, 4; **16**, 5; **17**, 6; **19**, 7; **20**, 12; **22**, 8; **24**, 4; **25**, 5; **32**, 21; **33**, 5.

12, *11: Radicals with cyano groups* by *H. D. Hartzler*

While the CN group itself is inert to attack by free radicals, it exerts a profound influence on the behaviour of the molecule to which it is attached; the CN group deactivates the latter towards abstraction by acceptor radicals, activates it towards abstraction by donor radicals, activates olefins towards radical attack and stabilizes α-cyanoalkyl radicals by electron delocalization.

Section II deals with hydrogen abstraction from nitriles and the selectivity of the reaction, e.g. in free-radical halogenation.

Section III discusses free radical addition of cyano esters and of halonitriles to olefins, including simple additions and cyclizations (Table 2).

Cyanobenzyl radicals are formed by dissociation of, e.g., tetraphenylsuccinonitrile: $Ph_2C(CN){-}C(CN)Ph_2 \rightleftharpoons 2Ph_2\dot{C}CN$. The primarily formed radicals may undergo recombination or isomerization to various products (Section IV).

Azonitriles (Section V) are used as initiators of free-radical reactions owing to their facile decomposition at low temperatures [$R_2C(CN){-}N{=}N{-}C(CN)R_2 \rightarrow 2R_2\dot{C}(CN) + N_2$] (Table 3). The radicals formed may combine mainly to dicyanoethane derivatives, $R_2C(CN){-}C(CN)R_2$, but various other reactions (disproportionation, reaction with the solvent, abstractions, oxidations, reductions) also occur. Cyanoolefins are especially reactive. They may undergo free-radical polymerization and reaction with aryldiazonium salts to yield α-halo-β-arylpropanonitriles (Meerwein reaction):

$$ArN_2^+ + CH_2{=}CHCN + CuCl_2^- \longrightarrow ArCH_2CHClCN$$

Acrylonitrile dimerizes to dicyanodimethylcyclobutanes and tetracyanoethylene and tetracyanoquinodimethane add the free radicals produced by decomposition of azobis(isobutyronitrile) (Section VII).

Cyanocarbenes, especially dicyanocarbene (Section VIII), add to olefins and insert in to C—H bonds in a non-discriminating manner. Some reactions of cyanonitrenes (NCN) are also mentioned.

Finally, in (Section IX) a few reactions of CN radicals (obtained by photolysis of ClCN or ICN) are described.

211 references up to 1968.
Complementary: none.
Relevant: **1**, 9; **13**, 3; **16**, 8; **19**, 19; **21**, 7; **31**, 10; **33**, 8; **36**, 9.

12, *12: The biological function and formation of the cyano group* by *J. P. Ferris*

Glycosides containing CN groups occur in many plants (Table 1). Their biosynthesis, hydrolysis and role in asparagine biosynthesis are described in Section II. Mandelonitrile [PhCH(CN)OH] is present in several arthropods and serves as a defensive agent (Section III). In various plants (*Lathyrus*, *Vicia* and other species), CN-containing compounds occur, which may cause serious neurotoxic effects ('lathyrisim') and occasionally death. These are β-cyanoalanine derivatives and their biosynthesis is discussed (Section IV).

Moulds and microorganisms produce cyanogenic substances either for defence or as intermediates in the synthesis of amino acids (Section V).

Cyanopyridine alkaloids occur in some plants, and their biosynthesis and enzymatic hydrolysis have been studied in detail. Indoleacetonitrile is a plant growth hormone occurring in cabbage and other plants (Section VII). It becomes active only after enzymatic hydrolysis to indoleacetic acid:

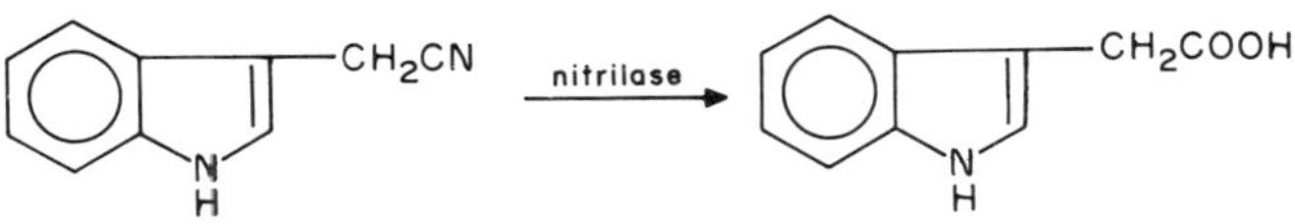

Section VIII deals with cyanide toxicity and with various detoxification mechanisms.

141 references up to 1968.
Complementary: **31**, 9.
Relevant: **2**, 7; **3**, 5; **4**, 9; **5**, 18; **7**, 1; **11**, 11; **13**, 14; **15**, 10; **16**, 13; **17**, 13; **18**, 13; **19**, 13; **20**, 6; **22**, 20; **24**, 14; **26**, 13; **28**, 17.

12, *13: Syntheses and uses of isotopically labelled cyanides* by *L. Pichat*

HCN (or its salts) labelled with ^{14}C or ^{13}C is prepared by reaction of CO_2 or carbonates by a variety of methods (reduction, direct exchange etc.; Sections II, III and IV). Organic nitriles have to be prepared by the use of ^{14}C-labelled alkali or Cu^{I} cyanides, or by transformation of ^{14}COOH into ^{14}CN. Cyanohydrin formation (Tables 1 and 2) is frequently used, and also the Strecker and Bücherer amino acid synthesis (Tables 3–5). Other routes involve reaction of alkali metal cyanides with halides (Tables 6 and 7), replacement of N_2^+ by CN (Sandmeyer reaction (Table 8), the nitrile–carboxyl exchange, e.g. $RCN + R'^{14}CO_2H \rightarrow R'CN + R^{14}CO_2H$ and reaction of cyanogen halides with labelled acids, e.g. $BrCN + R^{14}CO_2Na \rightarrow R^{14}CN + CO_2 + NaBr$. α-Deuteration occurs by base-catalysed H–D exchange in nitriles (Section V).

Section VI describes uses of nitriles, such as hydrolysis to carboxylic acids, reduction to labelled amines or aldehydes, preparation of ^{14}C-labelled ketones by Grignard reaction, formation of amidines and heterocycles and also some analytical applications.

229 references up to 1967.
Complementary: 5, 10.
Relevant: 13, 15; **17**, 12; **18**, 18; **19**, 17; **22**, 14; **23**, 16; **24**, 20; **26**, 11; **27**, 10; **28**, 8; **29**, 28; **33**, 18; **36**, 12.

12, *14: Nitrile oxides* by *Ch. Grundmann*

Nitrile oxides ($RC{\equiv}N \rightarrow O$) are isomeric with cyanates (ROCN) and isocyanates ($RN{=}C{=}O$). They are rather unstable and in order to use them it is often necessary to generate them *in situ* in the presence of a suitable reaction partner. The first member of the series, fulminic acid or formonitrile oxide ($HC{\equiv}N \rightarrow O$), is prepared in the form of highly explosive Ag or Hg salts. Higher nitrile oxides are synthesized from aldoximes by dehydrogenation ($ArCH{=}NOH \rightarrow ArC{\equiv}N \rightarrow O$) or by elimination of HX from hydroximic acid halides ($RCX{=}NOH \xrightarrow{\text{base}} RC{\equiv}N \rightarrow O + HCl$). Other routes involve nitroalkanes or fulminates (Table 1, Section II).

Section III describes the physical properties of nitrile oxides, which include rearrangement (to isocyanates), dimerization (to furoxanes), polymerization, reduction (to nitriles), additions of H_2O, HX, NH_3, N_3^-, S^-, SCN^-, CN^-, ROH [$ArC{=}N \rightarrow O + RO^- \rightarrow ArC(OR){=}NOH$, alkylhydroximic acids), RSH, RCOOH, RNH_2, Grignard reagents $\left(Ph\overset{+}{C}{=}N{-}\overset{-}{O} \xrightarrow{RMgX} PhCR{=}NOH\right)$, 1,3-dipolar additions with olefins, acetylenes, C=O or C=S compounds, etc., all leading to cyclic structures.

222 references up to 1968.
Complementary: 31, 18.
Relevant: none.

12, *15: Isonitriles* by *P. Hoffmann, D. Marquarding, H. Kliimann and I. Ugi*

Isonitriles can be described by various formulae: R—N=C, $R—\overset{+}{N}\equiv\overset{-}{C}$, RN⇄C. They are stable with a characteristic unpleasant smell.

Section II describes the syntheses of isonitriles by alkylation of cyanides (AgCN + RI → RNC + AgI), by the 'carbylamine' reaction ($RNH_2 + CHCl_3 + KOH \rightarrow\rightarrow RNC$), by dehydration of formamides ($RNHCHO \rightarrow RNC + H_2O$) and by some less general routes.

The reactions of isonitriles include C—N(R—N) single bond cleavage with rearrangement to cyanides; α-additions, e.g. $RNC + Cl_2 \rightarrow RN{=}CCl_2$, which may be followed by secondary reactions [e.g. the Passerini reaction in which α-acyloxycarboxamides are formed ($RCO_2H + R^1COR^2 + R^3N{=}C \rightarrow\rightarrow RCO_2CR^1R^2CONHR^3$) and four-component condensations of isonitriles with CO compounds in the presence of both amines and acids, forming a variety of products depending on the synthons and conditions (Table 1)], cycloadditions and cyclizations. Finally, Section IV treats isonitrile complexes with a wide choice of metals.

236 references up to 1969.
Complementary: 31, 2–4, 19, 20.
Relevant: 12, 16.I.I.A; **22**, 21.

12, *16: Rearrangement reactions involving the cyano group* by *J. Casanova, Jr*

I. Introduction (885); **II**. Rearrangement reactions directly involving the cyano group (886); **III**. Rearrangement reactions of the carbon skeleton to which the cyano group is attached (906); **IV**. Fragmentation reactions involving the cyano group (909); **V**. Rearrangement reactions involving intermediate cyano compounds: the von Richter reaction (936); **VI**. References (940).

Section II discusses the thermal rearrangement of isonitriles to nitriles ($ArN{\equiv}C \rightarrow ArC{\equiv}N$) and its scope and mechanism, electron-deficient rearrangements in which R or Ar groups migrate from metal to the cyanide carbon, with destruction of the CN group [$ArCN + PhMgBr \rightarrow ArC{\equiv}\overset{+}{N}{—}\overset{-}{M}g(Ph)Br \rightarrow ArC(Ph){=}NMgBr$], the tautomeric equilibrium between cyanides, ketenimines and ynamines ($RCH_2C{\equiv}N \rightarrow RCH{=}C{=}NH \rightarrow RC{\equiv}CNH_2$ and reactions involving cyanocarbene intermediates, obtained, e.g., by thermal decomposition of dicyanodiazomethane:

$$(NC)_2CN_2 \xrightarrow{\Delta} (NC)_2C: + N_2$$

Section III describes rearrangements of the carbon skeleton to which the CN group is attached. Thus dicyanocarbene yields 7,7-dicyanonorcaradiene with benzene, but thermolysis of the product yields two rearranged derivatives:

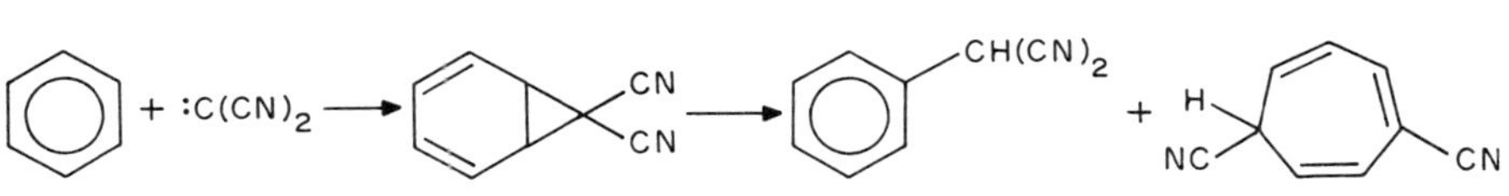

Section IV deals with fragmentations which involve the CN group and result in rearrangements of the carbon skeleton. These include rearrangements of nitrilium salts such as the Ritter reaction ($R^+ + N{\equiv}CR' \rightarrow RNHCOR'$) and the fragmentation of ketoximes and their derivatives, which involve unsaturated nitriles as intermediates and, depending on the structure of the substrate, may yield a wide variety of products (amides, unsaturated nitriles and unsaturated cyano esters, nitriles, isonitriles, acids, etc.). The common mechanistic feature of all these reactions is the intermediacy of nitrilium salts ($R'C{\equiv}N + R^+ \rightarrow [R'C{\equiv}\overset{+}{N}R \leftrightarrow R'C{=}NR]$.

The last section (V) treats the von Richter reaction, in which an aromatic nitro compound is treated with CN^-. The NO_2 group is lost and CN is attached to the ring, but not at the same position ('cine-substitution'):

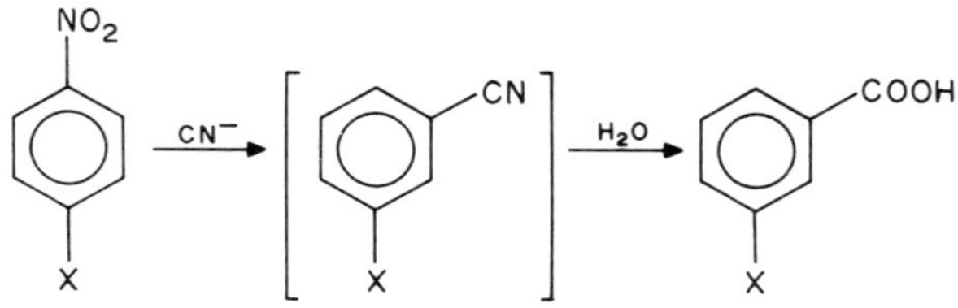

The mechanism of the reaction is discussed in detail.

136 references up to 1968.
Complementary: none.
Relevant: 1, 7; **2**, 15; **3**, 14; **4**, 10; **5**, 8; **8**, 9; **9**, 3; **10**, 3; **11**, 4; **13**, 13; **14**, 7; **15**, 8; **16**, 17; **17**, 8; **18**, 15; **9**, 18; **20**, 4; **23**, 13; **24**, 10; **26**, 15; **29**, 4;

The chemistry of the hydroxyl group (1971)

13, *1: Theoretical aspects of the chemistry of the hydroxyl group* by R. F. W. Bader

I. Introduction (1); **II**. A study of the O—H bond in OH^+, $OH^\cdot$ and OH^- (8); **III**. The molecular charge distributions of CH_3OH and CH_2FOH (33); **IV**. Acknowledgements (49); **V**. References (49).

Section I describes the charge distribution in molecules which in principle contains all the information necessary for a complete physical description of a system, and the relevant Hartree–Fock wave functions (Table I-1). Section II discusses the molecular charge distribution for various hydrides and CO species (OH^+, $OH^\cdot$, OH^-), using contour plots, total charge distributions and density difference distributions. It also treats the interpretation of the binding in terms of the forces exerted on the nuclei according to the Hellmann–Feynman theorem.

Section III deals with the molecular charge distributions of CH_3OH and of CH_2FOH, using a basis set of Gaussian-type atomic orbitals.

38 references up to 1969.
Complementary: 27, 9.
Relevant: 1, 1; **2**, 1; **3**, 1; **4**, 1; **5**, 1; **6**, 1; **8**, 1; **11**, 1; **12**, 1; **14**, 1; **15**, 1; **16**, 1; **17**, 1; **18**, 1; **19**, 1; **20**, 1; **22**, 1; **23**, 1; **24**, 1; **25**, 1; **26**, 1; **27**, 9; **28**, 1; **31**, 1; **32**, 1; **33**, 1; **36**, 2.

13, *2: Nucleophilic attack by hydroxide and alkoxide ions* by *C. A. Fyfe*

I. Introduction (52); **II**. Attack at saturated carbon (53); **III**. Attack on allylic systems (56); **IV**. Attack on carbon—carbon multiple bonds (61); **V**. Attack on aromatic systems (83); **VI**. Attack on aza-aromatic systems (113); **VII**. Acknowledgements (124); **VIII**. References (124).

Section II reviews briefly S_N2 and S_N1 reactions involving OH and OR ions. Section III describes substitutions on allylic systems, involving bimolecular attacks at α-carbons, attacks on the alkene moiety and also unimolecular reactions involving allyl carbonium ions.

Attacks on C═C bonds (Section IV) include attack by ^-OR ions on C═C leading to additions [$ArCH{=}CHNO_2 \xrightarrow{MeOH} ArCH(OMe)CH_2NO_2$], cleavage of C═C bonds by ^-OH or H_2O ($R^1R^2C{=}CR^3R^4 \rightarrow R^1R^2CO + CH_2R^3R^4$) and substitutions of labile groups:

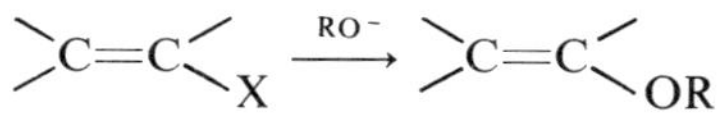

which may take place by addition–elimination and elimination–addition mechanisms. Similarly, C≡C bonds add alcohols under basic catalysis. The intermediates and transition states which may be involved are discussed.

Aromatic systems (Section V) are only attacked by nucleophiles if they contain strongly electron-attacking groups, although unactivated systems react also in the presence of catalysts and at high temperatures. Both 'two-stage' and 'synchronous' mechanisms are considered, on polynitroarenes and on polynitroanisoles, giving NMR, UV and thermodynamic data for these reactions. Results of molecular orbital calculations and their comparison with experimental data are also given.

Section VI treats attack on pyridine and its homologues, discusses kinetics and mechanisms, the evidence for intermediate complex formation and MO calculations. Pyridinium ions represent a special case in which, e.g., replacement of halogens is facile. Again the mechanism involves formation of an intermediate complex:

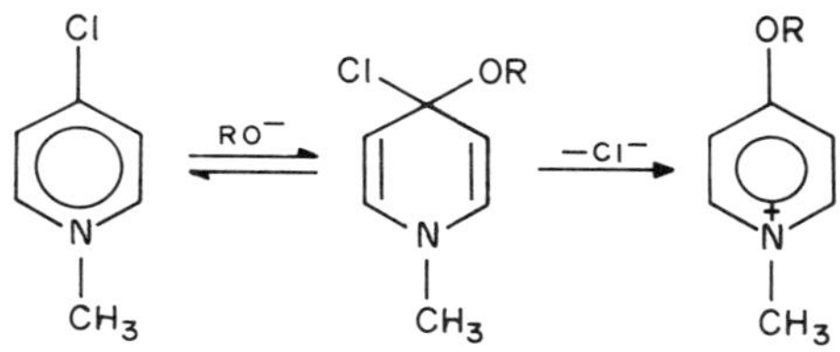

244 references up to 1969.
Complementary: 1, 8; **6**, 8; **29**, 27.
Relevant: 18, 16.

13, *3: Free radical and electrophilic hydroxylation* by *D. F. Sangster*

I. General (135); **II**. Radiolytic hydroxylation (136); **III**. Photolytic hydroxylation (160); **IV**. Chemical free-radical hydroxylation (163); **V**. Comparison of reagents (177); **VI**. References (181).

The chapter considers only direct hydroxylation of organic compounds by free-radical and electrophilic reagents.

In radiolytic hydroxylation OH radicals are formed together with other reactive species. HO$^{\bullet}$ may give addition, hydrogen abstraction, electron transfer and displacement reactions. In the case of benzenoid compounds, transient species and their reactions, kinetics, spectra, formation of hydroxy products, effects of additives and conditions are considered. With aliphatic compounds the main route is hydrogen abstraction leading to dimerization or the formation of peroxy, hydroperoxy or aldehyde compounds. Unsaturated compounds yield very complex product mixtures; the reactions of DNA bases, steroids and vitamins often involve hydroxylation as responsible for radiation damage. Other radiolytic reactions involve the hydroperoxyl radical, $HO_2^{\bullet}$ (Section II).

Section III discusses the photolytic production of OH radicals from H_2O or from H_2O_2 and their reactions (usually hydroxylation) with benzenoid and unsaturated compounds. Irradiation of various metal complexes (e.g. containing Fe^{3+}, $VO_2{}^{2+}$, Ce^{4+} and H_2O or OH^-) also produces OH radicals.

Section IV describes the production of OH radicals by chemical methods. The most important route involves H_2O_2 together with a metal ion of variable valency (e.g. Fenton's reagent, $H_2O_2 + Fe^{2+}$). These reagents hydroxylate benzenoid compounds producing phenols and are more effective in the presence of Cu^+ or V^{3+} instead of Fe^{2+}. They also react with olefins and with heterocyclic compounds. Similar $HO^{\bullet}$-producing systems are $Ti^{3+} + H_2O_2$, 1,2- or 1,4-dihydroxyarenes + Fe^{2+} and Udenfriend's reagent (Fe^{2+} + EDTA + ascorbic acid + O_2).

Several peracids undergo homolysis to free OH, including percarboxylic acids ($RCOOOH \rightarrow RCOO^{\bullet} + HO^{\bullet}$ by homolysis or $RCOOOH \rightarrow RCOO^- + OH^+$ by heterolysis), pernitrous acid ($H_2O_2 + HNO_2 \rightarrow H_2O + ONOOH \rightarrow ONO^{\bullet} + HO^{\bullet}$) and some other peracids and peresters.

Section V considers the identity of the reactive species (which is by no means always clear) and some discrepancies in the experimental results.

323 references up to 1969.
Complementary: none.
Relevant: **13**, 16, 17; **27**, 21, 23; **33**, 8.

13, *4: Formation of hydroxyl groups via oxymetallation, oxidation and reduction* by *I. R. L. Barker*

In principle, hydrogen bound to carbon can be oxidized to a hydroxyl group by free-radical reaction of the substrate with molecular oxygen. In various related reactions metallic compounds are involved.

Section II describes electrophilic additions of Hg^{II} salts with alkenes:

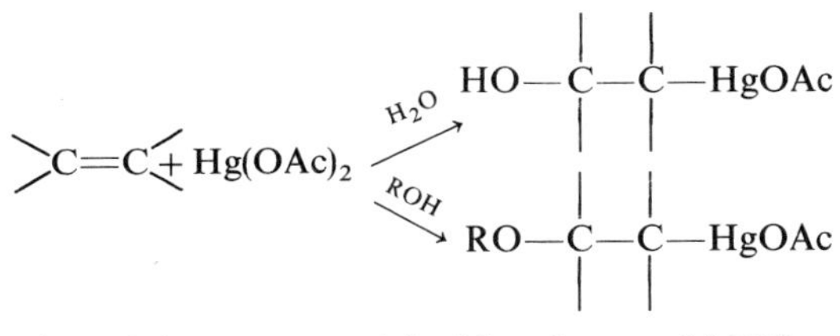

Subsequent reduction of the oxymercurial adduct by, e.g., $NaBH_4$ provides a saturated alcohol by substituting H for the HgOAc group. The mechanism and stereochemistry of the reaction are discussed. Thallium(III) salts react with olefins similarly, while Pd^{II} salts yield carbonyl compounds (e.g. methyl ketones from alk-1- and 2-enes and vinyl and allyl acetates in the presence of CH_3COOH. Allylic acetoxylation is the result of the reactions of SeO_2 in the presence of CH_3COOH.

Section III deals with reactions of molecular oxygen. Thus, singlet oxygen, in the presence or absence of initiators or sensitizers, produces peroxides or *endo*-peroxides (Table 1), from which alcohols may be obtained by a variety of reducing agents.

Enol forms of ketones give α-hydroperoxy ketones with O_2, which may be reduced to the corresponding hydroxy derivatives. The reaction occurs through a route involving carbanions. Autoxidation of metal alkyls and aryls of Al and B are important industrial processes and is used for the preparation of higher alcohols through the autoxidation and hydrolysis of the metal alkyls. The reaction is autocatalytic and probably a free-radical chain process.

Section V discusses reduction processes of oxygenated compounds to alcohols. These may involve one or two molecules of a carbonyl compound and produce alcohols or pinacols, respectively, using Li, Na, K, Ca, Mg, Fe or Sn in a wide variety of methods (Birch, Clemmensen, Bouveault–Blanc, etc.)

Other generally useful methods are catalytic hydrogenation and hydrogenolysis of aldehydes, ketones, esters, epoxides, peroxides, etc. Grignard and other organometallic reagents reduce carbonyl groups to alcohols, in addition to Meerwein–Ponndorf–Verley and other related reductions. Finally, the use of metal and organometal hydrides is reviewed.

294 references up to 1968.
Complementary: **2**, 11, 13.
Relevant: 33, 6, 8, 11.

13, *5: Electrochemistry of the hydroxyl group* by *H. Lund*

I. Introduction (254); **II.** Electrolytic preparation of alcohols and phenols (261); **III.** Electrolytic reactions of hydroxyl compounds (280); **IV.** References (290).

The introductory section reviews the main types of electrode reactions, i.e. type A in which electron transfer occurs directly between the electrode and the substrate (and which in turn is divided into type A1 when the reaction at controlled electrode potential can occur with 100% current efficiency and type A2 in which part of the current is consumed in the decomposition of the medium) and type B in which another species is involved in the electron transfer in addition to the substrate (e.g. solvated electrons).

Section II describes the electrolytic preparation of alcohols and phenols by reduction of acids (Table 1), amides (Tables 2 and 3), esters, acid chlorides, aldehydes and ketones. The mechanism and stereochemistry of the reduction are discussed (Table 4), and also the formation of alcohols by anodic oxidation (the Hofer–Moest modification of the Kolbe reaction).

Few OH compounds can be reduced electrolytically, but many of them can be oxidized anodically. Section III discusses briefly reductive removal of OH groups from certain α-hydroxyketones, triaryl alcohols and related compounds.

More details are given on the anodic oxidation of primary aliphatic alcohols through aldehydes to acids and of secondary aliphatic alcohols to ketones. Aromatic alcohols are oxidized to the corresponding aldehydes, the next step of oxidation being more difficult. Phenols can be oxidized to quinones (e.g. hydroquinone $\rightarrow$ quinone) by two-electron oxidation, but the reaction is often complicated by side-reactions such as substitution, ring fission and even rearrangements.

In one-electron oxidations the primarily formed radical often yields coupling products with another radical, and these may undergo further reactions, additions, polymerizations, etc.

106 references upto to 1970.
Complementary: 27, 8.
Relevant: 5, 2; **8**, 11; **16**, 15; **17**, 14; **19**, 12; **20**, 5; **21**, 5; **22**, 9; **23**, 10; **24**, 17; **25**, 13; **26**, 12; **27**, 8; **28**, 7; **29**, 8, 9; **31**, 7; **32**, 6; **34**, 1.

13, *6: Detection and determination of hydroxyl groups* by *S. Siggia, J. G. Hanna and T. R. Stengle*

I. Detection and determination of hydroxyl groups (296); **II**. Chemical methods (296); **III**. Physical methods (311); **IV**. References.

Qualitative analytical methods (Section II) described include derivatization as dinitrobenzoate and other esters, as urethanes and as xanthate. Esterifications in some favourable cases can be carried out quantitatively (using acyl halides, anhydrides, etc.); phenols and nitro alcohols are acidic enough to be titrated in non-aqueous media, potentiometrically or visually. Glycols can be quantitatively oxidized by periodic acid.

Physical methods (Section III) include IR spectroscopy (Table 1) and NMR methods involving direct analysis (Table 2) and the study of derivatives (esters etc.).

126 references up to 1968.
Complementary: **5**, 17; **25**, 10.
Relevant: **1**, 5; **2**, 8; **3**, 15; **4**, 3; **5**, 17; **8**, 3; **12**, 4; **14**, 3; **15**, 3; **16**, 3; **17**, 4; **18**, 5; **20**, 3; **22**, 5; **23**, 5; **24**, 5; **26**, 5; **28**, 3; **29**, 21; **30**, 14–20; **36**, 5.

13, *7: Acidity and inter- and intramolecular H-bonds* by *C. H. Rochester*

I. Introduction (328); **II**. Intermolecular H-bonding (328); **III**. Intramolecular H-bonding (353); **IV**. H-bonding and kinetics (367); **V**. The acidity of hydroxyl groups (369); **VI**. The protonation of hydroxyl groups (382); **VII**. References (383).

Section II discusses self-association of alcohols (Tables 1 and 2) and of phenols (Table 3); alcohols as hydrogen donors (Table 4); phenols as hydrogen donors (Table 6); the dependence of hydrogen bonding on structure, on steric effects (Table 7) and on solvent effects (Table 8) and the influence of hydrogen bonds on NMR and IR spectra.

Intramolecular bonding in alcohols and substituted phenols leads to cyclic structures and can be studied by IR, UV, Raman or NMR spectroscopy (Tables 10–12) (Section III).

The influence of hydrogen bonding on solution kinetics is discussed in Section IV. Section V deals with the acidity of OH groups in alcohols (Table 13), in carbohydrates (Table 14) and in *gem*-diols (Table 15), and with the ability of alcohols to undergo self-ionization (Table 16). This is followed by the description of the acidity of phenols (Table 17) and of the effect of substituents on the latter, of the acidity of phenols in excited states (Table 18) and of the acidity of some heteroaromatic OH compounds. Protonation of OH groups is briefly mentioned (Section VI).

375 references up to 1969.
Complementary: none.
Relevant: **2**, 9; **3**, 6; **4**, 4; **5**, 6; **7**, 6; **8**, 5; **11**, 3; **12**, 3, 10; **15**, 4; **16**, 5; **17**, 6; **18**, 8; **19**, 7; **20**, 12; **22**, 8; **23**, 6, 7; **24**, 4; **25**, 5, 6; **29**, 17; **31**, 17, 21; **32**, 21; **33**, 5.

13, *8: Directing and activating effects* by *D. A. R. Happer and J. Vaughan*

I. Introduction (394); **II**. Reactions of aliphatic systems (395); **III**. Reactions of aromatic side-chains (405); **IV**. Aromatic substitution reactions (418); **V**. References (446).

Section II discusses the influence of OH substituents on the reaction of aliphatic systems. These include electronic effects (Table 1), proximity effects in nucleophilic substitutions and in the hydrolysis of carboxyl derivatives and of phosphorus esters and aliphatic acid–base equilibria (Table 2).

The effects of ring-hydroxyl groups on the reactions of side-chains attached to the same aromatic system (Section III), including electronic effects involving solvents, different weak and strong $-R$ side-chains in addition to side-chains, have been studied in detail (Tables 3–9), and substituent constants for OMe, OH, and O^- substituents are discussed and compared. In continuation, proximity effects of OH groups are described, e.g. the abnormally high acidity of salicylic acid due to stabilization of the anion and the unexpectedly fast hydrolysis of aromatic OH-substituted esters.

Section IV deals with the mechanism of the electrophilic aromatic substitution in phenols; it discusses the mechanism and the *o*/*p* ratio in general and then various reactions separately, such as nitration (Tables 10–12), nitrosation, halogenation (Tables 13–15), sulphonation (Tables 16 and 17), diazonium coupling, Friedel–Crafts alkylation (Table 18), uncatalysed alkylation (Table 19), hydroxyalkylation (Table 20), haloalkylation, acylation, formylation (several methods, especially by dichloromethyl ether–$TiCl_4$, Table 21), ketone synthesis, carboxylation and briefly also hydrogen exchange, mercuration and hydroxylation. In conclusion, the comparative directing power of OH groups (Table 22) and the rates of electrophilic aromatic substitutions and nucleophilic aromatic substitutions (Table 23) are presented.

233 references up to 1968.
Complementary: none.
Relevant: **3**, 3; **4**, 5; **5**, 16; **6**, 8; **7**, 2, 3; **11**, 12; **12**, 5; **14**, 4; **15**, 5; **16**, 6; **18**, 9; **19**, 8; **23**, 11; **28**, 9; **29**, 6; **31**, 8; **36**, 15.

13, *9: Electrophilic attacks on the hydroxyl group* by *P. Salomaa, A. Kankaanperä and K. Pihlaja*

I. Introduction (454); **II.** Alkylation. Carbon atom as the electrophilic centre (455); **III.** Acylation. Carbon atom as the electrophilic centre (466); **IV.** Heteroatoms as the electrophilic centres (481); **V.** References (497).

The O atom of the hydroxyl group acts as an electron donor and is easily attacked by electrophiles.

Section II describes the nucleophilic reactivity of the OH group (Tables 1–3) in reactions of the type $ROH + RX \rightarrow ROR + HX$. The kinetics, mechanisms, equilibria and synthetic scope of alkylation with α-halo ethers, vinyl ethers, acetals and carbonyl compounds are discussed, in addition to alkylations with ketene acetals, ortho esters, alkenes, alkynes and carbenes ($R^1OH + R^2C\!: \rightarrow R^1OCHR$).

Section III deals similarly with acylations on the OH group, with C atoms acting as electrophilic centres (e.g. $RCOOH + R'OH \rightarrow RCOOR' + H_2O$). Conformational, steric and kinetic factors are considered, together with applications to syntheses, reactions with hindered acids, thermodynamic versus kinetic control in esterification, esterification with mixed anhydrides, autocatalytic reactions, acylations with ketenes and with isocyanates, intramolecular catalysis by neighbouring OH groups in hydrolysis and solvolysis of hydroxy esters and hydroxyamides.

Electrophilic attacks by heteroatoms are treated in Section IV. The reagents include inorganic oxyacids such as H_2SO_4 ($ROH + H_2SO_4 \rightarrow ROSO_3H + H_2O$) and their reactions with primary, secondary and tertiary alcohols (Tables 5–7). H_2SO_4 also forms diesters with alcohols [$ROSO_3 + ROH \rightarrow (RO)_2SO_2 + H_2O$]. Other acids treated are phosphoric, nitric and boric acids (Tables 8 and 9).

Apart from acids, other heteroatomic electrophiles which are considered include SO_3, sulphamoyl chlorides ($ClSO_2NR_2 + R'OH \rightarrow R'OSO_2NR_2$), phosphorus-containing electrophiles [H_3PO_3, PX_3, POX_3, $P(OR)_3$, etc.] and *O*-nitro and *N*-nitro compounds (e.g. $CH_3COONO_2 + ROH \rightarrow RONO_2 + CH_3COOH$).

214 references up to 1970.
Complementary: 3, 10; **5**, 9; **25**, 7.
Relevant: none.

13, *10: Oxidation and reduction of phenols* by *M. Lj. Mihailović and Ž. Čeković*

I. Introduction (506); **II**. Oxidation of phenols (506); **III**. Reduction of phenols (578); **IV**. References (581).

Section II discusses the mechanism of one-electron oxidation of phenols involving phenoxy radicals and quinone methides (**1** and **2**):

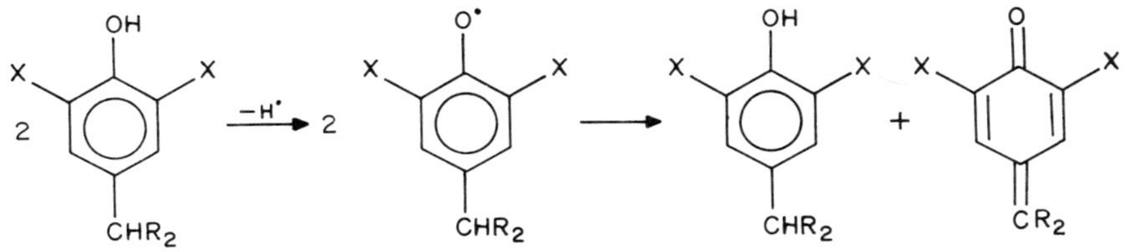

(**1**) (**2**)

The structure and characterization of phenoxy radicals are considered, and also oxidation potentials (Table 2). Oxidation by iron$^{(III)}$ salts, e.g. $K_3[Fe(CN)_6]$ and $FeCl_3$, may lead to a variety of products obtained by C—C dimerization of nuclear or of side-chain carbons, C—O dimerizations and also O—O dimerizations and CT complexes. A similar diversity of products is obtained in oxidations by $Pb(OAc)_4$ (Table 3) or by PbO_2. H_2O_2 oxidizes by $HO^{\bullet}$ or $HOO^{\bullet}$ radicals depending on the catalysts and conditions. Phenols inhibit and destroy peroxy radicals formed by autoxidation, and the kinetics and mechanism of these processes have been studied in detail. Other oxidants studied are acyl peroxides and peracids (which not only oxidize but usually also substitute the phenols at various positions); persulphate ($K_2S_2O_8$) oxidizes monohydric to dihydric phenols (Elbs reaction) via introduction of an OSO_3K group, which is in turn hydrolysed to OH; $NaIO_4$ removes oxidatively OR substituents in phenolic ethers, causes the formation of dimeric products with alkyl-substituted phenols and also yields various quinonoid products.

Molecular oxygen may lead to resin formation or to mixtures of quinones, dimers, linear polyphenylene ethers (in the presence of CuCl–pyridine catalyst) and various C—C and C—O coupling products.

Other oxidants mentioned are potassium nitrosodisulphonate [$ON(SO_3K)_2$], Ag_2O, halogens, $FClO_3$, CrO_2Cl_2 and organic oxidizing agents (high oxidation potential quinones, etc.). Electrochemical methods and some other oxidants are also mentioned briefly.

Section III describes the hydrogenation of the aromatic ring to cycloalkanols with or without the simultaneous hydrogenolysis of the OH group and formation of an aromatic hydrocarbon.

366 references up to 1969.
Complementary: 27, 12.
Relevant: 17, 7; **27**, 9.

13, *11: Displacement of hydroxyl groups* by *G. W. Brown*

I. Introduction (594); **II**. Displacement of halogen (595); **III**. Displacement by nitrogen (623); **IV**. Displacement by hydrogen (626); **V**. Displacement by oxygen (627); **VI**. Displacement by sulphur (627); **VII**. Displacement by carbon (628); **VIII**. Survey of displacements by halogen and specific hydroxyl functions (629); **IX**. References (632).

The chapter surveys the synthetic and mechanistic aspects of the replacement of OH groups by other functional groups.

Section II describes displacements by the halogens (ROH → RHal). Thus iodine can be introduced by direct halogenation ($ROH + P + I_2 \rightarrow RI + H_3PO_3$), by the use of HI, by phosphonium salts (Arbusov reaction) or via formation of *p*-toluenesulphonates and displacement of the latter by iodide ions (ROH → ROTs → RI).

Bromine and chlorine can be introduced by similar methods and also by the use of PX_5 and POX_3 and other P-containing reagents, by $SOCl_2$ and $SOBr_2$, by N-containing reagents such as chlorinated enamines ($Cl_2C{=}CClNEt_2$) and by organic [$COCl_2$, $(COCl)_2$] and inorganic (NaX, LiX, CaX_2, etc.) halides.

Fluorine cannot be introduced directly, but HF, fluoramines, SF_4 and $PhSF_3$ are useful reagents.

Section III discusses the displacement of OH by NH_2 and NR_2 groups (Bucherer reaction of naphthols to naphthylamines in the presence of $NHSO_3$ and NH_3) and by azide ions.

Benzylic alcohols are reduced to the parent hydrocarbons by catalytic or chemical hydrogenolysis. Ph_3COH is reduced to Ph_3CH by concentrated formic acid, and in other special cases P_2S_3 or $LiAlH_4$ can be used as reducing agents (Section IV).

For exchange with labelled OH groups, see **13**, 15 and with alkoxy groups see **3**, 10.

S-containing groups can be introduced by reaction of SH-containing compounds ($ArOH + HSCH_2COOH \rightarrow ArSCH_2COOH$) and also by the use of P_2S_5, thiols or thiocyanates (Section VI).

Cyanide ions attack OH groups in propargyl alcohols with formation of cyanoallenes, and some aromatic esters of allyl alcohols can be alkylated by Grignard reagents:

$$CH_2{=}CHCH_2OCOAr \xrightarrow{RMgX} CH_2{=}CHCH_2R$$

although the usual reaction is removal of the acidic proton (Section VII).

Section VIII surveys displacements on propargyl alcohols (Table 1), on allylic, homoallylic and allenic alcohols, on cyclopropanols and cyclopropyl carbinols and finally on phenols.

257 references up to 1969.
Complementary: none.
Relevant: 13, 8.

13, *12: The dehydration of alcohols* by *H. Knözinger*

I. Introduction (642); **II**. Dehydration in solution (642); **III**. Dehydration in the gas phase (662); **IV**. Concluding remarks (705); **V**. References (796).

Dehydration of alcohols may yield either olefins or ethers ($RCH_2CH_2OH \rightarrow RCH{=}CH_2$; $2ROH \rightarrow ROR$).

Section II treats the acid-catalysed dehydration of alcohols in solution, its kinetics and mechanism (Tables 1 and 2) and stereochemistry (Table 3) and also base-catalysed dehydration in concentrated alkaline media, most often used for the production of dienes.

Section III discusses dehydrations in the gas phase, both non-catalysed and catalysed (Table 4), homogeneous and heterogeneous. In the latter, the structure and selectivity of the catalysts (Table 5) and their modes of action, kinetics and mechanisms are considered for the formation of both olefins and ethers.

Processes on silica–alumina, molecular sieves, thoria, ion-exchange resins and metals are described. Finally, the effects of the structure of the substrate on its reactivity are discussed (Table 10).

386 references up to 1969.
Complementary: 3, 10; **27**, 11, 16.
Relevant: 1, 2, 3; **3**, 4.

13, *13: Uncatalysed rearrangements involving the hydroxyl group* by *E. N. Marvell and W. Whalley*

I. Introduction (719); **II**. Sigmatropic rearrangements (720); **III**. The enolene rearrangement (743); **IV**. References (750).

Section II discusses reactions taking place by sigmatropic rearrangements, including [1, 3]sigmatropic shifts:

and thermal and photochemical keto–enol interconversions. [1, 5]sigmatropic shifts include thermal and photochemical dienol–enone interconversions:

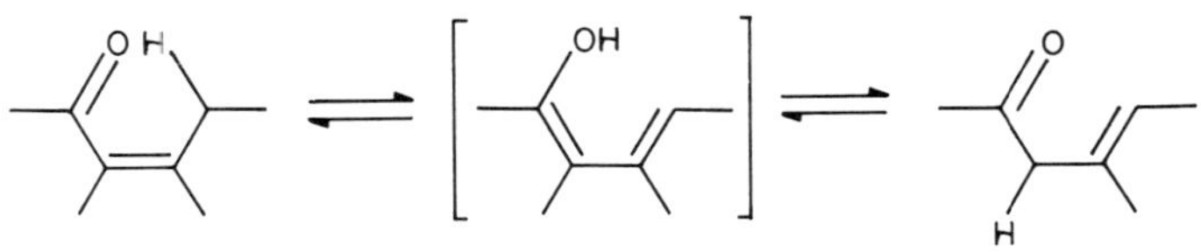

in which the dienol is only a transient species, however. The β-hydroxy olefin cleavage and the related decarboxylation of β,γ-unsaturated acids, the thermal aldol condensation and its reversal and the decarboxylation of β-keto acids also belong to this class. [1, 7]Sigmatropic shifts and [3, 3]sigmatropic shifts (e.g. the Cope rearrangement) are also discussed, considering the conservation of orbital symmetry rules of Woodward and Hoffmann.

Section III deals with the enolene rearrangement:

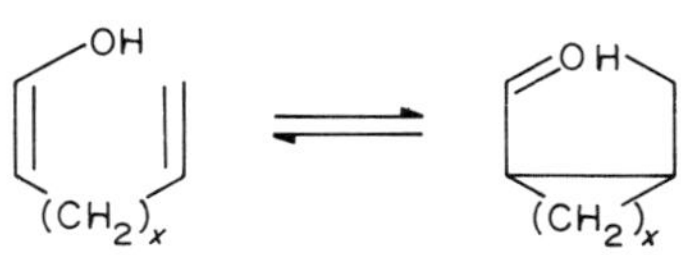

which is related to the Claisen reaction.

108 references up to 1970.
Complementary: **10**, 3; **27**, 1.
Relevant: **1**, 7; **2**, 15; **3**, 14; **4**, 10; **5**, 8; **8**, 9; **9**, 3; **10**, 3; **11**, 4; **12**, 16; **14**, 7; **15**, 8; **16**, 17; **17**, 8; **18**, 15; **19**, 18; **20**, 4; **23**, 13; **24**, 10; **26**, 15; **29**, 4.

13, *14: Biological formation and reactions of the hydroxyl group* by *S. K. Erickson, J. Schadelin, U. Schmeling, H.-H. Schott, V. Ullrich and Hj. Staudinger*

I. Introduction (756); **II**. Some concepts of enzymology and a list of abbreviations (757); **III**. Reduction of a carbonyl group (759); **IV**. Group transfer reactions (769); **V**. Aldol or ketol condensation reactions (766); **VI**. Addition of water at a double bond (768); **VII**. Isomerizations (769); **VIII**. Monooxygenation or mixed function oxygenation (722); **IX**. Acknowledgements (790); **X**. References (790).

The first two sections define the scope of the chapter and some general concepts. Section III deals with the reduction of C=O groups by alcohol dehydrogenase (Table 1). Section IV discusses reactions in which a group is transferred to an acceptor, resulting in the formation of an OH group. The enzymes which catalyse these processes are transferases (Table 2) and esterases (Table 3), each group including several sub-groups.

Section V deals with aldol or ketol condensations, catalysed by lyases (e.g. aldolase) which catalyse the reaction ketose-1-phosphate⇌dihydroxyacetone phosphate + aldehyde. Other condensations are also described (citrate, acetyl-CoA and sphingosine production).

Section IV discusses the reaction

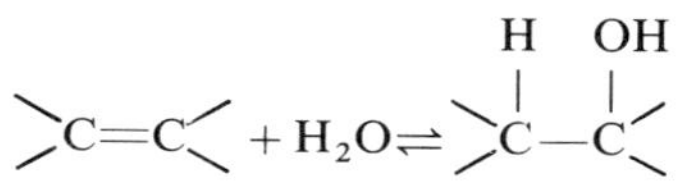

(Table 4) and Section VII isomerizations involving intramolecular oxidoreductions (triphosphate isomerase) or stereochemical changes (lactate racemase). Enzymatic hydroxylation (leading to dehydrogenation) occurs, utilizing O_2, catalysed by monooxygenases in mammals (Tables 6–9), in higher plants and in microorganisms (Table 10) (Section VIII).

229 references up to 1968.
Complementary: none.
Relevant: **2**, 7; **3**, 5; **4**, 9; **5**, 18; **7**, 14; **11**, 11; **12**, 12; **15**, 10; **16**, 13; **17**, 13; **18**, 13; **19**, 13; **20**, 6; **22**, 20; **24**, 14; **26**, 13; **28**, 17; **31**, 9.

13, *15: Syntheses and uses of* ^{18}*O-labelled hydroxylic compounds* by *S. Oae and S. Tamagaki*

I. Introduction (797); **II**. Oxygen exchange (799); **III**. Nucleophilic substitution (807); **IV**. Additions (818); **V**. Oxidations (820); **VI**. Reductions (822); **VII**. Rearrangements (824); **VIII**. References (830).

The main use of ^{18}OH-labelled compounds is for trace studies for the elucidation of reaction mechanisms.

Section II describes O exchange in alcohols (Tables 1 and 2) including mechanisms and stereochemistry, exchange reactions on solid surfaces (e.g. on Al_2O_3, Table 3) and also exchanges with phenols. Section III deals with nucleophilic substitutions ($RX + H_2{}^{18}O \rightarrow R^{18}OH + HCl$) and the influences of the leaving group and of the structure of the alkyl groups. Further, the hydrolysis of esters (Table 4) and of sulphonate, sulphate and phosphorus and other esters is treated.

Section IV discusses additions to C=C and C=O bonds and Section V oxidation with $KMnO_4$, peroxides, H_2O_2 and by O_2 enzymatically (Table 7).

Section VI treats reductions of carbonyl compounds to ^{18}O-labelled alcohols by using either $H_2{}^{18}O$ or labelled carbonyl compounds and finally Section VII deals with a variety of rearrangements involving ^{18}O-labelled esters, acyl halides, tertiary amine *N*-oxides, etc. (Tables 8 and 9).

93 references up to 1968.
Complementary: 27, 10.
Relevant: 5, 10; **12**, 13; **17**, 12; **18**, 18; **19**, 17; **22**, 14; **23**, 16; **24**, 20; **26**, 11; **27**, 10; **28**, 8; **29**, 28; **33**, 18; **36**, 12.

13, *16: Photochemistry of alcohols and phenols* by *H.-D. Becker*

Section II discusses the photolysis of alcohols, absorptions (Table 1), primary and secondary processes in the gas phase, photolyses in liquids (Tables 2–4) and low-temperature (75 K) studies. It also deals with photochemical dehydrogenation of alcohols by photoexcited compounds (Tables 5–7), by nitro and nitroso compounds or by photosensitized reductions of unsaturated compounds. Subsequently, photochemical addition reactions of alcohols to C≡C, C=C and C=N bonds and to *N*-heterocycles by free-radical routes are described and also ionic additions of alcohols to olefins, strained rings, nitrenes, carbenes, etc. Photosensitized autoxidation of alcohols (Table 8), photoinduced alcoholysis, acetalization, ketalization and esterification are also treated, with extensive reaction schemes and mechanistic discussions.

Section III describes photoenolization involving intramolecular (Table 9) or intermolecular hydrogen abstraction, reactions of enols with olefins (cycloadditions) or with molecular oxygen, and photoinduced rearrangements involving, e.g., the interchange of the positions of two atoms in a ring.

Section IV deals with the photolysis of phenols, electron ejection and homolysis, intermolecular proton transfer, intramolecular charge transfer, photoinduced reversible tautomerization of *ortho*-hydroxycarboxylic acids and carbonyl compounds (Table 11) and photochromic hydroxyl compounds (attributed to intramolecular hydrogen) transfer in the excited state).

Photolysis of bromo- and iodo-phenols (Table 12) usually leads to C—X bond cleavage and in phenolic acyl halides C—X cleavage, hydrogen abstraction and transfer and rearrangements occur.

Phenols are photochemically hydroxylated (Table 13) by H_2O_2 or O_2; they also undergo oxidative coupling ($2C_6H_5OH \rightarrow HOC_6H_4C_6H_4OH$, etc.) and other modes of oxidation. They also undergo intramolecular photoadditions to C=C bonds (Table 14) and sensitized additions to quinone methides (Table 15) and give various reactions with aromatic carbonyl compounds, intermolecular couplings and disproportionations.

305 references up to 1969.
Complementary: 27, 21.
Relevant: 2, 16; **3**, 8; **6**, 4; **8**, 12; **9**, 6; **11**, 5; **14**, 8; **15**, 9; **16**, 11; **17**, 9; **18**, 10; **19**, 20; **22**, 10; **23**, 9; **24**, 12; **25**, 11; **27**, 22; **28**, 5; **29**, 5, 6; **32**, 29; **33**, 21.

13, *17: The radiation chemistry of the hydroxyl groups* by *R. A. Basson*

The first three sections deal with the division of the subject, historical development, methodology and definitions.

Section IV discusses the stable end products obtained for primary alcohols, i.e. hydrogen, aldehydes and glycols (Tables 1 and 2) and (in low yields) hydrocarbons (Table 3) and some other minor products (Table 4). With branched alcohols, the product distribution is similar, although the differences are pronounced in the case of *t*-BuOH (Tables 5, 6).

Section V describes the intermediates occurring, including the yields, rate constants, structure and optical properties of solvated electrons (Tables 7–10), positive ions involved in ion–molecule reactions and in ion fragmentation (Tables 11, 12) and radicals (Tables 13 and 14). Finally, the concept of molecular products (e.g. H_2 and H_2O_2 formed in spurs) is explained.

Section VI deals with the general theory of reaction mechanisms, yields, dielectric constants and competitions between different processes and presents detailed general reaction schemes for primary and spur reactions.

Studies in vapour-phase radiolysis (Tables 16, 17) and in the solid state (Table 18) and the special characteristics of these are treated in Section VII. Sections VIII and IX deal with the effects of temperature (Table 19) and of the dose (Tables 20 and 21), respectively.

Linear energy transfer (LET) is the quantity used to describe the ionization density produced by a given radiation. The effect of LET on intermediates, products and yields is discussed in Section X (Tables 22–26), mainly with methanol and ethanol.

Section XI considers the effect of oxygen on the products (Table 27) and Section XII deals with the effect of acids, which increase the yields of H_2, glycols and some other products. Finally, Section XIII treats briefly some polyhydric alcohols (Table 30) and also some aryl-substituted alcohols and hydroxy acids.

144 references up to 1969.
Complementary: 27, 23.
Relevant: 9, 8; **10**, 6; **11**, 6; **15**, 9; **16**, 12; **17**, 10; **18**, 11; **19**, 10; **22**, 11; **25**, 12; **27**, 24; **29**, 7; **31**, 6; **32**, 9; **33**, 21; **36**, 8.

13, *18: Protection of the hydroxyl group* by *H. M. Flowers*

In polyfunctional compounds it is often necessary to protect one function, leaving others available for reaction, and to regenerate the original protected function at a later stage of the reaction.

Section II describes protection by converting OH groups to ethers which are stable. These can be alkyl, trityl, trimethylsilyl or tetrahydropyranyl ethers or also glycosides of sugars.

Protection as esters (Section III) can be made in the form of formates, trihaloacetates, acetates, benzoates, sulphonates, esters, tiglates, phenylcarbamates, carbalkoxy esters, carbonates, thiocarbonates or ortho esters. Inorganic esters (nitrates, sulphates, borates and phenylboronates) are also often prepared.

Section IV considers protection of OH groups by ring formation, i.e. epoxides, lactones and oxazolines.

Acid-catalysed formation of acetals from dihydroxy compounds with ketones and their cleavage are treated in Section V, and Section VI describes cases in which one OH group in a polyhydroxy compound can be protected selectively. Finally, indirect protection by chemical or biochemical conversion of an OH group to a carbonyl group (and, if necessary, protection of the latter) is described.

187 references up to 1968.
Complementary: none.
Relevant: 4, 11; **18**, 15.

13, *19: The mass spectra of hydroxyl compounds* by *R. G. Cooks*

I. Introduction (1045); **II**. General characteristics of the mass spectra of OH compounds (1046); **III**. Theoretical bases for the behaviour of OH compounds upon electron impact (1051); **IV**. Skeletal and hydrogen rearrangements (1059); **V**. Stereochemistry and the elimination of water (1064); **VI**. Compounds of natural origin (1067); **VII**. Derivatives of OH compounds (1071); **VIII**. New techniques for ion formation and analysis (1075); **IX**. References (1079).

Section II discusses the general features of the MS of OH compounds and the most characteristic fragmentation of aliphatic alcohols (α-cleavage:

$$R^1R^2R^3C\overset{+}{O}{}^{\bullet}H \longrightarrow R^2R^3C{=}\overset{+}{O}H + R^{1\bullet}).$$

as well as other routes in alcohols and phenols.

Section III treats the theoretical basis of the subject, including especially the quasi-equilibrium theory (QET) which is used as a basis for the discussion of rate processes.

Skeletal rearrangements which directly involve the OH group in bond-breaking and-making sequences and hydrogen migration (e.g., by ketonization of the C—OH) group) are treated in Section IV.

1, 4- and 1, 3-eliminations of H_2O from acyclic alcohols can be used for the elucidation of the stereochemistry of the reactions and has been used in the study of sterols, terpenols, etc. (Section V).

Section VI discusses natural products such as terpenes, steroids, saccharides, alkaloids, vitamins, colouring matters, etc., and some examples of structural determinations are considered.

To avoid some difficulties encountered in work with OH derivatives, in many cases derivatives such as trimethylsilyl ethers or acetates are used. These are easily prepared and usually parallels are observed between the behaviour of the OH compounds themselves and their derivatives as regards to the main reaction routes (α-cleavage, H_2O elimination and OH migration) (Section VII).

Finally, Section VIII describes new alternative techniques such as photoionization, field ionization and negative ion spectroscopy.

177 references up to 1970.
Complementary: **27**, 7.
Relevant: **9**, 7; **10**, 5; **16**, 14; **17**, 5; **18**, 6; **19**, 5; **22**, 7; **24**, 6; **25**, 4; **26**, 6; **29**, 3; **30**, 22; **31**, 3; **32**, 3; **33**, 4; **36**, 7.

13, *20: Hydroxide–alkoxide ion equilibria and their influence on chemical reactions* by *J. Murto*

I. Introduction (1088); **II**. Direct evidence for the existence of hydroxide–alkoxide equilibria (1089); **III**. Expressions for the equilibrium constants (1090); **IV**. Hydroxide–alkoxide equilibria in reaction kinetics (1091); **V**. Evaluation of hydroxide–alkoxide equilibrium constants (1100); **VI**. Relative reactivities of hydroxide and alkoxide ions (1109); **VII**. References (1112).

In the presence of alkali, two anion bases (OH^- and OR^-) are present in equilibrium in a water–alcohol mixture: $OH^- + ROH \rightleftharpoons RO^- + H_2O$. This is a proton transfer and usually attained instantaneously.

Section II describes the direct chemical and physical evidence available for the existence of the above equilibrium (reaction rates, products, conductances, etc.).

Section II reviews the thermodynamic equilibrium constant for the OH^-–RO^- equilibrium and the acid dissociation of alcohols ($ROH + H_2O \rightleftharpoons RO^- + H_3O^+$ (i.e. the autoprotolysis reaction).

Section IV considers the role of OH^-–RO^- equilibria in the reaction rates, thermodynamic functions of activation and reaction products of substitution reactions of the type $RX + OH^- \rightarrow ROH + X^-$, simultaneously with the reaction $RX + RO^- \rightarrow ROR + X^-$. Subsequently alkaline ester hydrolyses, eliminations and other reactions are treated similarly.

Section V describes the evaluation of OH^-–RO^- equilibrium constants by electrochemical methods, by indicator methods (Tables 1 and 2), by the reaction kinetic method and the calculation of the equilibrium constants and heats of the equilibrium reaction (Table 3). These values can be used in studies of reactivity correlations.

Section VI treats the problem of the relative reactivities of the OH^- and OR^- ions (Table 5). In many cases the reactivity of the OH^- ion is surprisingly low when compared with CH_3O^- and $C_2H_5O^-$, which often react several hundred times faster.

97 references up to 1968.
Complementary: none.
Relevant: none.

The chemistry of the azido group (1971)

14, *1: General and theoretical aspects* by *A. Treinin*

I. Introduction (2); **II.** Some structural properties of the azido group (14); **III.** Theoretical (34); **IV.** References (52).

Section I compares the properties of azides with those of the halides (Table 1) and justifies their inclusion in the group of pseudohalides, together with CN, NCO, SCN, etc. It gives a simplified model of the electronic structure (σ- and π-bonds and lone pairs), describes primitive Hückel-type calculations, the structure of the group and some relevant data on N atoms and N—N bonds (Table 5).

Section II discusses geometry (Table 6), thermodynamic data (Tables 7 and 8) dipole moments (Table 9), vibration spectra (Table 10), force constants (Table 11), internal rotation, electronic spectra (Tables 12–14), ORD and CD, quadrupole coupling and NMR (Table 15).

Section III describes the theoretical treatments of the azido group, the symmetry properties of the minimal basis set (Table 16), Walsh's correlation diagram, self-consistent Hückel molecular orbitals, semi-empirical self-consistent field methods and LCAO self-consistent field (Hartree–Fock) molecular orbitals.

110 references up to 1969.
Complementary: none.
Relevant: 1, 1; **2**, 1; **3**, 1; **4**, 1; **5**, 1; **6**, 1; **8**, 1; **11**, 1; **12**, 1; **13**, 1; **15**, 1; **16**, 1; **17**, 1; **18**, 1; **19**, 1; **20**, 1; **22**, 1: **23**, 1; **24**, 1; **25**, 1; **26**, 1; **27**, 9; **28**, 1; **31**, 27; **32**, 1; **33**, 1; **36**, 2.

14, *2: Introduction of the azido group* by *M. E. C. Biffin, J. Miller and D. B. Paul*

I. Introduction (58); **II.** Toxic and explosive properties of azides (61); **III.** Syntheses involving nucleophilic substitution reactions of azide ion (63); **IV.** Formation of azides by addition reactions (119); **V.** Synthesis of azides from diazotization and related reactions (147); **VI.** Miscellaneous syntheses (176); **VII.** References (177).

Section II describes the toxic and explosive properties of azides and draws attention to the safety measures which have to be taken in handling them.

Section III begins with a discussion of S_N reactions and the mechanistic aspects of these reactions involving N_3^- ions. The syntheses of aliphatic and alicyclic azides are treated (Table 1) followed by those of vinyl, allyl, acyl and carbamoyl azides. Epoxides with N_3^- yield azido alcohols:

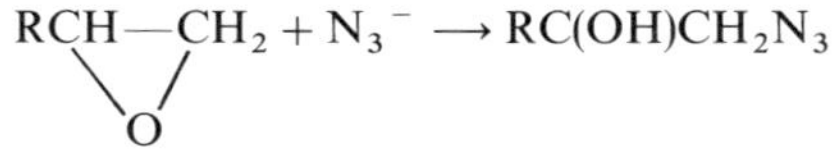

(Table 2) and formation of C—N_3 bonds is widely employed in the steroid (Section III.B.5) and carbohydrate (III.B.6) fields.

Aryl azides can be synthesized by S_NAr reactions (Table 4) in homocyclic, heterocyclic and non-benzenoid aromatic (e.g. tropone) systems.

Section IV deals with the addition of HN_3 to olefins [$CH_2{=}CHMe + HN_3 \rightarrow CH_3CH(N_3)Me$], conjugated olefins (Table 5), allenes, quinones, alkynes, ketenes and isocyanates. Trimethylsilyl azide and mercury$^{(II)}$ azide add to aldehydic carbonyl groups and to C=C double bonds, respectively, to yield azides (Tables 6 and 7).

Halogen azides give electrophilic addition to olefins (Tables 8 and 9), to conjugated olefins (Table 10) and to alkynes [$PhC{\equiv}CMe + IN_3 \rightarrow PhCI{=}C(N_3)Me$].

In solvents of low polarity and in the presence of light, halogen azides add to double bonds by a free-radical mechanism and the same occurs also in the presence of an organic peroxide–Fe^{2+} system (Fenton reaction).

Azides are also obtained from the reactions of diazonium compounds with a variety of N-containing nucleophiles (Section V):

$$PhN_2{}^+Br_3{}^- + NH_3 \longrightarrow PhN_3 + 3HBr$$

An analogous reaction occurs with hydrazine:

$$PhN_2{}^+Cl^- + H_2NNH_2 \longrightarrow [PhN{=}NNHNH_2] \longrightarrow PhN_3 + NH_3$$

and also with hydroazoic acid or with azides:

$$ArN_2{}^+ + N_3{}^- \longrightarrow ArN_3 + N_2$$

(Table 11). Aromatic nitroso compounds with azide ions also yield azides (Table 14) in a complicated reaction requiring two HN_3 molecules. Azides are also formed in moderate yields in a variant of the diazo transfer reaction (Table 16), and also by nitrosation of hydrazines:

$$ArNHNH_2 \xrightarrow{NO^+} ArN(NO)NH_2 \xrightarrow{-H_2O} ArN_3$$

(Table 18). Electrolytic and oxidative syntheses are briefly mentioned in Section VI.

483 references uo to 1968.
Complementary: none.
Relevant: none.

14, *3: Characterization and determination of organic azides* by *J. E. Gurst*

Qualitative analysis of azides is difficult and no generally valid method is available. Evolution of N_2 in the presence of H_2SO_4 has been reported, but is not reliable.

The toxic and explosive nature of many azides renders their quantitative determination difficult and most methods are based on evolution of N_2 by reaction with HI (Section II).

Section III discusses IR and Section IV UV spectroscopy. ORD and CD are described in Section V and mass spectrometry in Section VI.

82 references up to 1969.
Complementary: **32**, 3.
Relevant: **1**, 5; **2**, 8; **3**, 15; **4**, 3; **5**, 17; **8**, 3; **12**, 4; **13**, 6; **15**, 3; **16**, 3; **17**, 4; **18**, 5; **20**, 3; **22**, 5; **23**, 5; **24**, 5; **25**, 10; **26**, 5; **28**, 3; **29**, 21; **30**, 14–22; **36**, 5.

14, *4: Directing and activating effects of azido groups* by *M. E. C. Biffin, J. Miller and D. P. Paul*

Section II discusses polarization and polarizability effects, compares ionization constants of halo- and azido-substituted acids (Table 1) and gives σ, σ_m, σ_p, and σ_p^+ values (Tables 2 and 4) and ionization constants of azidobenzoic acids and azidoanilines (Table 3). The N_3 group may be either electron donating or attracting, depending on the demands of the approaching reagent. Electrophilic substitution on azidoarenes is discussed (Table 5) and activation with *o–p* direction is observed. Nucleophilic substitution in aromatic systems is weakly activated by N_3 (Tables 7 and 8).

Section III describes the effect of N_3 groups in additions to C=C double bonds of unsaturated azides and in elimination from iodo-substituted aliphatic azides to yield unsaturated azides.

In all cases the mechanisms are discussed in detail.

52 references up to 1969.
Complementary: none.
Relevant: **3**, 3; **4**, 5; **5**, 16; **6**, 8; **7**, 2, 3; **11**, 12; **12**, 5; **13**, 8; **15**, 5; **16**, 6; **18**, 9; **19**, 8; **23**, 11; **28**, 9; **29**, 16; **31**, 8; **36**, 15.

14, *5: Decomposition of organic azides* by *R. A. Abramovitch and E. P. Kyba*

Five major modes of decomposition are discussed and in each section alkyl, aryl and sulphonyl azides are treated separately.

Section II deals with acid-catalysed decomposition with protonic acids, when a mixture of products is obtained:

$$PhCH_2N_3 \xrightarrow{H^+} PhCH{=}NH + PhN{=}CH_2 + PhCH_2NH_2 + PhCH_2OH$$

Migration of protons and alkyl groups occurs simultaneously. Mechanisms and migratory aptitudes are treated for alkyl and aryl azides. Aromatic sulphonyl azides in the presence of benzene yield anilines and arenesulphonic acids.

Lewis acids have been less studied and nearly every individual azide gives different products.

Alkyl azides decompose thermally by a variety of routes, e.g.

$$CH_3N_3 \xrightarrow{\Delta} N_2 + CH_3N + CH_2 + HN_3$$

and with more complex azides the products are more varied and include aziridines and other heterocyclic compounds, migrations yielding rearranged products, etc.

Aryl azides give an even wider variety of products; thus, PhN_3 in the presence of aniline yields a ring-expanded product, 2-anilino-3*H*-azepine. However, the primary thermolytic process seems to be the formation of aryl nitrenes by loss of N_2 from aryl azides. In continuation, decompositions in the presence of unsaturated, saturated and aromatic compounds and assistance of azide decompositions by *ortho*-substituents (especially NO_2, leading to benzofuroxanes) are treated, followed by decompositions in the presence of nucleophiles and of radicals. The decomposition of sulphonyl azides in the presence of the same classes of compounds as with aryl azides is described (Section III).

Section IV describes decompositions with trivalent phosphorus compounds, usually yielding phosphinimines: $EtN_3 + Ph_3P \rightarrow EtN_3 \cdot PPh_3 \rightarrow EtN{=}PPh_3$. Analogous reactions take place with aryl azides and sulphonyl azides: $RSO_2N_3 + (R'O)_3P \rightarrow RSO_2N{=}P(OR')_3$.

Photolytic decompositions were studied both in the vapour phase and in solutions. Many reactions occur through nitrenes and involve rearrangements, formation of azomethine derivatives, azo compounds, heterocycles and, in the case of sulphonyl azides, polymeric materials, sulphonamides and their derivatives (Section V).

Section VI describes transition metal-catalysed decompositions, again leading to a wide variety of products (complexes, amino compounds, nitrene–metal–carbonyl complexes, aziridines, triazoles, tetrazoles) and, in the case of sulphonyl azides sulphonamides and also a variety of cyclic products.

270 references up to 1969.
Complementary: 32, 7, 8.
Relevant: none.

14, *6: Azides as synthetic starting materials* by *T. Sheradsky*

I. Introduction (332); **II.** Transformation of azido groups into other functional groups (333); **III.** Azides as starting materials in syntheses of heterocycles (358); **IV.** References (389).

Azides are most useful synthones, since they undergo facile electrophilic and nucleophilic attacks, cycloadditions, decomposition and reductions.

Section II describes the preparation of amines from azides by direct reductions ($RN_3 \xrightarrow{H} RNH_2 + N_2$), by acidolysis (e.g. $ArN_3 + HBr + AcOH \rightarrow ArNH_2$, when the formation of ring-brominated products is prevented by the addition of a bromine acceptor such as PhOH or acetone) and by thermolytic or photolytic nitrene formation followed by hydrogen abstraction from the environment. Azomethines are obtained by acidolysis yielding mixtures of the various possible (also rearranged) products:

$$RR^1CHN_3 \xrightarrow{H^+} RR^1C{=}NH + RCH{=}NR^1 + R^1CH{=}NR$$

or by rearrangement of nitrenes or by decomposition of adducts obtained by 1,3-dipolar cycloadditions with alkenes, followed by N_2 elimination from the primary 1,2,3-triazoline product.

Terminal vinyl azides yield nitriles on decomposition (via rearrangement of the primary nitrene product) and reaction with CO (catalysed by Ir or Rh complexes) gives isocyanates ($ArN_3 \rightarrow ArNCO + N_2$). Diazo compounds are formed by diazo transfer or by cleavage of azide adducts and azo compounds either by diazo transfer or by dimerization of nitrenes ($ArN_3 \rightarrow ArN \rightarrow ArN{=}NAr$). Syntheses of iminophosphoranes and -sulphuranes and of triazenes are also mentioned.

Section III discusses the syntheses of heterocycles from azides. Aziridines are obtained by additions of azides to C=C bonds, azirines by thermal or photochemical decomposition of azides, five-membered rings with one N atom by intramolecular nitrene insertion and some related methods and five-membered rings with two N atoms (imidazoles, pyrazoles) or containing both N and O atoms (oxazoles, isoxazoles, oxadiazoles) by related cyclizations.

1, 2, 3-Triazolines are obtained by reaction of azides with olefins or with sulphur ylides and 1, 2, 3-triazoles by addition of azides to olefins followed by elimination, by addition to C=C bonds, by nucleophilic attacks on azides followed by cyclization or by decomposition of *o*-azidoazobenzenes. Syntheses of tetrazoles and of six- and seven-membered rings are also treated briefly.

253 references up to 1969.
Complementary: none.
Relevant: 14, 5, 9, 10; **32**, 7, 8.

14, *7: Rearrangements involving azido groups* by *D. V. Banthorpe*

Throughout the chapter emphasis is put on the mechanistic aspects of the reactions.

Section II describes the Curtius rearrangement to yield amines [$RCON_3 \xrightarrow{(R'OH)} RNCO(+RNHCOOR') \xrightarrow{H+} RNH_2$]. This reaction can be carried out thermally, photochemically or under acid catalysis, and may involve nitrenes or, more probably, concerted migration and N_2 loss.

The Schmidt reaction includes interactions of HN_3 with various compounds and is closely related to the Curtius rearrangement. With carboxylic acids or ketones, amines are formed ($RCOR' + HN_3 \rightarrow RNHCOR' \rightarrow RNH_2 + R'COOH$). Various side-reactions may produce amides, tetrazoles or oxazoles. Under the same conditions cyanides yield tetrazoles and acetylenes give triazoles (Section III).

Acid-catalysed rearrangements usually produce azomethine compounds ($R_2CHN_3 \xrightarrow{H+} RCH{=}NR$). With cyclic azides, this often leads to ring expansion (Section IV), and very similar reactions also occur in thermal reactions of alkyl and aryl azides (Section V). Photochemical rearrangements (Section VI) probably involve triplet nitrene intermediates, e.g.

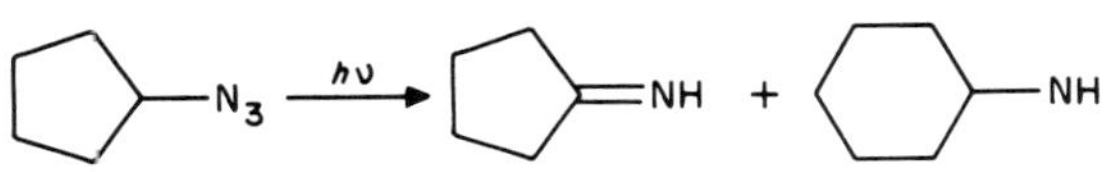

Section VII treats rearrangements in which azides are the products (e.g. from diazonium salts and HN_3) and finally the reactions of some acyclic and cyclic 1, 2, 3-triazenes which may be considered (with some licence) as substituted azides, e.g.

$$PhNHN{=}NPh \rightleftharpoons p\text{-}H_2NC_6H_4N{=}NPh$$

or skeletal rearrangement of *C*-aminotriazoles with ring opening and interchange of amino groups.

238 references up to 1969.
Complementary: **14**, 8; **32**, 7, 8.
Relevant: **1**, 7; **2**, 15; **3**, 14; **4**, 10; **5**, 8; **8**, 9; **9**, 3; **11**, 4; **12**, 16; **13**, 13; **15**, 8; **16**, 17; **17**, 8; **18**, 15; **19**, 18; **20**, 4; **23**, 13; **24**, 10; **26**, 15; **29**, 4.

14, *8: Photochemistry of the azido group* by *A. Reiser and H. M. Wagner*

Photoreactions of organic azides often involve nitrene intermediates. Section II and III describe the spectra (Table 1) and excited states of alkyl, aryl and acyl azides, the quantum yield (Tables 2 and 3) and the mechanisms of direct and sensitized photolysis.

Section IV discusses the electronic structure and spectra of nitrenes (Table 4), their general reactions such as recombination to form azo compounds, insertion into C—H, O—H and N—H and hydrogen abstraction, addition to C=C bonds, 1,3-cycloaddition and electrophilic attack on non-bonding electron pairs, describing in each case the mechanisms and products of these processes.

Section V deals with photoreactions of alkyl, vinyl, acyl and aryl azides and gives detailed schemes relating to the mechanism and the products of the reactions. Among acyl azides, azidoformates, acetyl, pivaloyl and benzoyl azides are treated in detail, including cases in which Curtius rearrangement to isocyanates does or does not occur.

144 references up to 1968.
Complementary: none.
Relevant: **2**, 16; **3**, 8; **6**, 4; **8**, 12; **9**, 6; **11**, 5; **13**, 16; **15**, 9; **16**, 11; **17**, 9; **18**, 10; **19**, 20; **22**, 10; **23**, 9; **24**, 12; **25**, 11; **27**, 21, 22; **28**, 5; **29**, 5, 6; **32**, 29; **33**, 21.

14, *9: Acyl azides* by *W. Lwowski*

Section I describes the preparation and properties of alkanoyl, aroyl, alkenoyl, thiocarbonyl and imidoyl azides, the last two displaying ring–chain tautomerism:

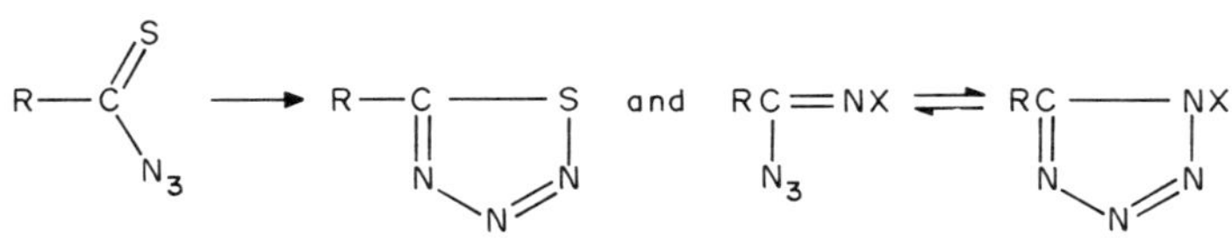

In both cases the cyclic tautomers are more stable, although some can also be isolated in the open-chain form, such as azidoximes [RC(N)$_3$=NOH], cyanoguanyl azides [RR′NC(N$_3$)=NCN], hydrazidic azides [RC(N)$_3$=NNHAr] and azidobenzalazines [ArC(N$_3$)=NN=CHAr]. Subsequently, the preparation, reactions, dissociation, Curtius rearrangement and nitrene reactions of carbamoyl azides of the type RR′NCON$_3$ are treated, and briefly also azidoformates ($ROCON_3$) and mercaptocarbonyl azides ($RSCON_3$).

Section II discusses reactions which are common to all classes of acyl azides. These include 1,3-dipolar additions to double and triple bonds yielding triazolines and triazoles, respectively, additions by nucleophilic attack on the terminal N atom of acyl azides yielding iminophosphoranes with phosphines ($R'N_3 + R_3P \rightarrow R_3P{=}NR'$), related attacks on the same N atom by ylids or by Grignard compounds, nucleophilic displacement of the whole N_3 group by alkoxycarbonyl groups (e.g. in peptide synthesis) or by carbanions, reactions with acids (Curtius, etc.) and radical-induced decompositions which may occur by a chain mechanism and yield amides by dehydrogenation of alcohols present as solvents ($RCON_3 + Me_2CHOH \longrightarrow RCONH_2 + Me_2CO$).

Finally, the formation of nitrenes from acyl azides is discussed in detail, dealing with the dissociation of acyl nitrenes, with insertion into C—H, O—H and N—H bonds, with additions to C=C, C≡C, C=N, N=C and SO groups and to aromatic and heterocyclic systems.

369 references up to 1970.
Complementary: 14, 8; **32**, 8.
Relevant: 14, 2.

14, *10: The Chemistry of vinyl azides* by *G. Smolinsky and C. A. Pryde*

Vinyl azides can be prepared by addition HN_3 to acetylenediacarboxylic esters [$RO_2CC{\equiv}CCO_2R + HN_3 \rightarrow RO_2CC(N_3){=}CHCO_2R$] or, better, by additions of IN_3 or BuN_3 to alkynes [$RC{=}CR' + IN_3 \rightarrow RCI{=}C(N_3)R'$]. The most general synthesis is the elimination of HX from azidohaloalkanes ($XCH_2CH_2N_3 \xrightarrow{KOH} CH_2{=}CHN_3$). Other methods include the addition of N_3^- to β-chlorovinyl ketones, followed by elimination of Cl^- and treatment of phosphoranes with acetyl azide or with sodium azide (Section II).

The most important reactions of vinyl azides (Section III) involve loss of N_2 and reorganization of the molecular residue. Thus, vinyl azides yield aziridines by pyrolysis or photolysis:

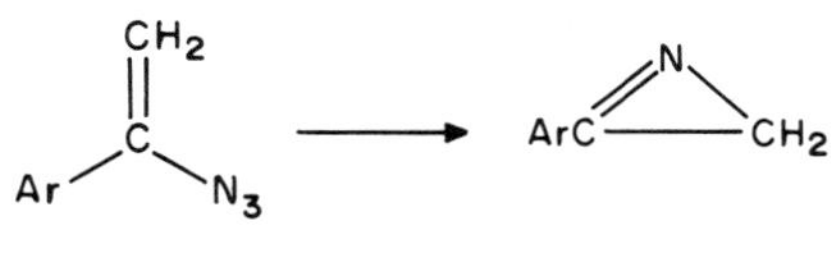

but other cyclic products, e.g. indoles, are also frequently obtained. Vinyl azides containing strongly electron-withdrawing substituents (e.g. CF_3) at the double bond are unstable and decompose at much lower temperatures (room or below) either to azirines or to other products. Vinyl azides were postulated to be intermediates in various reactions forming furoxans, nitriles, azidoketenes, etc. Reactions in which all three N atoms of the azide group are retained also occur and may yield polymers in addition to triazoles.

Section IV deals with the mechanisms of the reactions discussed under various conditions and emphasizes the nitrene route.

Finally, Section V treats the rearrangements of azidoquinones, in which expulsion of N_2

is usually followed by C—C bond rupture and often involve either ring contraction or expansion.

49 references up to 1969.
Complementary: 32, 8.
Relevant: 14, 7.

The chemistry of acyl halides (1972)

15, *1: General and theoretical aspects* by *M. Simonetta and P. Beltrame*

Section II describes the geometry of acyl halides, including data from X-ray diffraction, electron diffraction (Table 1) and microwave spectroscopy (Table 2).

Section III discusses thermodynamic properties and bond energies (Tables 3 and 4), ionization and appearance potentials (Table 5), dipole moments (Table 6) and polarity.

Section IV deals with rotational (Table 7), vibrational (Tables 8–13), electronic (Tables 14 and 15), NMR (Tables 16–18) and mass spectra.

Section V deals wtih theoretical aspects and the results of various MO calculations.

147 references up to 1970.
Complementary: 25, 1.
Relevant: 1, 1; **2**, 1; **3**, 1; **4**, 1; **5**, 1; **6**, 1; **8**, 1; **11**, 1; **12**, 1; **13**, 1; **14**, 1; **16**, 1; **17**, 1; **18**, 1; **19**, 1; **20**, 1; **22**, 1; **23**, 1; **24**, 1; **25**, 1; **26**, 1; **27**, 1; **28**, 1; **31**, 27; **32**, 1; **33**, 1; **36**, 2.

15, *2: Preparation of acyl halides* by *M. F. Ansell*

Acyl halides are most frequently prepared from the corresponding acids or anhydrides with a variety of reagents. Even when starting from acids, anhydrides are involved:

$$RCOOH + SOCl_2 \longrightarrow RCOCl$$

$$RCOCl + RCOOH \longrightarrow (RCO)_2O \xrightarrow{SOCl_2} RCOCl$$

The reagents include thionyl chloride (Tables 1 and 2), PX_3, PX_5 and POX_3, high-boiling acyl halides (with continuous removal of the formed lower boiling product, sulphonyl halides, α-chloro ethers, halogen acids, imidoyl halides (Table 3), benzotrichloride, SF_4 and sometimes $SiCl_4$.

Acyl derivatives may also be starting materials (Section III). Thus new acyl halides may be prepared by halogen exchange from acyl chlorides (e.g. $RCOCl + HBr \rightarrow RCOBr$), from acyl hydrazides with HCl or $SOCl_2$ ($RCONHNH_2 \xrightarrow[\text{several steps}]{SOCl_2} RCOCl$) and also from esters in special cases [$RCOOC(=CH_2)CH_3 \rightarrow RCOX + Me_2CO$].

The aldehydic proton can be replaced by halogen in free-radical reactions ($ArCHO + Cl_2 \rightarrow ArCOCl$) (Section IV) and halocarbonyl (COX) groups can be directly introduced into alkanes, alkenes or arenes with oxalyl chloride or $COCl_2$.

$$CH_3CH{=}CH_2 + COCl_2 \longrightarrow ClCH_2CH(CH_3)COCl$$

or by a radical chain reaction involving CCl_4 and CO (Section V). Ketenes yield acyl halides with halogen acids or haloacyl halides with molecular halogens (both by addition to the C=C double bond) and similar additions or alternatively from diketene and HX or X_2 (Section VI).

Some less general methods are briefly mentioned in Section VII.

163 references up to 1970.
Complementary: none.
Relevant: 5, 9; **8**, 13; **26**, 8.

15, *3: Detection, determination and characterization* by *H. Weiler-Feilchenfeld*

Section I describes chemical methods of detection (e.g. hydrolysis, hydroxamic acid test), separation, characterization (especially by solid derivatives and by gas chromatography) and determination by titration (conversion to amides, hydrolysis, esterification), by spectrophotometric methods and by gas chromatography.

Section II deals with physical methods, such as IR and UV spectroscopy (Tables 1–9), and also with electric dipole moments (Table 10), nuclear quadrupole resonance, magnetic susceptibility and the Faraday effect.

190 references up to 1969.
Complementary: 25, 10.
Relevant: 1, 5; **2**, 8; **3**, 15; **4**, 3; **5**, 17; **8**, 3; **12**, 4; **13**, 6; **14**, 3; **16**, 3; **17**, 4; **18**, 5; **20**, 3; **22**, 5; **23**, 5; **24**, 5; **25**, 10; **26**, 5; **28**, 3; **29**, 21; **30**, 14–22; **36**, 5.

15, *4: Acid–base behaviour and complex formation* by *D. P. N. Satchell and R. S. Satchell*

Acyl halides may behave either as Lewis bases or Lewis acids, depending on their structure and the nature of the substituents in the molecule.

Section II discusses basic behaviour, types of complexes formed with acids (Table 1), their IR (Table 2), NMR and X-ray spectra and the structure and stability of the complexes (Table 3) with HX, X_2 or metal ions.

Acid behaviour is described in Section III, including isolatable complexes (Table 4) and the role of complexes as intermediates in basic systems, e.g. in the course of the tertiary amine-catalysed preparation of acid anhydrides from acyl halides and carboxylic acids.

Some acyl halides are used as solvents ($COCl_2$, CH_3COCl, C_6H_5COCl and C_6H_5COF, Table 5). These dissolve various metal halides (Table 6) in the form of ionic species. Tertiary nitrogen bases (Et_3N, pyridine) also dissolve in acyl halides, possibly but not necessarily by complex formation. Reactions between acids and bases in acyl halide solutions have been studied (Section IV).

85 references up to 1970.
Complementary: none.
Relevant: 2, 9; **3**, 6; **4**, 4; **5**, 6; **7**, 6; **8**, 5; **11**, 3; **12**, 3, 10; **13**, 7; **16**, 5; **17**, 6: **18**, 8; **19**, 7; **20**, 12; **22**, 8; **23**, 6, 7; **24**, 4; **25**, 5, 6; **29**, 17; **31**, 17, 21; **32**, 21; **33**, 5.

15, *5: Directing and activating effects of COX groups* by *P. H. Gore*

I. Introduction (138); **II**. Directing and activating effects of COX groups in aliphatic carbonyl halides (141); **III**. Directing and activating effects of COX groups in aromatic carbonyl halides (157); **IV**. Effects of COX substituents on the aryl groups in ω-arylalkane carbonyl halides (166); **V**. Substituent effects of COX groups in nuclear magnetic resonance spectroscopy (169); **VI**. References (172).

In the COX groups the carbonyl oxygen powerfully attracts electrons by inductive ($-I$) and mesomeric ($-M$) effects. These are modified by the $-$I and $+$M effects of the halogen atoms. In benzoyl halides electrophilic substitution will occur predominantly at the *meta* positon, with deactivation.

Section II describes fluorination (vapour-phase and electrolytic), chlorination [ionic in the presence of halogen carriers such as I_2 or PCl_5, or free-radical and mainly photolytic (Tables 2–11), ionic and free-radical brominations (Tables 12–14), chlorocarbonylations, sulphoxidations and sulphochlorinations (Table 15) and eliminations and solvolyses (Table 16).

Section III discusses reactions in aroyl halides, such as light-catalysed addition of Cl_2 to PhCOCl to yield $C_6H_5Cl_6COCl$, nitrations, chlorinations and brominations of aroyl chlorides, side-chain reactions such as Friedel–Crafts acylations of arenes with substituted benzoyl halides (Table 17) and the kinetics of solvolytic reactions of phthaloyl, isophthaloyl and terephthaloyl chlorides (Table 18) and of the reductions of COX-substituted nitrobenzenes (Table 19).

Section IV deals with substitutions and self-condensation in ω-arylalkane carbonyl halides such as the oligomerization of PhCH, COCl, catalysed by $AlCl_3$:

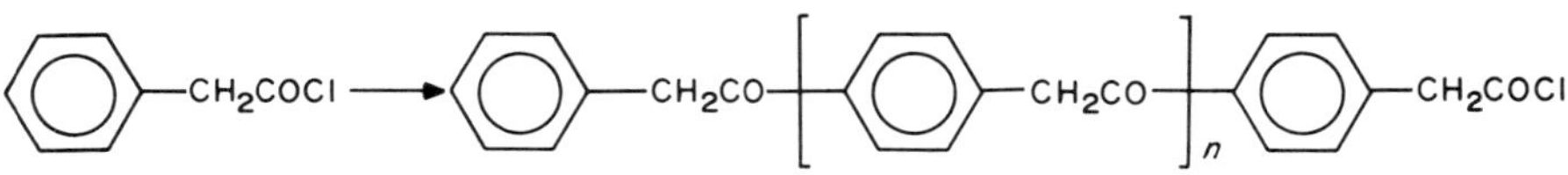

Effects of the COX groups in NMR spectra are treated in Section V (Tables 20–23).

126 references up to 1971.
Complementary: 5, 16.
Relevant: 3, 3; **4**, 5; **5**, 16; **6**, 8; **7**, 2, 3; **11**, 12; **12**, 5; **13**, 8; **14**, 4; **16**, 6; **18**, 9; **19**, 8; **23**, 11; **28**, 9; **29**, 16; **31**, 8; **36**, 15.

15, *6: Mechanisms of substitution at the COX group* by *A. Kivinen*

I. Introduction (178); **II**. Reactions with water, alcohols and phenols (179); **III**. Reactions with amines (210); **IV**. Solvolytic reactions catalysed by tertiary amines (219); **V**. Comparison of reactivities of COF, COCl, COBr and COI groups (221); **VI**. Acknowledgement (221); **VII**. References (222).

Reactions of COX groups of the type RCOX + Y → RCOY + X are widely employed in organic synthesis.

Section II discusses the scope, rates and mechanisms of reactions acyl chlorides with H_2O, ROH and ArOH, including solvent, substituent and salt effects (Tables 1–10), heat capacities of activation, the effect of pressure (Table 11), oxygen exchange (Table 12), kinetic solvent and secondary deuterium isotope effects (Tables 13 and 14), reactions in special solvents (CH_3CN, CH_3NO_2, SO_2, HCOOH), acid catalysis and reactions with HO^-, RO^- and ArO^- ions. Subsequently, reactions of acyl fluorides, bromides and iodides (Table 15) are treated and finally the mechanisms of the unimolecular solvolysis (involving probably the route $RCOX \xrightarrow{\text{slow}} RC^+{=}O \xrightarrow[H_2O]{\text{fast}}$ products) and of the more frequent bimolecular route (involving either a synchronous or an addition–elimination mechanism) are discussed.

Section III treats in detail the reactions of acyl chlorides with amines, including kinetics (Tables 16–18), substituent effects, isotope effects, acid catalysis (Table 19) and the effect of the basicity of amines. Acyl fluorides, bromides and iodides (Tables 20 and 21) are treated more briefly and so also are the mechanisms of these reactions.

Tertiary amines strongly catalyse the solvolyses of acyl halides. Formation of a quarternary amine adduct occurs. The catalytic effect is exerted by both tertiary aliphatic (Et_3N, etc.) and aromatic (pyridine) amines (Section V).

281 references up to 1970.
Complementary: none.
Relevant: 5, 9; **10**, 1.

15, *7: Reduction* by *O. H. Wheeler*

I. Introduction (231); **II**. Catalytic reduction (232); **III**. Metal hydride reduction (235); **IV**. Reduction by dissolving metals (243); **V**. Miscellaneous methods (246); **VI**. References (247).

Acyl halides may be reduced to alcohols ($RCOX \rightarrow RCH_2OH$) or to the corresponding aldehydes (RCOX → RCHO) and by bimolecular reductions also to acyloins (RCOCHOHR′) or to glycols [RCH(OH)CH(OH)R′].

Section II discusses the catalytic hydrogenation to aldehydes (Rosenmund reduction), which in some cases may also produce alcohols or hydrocarbons as by-products:

$$RCOX \xrightarrow{H_2} RCHO \xrightarrow{H_2} RCH_2OH \xrightarrow{H_2} RCH_3$$

and in some cases may give decarbonylation [ArCOCl → ArH, when $Ar = CH_3OC_6H_4$— or 3,4,5-$(CH_3O)_3C_6H_2$—].

$LiAlH_4$, $NaBH_4$ and similar metal hydrides yield the corresponding carbinols whereas, e.g., $LiAl(OBu)_3H$ yields only the aldehyde; other selective catalysts have been discovered and studied (Section III).

The use of metals dissolving in various hydrogen donors is limited by the occurrence of unwanted side-reactions, and in most cases is used only for bimolecular reductions (Section IV).

Some special and non-general methods are mentioned in Section VI.

160 references up to 1968.
Complementary: none.
Relevant: 2, 11; **9**, 4; **12**, 7; **19**, 15.

15, *8: Rearrangements involving acyl halides* by *D. V. Banthorpe and B. V. Smith*

Half-ester chlorides (i.e. compounds which have both a COX and a COOR functional group, e.g. $RO_2CCH_2CR_2COCl$) are often used for the preparation of keto esters by treatment with organozinc or organocadmium compounds:

$$R_2Cd + ClCOCH_2CO_2Et \rightarrow RCOCH_2CO_2Et$$

The usefulness of this reaction is limited by rearrangement of the half-ester chloride. Thus, from $ClOCCRR'CH_2CH_2CO_2Me$ derivatives of its rearranged isomer $MeO_2CCRR'CH_2CH_2COCl$ are obtained. The mechanism is not clear, although a cyclic intermediate is probably involved (Section II).

Unsaturated acyl halides undergo positional isomerization by migration of the double bond, e.g.

$$\underset{}{MeCH{=}CHCOCl} + ROH \xrightarrow{R_3N} \underset{\text{(major)}}{CH_2{=}CHCH_2CO_2R} + \underset{\text{(minor)}}{MeCH{=}CHCO_2R}$$

Geometric isomerizarion interconverts the dichloride of maleic acid to that of fumaric acid and *vice versa* (Section III). Reductions of diacid chlorides often occur with rearrangements, e.g. (Section IV).

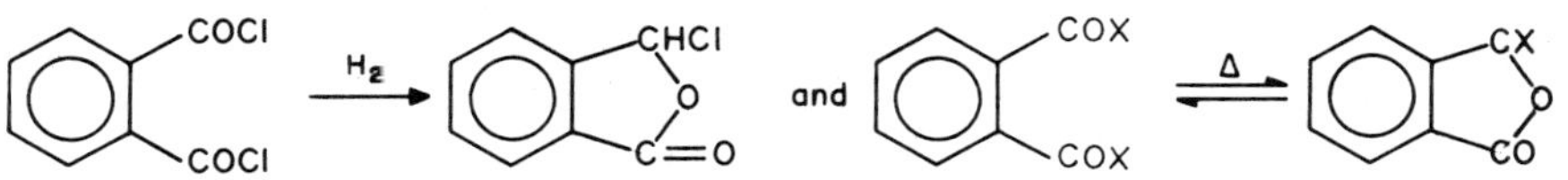

Rearrangements occuring during Friedel–Crafts reactions of dibasic acid chlorides are discussed in Section V, e.g. phthaloyl chloride gives products derived from the rearranged

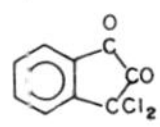 form. Other reactions of acyl halides occur with hydride or alkyl shifts, e.g.

$$Me_3CCOCl \xrightarrow{AlCl} Me_3C\overset{+}{C}O \longrightarrow Me_2\overset{+}{C}COMe \xrightarrow{H^-} Me_2CHCOMe$$

Rarrangement of the attacked molecule may occur during acylation of n-alkanes and in cyclizations, e.g. in Friedel–Crafts reactions (Section VI) may be accompanied by rearrangements:

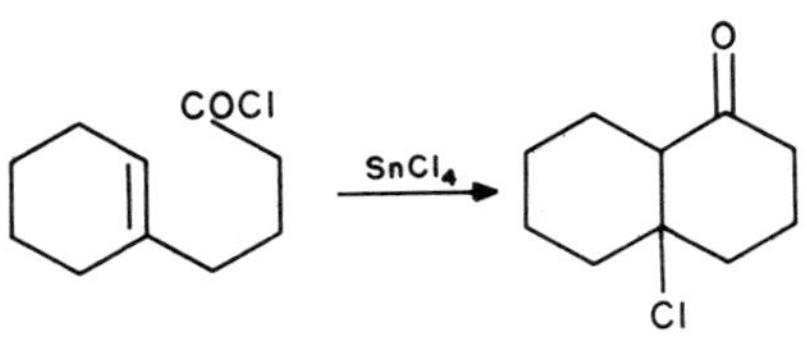

Section VII treats some special cases, e.g. rearrangements in the course of the reaction of Grignard reagents with acyl halides:

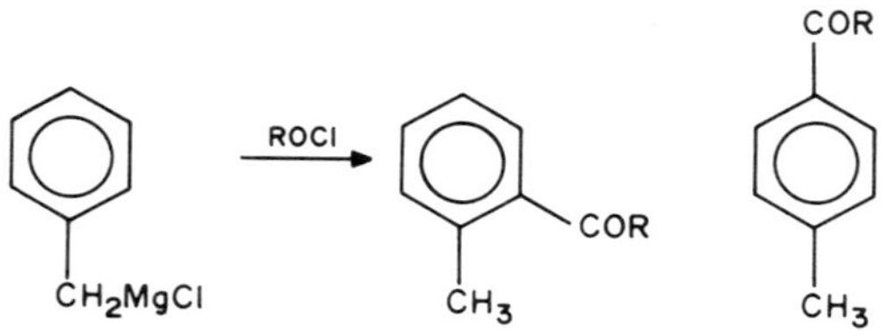

and some other reactions in which rearrangements occur with the elimination of CO or CO_2.

234 references up to 1970.
Complementary: none.
Relevant: 1, 7; **2**, 15; **3**, 14; **4**, 10; **5**, 8; **8**, 9; **9**, 3; **11**, 4; **12**, 16; **13**, 13; **14**, 7; **16**, 17; **17**, 8; **18**, 15; **19**, 18; **20**, 4; **23**, 13; **24**, 10; **26**, 15; **29**, 4.

15, *9: Photochemistry and radiation chemistry of carbonyl halides* by *U. Schmidt and H. Egger*

I. Introduction (293); **II**. Electron absorption spectra of acyl halides (294); **III**. Photochemical and radiation-chemical formation of acid halides (296); **IV**. Photolysis, photochemical oxidation and reduction of acyl halides (298); **V**. Photolysis of acyl halides in the presence of hydrogen donors, alkanes and alkenes (302); **VI**. Acknowledgement (311); **VII**. References (311).

The primary photochemical step with acyl halides is probably the formation of an acyl–hydrogen radical pair, but very complex mixtures may arise by secondary reactions.

Section II describes electron absorption spectra and Section III deals with formation of acyl chlorides photochemical, radiation-chemical and photo-oxidation routes.

Photolysis, photo-oxidation and reduction of phosgene ($COCl_2 \xrightarrow{h\nu} CO + Cl_2$; $COCl_2 + H_2 \xrightarrow{h\nu} CO + 2HCl$), of oxalyl halides, of acyl halides (e.g. $CH_3COCl \xrightarrow{h\nu} CH_3Cl + CO + CH_4 + CH_2{=}CHCl$ + polymers) and aroyl bromides are discussed in Section IV.

Photolysis of acyl halides and cyanides in the presence of hydrogen donors, alkanes and alkenes is the subject of Section VI. E.g. for the reaction of CH_3COCl in the presence of cyclohexane, the products isolated were cyclohexyl methyl ketone, cyclohexyl chloride, bicyclohexyl, HCl and small amounts of chlorine.

40 references up to 1970.
Complementary: none.
Relevant: 2, 16; **3**, 8; **6**, 4; **8**, 12; **9**, 6, 8; **10**, 6; **11**, 5, 6; **13**, 16, 17; **14**, 8; **16**, 11, 12; **17**, 9, 10; **18**, 10, 11; **19**, 10, 20; **22**, 10, 11; **23**, 9; **24**, 12; **25**, 11, 12; **27**, 21–24; **28**, 5; **29**, 5–7; **31**, 6; **32**, 9, 29; **33**, 21; **36**, 8.

15, *10: Biological reactions of carbonyl halides* by *S. Cohen*

I. Introduction (313); **II**. Actual and potential carbonyl halides (314); **III**. Potential reaction sites in biological media (315); **IV**. Toxicological effects due to generation of acid (318); **V**. Phosgene and related compounds (320); **VI**. Halides of toxic carboxylic acids (326); **VII**. Fluoroalkenes as latent carbonyl halides (327); **VIII**. Fine biological effects of carbonyl halides (331); **IX**. Selectivity versus reactivity (337); **X**. Concluding notes (342); **XI**. References (343).

Compounds which may be converted into acyl halides are of the type RCX_3 ($\xrightarrow{H_2O}$ RCOX) and highly electrophilic olefins

$$CFBr{=}CFBr \xrightarrow{O_2} CFBr_2COF \text{ or } CHF{=}Br_2 \xrightarrow{O_2} CHBrFCOBr$$

(Section II). The potential reaction sites and reactive groups in biological media which can be attacked are tabulated (Table 1, Section III). Many toxicological effects are due to the generation of acids from acyl halides, the acids causing weak pulse, shallow breathing, convulsions and collapse. The lungs, owing to their very large surface area (80 m^2 in human adults), are especially vulnerable (Section IV, Table 2). The effects of phosgene, diphosgene ($COCl_2$, $ClCOOCCl_3$) and related compounds are described in Section V (Table 3–5) and of other toxic acyl halides in Section VI (Table 6). Fluoroalkenes which can generate acyl halides and hence acids are described in Section VII (Tables 7 and 8) and the acylations and other reactions of biological systems with acyl halides in Section VIII (Tables 9–13).

Acyl halides are usually unselectively reactive, but in some cases different sites in biological systems can be selectively acylated or alkylated (Section IX).

129 references up to 1970.
Complementary: 5, 18.
Relevant: 2, 7; **3**, 5; **4**, 9; **7**, 4; **11**, 11; **12**, 12; **13**, 14; **16**, 13; **17**, 13; **18**, 13; **19**, 13; **20**, 6; **22**, 20; **24**, 14; **26**, 13; **28**, 17; **31**, 9.

15, *11: Thiocarbonyl halides* by *K. T. Potts and C. Sapino*

I. Introduction (350); **II**. Preparation of thiocarbonyl halides (350); **III**. Physical properties of thiocarbonyl halides (358); **IV**. Chemical properties of thiocarbonyl halides (362); **V**. Thioaroyl chloride *S*-oxides (375); **VI**. References (377).

Thioaroyl halides can be prepared from thio acids and chlorinating agents [$ArCS_2H \xrightarrow{SOCl_2} ArC({=}S)Cl$] whereas thioacyl halides are usually prepared from various haloethylenes with sulphur [e.g. $CFCl{=}CF_2 + S \xrightarrow{\Delta} CClF_2C({=}S)F$] or from perfluoroalkylmercury compounds with sulphur. Thiophosgene ($CSCl_2$) is obtained from reduction of perchloromethyl mercaptan [$Cl_3CSCl \rightarrow C({=}S)Cl_2$] and other special routes provide $C({=}S)F_2$ and $C({=}S)Br_2$. *N*, *N*-Disubstituted thiocarbamoyl chlorides are obtained from the corresponding thioformamides (Section II) [$R_2NC({=}S)H + S_2Cl_2 \rightarrow R_2NC({=}S)Cl$].

Section III deals with physical properties (b.p., d^4_{20} and m^{25}_4, Table 1) with UV (Table 2), IR and NMR spectra. Chemical properties (Section IV) include reactions with amines, alcohols (Table 3), phenols, thiols and thiophenols, cyanides, azides, diazoalkanes, heterocyclic, aromatics (i.e. thioacylations), dimerizations and cycloadditions. Finally, thioaroyl chloride *S*-oxides [$ArC(Cl){=}S{=}O$] are treated briefly (Section V).

93 references up to 1970.
Complementary: 25, 18.
Relevant: 5, 15; **11**, 8, 9.

15, *12: Chloroformate esters and related compounds* by *N. Kevill*

Chloroformic acid (ClCOOH) is extremely unstable and its derivatives should be handled with caution owing to their lachrymatory properties and instability. They are nearly always prepared from alcohols with phosgene, preferably in the presence of a tertiary amine (Section II) ($ArOH \rightarrow ArONa \xrightarrow{COCl_2} ArOCOCl$).

Thermal decomposition (Section III) in the liquid or gaseous phase leads to elimination of small, stable molecules with formation of halides or olefins:

$$ROCOCl \rightarrow RCl + CO_2$$
$$R_2CHCR_2OCOCl \rightarrow R_2C{=}CR_2 + CO_2 + HCl$$

The scope, kinetics and mechanisms of these reactions are discussed in detail.

Section IV deals with the meagre data available for the free-radical photolysis of ethyl and phenyl chloroformate.

Section V describes IR and electron diffraction data, dipole moments (Table 1), molar Kerr constants and conformations, indicating a *cis* arrangement of the alkyl group and the carbonyl oxygen.

Nucleophilic substitutions in ROC(=O)Cl are mostly similar to those in other esters and the $B_{AL}1$, $B_{AL}2$, $B_{AC}1$ and $B_{AC}2$ mechanisms operate usually, but can be complicated by CO loss (Table 2). Hydrolytic (Table 3) and solvolytic reactions are described, including kinetics, mechanisms and stereochemistry. Non-solvolytic reactions ($ROCOCl + Y^- \rightarrow ROCOY + Cl^-$) are considered, with Y^- being N_3^-, F^-, I^-, No_3^-, ArO^-, etc. (Table 4), and reactions in which solvolysis competes with decomposition (Tables 5 and 6).

Friedel-Crafts reactions may occur with loss of CO_2 ($EtOCOCl + ArH \rightarrow ArEt + CO_2 + HCl$), but aromatic chloroformates yield the esters ($ArOCOCl + ArH \rightarrow ArCOOAr$). Silver ion catalysed reactions of chloroformates afford routes to aliphatic carbonium ions of high energy, which are difficult to prepare in other ways (Tables 7 and 8). Various mechanisms have been proposed for the reaction $ROCOCl + Ag^+ \rightarrow R^+ + CO_2 + AgCl$ and these are discussed in detail.

Section VIII describes the preparation and chemical and physical properties of fluoroformates, bromoformates and iodoformates and Section IX deals similarly with chloroglyoxalates (ROCOCOCl).

267 references up to 1969.
Complementary: none.
Relevant: none.

15, *13: The acyl hypohalites* by *D. D. Tanner and N. J. Bunce*

Acyl hypohalites (RCOOHal) derive their main reactivity from the O—Hal bond. Only a very few of them have been isolated; most of them have been postulated intermediates on the basis of further reactions which occur.

Section II describes perfluoroacyl hypofluorites R_FCO_2F, their free-radical reactions and physical properties. Only the lower members of the series (up to C_4) are isolable since their stability decreases as the MW rises.

Section III discusses acyl hypochlorites and hypobromites, of which only acetyl hypochlorite has been isolated $[(CH_3CO)_2O + Cl_2 \rightarrow 2CH_3CO_2Cl]$. Another route is from metal carboxylates ($RCO_2M + X \rightarrow RCO_2X + MX$). This system is employed for various reactions of the postulated hypohalite, which include electrophilic halogenation of aromatic compounds (a route which offers no advantage over more conventional ones), addition to olefins giving 2-haloalkyl esters (the Prevost reaction) and the Hunsdiecker decarboxylation ($RCO_2M + X_2 \rightarrow RCO_2X \rightarrow RX \rightarrow CO$), which is of very wide applicability (except for tertiary halides) and has been the subject of detailed synthetic and mechanistic studies.

Acyl hypohalites are probably involved in the LiCl-catalysed fast decomposition of peroxides $[(PhCO_2)_2 + Cl^- \rightarrow PhCO_2Cl + PhCO_2{}^-]$ and in the halogenation of alkanes by mixtures of hydrohalic acids and benzoyl peroxide. Similar reactions were studied with valeryl peroxide and Li or Cu salts.

Acyl hypohalites are probably formed from acids and halogen monoxide ($X_2O + RCO_2 \rightleftharpoons HOX + RCO_2X$) or with hypohalites ($HOCl + CH_3COOH \rightarrow CH_3CO_2Cl + H_2O$) and possibly also by other routes, e.g. from peracetic acid and bromine.

Section IV describes the reaction of I_2 with metal carboxylates (e.g. $RCO_2Ag + I_2 \rightarrow RCO_2I + AgI$), where the presence of hypohalites as intermediates is postulated mainly by analogy and not by direct evidence. Other active species have been proposed by different stoichiometries (e.g. $2RCO_2Ag + I_2 \rightarrow AgI + RCO_2Ag{\cdot}RCO_2I$) ('Simonini complex'). Acyl hypoiodites are obtained from positive iodine compounds and acids ($Me_3COI + RCO_2H \rightarrow [RCO_2I] + Me_3COH$), the actual products being CO_2 and an alkyl halide ($RCO_2I \rightarrow RI + CO_2$). I_2 reacts similarly with benzoyl peroxide when from the postulated intermediate $PhCO_2I$ an excellent yield of PhI can be obtained.

Section V treats briefly acyl cyanates, acyl nitrites and acyl nitrates, which can be regarded as analogues of acyl hypohalites.

226 references up to 1969.
Complementary: none.
Relevant: 32, 11.

The chemistry of the carbon–halogen bond. Parts 1 and 2 (1973)

16, *1: General and theoretical aspects of the carbon—halogen bond* by *G. H. Wagnière*

I. General introduction (2); **II**. Aliphatic carbon—halogen compounds (14); **III**. Aromatic carbon—halogen compounds (37); **IV**. References (43).

Section I reviews advances in the study of molecular properties, *ab initio* and semi-empirical methods of calculation, the uses of nuclear quadrupole, Mössbauer and photoelectron (PE) spectroscopy and a comparison of the electronic properties of F_2, Cl_2, Br_2 and I_2 (Tables 1a–e).

Section II discusses alkyl halides, including the electronic properties of halomethanes (Tables 2–7), the inductive effect and dipole moments of alkyl halides (Tables 8 and 9) and barriers to internal rotation (Tables 10 and 11). In continuation, halogenated ethylenes are similarly discussed, emphasizing PE spectra (Table 12) and electronic spectra (Table 13). Finally, the PE spectra of halogenated acetylenes (Table 14) and their electronic structures are treated (Table 15).

Section III deals with the UV spectra of halobenzenes (Tables 16 and 17) and with their PE spectra.

151 references up to 1972.
Complementary: 32, 1, 3.
Relevant: 1, 1; **2**, 1; **3**, 1; **4**, 1; **5**, 1; **6**, 1; **8**, 1; **11**, 1; **12**, 1; **13**, 1; **14**, 1; **15**, 1; **17**, 1; **18**, 1; **19**, 1; **20**, 1; **22**, 1; **23**, 1; **24**, 1; **25**, 1; **26**, 1; **27**, 9; **28**, 1; **31**, 27; **33**, 1; **36**, 2.

16, *2: Structural chemistry of the C—X bond* by *J. Trotter*

I. Introduction (49); **II**. Carbon—chlorine bonds (53); **III**. Carbon—Bromine and carbon—iodine bonds (59); **IV**. Carbon—fluorine bonds (59); **V**. Conclusion (60); **VI**. References (61).

The reduction in C—X bond distances in alkanes, alkenes and alkynes is due to increasing π-bond character and to other effects (Table 1 and 2), although hybridization of the C atom is the major influence.

Section II discusses C—Cl bonds in saturated (Table 3), aromatic (Table 4) and ethylenic and acetylenic compounds. Section III deals briefly with C—Br and C—I bonds and Section IV with C—F bonds (Table 5). In the Conclusion (Section V) the results are summarized in Table 6.

78 references up to 1971.
Complementary: 32, 24.
Relevant: 17, 2; **18**, 2; **19**, 2; **20**, 2; **21**, 1; **22**, 2; **23**, 3; **24**, 2; **26**, 2; **27**, 4; **28**, 2; **29**, 1; **30**, 1; **31**, 24; **33**, 12; **34**, 10.

16, *3: Analysis of organic halogen compounds* by *J. Zabicky and S. Ehrlich-Rogozinski*

I. Introduction (65); **II**. Elemental analysis (68); **III**. Molecular analysis (105); **IV**. Functional analysis by chemical methods (115); **V**. Functional analysis by physical methods (156); **VI**. Acknowledgements (198); **VII**. References (198).

Section II describes methods of elemental analysis, including X-ray spectra (Table 2), electron microprobe analyser; nuclear activation (Tables 3–5), emission spectra, mass spectra (Tables 6 and 7), conversion of the halogen into X_2 or X^- ions by combustion (Table 8), fusion (Table 9) and similar methods, followed by titration, precipitation, spectrocolorimetry, chromatography or electrometric analysis of the 'mineralized' halogen (Tables 10–13).

Section III discusses chromatographic methods (Tables 14–17) and some spectral properties used to separate and determine halogen-containing compounds.

Section IV deals with derivatization (Table 18) and displacement of halogens by chemical methods (hydrolysis, ether or ester formation, alkylation, etc.), displacement via organometallic derivatives, elimination of HX from various types of halides, substitution of X by H in various types of halo compounds and reductive eliminations, oxidations and other reactions.

Section V treats physical methods. These include acid–base properties of halogenated carboxylic acids and amines (Tables 20 and 21), polarography of various types of halides (Tables 22–26), mass spectrometry, dipole moments, UV spectra (Table 27), optical rotation (Table 28), ORD (Table 29), IR spectra (Tables 30–33), NMR (Tables 34–38) and NQR (Table 39).

893 references up to 1971.
Complementary: 32, 3, 4.
Relevant: 1, 5; **2**, 8; **3**, 15; **4**, 3; **5**, 17; **8**, 3; **12**, 4; **13**, 6; **14**, 3; **15**, 3; **17**, 4; **18**, 5; **20**, 3; **22**, 5; **23**, 5; **24**, 5; **25**, 10; **26**, 5; **28**, 3; **29**, 21; **30**, 14–22; **36**, 5.

16, *4: Mass spectrometry and the carbon—halogen bond* by *A. G. Loudon*

I. Introduction (223); **II**. Data selection, presentation and errors (226); **III**. The mass spectra of the haloalkanes (227); **IV**. The haloethylenes (243); **V**. Influence of remote π-electrons on the fragmentation of carbon—halogen bonds (245); **VI**. Aromatic and benzylic halides (250); **VII**. Negative ion formations (process 2) (256); **VIII**. Ion pair formation (process 3) (259); **IX**. Conclusion (260); **X**. References (260).

The Introduction reviews some general and relevant points of mass spectrometry. Section II deals with the selection, presentation and the errors in the intensity and energetic data included in the chapter.

Section III discusses the nature of the ground state of the molecular ions of haloalkanes, their ionization potentials (Tables 1 and 2), fragmentation processes (Tables 3–10) in halo-, polyhalo- and perhaloalkanes and the intensities of the molecular ions in these (Table 11).

Section IV describes the MS of haloethylenes (Tables 12–14) and Section V deals with the influence of remote π-electrons on the fragmentation of C—X bonds (Tables 15 and 16). Section VI discusses halobenzenes (Tables 17–19) and benzylic halides (Tables 20 and 21) and neighbouring group participation in the expulsion of aromatic halogens (e.g. in *ortho*-substituted haloacetanilides and halophenylurea). Negative ion formation (Table 23) from fluoro compounds and ion-pair formation from various alkyl halides (Table 24) are treated in Sections VII and VIII.

94 references up to 1971.
Complementary: 32, 3.
Relevant: 9, 7; **10**, 5; **13**, 19; **17**, 5; **18**, 6; **19**, 5; **22**, 7; **24**, 6; **25**, 4; **26**, 6; **27**, 7; **29**, 3; **30**, 22; **31**, 3; **33**, 4.

16, *5: Hydrogen-bonding and complex-forming properties* by *J. W. Smith*

I. Introduction (265); **II**. Hydrogen-bonding (267); **III**. Complex formation with Lewis acids (292); **IV**. Complex formation arising from the polarizability of carbon—halogen bonds (295); **V**. Hydrogen-bonding by polyhalogenated hydrocarbons (297); **VI**. References (298).

Various forms of hydrogen bond and complex formation arise from the presence of the lone-pair electrons on the halogen atom in organic halides.

Section II discusses the factors influencing the formation of intramolecular hydrogen bonds with a proton-donor group. Evidence for hydrogen bond formation can be derived from b.p. and m.p. data in various halophenols, haloanilines and related compounds (Tables 1 and 2) and also from viscosity data (Table 3) and IR data in the vapour state and in various solvents (Tables 4–7). NMR spectra were also studied (Tables 8 and 9). Hydrogen bonds are also formed in *o*-halothiophenols and in *o*-halobenzaldehydes; however, in *o*-halocarboxylic acids the intermolecular hydrogen bonding competes favourably with the intramolecular type. In halohydrins and related compounds intramolecular hydrogen bonding was studied by IR (Table 10) and Raman spectra and by electron diffraction.

In *m*- and *p*-halophenols intermolecular binding between OH groups in two molecules dominates, with hardly any involvement of the halogen, with the probable exception of fluoro compounds.

Section III describes complex formation of halides with Lewis acids ($AlCl_3$, $GaCl_3$, BF_3, etc.) and their role in the halogen interchange between alkyl halides and boron trihalides.

Section IV deals with complex formation between perhaloalkanes and aniline and phenol (and less also with ethers and ketones). In these cases it is postulated that the complex is formed with a polarized C—X bond.

The presence of the C—X bond increases the acidity of hydrogen atoms bound to the same carbon atom owing to the powerful electron withdrawal by the halogen. In some cases ($CHCl_3$, $CHBr_3$, CCl_3CHCl_2 and especially CHF_3) the acidity of the C—H bond is sufficient for the formation of intermolecular hydrogen bonds with the halogens of molecules of the same type or with solvents.

73 references up to 1971.
Complementary: 32, 21.
Relevant: 2, 9; **3**, 6; **4**, 4; **5**, 6; **7**, 6; **8**, 5; **11**, 3; **12**, 3, 10; **13**, 7; **15**, 4; **17**, 6; **18**, 8; **19**, 7; **20**, 12; **22**, 8; **23**, 6, 7; **24**, 4; **25**, 5, 6; **29**, 17; **31**, 17, 21; **32**, 12; **33**, 5.

16, *6: Directing, activating and deactivating effects* by *G. Modena and G. Scorrano*

The Introduction reviews electronegativities, ionization potentials and electron affinities (Table 1), pK_a values of aliphatic halo acids (Table 2), some NMR data (Table 3), pK_a values of halophenols, -anilines and -benzoic acids (Table 4), bond dissociation energies (Table 5), σ-substituent constants (Table 6), C—Hal distances in halomethanes (Table 7) and hydrogen bonds between phenol and cyclohexyl halides (Table 8).

Section II discusses the effects of halogens on the stability (Tables 9 and 10), internal rotation (Table 11), NMR data (Tables 12–14), pK_{R+} values of phenylcarbonium ions (Table 15) of a variety of alkyl and aryl halides. In reactions involving carbonium ions, the effects of α- and β-halogens on electrophilic addition to halogenoethylenes (Table 18) and on solvolysis (Tables 19–22) are considered. Halocarbonium ions are also involved in halogen addition to C=C double bonds (Tables 23–28) and to acetylenes (Table 29). Remote (further than β) halogen atoms also influence additions to alkenes (Tables 31 and 32) and solvolyses (Tables 33 and 34). Positive ions are also formed as intermediates in aromatic electrophilic substitution and the influence of halo substituents is discussed (Tables 35–37).

The influence of halogens on carbanion stability is a balance of the opposing inductive and lone pair–lone pair repulsion effects. This is illustrated by H-exchanges and pK_a values in haloforms, haloacids, haloarenes, etc. (Tables 38–49). Halo substitutents influence the reactivity of alkyl anions (e.g. in decarboxylations of trihaloacetic acids (Table 50):

$$X_3CCOO^- \rightarrow X_3C^- + CO_2;\ X_3C^- + H_2O \rightarrow X_3CH + OH^-$$

or additions of dinitromethide ions [Table 51: $FC(NO_2)_2^-$ adds about 40 000 times faster to methyl acrylate than $MeC(NO_2)_2^-$]. Similar effects are observed on vinyl anions (Table 53) and on aryl anions (in nucleophilic additions to benzynes) (Tables 54–58) (Section III).

Section IV deals with effects on radicals, which are also stabilized by halogens (Tables 59–63) and influence the rates of homolytic processes such as halogenations in the gas phase (Tables 64–72), homolytic additions to double and triple bonds via radical chain processes (Tables 73–76), the stereochemistry of these reactions (Tables 77–79) and also homolytic aromatic substitutions (Tables 81–85).

281 references up to 1971.
Complementary: none.
Relevant: **3**, 3; **4**, 5; **5**, 16; **6**, 8; **7**, 2, 3; **11**, 12; **12**, 15; **13**, 8; **14**, 4; **15**, 5; **18**, 9; **19**, 8; **23**, 11; **28**, 9; **29**, 16; **31**, 8; **36**, 15.

16, *7: Heterolytic mechanisms of substitution involving carbon—halogen bonds* by *P. B. D. de la Mare and B. E. Swedlund*

The Introduction reviews the classification of reactions and of mechanisms which are relevant to the mechanisms of the reactions of C—X groups.

Section II discusses nucleophilic replacement of halogens at saturated centres, starting with bimolecular (S_N2) processes, their kinetic criteria, relative reactivities of halogens (Tables 1 and 2), influence of α- and β-alkylation (Table 3), kinetic isotope effects with various nucleophiles (Tables 4 and 5); structural effects (Tables 6 and 7) and modified transition states. Further, intramolecular processes, bimolecular radical-ion processes, S_N2' processes (i.e. involving rearrangement, Table 10), unimolecular (S_N1) reactions, the $S_N2(C^+)$ mechanism, neighbouring group participation (Table 11), electrophilically assisted processes and mixed and multi-stage reactions are considered.

Subsequently, reactions are discussed in which the halogen acts as the nucleophile at saturated centres (halogen exchange; reaction of ethers and alcohols with HX; replacement of OH by X). The analogous reaction at unsaturated centres is discussed in detail, including the bimolecular mechanism, the influence of halogens as leaving groups (Table 14), isotope effects (Table 15) and electrophilic catalysis. The unimolecular mechanism, the elimination–addition sequence, replacements and rearrangements and various modes of reactions of carbonyl halides (Table 16) are considered. In nucleophilic replacement by halogen at unsaturated centres, the routes involving second-order processes, aryl cations (Table 17) and S_N1 reactions are treated.

Section III describes electrophilic replacement by halogen at unsaturated centres, including aromatic fluorination (Table 18), chlorination, bromination and iodination. The various neutral and positively charged species are dealt with in addition to the reaction sequences occurring [coordination – heterolysis; displacement with rearrangement; addition–elimination; intramolecular rearrangement of the carbonium ion; rate-determining proton loss; indirect substitution (in which the primary attack by the electrophile is at one site in the molecule, but its final attachment is at another site); and finally displacements of groups other than H].

Replacement of halogens at unsaturated centres (protodehalogenation, nitrodehalogenation, etc.) is less well studied and documented. Electrophilic replacement by halogens at saturated centres can occur if the halogen can be introduced adjacent to an electron-withdrawing substituent; however, halogens can replace more easily groups in sulphides or in sulphenates.

Some reactions which may involve electrophilic replacement of halogen at saturated centres are metallation of halides, iodine exchange and halogen–metal exchange.

383 references up to 1972.
Complementary: **32**, 12, 15, 16, 20.
Relevant: none.

16, *8: Homolytic mechanisms of substitution* by *E. S. Huyser*

I. Introduction (550); **II**. Enthalpic and kinetic aspects (555); **III**. Chlorinations (564); **IV**. Brominations (585); **V**. Fluorination (602); **VI**. Iodination (603); **VII**. References (604).

The introduction reviews the general mechanisms involved in free-radical halogenation and the use of molecular halogens and other species in these reactions.

Section II deals with bond dissociation energies (Table 1), reaction enthalpies, kinetic aspects and rate laws for free-radical halogenation.

Section III describes chlorinations with Cl_2 of alkanes (Table 2), alkylaromatics (Table 4), alkyl halides and acid derivatives (Table 5), alcohols, amines, ethers and aldehydes. Other reagents, such as SO_2Cl_2 (Table 6), *tert*-butyl hypochlorite (Table 7), Cl_3CSO_2Cl, Cl_3CSCl, PCl_5, $PhICl_2$, $CuCl_2$ and *N*-chloroamines, are considered (Table 8). Section IV treats similarly brominations of alkanes, alkylaromatics and alkyl halides (Table 10) by Br_2 and also by NBS and related compounds, by polyhaloalkanes (e.g. $BrCCl_3$, which can introduce either $Br^\bullet$ or $^\bullet CCl_3$ radicals), by *t*-BuOCl, by BrCl and by Cl_3CSO_2Br. Fluorination is treated in Section V and iodination in Section VI.

82 references up to 1971.
Complementary: 32, 22.
Relevant: none.

16, *9: Elimination reactions in solution* by *R. A. More O'Ferrall*

Sections I and II review the literature and the principal mechanisms of elimination and their notation.

Section III discusses the *E*2 mechanism, methods of determining transition state structures, Hammett ρ values (Table 1) and Brønsted exponents (Table 2), interpretations of structural changes in the *E*2 transition state and the factors affecting these (Tables 3 and 4), the Hofmann and Saytzeff rules in alkyl halides and alkyl onium ions (Table 5) and the stereochemistry of *E*2 eliminations in small and medium rings (Tables 7–9) and in open-chain substrates. There follows a discussion of *E*2C eliminations, involving the transition

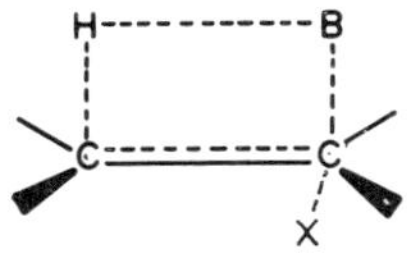

state, its stereochemistry, substituent effects (Table 10), orientation and *cis–trans* product ratios (Table 11), H-isotope effects and effects of the leaving group and some criticisms of this mechanism.

Section IV deals with the *E*1cB mechanism, which involves a carbanion intermediate. The methods of diagnosing this mechanism include stereochemistry, isotope effects and exchange, base catalysis and the element effect. Factors favouring the *E*1cB mechanism are described, together with the role of ion pairs in aprotic and in hydroxylic solvents and the links and borderline region between the *E*2 and *E*1cB routes.

Section V deals with the *E*1 mechanism, which proceeds by way of a carbonium ion intermediate, its diagnosis and scope and effects of ion pairing, stereochemistry in open-chain and cyclic substrates and the route involving rate-determining attack on an ion pair.

244 references up to 1972.
Complementary: 1, 2; **32**, 23.
Relevant: 21, 4; **32**, 5.

16, *10: Pyrolysis reactions involving carbon—halogen bonds* by *K. W. Egger and A. T. Cocks*

Section I reviews the main experimental methods (static, flow and shock tube) and general aspects of thermochemical kinetics, activation parameters, bond dissociation energies (Table 1) and stabilization energies (Table 2) for gas-phase radical and molecular reactions.

Section II deals with unimolecular thermal reactions involving cleavage of C—X bonds, including molecular HX elimination and radical reactions of halomethanes and four-centre eliminations of HX and X_2 from ethane and higher hydrocarbons (Tables 4 and 5), from cyclopropyl halides (Table 6), from higher cycloalkyl halides (Table 7), from vinyl halides (Table 8), from allyl and benzyl halides (Tables 9 and 10), from aromatic halides (in some cases leading to benzynes) (Table 11) and from acyl and benzoyl halides and halogenated ketones (Table 12), fluoroacetic acids, chloroformates and chloro ethers.

Section III summarizes the molecular reaction paths, including three- and four-centre (Tables 13–15) and also five- and six-centre mechanisms.

216 references up to 1972.
Complementary: 1, 3.
Relevant: 27, 11; **32**, 7.

16, *11: Photochemistry of the C—X group* by *P. G. Sammes*

I. Introduction (747); **II**. Ultraviolet properties of organic halides (748); **III**. Aliphatic halides (751); **IV**. Halocarbonyl compounds (763); **V**. Aromatic halides (772); **VI**. Aromatic substitution reactions (782); **VII**. References (789).

Section II reviews the UV properties of alkyl (Table 1) and aryl (Table 2) halides and of haloketones (Table 3).

Aliphatic halides usually undergo C—X bond homolysis under the influence of light. Detailed photolytic schemes are given for various alkyl halides, for the addition of polyhaloalkanes to C=C and C≡C bonds (Table 5) and for the reactions of vinylic and allylic halides (Section III).

Section IV deals similarly with acyl halides, α-haloketones and chloroacetamides.

Aromatic halides (Section V), especially iodides, undergo facile C—X bond cleavage and the aryl radical so formed may undergo a variety of reactions, including self-couplings and arylations (Table 6) and photocyclizations through intramolecular arylations. The last route is used extensively also for the preparation of alkaloids.

Section VI discusses homolytic aromatic substitutions, nucleophilic aromatic substitutions enhanced in excited states and some studies of haloheteroarenes.

197 references up to 1971.
Complementary: 32, 29.
Relevant: 2, 16; **3**, 8; **6**, 4; **8**, 12; **9**, 6; **11**, 5; **13**, 16; **14**, 8; **15**, 9; **16**, 11; **17**, 9; **18**, 10; **19**, 20; **22**, 10; **23**, 9; **24**, 12; **25**, 11; **27**, 21, 22; **28**, 5; **29**, 5, 6; **33**, 21.

16, *12: Radiation chemistry of the carbon—halogen bond* by *R. E. Bühler*

I. Introduction (796); **II**. Electron attachment (798); **III**. Electron scavenging processes (816); **IV**. Ionic reactions (832); **V**. Excited states and energy transfer (834); **VI**. Radical reactions (837); **VII**. Radiolysis of selected halocarbons (849); **VIII**. References.

After a brief review of radiolytic processes in general, Section II discusses electron affinities (Table 1A) and vertical attachment energies (Table 13), dissociative attachment cross-sections and resonance energies (Tables 2–4) of various alkyl and aryl halides, short- and long-lived anion states (Table 5) and potential energy diagrams and E_{oct} values (Table 6).

Electron scavenging processes in the gas phase (Table 7), in non-polar liquids (Tables 8A, 8B and 9) and in polar solvents (Tables 10–12) are discussed, including competition studies, absolute rate constants, products and reactions in organic glasses (Section III).

Section IV describes ionic reactions in which the negative halocarbon ions are formed by electron attachment or by electron transfer from precursor anions or form radicals in solutions. Section V deals with excited states and energy transfer in radiation chemistry, and with various reactions of the excited RX* molecules.

Section VI treats reactions of radicals produced through dissociative electron attachment, including perfluoro radical reactions (Table 14), radical reactions in irradiated aqueous solutions (e.g. with OH radicals, Table 15), with reactions of H atoms with halocarbons in organic solvents or in the gas phase (Table 16) and with the reaction of halogen atoms involved in the reactions of halocarbons (usually occurring via halogen charge-transfer complexes). In continuation, chain reactions involving halocarbons with olefins and with saturated hydrocarbons are treated.

Section VII describes the radiolysis of CCl_4, $CHCl_3$ (Table 17) and alkyl iodides (Table 19) in detail, presenting mechanistic schemes and radical and product yields for all discernible stages of the reactions.

288 references up to 1971.
Complementary: 32, 9.
Relevant: 9, 8; **10**, 6; **11**, 6; **13**, 17; **15**, 19; **17**, 10; **18**, 11; **19**, 10; **22**, 11; **25**, 12; **27**, 23, 24; **29**, 7; **31**, 6; **33**, 21; **36**, 8.

16, *13: The biochemistry of carbon—halogen compounds* by *S. Doonan*

List of abbreviations (866); **I**. Introduction (867); **II**. Fungal metabolites containing chlorine (869); **III**. Thyroid hormones (880); **IV**. Metabolism and toxicity of synthetic halogen-containing compounds (892); **V**. References (910).

Few covalent organic halogen compounds occur in nature, but their use as chemotherapeutic agents, pesticides and herbicides is extensive.

Section II reviews fungal metabolites which contain Cl (Fig. 2) almost all with the Cl being bound to a five- or six-membered ring. Especially important are those having antibiotic activity (chloramphenicol, griseofulvin, tetracyclines). The biosynthesis of these is treated briefly.

A series of iodine-containing compounds occurs in the thyroid gland. These hormones are essential for the growth and development of embryos and for the maintenance of normal body functions in adult animals. Most important among these are thyroxine analogues (Fig. 4 and Table 1). The biosynthesis, release, control and metabolism of thyroid hormones are discussed (Section III).

Synthetic halogen-containing compounds are dealt with in Section IV. These include various drugs, such as the muscle-relaxant chlorobenzoxazolinone, the hypnotic chloral hydrate, the local anaesthetic chloroprocaine and the tranquillizer chloropromazine (Table 2). The metabolism of these and of some simple chloro compounds (CCl_4, $CHCl_3$, haloethanes, haloethylenes, etd.) is discussed. Among F-containing compounds, the toxicities of CH_2FCOOH (Table 3) and of compounds which metabolize to CH_2FCOOH (Table 4) are high, since they interfere with the metabolic pathway of fatty acids. 5-Fluorouracil interferes in the process of DNA synthesis and is used in the treatment of some tumours.

The metabolic fate of Cl-containing pesticides may involve pathways without dechlorination (e.g., in the case of aldrin, which is converted by epoxidation into dieldrin, the latter being the principal toxic product) or with dechlorination (e.g., DDT, lindane).

169 references up to 1971.
Complementary: none.
Relevant: 2, 7; **3**, 5; **4**, 9; **5**, 18; **7**, 4; **11**, 11; **12**, 12; **13**, 14; **15**, 10; **17**, 13; **18**, 13; **19**, 13; **20**, 6; **22**, 20; **24**, 14; **26**, 13; **28**, 17; **31**, 9.

16, *14: Perchloro-, perbromo- and periodo-compounds* by *T. Chivers*

I. Introduction (918); **II**. Conjugated cyclic halocarbons (920); **III**. Perhalo-aromatic and -heteroaromatic compounds (926); **IV**. Perhalocarbon radicals, carbonium ions and carbanions (935); **V**. Chlorocarbon and bromocarbon derivatives of metals and metalloids (937); **VI**. Dihalocarbenes, tetrahalobenzynes and perhalohetarynes (950); **VII**. Structural and spectroscopic studies (954); **VIII**. Acknowledgements (966): **IX**. References (966).

The chapter treats perhalo derivatives containing Cl, Br and I but not F; some general properties are given in the Introduction.

Section II describes perhalo derivatives of 3-, 4-, 5-, 7- and 8-membered conjugated cyclic compounds and their relevant positive and negative ions and salts, with their preparation and physical properties and reactions.

Section III treats similarly the perhalo derivatives of benzene, toluene, the xylenes, pyridine (Table 3) and some other heterocyclic compounds (Table 4). These usually undergo aromatic nucleophilic substitution and in the case of bromides and iodides, also protodehalogenations.

The three species $(C_6Cl_5)_3C^+$, $(C_6Cl_5)_3C^-$ and $(C_6Cl_5)_3C^{\cdot}$ are exceptionally inert, probably owing to the complete steric shielding of the central C atom by the *o*-chloro substituents. This is also true for carbanions of the type $(C_6Cl_5)_2\bar{C}Cl$ and the analogous radical, but not for simple anions such as $^-CCl_3$ and $^-CBr_3$, the alkali metal derivatives of which decompose at low temperatures (Section IV).

Section V deals with perchloro and perbromocarbon derivatives of Li (Table 5) and Mg (Table 6) and their preparation and thermal stability. This is followed by treatment of derivatives of Si and Ge which can be prepared by direct halogenation $[Me_2SiCl_2 \xrightarrow{Cl_2} (CCl_3)_2SiCl_2]$ and of σ-bonded derivatives of transition metals [e.g. $(PhEt_2P)_2Co(C_6Cl_5)_2$] and of π-bonded derivatives of the same (e.g. the π-complex of Pt with $Cl_2C{=}CCl_2$) and of mercury derivatives (CCl_3HgBr, C_6Cl_5HgCl, etc.).

Section VI discusses trihalomethylmercury compounds and dihalocarbene precursors ($PhHgCCl_2Br \rightarrow CCl_2$) and their reactions, the elimination of metal halide from perhalo metal derivatives to obtain tetrohalobenzyne intermediates:

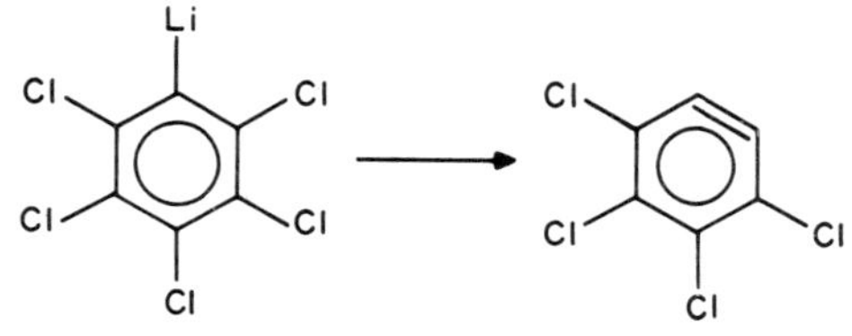

and with the analogously obtained perhalohetero arynes.

Section VII deals with X-ray, NQR, MS (Table 7), ESR, IR, UV (Table 8), dipole moment and electrochemical studies.

345 references up to 1971.
Complementary: none.
Relevant: none.

16, *15: Electrochemistry of the carbon—halogen bond* by *J. Casanova and L. Eberson*

I. Introduction (980); **II**. General reaction types (980); **III**. Cathodic reactions (982); **IV**. Anodic reactions (1036); **V**. Acknowledgements (1040); **VI**. References (1041).

Most electrochemical reactions involve cathodic reductions of halides, oxidative anodic reactions being almost completely confined to iodo compounds (Sections I and II). Section III discusses the ease of reduction of halogen types (I > Br > Cl > F), the trends in various series [e.g. Table 1 for CH_3X to CX_4; Table 2 for alkyl halides containing HO, CO, $CH_2{=}CH$ and CN groups; Table 3 for phenacyl derivatives ($PhCOCH_2X$ where X = halogen or various other groups]. Among monohalides the nature of the R groups is less influential (Table 4); the influence of various substituents in $R^1R^2C{=}CR^3Br$ is shown in Table 5 and of various ring substituents in ArI, ArBr and ArCl in Table 6. Subsequently cathodic reactions of geminal polyhalides, e.g. mono-, di- and trihaloacetic acids (Tables 7 and 8), 1,2-dihalides (Table 9), 1,ω-dihalides (Table 10) and bicyclic bridgehead dihalides (Table 11) are treated. *p*-Bis(halomethyl)benzenes of the types $XCH_2C_6H_4CH_2X$ to $X_3CC_6H_4CX_3$ give polymers $(CH_2C_6H_4CH_2)_n$ to $(X_2CC_6H_4CX_2)_n$ (Tables 12 and 13).

The mechanisms are treated, including the sequence of electron transfers and chemical steps, analogy to chemical mechanisms, LFE correlations and the evidence for carbanion and radical intermediates. The stereochemistry of the electroreductions of aliphatic, alicyclic and cyclopropyl monohalides and of cyclopropyl geminal dihalides is treated (Tables 14–16), together with the proton-donating properties of solvents (Table 17) and the formation of organometallics with the electrode material during the reaction (Table 18).

Section IV describes anodic reactions, e.g. the oxidation of alkyl iodides (Table 20) and a very few cases of reactions with bromides and perfluorinated aromatic compounds (e.g. $C_6F_5NH_2$, which gave a poor yield of a mixture of $C_6F_5N{=}NC_6F_5$ and octafluorophenazine).

203 references up to 1972.
Complementary: 32, 6.
Relevant: 5, 2; **8**, 11; **13**, 5; **17**, 14; **19**, 12; **20**, 5; **21**, 5; **22**, 9; **23**, 10; **24**, 17; **25**, 13; **26**, 12; **27**, 8; **28**, 7; **29**, 8, 9; **31**, 7; **34**, 1.

16, *16: Thermochemistry of organic halides* by *R. Shaw*

I. Introduction (1049); **II**. Types of thermochemical data and where to find them (1050); **III**. Thermochemistry and kinetics (1053); **IV**. Some current topics in organic halogen thermochemistry (1054); **V**. Acknowledgement (1068); **VI**. References (1068).

In this chapter the term 'thermochemistry' is used to include entropy and heat capacity and hence is synonymous with 'chemical thermodynamics.' Section II defines group additivity values and quotes the sources of critically evaluated reviews and of original data.

Section III discusses briefly the relation between thermochemistry and kinetics. Section IV deals with the strength of the F—F bond, rotating bomb calorimetry, the heat of formation of CCl_4, chlorofluoromethanes (Tables 1–5), CF_3OOCF_3 and the CF_3O radical, chloroolefins and a large body of recent results on $\Delta H_f{}^0$, S^0 and $C_p{}^0$ for a variety of organic halogen compounds (Table 6).

67 references up to 1972.
Complementary: 16, 10.
Relevant: 18, 3; **19**, 3; **20**, 11; **21**, 3; **22**, 6; **23**, 4; **24**, 3; **25**, 2; **26**, 4; **27**, 9; **28**, 4; **29**, 24; **30**, 2; **31**, 2; **33**, 3; **36**, 4.

16, *17: Rearrangements involving halides* by *C. Rappe*

I. Introduction (1072); **II**. Isomerizations without skeletal rearrangement (1072); **III**. Isomerizations with skeletal rearrangement (1077); **IV**. Rearrangements during substitution without skeletal rearrangement (1079); **V**. Rearrangements during substitution with skeletal rearrangement (1080); **VI**. Rearrangements with elimination of hydrogen halide (1082); **VII**. Rearrangements involving three-membered rings (1084); **VIII**. References (1110).

The chapter classifies rearrangements according to several types, and in each according to whether they do or do not undergo skeletal rearrangements.

Section II deals with isomerizations of aryl halides and of α-haloketones without skeletal rearrangement, such as aromatic halogen migrations, or migration of Br from the α- into the γ-position in bromoacetoacetate.

Section III discusses isomerization with skeletal rearrangement, e.g. the conversion of bromocyclooctatetraene to *trans*-β-bromostyrene.

Section IV treats rearrangements which occur during substitutions without skeletal rearrangement and Section V deals with similar substitutions which do involve rearrangements, e.g. the frequent carbonium-ion rearrangements during S_N1 substitutions.

Section VI describes rearrangements with elimination of HHal, e.g. $Ar_2C{=}CHX \rightarrow ArC{\equiv}CAr + HX$, or the overall reaction.

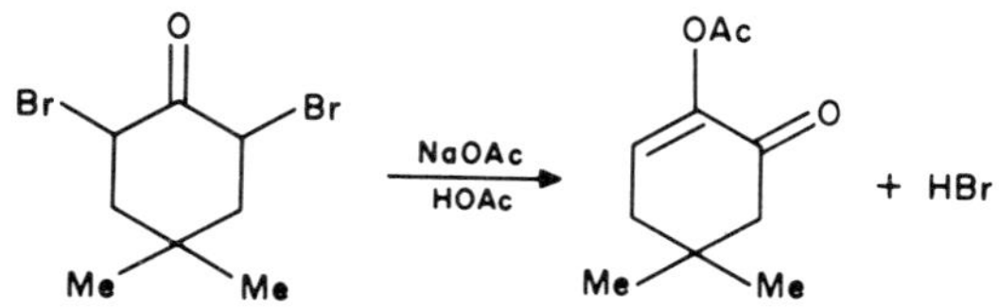

Section VII deals with the Favorsky rearrangement of α-haloketones in the presence of bases, in general.

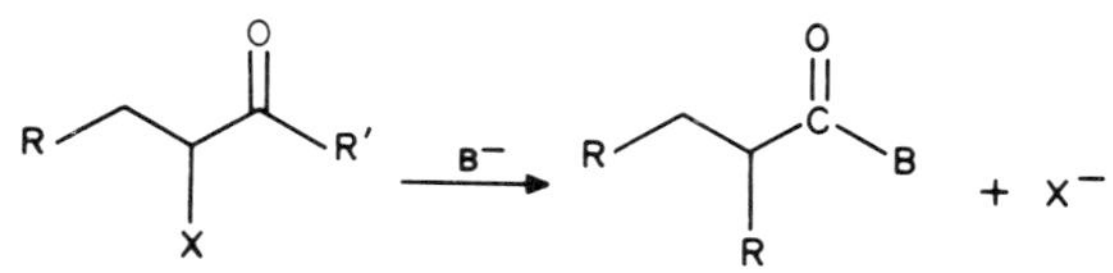

The scope of the reaction with various mono-, di- and polyhaloketones is presented and the mechanisms (especially those involving a cyclopropanone intermediate) are discussed. Other Favorsky-like rearrangements are treated briefly, and also rearrangements of α-haloketals, α-haloepoxides, halogenated allene oxides and α-haloamides are treated. Finally, the Ramberg–Bäcklund rearrangement of α-monohalosulphones with at least one α'-proton:

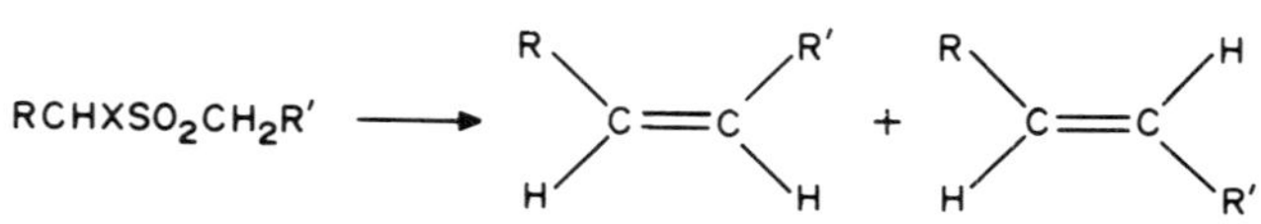

is discussed, including the scope, synthetic utility (as a general olefin synthesis) and mechanism of the reaction.

147 references up to 1971.
Complementary: none.
Relevant: **1**, 7; **2**, 15; **3**, 14; **4**, 10; **5**, 8; **8**, 9; **9**, 3; **10**, 3; **11**, 4; **12**, 16; **13**, 13; **14**, 7; **15**, 8; **17**, 8; **18**, 15; **19**, 18; **20**, 4; **23**, 13; **24**, 10; **26**, 15; **29**, 4.

The chemistry of the quinonoid compounds (1974)

17, *1: Theoretical and general aspects* by *G. J. Gleicher*

I. Introduction (1); **II.** Theoretical treatment of quinones (2); **III.** Physical properties of quinones (6); **IV.** Quinododimethanes (26); **V.** Acknowledgements (30); **VI.** References (30).

Section II discusses the various quantum chemical calculations applied to quinones. These are mainly Hückel MO calculations, taking into account electronic factors associated with the heteroatom and also other more advanced MO techniques.

Section III deals with physical properties, such as thermochemistry, heats of combustion and resonance energies (Tables 1 and 2), bond lengths (Tables 3 and 4), dipole moments, magnetic susceptibilities (Table 5), reduction potentials (Table 6) and IR (Table 7), UV and NMR spectroscopy.

Section IV gives a short summary of the theory and properties of quinododimethanes (= quinodimethanes).

224 references up to 1971.
Complementary: 40, 1.
Relevant: 1, 1; **2**, 1; **3**, 1; **4**, 1; **5**, 1; **6**, 1; **8**, 1; **11**, 1; **12**, 1; **13**, 1; **14**, 1; **15**, 1; **16**, 1; **17**, 1; **18**, 1; **19**, 1; **20**, 1; **22**, 1; **23**, 1; **24**, 1; **25**, 1; **26**, 1; **27**, 9; **28**, 1; **31**, 27; **32**, 1; **33**, 1; **36**, 2.

17, *2: The structural chemistry of quinones* by *J. Bernstein, M. D. Cohen and L. Leiserowitz*

I. Introduction (38); **II.** Molecular geometry of quinones (39); **III.** Quinone packing modes (83); **IV.** Acknowledgement (105); **V.** References.

Section I considers the significance of X-ray crystallographic data. Section II treats the molecular geometry of various groups of quinones separately, including 1,4-benzoquinones (Tables 1 and 2) and their molecular complexes (Table 3) and 'modified' benzoquinones in which the C=O group has been replaced by another function [C=C(CN)$_2$, C=NH, etc., Table 4]; crystallographic constants (unit cell dimensions, space groups) are given in Table 5 for all benzoquinones treated. Data for naphthoquinones (1,4-, 1,2- and variously substituted) are given in Tables 6–9, for anthraquinones in Tables 10–12 and for larger five-ringed quinones in Table 13.

Section III describes quinone packing modes and aspects of the molecular packing in the crystal, including the arrangement of molecules within the stack and the lateral contacts between molecules of neighbouring stacks, such as C—H···O contacts, halogen–carbonyl contacts (Table 14) and halogen–halogen contacts.

168 references up to 1972.
Complementary: 40, 20.
Relevant: 16, 2; **18**, 2; **19**, 2; **20**, 2; **21**, 1; **22**, 2; **23**, 3; **24**, 2; **26**, 2; **27**, 4; **28**, 2; **29**, 1; **30**, 1; **31**, 24; **32**, 24; **33**, 12; **34**, 10; **36**, 2.

17, *3: Synthesis* by *R. H. Thomas*

I. Introduction (111); **II**. Oxidative methods (112); **III**. Cyclization methods (136); **IV**. Condensation methods (143); **V**. Miscellaneous (146); **VI**. Annellation methods (149); **VII**. References (154).

The chapter is restricted to methods of preparation of quinones from non-quinonoid precursors and to annelation methods in which an additional ring is built on to an existing quinone.

Section II describes the most general method, which is oxidation. A wide variety of oxidants give quinones from monohydric phenols, catechols, quinols, catechol and quinol ethers, aromatic amines, aminophenols and polycyclic aromatic hydrocarbons.

Section III deals with cyclization methods, e.g. the intramolecular Friedel–Crafts ring closure of *o*-benzoylbenzoic acids to 9, 10-anthraquinones and the Scholl reaction for the synthesis of higher polycyclic quinones:

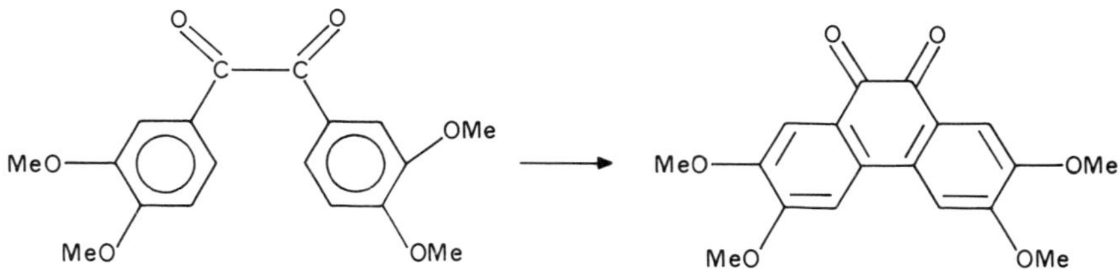

Section IV deals with methods for the condensation of quinols with cyclic anhydrides (e.g. quinol + maleic anhydride → → naphthazarin) and of dialdehydes with cyclohexane-1, 4-dione (e.g. the latter with two molecules of phthalaldehyde yields pentacenequinone).

Section V reviews some other approaches to quinone synthesis, starting from *o*-bisarylacetylenic ketones, triketoindanes and other ketones and keto esters.

Section VI discusses annellation methods through Diels–Alder reactions (e.g. cycloaddition of a conjugated diene to a quinone, aromatization of the adduct to a quinol and oxidation of the quinol to a quinone) or through 1, 3-dipolar cycloadditions of, e.g., diazoalkanes or azides to quinones and subsequent oxidation of the product.

222 references up to 1971.
Complementary: 13, 10; **40**, 8.
Relevant: none.

17, *4: Identification and determination of quinones* by *St. Berger and A. Rieker.*

I. Introduction (164); **II**. NMR spectra of quinones (195); **III**. IR spectra of quinones (186); **IV**. UV spectra of quinones (195); **V**. Identification of quinones by polarography and ESR spectroscopy (204); **VI**. Chemical methods (214); **VII**. Quantitative determination of quinones (219); **VIII**. Distinction between *p*- and *o*-quinones (222); **IX**. Acknowledgement (223); **X**. References (223).

Section II describes the chemical shifts, coupling constants and applications of the PMR spectra of benzoquinones, anthraquinones (Table 5) and phenanthraquinones (Table 6) and with the ^{13}C NMR spectra of the same quinones (Table 7).

Section III discusses IR spectra of benzoquinones (Tables 9 and 10) and of some condensed quinones (Table 11).

Section IV deals with UV spectra of benzoquinones and also of other polycyclic compounds with *p*-quinonoid structures (Tables 12 and 14) and their *o*-quinonoid analogues (Table 13 and 15).

Section V considers the polarography (Table 16) and ESR spectra (Tables 17 and 18) of quinones, while Section VI treats chemical methods, colour tests (Tables 19 and 20), reductions and reoxidations, reactions with amines (to form quinone imines) and reactions with C–H acids (Michael addition, used for a characteristic colour test, Table 20). The last section considers the possibilities of distinguishing between *o*- and *p*-quinones.

171 references up to 1971.
Complementary: 40, 2.
Relevant: 1, 5; **2**, 8; **3**, 15; **4**, 3; **5**, 17; **8**, 3; **12**, 4; **13**, 6; **14**, 3; **15**, 3; **16**, 3; **18**, 5; **20**, 3; **22**, 5; **23**, 5; **24**, 5; **25**, 5; **26**, 5; **28**, 3; **29**, 21; **30**, 14–22; **32**, 1.

17, *5: Mass spectra of quinones* by *K.-P. Zeller*

I. Introduction (231); **II**. The structure of $[M-CO]^{+\cdot}$ and $[M-2CO]^{+\cdot}$ ions (233); **III**. The $[M+2]$ peak (236); **IV**. Benzoquinones (237); **V**. Naphthoquinones (243); **VI**. Anthraquinones (250); **VII**. References (255).

One of the characteristic features of the MS of quinones is the appearance of the $[M-CO]^{+\cdot}$ and $[M-2CO]^{+\cdot}$ ions. Various interpretations have been proposed for the structure of these ions (Section II). The second characteristic feature of the MS of quinones with high redox potential and especially of *o*-quinones is the $[M+2]^{+\cdot}$ peak (Section III, Table 1).

Subsequent sections consider in detail the MS of benzoquinones, naphthoquinones and anthraquinones, describe the fragmentation paths observed, reproduce typical mass spectra and compare the behaviours of the various quinones.

38 references up to 1972.
Complementary: 40, 3.
Relevant: 9, 7; **10**, 5; **13**, 19; **16**, 4; **18**, 6; **19**, 5; **22**, 7; **24**, 6; **25**, 4; **26**, 6; **27**, 7; **29**, 3; **30**, 22; **31**, 3; **32**, 3; **33**, 4; **36**, 7.

17, *6: Quinone complexes* by *R. Foster and M. I. Foreman*

I. Introduction (258); **II**. Electron donor–acceptor complexes (258); **III**. Hydrogen-bonded complexes (304); **IV**. Metal complexes (314); **V**. Acknowledgement (324); **VI**. References (325).

Section II describes the general features of electron donor–acceptor complexes (EDAC), also termed charge-transfer complexes, in inert solvents and in the vapour phase and discusses the energy of the charge-transfer transition (Tables 1–3) and equilibrium constants (Tables 4 and 5). It also deals with electron affinities of quinones (Tables 6–11), which are ready acceptors of electrons since they possess low-lying unoccupied electronic energy levels.

Solid EDAC are easily prepared from their components. Their crystal structures

(Table 12), electronic and IR (Tables 13–15) absorption spectra and electrical properties as organic semiconductors are discussed. Quinhydrone systems and intramolecular complexes are treated and also the involvement of EDAC in organic reactions.

Section III discusses hydrogen-bonded complexes, their crystal structure, IR spectra (Tables 16 and 17), photochemistry and polarography. Section IV treats metal-bonded *p*- and *o*-quinone complexes and organometallic EDAC.

310 references up to 1971.
Complementary: none.
Relevant: **2**, 9; **3**, 6; **4**, 4; **5**, 6; **7**, 6; **8**, 5; **11**, 3; **12**, 3, 10; **13**, 7; **15**, 4; **16**, 5; **18**, 8; **19**, 7; **20**, 12; **22**, 8; **23**, 6, 7; **24**, 4; **25**, 5, 6; **29**, 17; **31**, 17, 21; **32**, 21; **33**, 5.

17, *7: Quinones as oxidants and dehydrogenating agents* by *H.-D. Becker*

The H-transfer for a donor molecule AH_2 to a quinone Q occurs in two stages: $AH_2 + Q \rightarrow AH^+ + OH^- \rightarrow A + QH_2$.

Section II describes the general features of the reaction, the influence of various factors on the oxidation potential (Tables 1–5) and the kinetics and the mechanism of the process.

Section III deals with the dehydrogenation of hydrocarbons by chloranil or DDQ, the oxidation of allyl and propargyl alcohols (Table 6) to the corresponding carbonyl compounds, the dehydrogenation of ketones (introducing double bonds into the molecule), lactones, lactams, enols, enol ethers and esters and the Oppenauer oxidation of hydroxysteroids to ketosteroids.

Section IV describes the conversion of suitable activated $—CH_2—CH_2—$ groups into $—CH{=}CH—$ (e.g., $Ar_2CHCH_2Ar \rightarrow Ar_2C{=}CHAr$) and the aromatization of hydroaromatic compounds (Tables 7 and 8), sometimes accompanied by migration of alkyl groups. High potential quinones are also effective in dehydrogenations leading to non-benzenoid aromatic and related compounds.

Section V deals with the oxidation of phenalenes, arylalkanes and arylalkenes (Table 9), benzyl alcohols and benzyl ethers (Tables 10–12) and also with the dehydrogenation of aromatic dihydroxy compounds (to quinones), of monohydric phenols (often leading by coupling to biphenols) and the oxidative conversion of 2-isoprenylphenols into chromenes.

Section VI considers the oxidation and dehydrogenation of aliphatic and aromatic amines, conversions of amines into ketones (Table 14) and oxidations of hydroxylamines, nitronic acids, hydrazines and nitrogen heterocycles (Table 15) to their aromatic analogues.

Section VII deals with the dehydrogenation of O- and S-containing heterocycles and Section VIII with the oxidation of Grignard, organolithium and metallocene compounds and with the oxidative–reductive addition to quinones by organic Si and P compounds and metal complexes. Finally, Section IX considers the H-transfer process in strongly acidic media (H_2SO_4, CF_3COOH) and the involvement of mono- and diprotonated conjugate acids (QH^+ and QH_2^{2+}) of quinones.

262 references up to 1971.
Complementary: 40, 23.
Relevant: 13, 10; **18**, 17; **27**, 12, 13.

17, *8: Rearrangements of quinones* by *H. W. Moore and R. J. Wikholm*

I. Introduction (425); **II**. Rearrangements of hydroxyquinones (426); **III**. Rearrangements of quinones upon reaction with hydrazoic acid and organic azides (446); **V**. Miscellaneous rearrangements of quinones (459); **VI**. Rearrangements resulting in the formation of quinones (460); **VII**. References (462).

The chapter deals with rearrangements in which the quinonoid rings are involved (and not with those which affect only substituents).

Section II discusses oxidative rearrangements of 2-hydroxy-3-alkyl(or alkenyl)-1, 4-naphthoquinones, in which the alkyl (or alkenyl) group is tranformed to the next lower homologue and, in addition, the alkyl and OH groups exchange places (Hooker oxidation, Table 1):

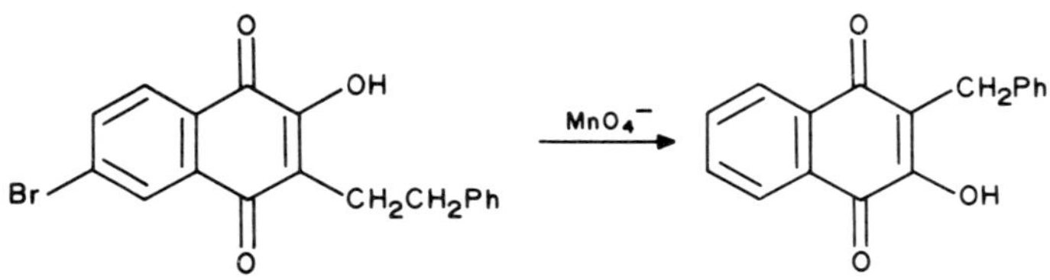

The scope and mechanism of this and related reactions are discussed in detail, including cases in which the quinonoid ring undergoes contraction to produce a derivative of a five-membered cyclic ketone.

Some highly substituted (hindered) hydroxyquinones undergo base-induced rearrangements to indenonecarboxylic acids, e.g.

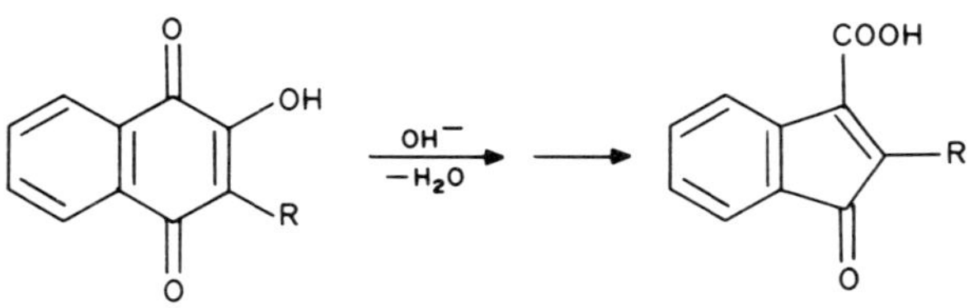

Section III describes the ring expansion of substituted benzo- and naphthoquinones to 2, 5-*H*-azepine-2, 5-diones by action of NaN_3 and ring contraction under the influence of organic azides.

Section IV deals with azidoquinones. These undergo acid-catalysed rearrangements to γ-cyanoalkylidene-Δ^{α}, β-butenolides:

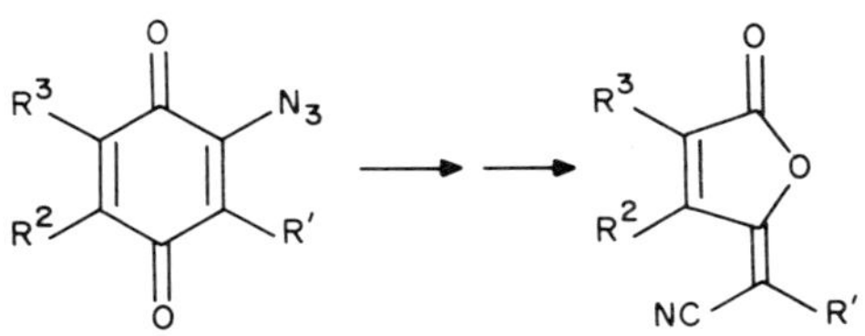

The thermal rearrangement of the same azidoquinones leads to the formation of 2-cyanocyclopentene-1, 3-diones:

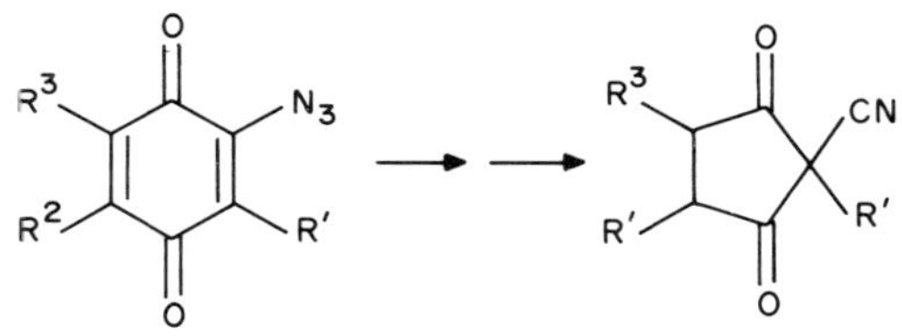

The scope and mechanism of both the above-mentioned reactions are discussed in detail. In conclusion, the rearrangement of 2, 3-diazidohydroquinone derivatives to 1, 4-dicyanobuta-1, 3-diene derivatives is treated.

Section V deals with a unique thermal rearrangement of perezone (a natural hydroxyquinone) and with the Baeyer–Villiger oxidation of 3-amino-1, 2-benzoquinones to the corresponding hydroxypicolinic acids.

Finally, Section VI describes some (rather uncommon) ring expansions of five-membered rings which lead to the formation of quinones.

103 references up to 1972.
Complementary: none.
Relevant: **1**, 7; **2**, 15; **3**, 14; **4**, 10; **5**, 8; **8**, 9; **9**, 3; **10**, 3; **11**, 4; **12**, 16; **13**, 13; **14**, 7; **15**, 8; **16**, 17; **18**, 15; **19**, 18; **20**, 4; **23**, 13; **24**, 10; **26**, 15; **29**, 4.

17, *9: Photochemistry of quinones* by *J. M. Bruce*

I. Introduction (466); **II**. Spectra and excited states (468); **III**. Cycloaddition reactions (478); **IV**. Addition to benzene (493); **V**. Reactions involving abstraction of hydrogen from substrates (494); **VI**. Miscellaneous systems (507); **VII**. Reactions involving substituents (509); **VIII**. 4, 4′-Diphenoquinones (527); **IX**. Quinone methides and quinone imines (528); **X**. References (530).

Section II describes the main features of the absorption spectra of *p*- and *o*-quinones including the spectra of hydrogen-bonded and complexed quinones and gives evidence on the nature and properties of excited states, of which the μ, π^* triplet state of the carbonyl group seems to be the most important and is expected to show electrophilic character.

Section III discusses photodimerization of quinones and photoaddition to quinones of alkenes, dienes, trienes, alkynes, SO_2 and an iridium compound.

Section IV mentions briefly the (largely unexplained) reactions of quinones with benzene (e.g. 1, 4-benzoquinone + benzene $\xrightarrow{h\nu}$ *p*-phenoxyphenol).

In general, quinones are reduced to hydroquinones in the presence of hydrogen donors. Thus, hydrocarbons yield olefins or dehydro-dimers; an α-H atom is usually abstracted from ethers leading to free-radical reactions; alcohols give the corresponding aldehydes or ketones; with aldehydes the dominant products are either ketones (i.e. acylhydroquinones) or hydroquinone monoesters; anhydrides give hydroquinones acylated either at the O atom or at the α-C atom and in the presence of water the main reaction is mono- or polyhydroxylation (Section V).

Various quinones may act as photosensitisers (e.g. in decarboxylations, dimerizations) and are also involved in biological electron transfer (Section VI).

Section VII discusses reactions (cycloadditions, H-abstractions) in which side-chains are involved. These include the formation of radical, diradical, spiro or zwitterionic intermediates from saturated substituents; from olefinic and benzenoid substituents; side-chains bearing OH groups which can act as reactive centres; CH_3O groups may react through zwitterionic or spirooxirane routes and may also be replaced photolytically by NH_2:formyl, ester, halogen, amino, diazonium and azido substituents are also mentioned.

Sections VIII and IX deal briefly with 4,4′-diphenoquinones, with quinone methides and with quinones imines.

278 references up to 1973.
Complementary: 40, 6, 13, 20.
Relevant: 2, 16; **3**, 8; **6**, 4; **8**, 12; **9**, 6; **11**, 5; **13**, 16; **14**, 8; **15**, 9; **16**, 11; **18**, 10; **19**, 20; **22**, 24; **23**, 9; **24**, 12; **25**, 11; **27**, 21, 22; **28**, 5; **29**, 5, 6; **32**, 29; **33**, 21.

17, *10: Radiation chemistry of quinones* by *J. H. Fendler and E. J. Fendler*

Section II reviews the fundamentals of radiation chemistry (sources, *G*-values and the use of scavangers in order to deal only with a single primary species (hydrated electron, H atom or OH radical, Table 1).

Section III describes in detail the aqueous radiation chemistry of simple aromatic quinones, since to an extent they can be used as models for the behaviour of more complex quinones. Rates and equilibrium constants (Table 2) and transient spectra from pulse radiolysis and various mechanistic schemes are presented. The formation of semiquinones from *p*-hydroxy-substituted aromatic compounds is treated (e.g. *p*-nitrophenol, Table 3), then some complex quinones and dyes (Table 4), followed by pulse and radiolytic investigations of electron-transfer processes relevant to biological systems.

Section IV deals with non-aqueous solutions, the radiolysis of duroquinone and ubiquinone in methanol (Table 5), the radiolysis of *p*-benzoquinone, duroquinone and ubiquinone in cyclohexane or in benzene (Table 6) and irradiations in the solid state. Section V is an extensive listing of the absorption spectra of quinones, hydroquinones and their transients.

92 references up to 1971.
Complementary: 40, 14.
Relevant: 9, 8; **10**, 6; **11**, 6; **13**, 17; **15**, 9; **16**, 12; **18**, 11; **19**, 10; **22**, 11; **25**, 12; **27**, 23, 24; **29**, 7; **31**, 6; **32**, 9; **33**, 21; **36**, 8.

17, *11: Fragmentation reactions of quinones* by *P. Hodge*

The chapter treats reactions of the title compounds, in which two or more C—C bonds of the quinone ring are broken and fragments are formed.

Section II discusses hydrolytic reactions hydroxybenzoquinones leading either to ring opening or to ring-contraction:

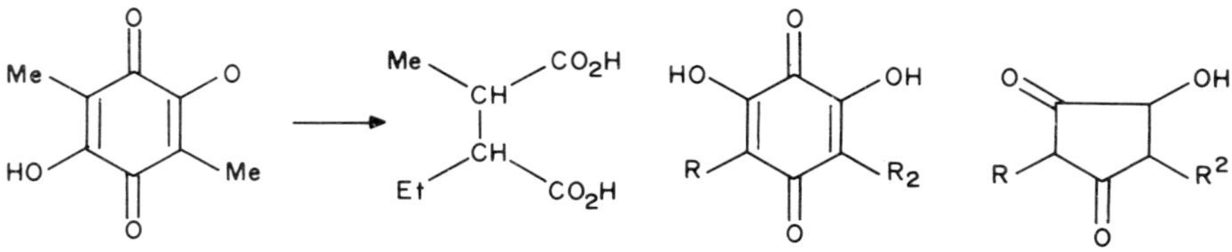

Oxidations of quinones are considered, in which all or most of the C atoms of the original ring are present also in the product or, alternatively, under different conditions, in which the carbon skeleton is extensively degraded (e.g. in ozonolysis). Some reactions of quinones with NaN_3 (Schmidt reaction, leading mainly to azepin-2, 5-diones) are treated.

Section III treats analogously hydrolytic, oxidative and Schmidt reactions of naphthoquinones and a Beckmann fragmentation of 1, 2-naphthoquinone-1-oxime.

Section IV deals with anthraquinones, which are more stable and present fewer fragmentation routes. Hydrolytic reactions usually occur only under drastic conditions, using concentrated bases or potassium *tert*-butoxide reagent. Conditions, yields and major acid products obtained by the latter are presented in Table 1.

Anthraquinones are also resistant to oxidations, and often β-CH_3 groups can be oxidized to β-COOH without breakdown of the quinone ring. However, hydroxyanthraquinones are much more reactive, some can even be oxidized by air in the presence of alkali and many yield phthalic acid derivatives. Schmidt reactions occur with NaN_3.

Section V treats phenanthrenequinones. The 9, 10-quinones undergo benzilic acid rearrangement with alkali and are readily oxidized to diphenic acids by chromic acid, H_2O_2, $KMnO_4$, etc. They also react with NaN_3 and their monooximes undergo Beckmann rearrangements.

119 references up to 1973.
Complementary: none.
Relevant: **2**, 15; **14**, 7; **21**, 8.

17, *12: Syntheses and uses of isotopically labelled quinones* by *M. Zielinski*

I. Theoretical introduction (619); **II**. Syntheses of labelled quinones (631); **III**. Tracer applications of labelled quinones (659); **IV**. Isotope effect studies with quinones (666); **V**. Acknowledgement (675); **VI**. References (679).

Section I reviews the absolute rate theory of isotope effects in two- and three-centre reactions, tunnelling in isotopic chemical reactions, relative T and D isotope effects, the experimental methods of determining kinetic isotope effects (chemical and isotopic competitive methods) and tracer studies with isotopes.

Section II describes the syntheses of *p*-benzoquinones, labelled with ^{13}C at position 1 or at positions 1, 3, and 5 and also the synthesis of 2-*tert*-[β-^{13}C]butylhydroquinone. This is followed by syntheses of ^{14}C-labelled quinones, including tetramethyl-*p*-

[^{14}C]benzoquinone, some $^{14}CH_3$-substituted and ring-substituted 1,4-naphthoquinones, vitamin K, labelled with ^{14}C either in the methyl group or in the long side-chain, and anthraquinones labelled in the 9-position. Subsequently, biosyntheses of ^{14}C-labelled phythoquinones and naphthoquinones are treated and finally syntheses of *o*-[^{14}C]benzoquinone diacetate and of uniformly labelled 3,3′,5,5′-tetra-*tert*-butyldiphenoquinone.

Section IIC discusses syntheses of T- and D-labelled benzoquinones, naphthoquinones, anthraquinones, vitamin K, ubiquinones and DL-α-tocopherol (the latter two being labelled with both ^{3}H and ^{14}C).

Section IID describes ^{18}O exchange in benzoquinones and the syntheses of ^{18}O-labelled naphthoquinones and some ^{17}O-labelled quinones. Section IIE deals with the synthesis of some ^{14}C-, ^{3}H-, ^{131}I- and ^{82}Br-labelled quinones and quinone precursors used in metabolic investigations (mostly variously substituted 1,4-naphthoquinol derivatives).

Section III discusses H–D exchange in methylquinones, hydrogen bonding in benzohydroquinone and its role in quinone–hydroquinone exchange reactions the similar hydrogen exchange between duroquinone and durohydroquinone, T-shifts in the naphthalene ring during oxidation of the latter to naphthoquinone, ^{18}O studies of the oxidative fission of hydroquinone ethers with AgO and finally tracer studies performed on the diketone–phenol rearrangement.

Section IV describes additional applications of isotopes in mechanistic investigations involving quinones. Among these are the oxidation of leuco-triphenylmethane dyes by quinones (Tables 1–5), the oxidation of ^{3}H-labelled tetrahydronaphthalene to ^{3}H-labelled naphthalene by 2,3-dichloro-5,6-dicyanoquinone and the ^{14}C isotope effect in the Friedel–Crafts condensation of *o*-benzoylbenzoic acid (labelled with ^{14}C in the carboxyl group) to yield 9-^{14}C-labelled anthraquinone.

188 references up to 1972.
Complementary: 40, 19.
Relevant: 5, 10; **12**, 13; **13**, 15; **18**, 18; **19**, 17; **22**, 14; **23**, 16; **24**, 20; **26**, 11; **27**, 10; **28**, 8; **29**, 28; **33**, 18; **36**, 12.

17, *13: Biological reactions of quinones* by *R. Bentley and I. M. Campbell*

I. Introduction (683); **II.** The biosynthesis of quinones (685); **III.** The functional significance of quinones (715); **IV.** Epilogue (729); **V.** References (730).

The chapter deals only with quinonoid materials which biological systems produce themselves and with the functional roles these materials play in the system that produces them.

Section II reviews the pathways used by cells to make quinonoid materials. Fungi, plants, insects, sea urchins and *Streptomyces* are able to synthesize quinones from simple aliphatic acids by the 'polymalonate' condensation, which involves CoA and leads to the formation of aromatic hydroxy acids (e.g. orsellinic acid), from which by a variety of substitution and oxidation steps the quinones themselves are produced.

Many quinones are formed in biological systems from shikimic acid (which in its turn is derived from carbohydrates), from which quinones are obtained by the critical intermediacy of either *p*-hydroxybenzoate, homogentisate or phenylpyruvate. A different pathway leading to naphthoquinones or anthraquinones in bacteria and plants is through succinylbenzoate; in fungi or plants, benzoquinones and naphthoquinones are fully derived from terpenes and hence from mevalonate.

Quinones often function as intermediates in the biosynthesis of other secondary metabolites (e.g. tetracyclines and aflatotoxins). Finally, some polymeric quinones and their biosyntheses are treated.

Section III discusses the functional significance of quinones and their most important biochemical reaction, namely their reversible reduction to the corresponding hydroquinone, which constitutes one of the elements of the electron transport chain found in both photosynthesis and respiration. The discussion of quinones which are primary metabolites includes the photosynthesis in encaryotes (animals, plants, fungi, algae whose cells possess nuclei) and in procaryotes (bacteria and blue–green algae whose cells lack nuclei) and is followed by discussion of respiration in encaryotes and in procaryotes. In conclusion, quinones that are secondary metabolites are considered. Unlike the compounds considered in Section IIIA, which are very similar to each other (plastoquinones, ubiquinones, menaquinones), those in Section IIIB are different in each species and even in each strain. Among these are defensive secretions of various beetles, some polymeric quinones used as structural materials and secondary metabolites whose functional role is not yet clear.

205 references up to 1972.
Complementary: 17, 15.I; **40**, 22.
Relevant: 2, 7; **3**, 5; **4**, 9; **5**, 18; **7**, 4; **11**, 11; **12**, 12; **13**, 14; **15**, 10; **16**, 13; **18**, 13; **20**, 6; **22**, 20; **24**, 14; **26**, 13; **28**, 17; **31**, 9.

17, *14: Electrochemistry of quinones* by *J. Q. Chambers*

The chapter emphasizes primarily the mechanistic aspects of quinone and hydroquinone electrode reactions.

Section II discusses half-wave potentials in aqueous and non-aqueous solutions, substituent effects on reversible half-wave potentials of quinone–hydroquinone couples, MO correlations to interpret potentiometric E_0 values and some spectroscopic correlations between charge-transfer absorption bands and polarographic $E_{1/2}$ values.

Section III describes the polarographic behaviour of quinones in non-aqueous solvents, proton donor effects, in processes such as protonations of $Q^{\cdot -}$ and Q^{2-}, disproportionations, preprotonation reactions, reductions of hydroxyquinones and C—O bond scissions in quinones. Oxidation of various hydroquinones, QH^-, QH_2 and tocopherols is dealt with, and also the striking effects of small amounts of metal ions on current–potential curves.

Section IV deals with electrochemical kinetics at solid and at Hg electrodes and with chemical reactions coupled to the electron-transfer steps. An extensive appendix (Section V) gives half-wave potentials of several hundred quinones in non-aqueous solvents.

211 references up to 1972.
Complementary: 40, 12.
Relevant: 5, 2; **8**, 11; **13**, 5; **16**, 15; **19**, 12; **20**, 5; **21**, 5; **22**, 9; **23**, 10; **24**, 17; **25**, 13; **26**, 12; **27**, 8; **28**, 7; **29**, 8, 9; **31**, 7; **32**, 6; **34**, 1.

17, *15: Polymeric quinones* by *A. S. Lindsey*

behaviour of polyquinones (830); **VI**. Electronic properties of polymeric quinones (841); **VIII**. Reference (847).

Section II describes tannins, lignins, humic acid, coal, melamins and some other natural quinone polymers.

Section III discusses fused polynuclear polyquinones and their preparation, conjugated polyquinones, their synthesis and interesting properties (photodynamism, photochromism, semiconductors) and quinonoid polymers in which fused rings ('ladders') possess quinonoid structures and special characteristics (free electrons, semiconductivity, high thermal stability).

Section IV deals with polymers which contain directly linked quinone groups or in which quinones are attached to a vinyl-type polymer (Table 1), with quinones which are polymerized by condensation with formaldehyde (Table 2) and with polymer-supported quinones. Section V treats the electrochemical behaviour of the polymeric quinone–hydroquinone system (Table 3) and of polyquinone chain segments (Table 4) and the reaction kinetics of polymeric quinones.

Section VI deals with electronic spectra, semiconductivity and photoconductivity and catalytic properties of polymeric quinones.

192 references up to 1971.
Complementary: 40, 18.
Relevant: none.

17, *16: Non-benzenoid quinones* by *T. A. Turney*

I. Introduction (858); **II**. Even-membered rings (860); **III**. Odd-membered rings (867); **IV**. Other dicarbonyl systems (870); **V**. Acknowledgement (874); **VI**. References (874).

The title compounds are defined as 'any dicarbonyl species whose two-electron reduction product would generate a cyclic non-benzenoid structure containing conjugated double bonds.'

Section II discusses even-membered rings such as cyclobutenequinone and squaric acid:

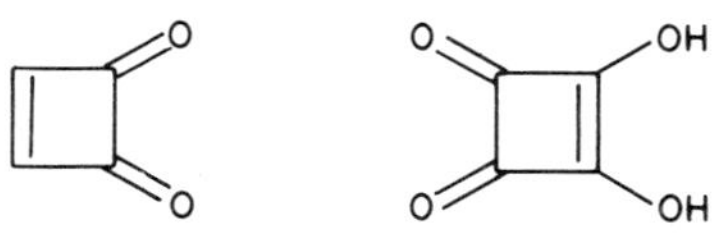

and their derivatives, and also cyclooctatrienequinone.

Section III deals with odd-membered rings, such as cyclopropanetriquinone, cyclopentenequinone and cycloheptadienequinone:

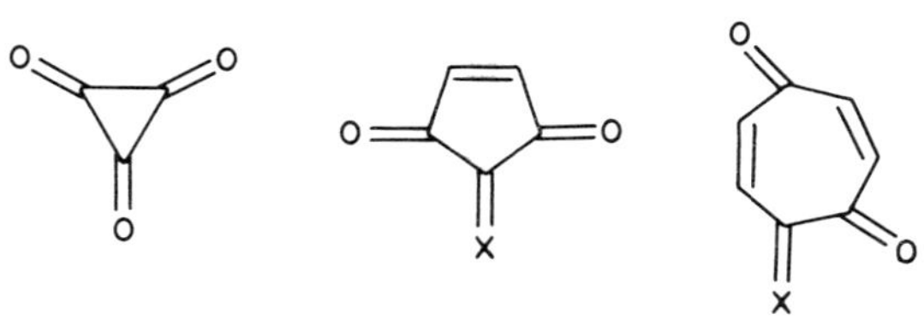

Section IV describes some other 1, 2-dicarbonyl systems such as acenapththenequinone and pyracyloquinone and also a 4,8-system, dibenzo[*cd*, *gh*]pentaleno-4,8-quinone. In each section synthetic methods are indicated and chemical and physical properties of interest are recorded.

39 references up to 1971.
Complementary: 40, 26.
Relevant: 40, 24.

17, *17: The addition and substitution chemistry of quinones* by *K. T. Finley*

The chapter is divided according to reaction types and subdivided according to the nature of the non-quinonoid reactant.

Section II describes nucleophilic (generally 1,4-) additions with the primary hydroquinone product often being oxidized to a new, substituted quinone by air, added oxidant or the quinone starting material itself:

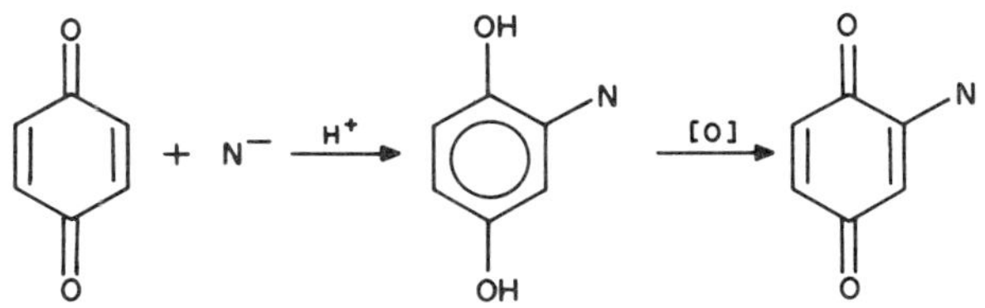

S-Containing nucleophiles include sulphinic acids, thiols, thioglycolic acid, *S*-heterocycles, thiourea, etc. (Tables 1–3). *N*-Substituted quinones are obtained from amines, anilines, pyrroles and other heterocycles, etc. (Tables 4 and 5). With both N- and S-containing nucleophiles, the method of preparing the quinone *in situ* by oxidation of the corresponding dihydroxy, aminohydroxy or other quinone-producing compounds (nascent quinones) is described.

The Thiele acylation of quinones with Ac_2O leads to both addition and esterification:

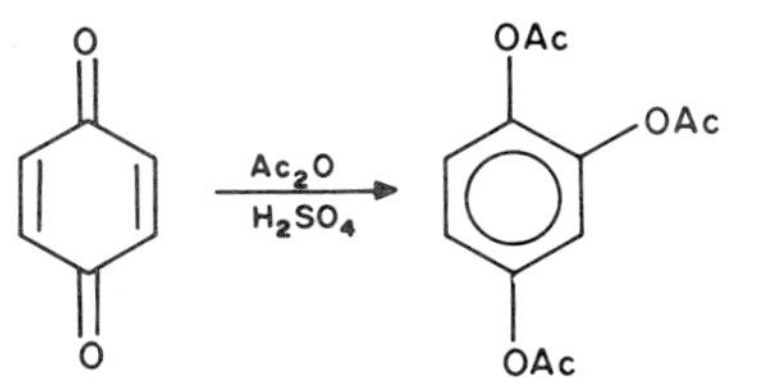

Although the reaction is properly considered electrophilic (by $CH_3\overset{+}{C}O$), it leads to product orientation predicted by nucleophilic considerations. The mechanism and scope of the reaction are discussed in detail (Tables 6 and 7). Subsequently, the addition of halogens, hydrogen halides, HCN, HN_3, H_2SO_3 and aryl phosphorus compounds is described. O-Containing nucleophiles are discussed, including alcohols, H_2O and OH^-, leading to alkoxy- or hydroxy-substituted hydroquinones or quinones. The epoxidation of the quinonoid C=C bond is treated.

Section III discusses features of quinone chemistry which relate chiefly to the C=O group. Among these are formation of mono- and dioximes, hydrazones and interaction with tertiary phosphines $(RO)_3P$, which may lead either to simple reduction of the quinone to the corresponding hydroquinone or to hydroquinone ethers.

Section IV deals with the fate of quinones as inhibitors in radical polymerization and the related issue of the copolymerization of quinones with other monomers (styrene, methyl methacrylate), with the reaction of single radicals with quinones (Table 8), with free-radical alkylation of quinones and acyl peroxide alkylations [in which, e.g., $Pb(OAc)_4$ introduces a CH_3 group into the quinone nucleus, a reaction which had great importance in the synthesis of a very large series of potential antimalarial drugs], with the reductive methylation of quinones (Table 9) and with hydroboration:

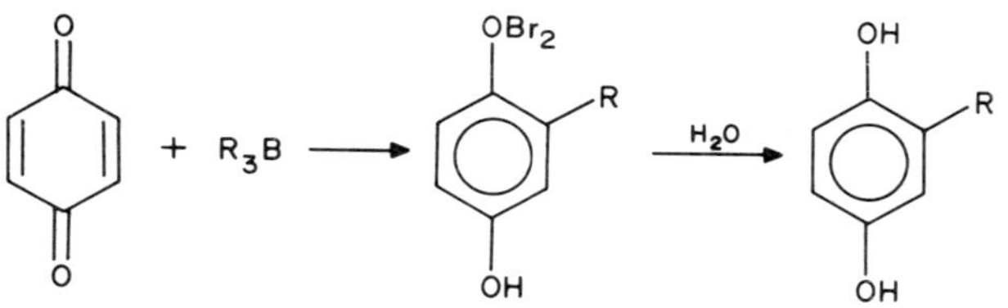

Section V deals with various cycloadditions. In the Diels–Alder reaction the quinonoid C=C bond acts as the dienophile:

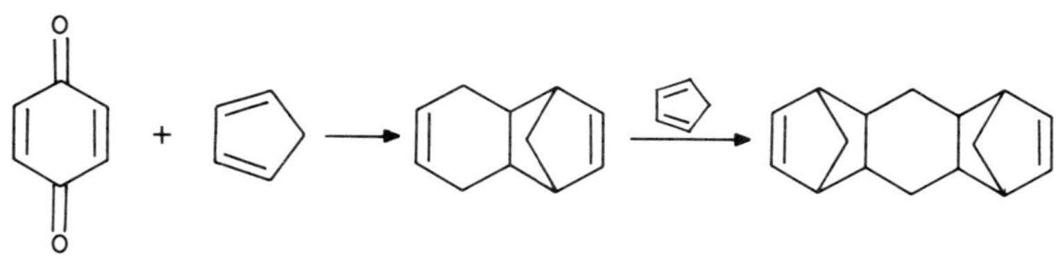

The mechanism of and scope of the reaction are described in detail (Tables 10–15) and also the use of 1, 2-quinones in the reaction.

Cycloaddition of diazo compounds occurs cleanly:

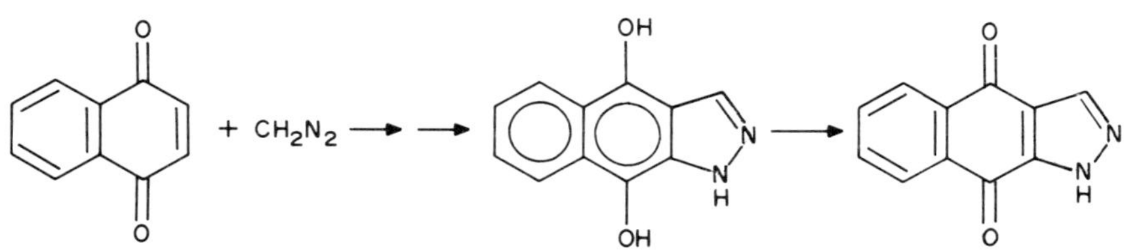

Enamines produce variously substituted indoles (Nenizetscu condensation, Table 16). Some other less general reactions, such as the oxidation of tertiary amines by quinones and the cycloaddition of ketenes, indoles and isocyanates with quinones, are mentioned briefly.

Section VI deals with the electrophilic arylation of quinones, especially via diazonium intermediates:

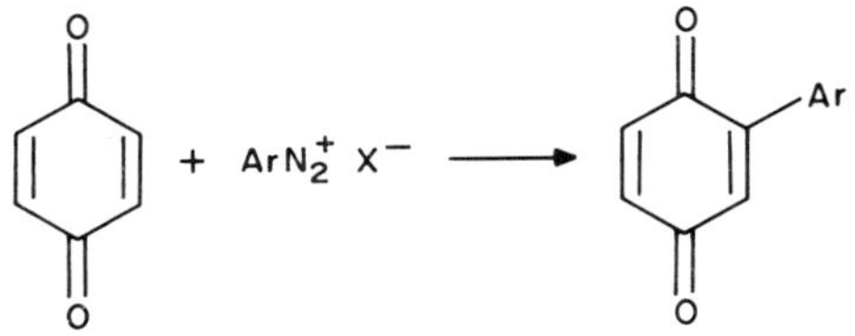

Section VII describes the reactions of quinones with active methylene anions. These reactions give well defined products only if the quinones are extensively substituted (e.g. duroquinone), and probably involves the methylene tautomer:

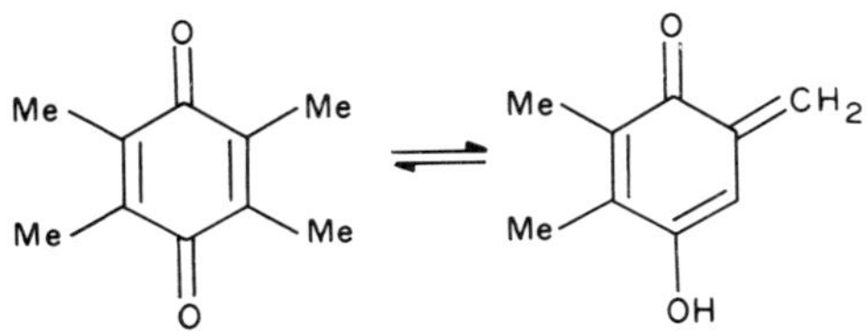

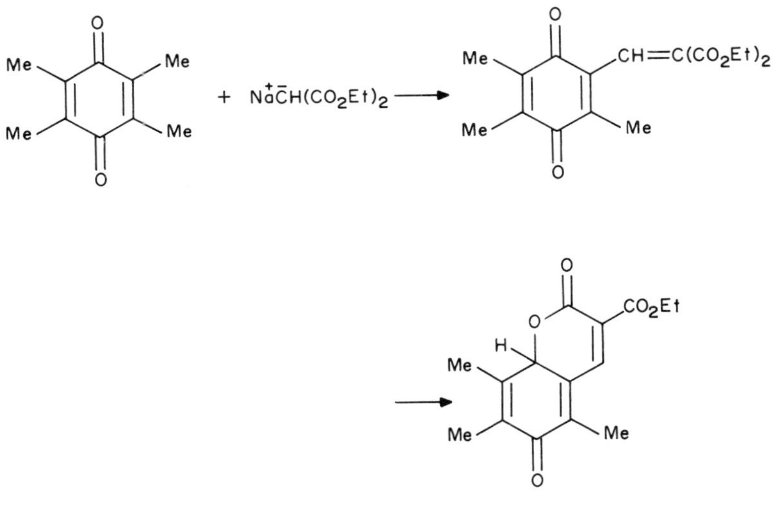

The reaction has been extended to naphthoquinones and to variously substituted benzo- and naphthoquinones.

Section VIII describes the substitution of a labile group on the quinone nucleus by various agents. Thus, halogen is replaced by a variety of N-containing groups, e.g.

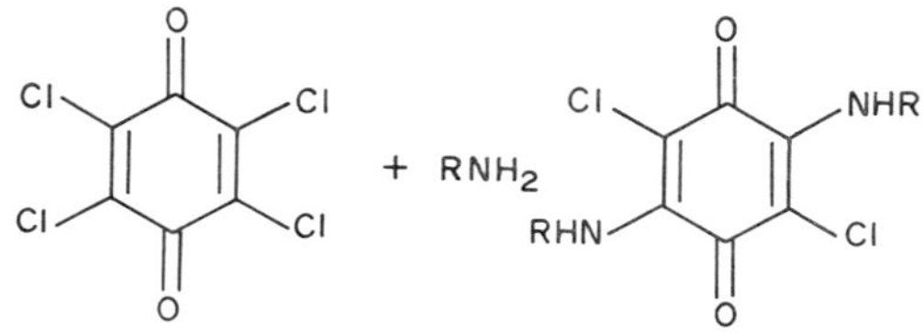

This substitution competes with the addition described in Section II.C (Table 18); moreover, ring-bound alkyl groups of the quinone can also be displaced by amines and in some cases hydrogen atoms of such alkyl substituents are substituted by the amine (Tables 19 and 20). The mechanisms and the synthetic uses of these reactions are surveyed in detail. Halogen atoms on quinone rings can be substituted by ethyleneimine:

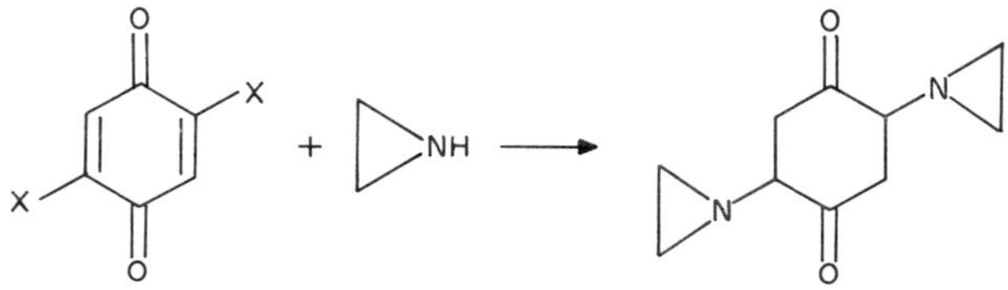

The reaction has many synthetic uses and the ethyleneimino group is ring-opened (without rearrangement) by HCl to give the 2-chloroethylamino derivative. Finally, the section describes displacement of ring halogen atoms by various oxygen and sulphur nucleophiles (ROH, ArOH, RSH, HSO_3^-, etc.) and the removal of halogens and other groups from the nucleus and their replacement by hydrogen.

681 references up to 1972.
Complementary: 40, 11.
Relevant: 1, 11; **21**, 6.

17, *18: Quinone methides* by *R. Gompper and H.-U. Wagner*

I. Introduction (1145); **II.** Preparative methods (1150); **III.** Synthesis by electrophilic aromatic substitution of phenols and subsequent elimination (1154); **IV.** Synthesis from aromatic hydroxy-aldehydes and -ketones (1158); **V.** Synthesis by elimination and deprotonation (1160); **VI.** Synthesis by electrophilic aromatic substitution of phenols and subsequent oxidation (1161); **VII.** Synthesis from aromatic hydroxy-aldehydes by oxidation of derivatives (1163); **VIII.** Synthesis by oxidation of (hydroxylaryl) methyl compounds (1164); **IX.** Synthesis from quinones (1164); **X.** Photochemical syntheses (1166); **XI.** General reactions (1168); **XII.** Cycloaddition reactions (1172); **XIII.** Acknowledgement (1174); **XIV.** References (1175).

Replacement of a carbonylic oxygen of a quinone leads to a quinone methide, 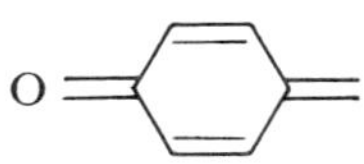 CH_2. The Introduction gives a short theoretical treatment of these (HMO description, charge distributions, reactivity).

Section II surveys the types of preparative methods available. These include chloromethylation of phenol and subsequent elimination of HCl and related reactions

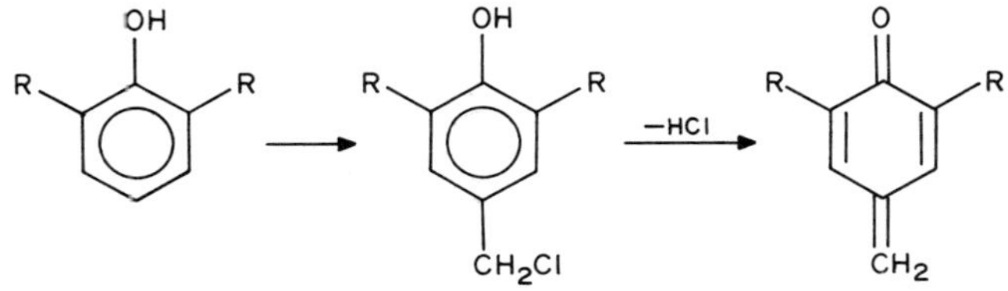

(Section III): syntheses starting from aromatic *p*-hydroxy aldehydes and ketones (Section IV); by H_2O eliminations (V); by substitution and oxidation of phenols (VI); from hydroxyaldehydes (VII); from (hydroxyaryl)methyl compounds (VIII); by carbonyl–methylene condensations from quinones (IX); and through some photosynthetic processes (X).

Section XI deals with some addition, substitution and other reactions and Section XII with cycloadditions of quinone methides.

138 references up to 1971.
Complementary: none.
Relevant: 40, 21.

The chemistry of the thiol group (1974)

18, *1: General and theoretical aspects of the thiol group* by *I. G. Csizmadia*

I. Physical properties (2); **II**. Theory (44); **III**. Calculations of molecular wave functions and energies (75); **IV**. Analysis of electron distribution (91); **V**. Concluding remarks (104); **VI**. Acknowledgement (107); **VII**. References (108).

The main purpose of Section I is to show that there is no fundamental difference between the corresponding O and S compounds. Thus, standard states and relative energies (Tables 2 and 3 and Figs. 1 and 2) are compared, in addition to bond energies and molecular vibrations (Tables 4–8), electronic excitations (Table 9), ionization potentials and electron affinities (Tables 10–14) and proton, hydrogen and hydride affinities (Tables 15–19). Subsequently stereochemistry and dipole moments are considered.

Section II discusses some fundamental concepts of quantum chemistry, including the Schrödinger equation, the variation theorem, the principles of constructing many-electron wavefunctions, the non-empirical SCF–MO theory (Table 23) and its applications, the concept of localized MO and the notion of d-orbital participtation (Tables 24 and 25).

Section III reviews calculations of molecular wavefunctions and energies and demonstrates that accurate Hartree–Fock type molecular calculations are technically feasible even for large systems such as RSH. Studies regarding HS, H_2S and H_3S are first described (Tables 26–29), followed by studies on CH_3SH (Tables 30–32) and on other structures involving the SH group (Table 33).

Section IV suggests that the magnitude of the electron density is the major factor which determines the difference in the physical and chemical properties of alcohols and thiols. Charge distributions and dipole moments (Table 34), electron density contours and the sizes and shapes of electron pairs (Tables 35 and 36) and of functional groups (Table 37) are discussed.

38 references up to 1973.
Complementary: none.
Relevant: 1, 1; **2**, 1; **3**, 1; **4**, 1; **5**, 1; **6**, 1; **8**, 1; **11**, 1; **12**, 1; **13**, 1; **14**, 1; **15**, 1; **16**, 1; **17**, 1; **19**, 1; **20**, 1; **22**, 1; **23**, 1; **24**, 1; **25**, 1; **26**, 1; **27**, 9; **28**, 1; **31**, 27; **32**, 1; **33**, 1; **36**, 2.

18, *2: Structural chemistry of the thiol group* by *I. C. Paul*

I. Introduction (111); **II**. Molecular dimensions and conformational information obtained from X-ray crystallographic studies (113); **III**. A discussion of some structures that have been reported to contain the thiol group (119); **IV**. Relative occurrence of thiol: thione tautomers in the solid state (123); **VI**. Magnetic resonance investigations of conformation on molecules containing thiol groups (131); **VII**. Hydrogen-bonding properties of sulphur (133); **VIII**. Acknowledgement (146); **IX**. References (146).

The chapter reviews the available information on the structure of thiols in the solid, liquid and gaseous phases.

Section II describes molecular dimensions (Table 1) and conformations (Table 2) obtained by X-ray studies.

Section III warns that in many cases when thiol groups were postulated in a structure actually the thione tautomer is present, and the latter tautomer almost always dominates in the solid state (Section IV) and often also in the liquid and in solution.

Section V deals with electron diffraction and with microwave studies, which give very accurate information in the case of small molecules in the gas phase (dipole moments, conformations, rotational barriers, etc.).

Section VI treats briefly NMR studies and Section VIII the hydrogen bonding properties of S both as donor S—H···B and as acceptor A—H···S (Table 3). Several stereoscopic views of SH-containing molecules illustrate the hydrogen bond in the crystals.

89 references up to 1973.
Complementary: none.
Relevant: 16, 2; **17**, 2; **19**, 2; **20**, 2; **21**, 1; **22**, 2; **23**, 3; **24**, 2; **26**, 2; **27**, 4; **28**, 2; **29**, 1; **30**, 1; **31**, 24; **32**, 24; **33**, 12; **34**, 10; **32**, 3.

18, *3: Thermochemistry of thiols* by *R. Shaw*

I. Introduction (151); **II**. Measured data (152); **III**. Estimation by group additivity (152); **IV**. Kinetics and thermochemistry (157); **V**. Acknowledgement (161); **VI**. References (161).

Sections I and II review the concepts and the published experimental data (Table 1) for thiols. Section III describes group additivity and compares observed and estimated heats of formation (Table 2) and heat capacities (Table 3).

Section IV discusses the use of thermochemistry to help to understand kinetics and the relationship between the two. Thus, heats of formation can be used to calculate S—H bond strengths (Table 4). Finally, the strengths of O—H and S—H bonds in various compounds are compared.

24 references up to 1972.
Complementary: none.
Relevant: 16, 16; **19**, 3; **20**, 11; **21**, 3; **22**, 6; **23**, 4; **24**, 3; **25**, 2; **27**, 9; **28**, 4; **29**, 24; **30**, 2; **31**, 2; **33**, 3; **36**, 4.

18, *4: Preparation of thiols* by *J. L. Wardell*

I. Introduction (164); **II**. Formation from alkenes (169); **III**. Formation from alcohols (179); **IV**. Formation from halides and hydrogen sulphide (180); **V**. Using phosphorothiolate ion (185); **VI**. Formation via iso-thiuronium salts: use of thiourea (186); **VII**. Formation via Bunte salts; using thiosulphate (192); **VIII**. Formation via xanthates and related esters (194); **IX**. Sulphur insertion reactions of organometallic compounds (211); **X**. Formation and reduction of sulphonyl chlorides and related compounds (216); **XI**. The formation and conversion of disulphides to thiols (220); **XII**. The formation of thiocyanates and their conversion to thiols (230); **XIII**. Dealkylation of sulphides: carbon—sulphur bond cleavage (235); **XIV**. Thiol formation from the ring opening of heterocyclic compounds (246); **XV**. Thiols from aldehydes and ketones (251); **XVI**. Formation from carboxylic acid derivatives; **XVII**. Miscellaneous methods (256); **XVIII**. References (258).

Usually different methods are used for the formation of alkanethiols (Table 1) and of aromatic thiols.

Section II deals with syntheses starting from alkenes. Addition of H_2S to C=C double bonds yields thiols, but the reaction often continues to the sulphide:

$$\mathrm{C{=}C \quad + H_2S \longrightarrow HS\overset{|}{\underset{|}{C}}{-}\overset{|}{\underset{|}{C}}H \longrightarrow H\overset{|}{\underset{|}{C}}{-}\overset{|}{\underset{|}{C}}{-}S{-}\overset{|}{\underset{|}{C}}{-}\overset{|}{\underset{|}{C}}H}$$

Mechanisms, rates and stereochemistry are discussed (Table 2). Thiolacetic acid [$CH_3C(O)SH$] also adds to C=C (Table 3) and the thiol is obtained after hydrolysis. Section III describes the preparation of thiols from alcohols, using either H_2S or P_2S_5. Section IV deals with the synthesis using H_2S and alkyl, heterocyclic or aryl (Table 4) halides.

Other syntheses involve alkyl halides and the phosphorothiolate ion ($PSO_3{}^{3-}$) (Table 5, Section V), the use of isothiuronium salts (Section VI) from alkyl halides (Table 6) and the preparation of aromatic thiols through *S*-arylisothiuronium salts (obtained from aryldiazonium salts and thiourea (Table 7):

$$\mathrm{ArNH_2 \longrightarrow ArN_2{}^+ \xrightarrow{S{=}C(NH_2)_2} ArSC(NH_2){=}NH{\cdot}HX \longrightarrow ArSH}$$

Section VII deals with preparations from halides and thiosulphate via Bunte salts ($RX + S_2O_3{}^{2-} \rightarrow RSSO_3{}^- \rightarrow RSH$) (Table 8). Thiosulphate reacts similarly with quinones and, after reduction and hydrolysis, yields aromatic thiols.

Section VIII deals with the production of thiols from alkyl halides (Table 9) via xanthates ($RX \rightarrow {}^-SCOOR' \rightarrow RSCOOR' \rightarrow RSH$) or from aromatic amines, again via xanthates (Table 10) ($ArNH_2 \rightarrow ArN_2{}^+ \rightarrow ArSCOOR' \rightarrow ArSH$).

Like xanthates, trithiocarbonates are also easily reduced to thiols ($RCl + Na_2CS_3 \rightarrow NaSCSSR \rightarrow RSH$) (Table 11). Thiols are formed from phenols with dialkylthiocarbonyl chloride (Table 12). Thioesters are prepared by several routes (rearrangement of thiolcarboxylates, addition of thiolcarboxylic acids to alkenes, replacement of halide by thiolcarboxylate) and are easily hydrolysed to thiols (Table 13). A similar route from halides to thiols (with alkyl and aralkyl, Table 14) is through dithiocarbamates.

Section IX describes S insertion reactions into organometallic compounds (Table 15), e.g. $RMgCl + S_8 \rightarrow RSH$ or $ArLi + S_8 \rightarrow ArSH$.

Sulphonyl chlorides can be reduced to thiols, especially aromatic thiols, by $LiAlH_4$, Zn–H^+, etc. (Tables 16 and 17; Section X).

Section XI describes the reduction of disulphides to thiols by $LiAlH_4$, $NaBH_4$, metal + acid, glucose, electrolysis, etc. (Table 18).

Thiocyanates can be prepared by nucleophilic substitution of halides and also by direct thiocyanation of reactive arenes or via the Gatterman or Sandmeyer methods. The products are easily reduced to thiols (Table 19, Section XII).

Section XIII discusses the preparation of thiols from sulphides by C—S bond cleavage (Tables 20 and 21), e.g. by the action of various metal–amine systems: $\mathrm{ArSR \xrightarrow[ii.\ H_3O^+]{i.\ metal-amine} ArSH}$. The direction of the cleavage in the case of unsymmetrical sulphides and its mechanism are described.

Thiols are obtained by ring opening of three-membered heterocycles with H_2S. The reaction is discussed for aziridines (Table 22), thiiranes (Table 23) and epoxides.

Section XV treats the catalytic reduction of aldehydes and ketones by H_2S to give thiols (or, in various circumstances, also enethiols, *gem*-hydroxythiols and *gem*-dithiols). Section XVI mentions briefly methods starting from acyl halides or from thioacids and Section XVII the direction reaction of sulphur with arenes, heteroarenes and alkenes and, finally, the reaction of aromatic compounds containing strong electron-donating groups with sulphur mono- and dichloride and subsequent transformation of the introduced S-containing group into SH.

404 references up to 1972.
Complementary: none.
Relevant: 27, 13.

18, *5: Detection and determination of thiols* by *A. Fontana and C. Toniolo*

Section II describes qualitative tests, spot tests on chromatograms and identification through derivatives. Section III deals with the reaction $2RSH \rightarrow RSSR + 2H^+ + 2e$ and the problems occurring in its practical use (over-oxidation, lack of specificity, etc.).

Thiols with heavy metals give mercaptide derivatives. The method can be used in electrometric procedures or in titrations of the thiols by inorganic mercury salts, alkylmercury compounds, *p*-chloromercury benzoate, silver ions and other mercaptide-forming agents which contain heavy metals salts in large organic molecules (Section IV).

Section V discusses colorimetric procedures, used especially for the determination of the cysteine content of proteins.

Section VI deals with alkylating agents (carboxymethylation and addition to double bonds) and Section VII with radiochemical methods (involving isotopes of C, S and Hg). Section VIII treats the determination of total S, sulphides and disulphides. Section X describes UV (Tables 1 and 2), NMR (Table 4) and ESR studies. Table 5 (Section X) compares analytical results for SH obtained by four different methods.

288 references up to 1972.
Complementary: 27, 7, 8.
Relevant: 1, 5; **2**, 8; **3**, 15; **4**, 3; **5**, 17; **8**, 3; **12**, 4; **13**, 6; **14**, 3; **15**, 3; **16**, 3; **17**, 4; **20**, 3; **22**, 5; **24**, 5; **25**, 10; **26**, 5; **28**, 3; **29**, 21; **30**, 14–22; **36**, 5.

18, *6: The mass spectra of thiols* by *C. Lifshitz and Z. V. Zaretskii*

Section II describes the general characteristics of the MS of aliphatic, alicyclic, aromatic and heterocyclic thiols and SH-containing amino acids and peptides. Fragmentation schemes are presented for each group.

Section III discusses ionization potentials (Table 1), appearance potentials and heats of formation (Table 2), structures of various S-containing ions, bond energies, activation energies and fragmentation pathways.

Negative ions (CH_3S^-, CH_2S^-, HS^-, S^-, etc.) and dissociative electron-capture processes and energetic considerations are presented in Section IV.

Section V deals with ion–molecule reactions, the reactions of positive ions and their rate constants, reactions of negative ions and proton affinities and gas-phase basicities and acidities.

59 references up to 1972.
Complementary: **27**, 7.
Relevant: **9**, 7; **10**, 5; **13**, 19; **16**, 4; **17**, 5; **19**, 5; **22**, 7; **24**, 6; **25**, 4; **26**, 6; **29**, 3; **30**, 22; **31**, 3; **32**, 3; **33**, 4; **36**, 7.

18, *7: The optical rotatory dispersion and circular dichroism of thiols* by *C. Toniolo and A. Fontana*

Section II reviews the UV absorption of thiols and thioethers. Optical dissymmetry effects (ODE) of monochromophoric thiols (Table 1) are compared with those of alcohols (Table 2), followed by monochromophoric open-chain thioethers (Table 3). Section IV discusses the ODE of monochromophoric thiiranes (Table 4), Section V that of thiolanes and thianes (Table 5) and Section VI that of 1, 3-dithiolanes (Table 6).

Section VII deals with the ODE of thiols containing additional chromophores (Table 7) and with open-chain thioethers also containing additional chromophores (Table 8).

68 references up to 1973.
Complementary: **27**, 6.
Relevant: **8**, 4; **19**, 6; **20**, 2; **22**, 4; **25**, 3; **26**, 3; **27**, 6; **28**, 10; **29**, 23; **31**, 1; **36**, 17.

18, *8: Acidity and hydrogen bonding* by *M. R. Crampton*

Section I describes self-association in H, S, aliphatic thiols and thiophenols (dimers by NMR; Table 1) and other compounds. This is followed by a description of thiols as hydrogen bonding acids (Tables 2 and 3) and of intramolecular hydrogen bonding in *ortho*-substituted thiophenols, enethiols (thiocarbonyl tautomers) and related compounds.

Section II discusses the acidity of aliphatic thiols (Table 4), H_2S, some aminothiols (Table 5), some thio and dithio acids (Table 6), thiophenols (Table 7) and heteroaromatic thiols. Finally, deuterium isotope effects on thiol ionization, the thermodynamics of the ionization (Table 8) and the carbon basicities of sulphur bases (Table 9) are treated.

180 references up to 1973.
Complementary: none.
Relevant: **2**, 9; **3**, 6; **4**, 4; **5**, 6; **7**, 6; **8**, 5; **11**, 3; **12**, 3, 10; **13**, 7; **15**, 4; **16**, 5; **17**, 6; **19**, 7; **20**, 12; **22**, 8; **23**, 6, 7; **24**, 4; **25**, 5, 6; **29**, 17; **31**, 17, 21; **32**, 21; **33**, 5.

18, *9: Directing and activating effects* by *G. Maccagnani and G. Mazzanti*

I. Introduction and general features (417); **II**. Polar effects (419); **III**. Correlation between structure and reactivity (428); **IV**. Electrophilic aromatic substitution reactions (431); **V**. Proximity effects (437); **VI**. References (449).

The chapter deals with the electronic and proximity effects of the thiol group. Section II discusses the polar effects of the SH group, its inductive ($-I$) effect in saturated systems and the relative position of the SH group compared with other representative substituents regarding dipole moments (Table 1), acidities (Table 2) and electronegativities (Table 3). Table 4 shows PMR chemical shifts of various S-containing compounds. Resonance ($+R$) effects in aromatic and unsaturated systems are discussed and especially their acid-strengthening effect (Tables 5 and 6).

Section III deals with LFER, including various σ values (Table 7), electrophilic substituent constants (Table 8) and Taft's σ_I, σ_R treatment.

Section IV considers electrophilic aromatic substitutions, when attack on the S atom competes with ring substitution and protection of the SH group is often necessary. The processes described are protodesilylation (Table 9), Friedel–Crafts alkylation, tritylation and acylation.

Section V deals with neighbouring group participation by the SH group in nucleophilic substitutions, other proximity effects, acid–base equilibria and conformational equilibria (Tables 10–12).

147 references up to 1972.
Complementary: none.
Relevant: **3**, 3; **4**, 5; **5**, 16; **6**, 8; **7**, 2, 3; **11**, 12; **12**, 5; **13**, 8; **14**, 4; **15**, 5; **16**, 6; **19**, 8; **23**, 11; **28**, 9; **29**, 16; **31**, 8; **36**, 15.

18, *10: Photochemistry of thiols* by *A. R. Knight*

I. Introduction (455); **II**. Gas phase photodecomposition of thiols (458); **III**. Condensed phase photolysis (471); **VI**. References (478).

Section II discusses the gas-phase photolysis of CH_3SH and of C_2H_5SH and presents detailed mechanisms for both processes. It also describes the energy partitioning at various wavelengths between RS and H fragments (Table 1) and kinetic and mechanistic interpretations of the process.

Section III deals with the photolysis of liquid thiols, with thiols as sources of H atoms and of thiyl radicals in solutions, and with some other studies including aliphatic and aromatic thiols in solutions and also in the solid state.

45 references up to 1973.
Complementary: **27**, 22.
Relevant: **2**, 16; **3**, 8; **6**, 4; **8**, 12; **9**, 6; **11**, 5; **13**, 16; **14**, 8; **15**, 9; **16**, 11; **17**, 9; **19**, 20; **22**, 10; **23**, 9; **24**, 12; **25**, 11; **27**, 21; **28**, 5; **29**, 5, 6; **32**, 29; **33**, 21.

18, *11: The radiation chemistry of thiols* by *J. E. Packer*

Section II discusses the radiolysis of thiols in oxygen-free aqueous solutions involving OH radicals (Table 1), aquated electrons (Table 2) and H atoms. The products are disulphide, H_2 and H_2S and the mechanism is considered. Pulse radiolysis studies show the presence of the transient RS SR (Table 3). In confirmation, the reaction of OH, H and e_{aq}^- with disulphides, with large molecules of biological interest (S-containing enzymes) and with thiolactones are described in addition to reactions with secondary radiation-produced radicals.

Section III deals with oxygenated aqueous solutions, presenting products and yields (Tables 4 and 5). The effect of oxygen is felt in the competition for primary radicals, in the presence of HOO· which reacts with RSH and in the reactions of RSSR, thiyl and alkyl radicals with O_2. The mechanisms for cystine, other thiols and disulphides are considered. Section IV deals with thiols irradiated in the pure liquid state and Section V with reactions and ESR studies of thiols in the solid state and in frozen solutions and glasses.

Section VI describes the phenomenon of protection of mammals against the harmful effects of ionizing radiation, considering the mechanism of the protection and studies on model substrates in solutions and also in the solid state. The mechanisms of 'competition scavenging' and 'repair' are considered; in both the thiol is thought to prevent or reduce damage caused by attack of free radical precursors on the biological solute ('indirect action'), since thiols with OH, H or e_{eq}^- yield thiyl radicals and these are relatively unreactive towards biological molecules and thus the damage is reduced.

Section VII mentions briefly radiation-induced addition of thiols to olefins.

107 references up to 1972.
Complementary: 27, 24.
Relevant: 9, 8; **10**, 6; **11**, 6; **13**, 17; **15**, 9; **16**, 21; **17**, 10; **19**, 10; **22**, 11; **25**, 12; **27**, 23; **29**, 7; **31**, 6; **32**, 9; **33**, 21; **36**, 8.

18, *12: Synthetic uses of thiols* by *R. K. Olsen and J. O. Currie, Jr*

Thiols can be converted into a wide variety of bivalent organosulphur compounds.

Section II describes the formation of dithioacetals, widely used for the protection of carbonyl compounds, and their hydrolysis to the parent CO compound (Table 1). Dithioacetals can be reduced (Raney Ni, etc.) to the corresponding methylene compounds or to olefins. Dithioacetals are extensively used as methylene blocking groups, making it possible to carry out alkylations, decarbonylations, formation of dicarbonyl compounds, ketone transpositions and selective C—C bond cleavages at other positions in the molecule.

2-Lithio-1, 3-dithianes are easily prepared: They undergo a variety of reactions with electrophiles and can be hydrolysed to carbonyl compounds or desulphurized to the corresponding methylene derivatives:

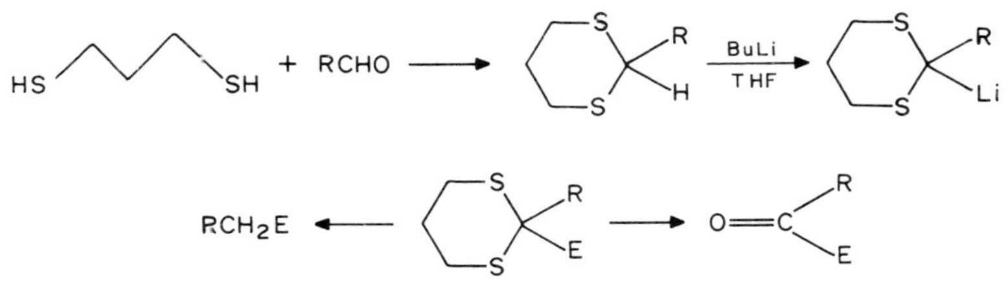

The reactions of 2-lithio-1,3-dithianes with alkyl and aryl halides, epoxides, aldehydes, ketones, acylating agents and oxidants (I_2, Cu salts, etc., leading to dimerization) are described.

Section III discusses the preparation of monothioacetals by C=O condensations with $HOCH_2CH_2SH$ or $HOCH_2CH_2CH_2SH$, and the regeneration of the C=O group by the use of Raney Ni and by other methods. Section IV deals with the similar formation and removal of thiazolidines from C=O compounds with $HSCH_2CH_2NH_2$.

Section V considers the formation of thioenol ethers from carbonyl groups:

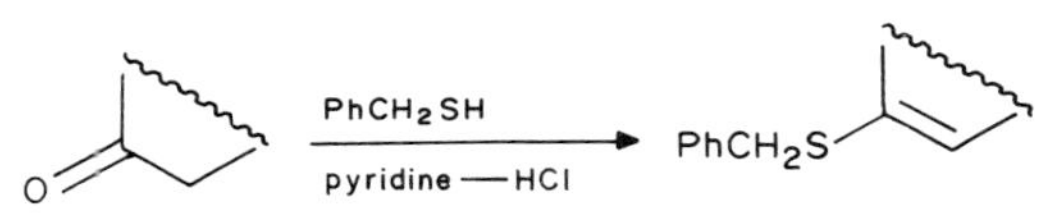

especially in ketosteroids and related compounds. These are reconverted to the parent compound on dilute acid hydrolysis.

Thiols are also used as methylene blocking groups when under *p*-toluenesulphonic acid catalysis they give thioenol ethers:

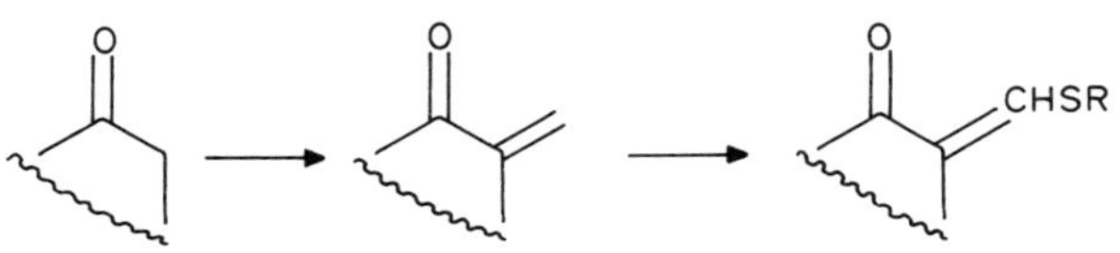

so that the ketones can be alkylated (Table 2) without interference by the active methylene site. Similar procedures are used in monoalkylations, geminal alkylations (Table 3 and the preparation of α, β-unsaturated aldehydes (Table 4).

Section VI deals with the Stevens rearrangement of sulphonium salts (allyl and non-allyl, Table 5) and subsequent S extrusion from the products. Extrusion of SO_2 by pyrolysis of sulphones ($RSO_2R' \xrightarrow{\Delta} R - R' + SO_2$) and the Ramberg–Bäcklund reaction are also treated. Finally (Section VII) some less general synthetic uses of thiols are considered, such as methylation, blocking of α-CH_2 groups in esters, cleavage of methyl esters and ethers, dehalogenations and the use of α-sulphenyl carbanions, are described.

198 references up to 1972.
Complementary: none.
Relevant: none.

18, *13: Biochemistry of the thiol group* by *A. L. Fluharty*

I. Introduction (590); **II**. Thiol metabolism (591); **III**. Biological thiols and their function (608); **IV**. Conclusion (662); **V**. References (664).

The SH function is relatively reactive under biological conditions (moderate temperature and netural pH).

Section II discusses various aspects of thiol metabolism. Among these are reduction of sulphate to sulphite and to sulphide, assimilation of sulphide, oxidation and desulphuration of cysteine, cysteine–cystine interconversion, trans-sulphuration via cystathione and thiol formation by cysteine incorporation. In all cases the role of enzymes and the metabolic pathways are presented in detailed schemes.

Section III describes the most important biological thiols. Glutathione is probably the most ubiquitous single thiol, and its biosynthesis, degradation, role in keeping cells in a reduced state, involvement in electron transport, use as an enzyme cofactor and importance in mercapturic acid formation and its detoxification are considered.

The significance of methionine and *S*-adenosylmethionine in methylations, transmethylations and other sulphonium ion alkylations are discussed. This is followed by treatment of pantetheine cofactors, the biosynthesis of coenzyme A and its thioesters and the reactions of the latter, and by consideration of phosphopantetheine proteins and lipoic acid and its derivatives.

Thiol proteins are discussed, including thioester enzyme intermediates (e.g. glyceraldehyde phosphate dehydrogenase), persulphide enzyme intermediates (rhodanase, which catalyses—SCN formation from thiosulphate and cyanide) and thiols which participate in the binding of substrates or coenzymes at active sites (e.g. metals in metalloproteins), and the section deals also with the structural importance of thiols and disulphides in proteins.

Lastly, dithiol and polythiol proteins are described, including thioredoxins, dithiol–flavin enzymes, polythiol metal-binding centres and iron–sulphur redox proteins ('non-haem iron proteins').

102 references up to 1973.
Complementary: none.
Relevant: 2, 7; **3**, 5; **4**, 9; **5**, 18; **7**, 4; **11**, 11; **12**, 12; **13**, 14; **15**, 10; **16**, 13; **17**, 13; **19**, 13; **20**, 6; **22**, 20; **24**, 14; **26**, 13; **28**, 17; **31**, 9.

18, *14: Protection of the thiol group* by *Y. Wolman*

I. Introduction (669); **II**. Disulphides as a protecting group (670); **III**. Thioethers (671); **IV**. Thioesters (677); **V**. Semithioacetals (680); **VI**. Heterocycle rings (682); **VII**. Acknowledgements (682); **VIII**. References (682).

Conversion of SH groups to more stable compounds enables reactions to be carried out at other sites in the molecule.

Disulphides are obtained by oxidation (RSH → RSSR) and can be easily regenerated (Section II). Benzilic, diphenylmethyl and triphenylmethyl ethers are often used as protecting groups, and also picolyl, acetamidomethyl β,β,β-trifluoroacyl-α-acylaminoethyl and β, β-diethoxycarbonylethyl derivatives (Section III). Among thioesters, acetyl, benzoyl, benzyloxycarbonyl and urethane derivatives are useful (Section IV).

Semithioacetals employed include tetrahydropyranyl, benzylthiomethyl, phenylthiomethyl and isobutyloxymethyl derivatives (Section V). β-Aminothiols (cystein and its derivatives) react with carbonyl compounds to form thiazolidine derivatives (Section VI).

In each section, the syntheses and decompositions of the protected derivatives are discussed and documented.

71 references up to 1973.
Complementary: none.
Relevant: 4, 11; **13**, 18.

18, *15: Rearrangements involving thiols* by *T. Sheradsky*

I. Introduction (686); **II**. Group migrations from and on to thiols (686); **III**. Rearrangements of the *O*-thioacyl system (698); **IV**. The thio-Claisen rearrangement (702); **V**. Internal additions, eliminations and ring-chain tautomerisms (706); **VI**. Miscellaneous rearrangements (715); **VII**. References (717).

Section II discusses alkyl migrations from sulphur to carbon (e.g. $PhCH_2SCH_2PH \rightarrow PhCH{=}CHPh$) and from S to O and also aryl migrations (Smiles rearrangement):

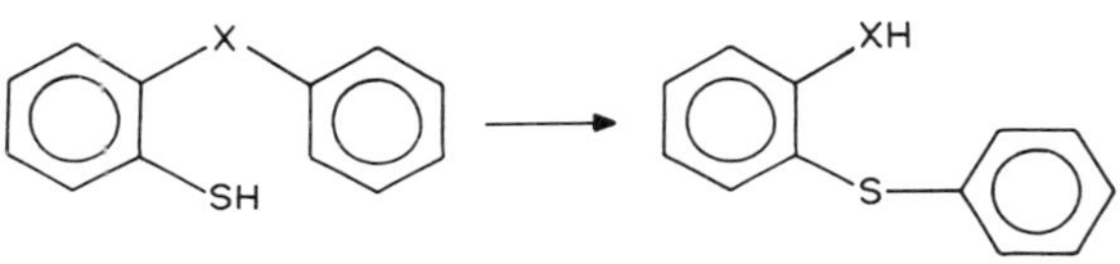

acyl migrations from S to O and from S to N (e.g. $CH_3COSCH_2CH_2NH \rightarrow HSCH_2CH_2NHCOCH_3$) and migration of cyano and amidino groups.

Section III describes rearrangements of *O*-thioacyl systems to *S*-acyl systems ($ArOCOAr \rightarrow ArOCSAr$), occurring either through a cyclic transition state or through dissociation and return. Section IV deals with the thio-Claisen rearrangement of allyl and prop-2-ynyl aryl sulphides.

Section V treats intramolecular additions to double and to triple bonds (Table 1), cyclization and ring-chain tautomerism of cyanothiols, mercaptoaldehydes and mercapto ketones, ring openings of cyclic sulphides to unsaturated thiols:

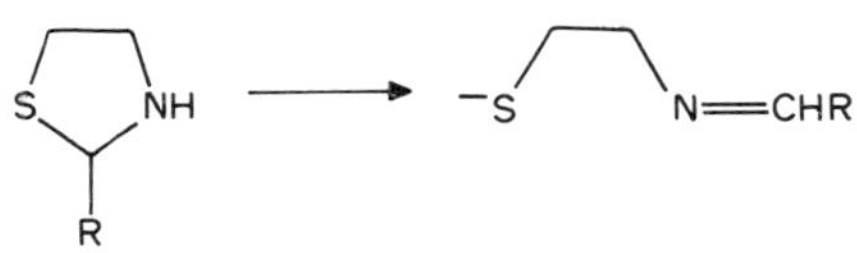

and some other non-general reactions.

119 references up to 1972.
Complementary: none.
Relevant: 1, 7; **2**, 15; **3**, 14; **4**, 10; **5**, 8; **8**, 9; **9**, 3; **10**, 3; **11**, 4; **12**, 16; **13**, 13; **14**, 7; **15**, 8; **16**, 17; **17**, 8; **19**, 18; **20**, 4; **23**, 13; **24**, 10; **26**, 15: **29**, 4.

18, *16: Thiols as nucleophiles* by *M. E. Peach*

I. Introduction (722); **II**. Substitution reactions (725); **III**. Addition reactions (760); **IV**. References (775).

Thiols may act as nucleophiles in substitutions ($RSH + AB \rightarrow RSA + HB$) or in additions $\left(RSH + \;\rangle C{=}C\langle\; \rightarrow RS\overset{|}{\underset{|}{C}}-\overset{|}{\underset{|}{C}}H\right)$.

Section II discusses the various types of substitutions with RSH or RS^-. In aliphatic halo acids, halo ketones and sulphonates, the halo or sulphonyl group is displaced by thiolate: $ClCH_2CO_2H \rightarrow RSCH_2CO_2H$. In benzyl halides and alicyclic halides or toluenesulphonates the substituent is replaced with thiolate and the same is true for alkyne, alkene and imine derivatives.

In the reactions of aromatic compounds the displaced group is usually halide, and in polyhalides all halogens can be replaced:

$$C_6X_6 + RS^- \rightarrow C_6X_5SR \rightarrow \rightarrow \rightarrow C_6(SR)_6$$

Such reactions are described with mixed hexahalobenzenes and with miscellaneous polyhalogenated arenes, monohalogenobenzenes, halogeno heteroarenes and aromatic compounds containing other substituents ($SO_2CH_3 > NO_2 > F > Cl$). Thiols can also be used for demethylation: $R_2\overset{+}{N}Me_2Cl \xrightarrow{ArS^-} R_2NMe$.

Subsequently the preparation and reactions of thiol derivatives of main group elements are described, such as $(t\text{-BuS})_2Be_3(OBu\text{-}t)_4$, $B(SC_6F_5)_3$, $MeSSiH_3$, $(RS)_3P$, $(RS)_3PO$, $PhAs(SBu)_2$ and ArSSSAr. Transition metal derivatives of simple (monofunctional) thiols are discussed next, including compounds such as $Ni(SR)_2$, $Hg(SR)_2$ and $[Co(SR)_2]_n$ and also complex ions of the type $[M(SR)_2]^-$ and $[M(SR)_4]^-$ (where M = Co, Pd, Pt, Zn, Cd, Hg, Cu, Agor Au) and organometallic transition metal complexes, e.g. $(\pi\text{-}C_5H_5)_2Ti(SR)_2$. Finally, metal (Th, Ir, Cr, Mo, W, etc.) carbonyl complexes containing thiol groups are considered.

Section III describes addition reactions of thiols or thiolate ions. Most of these occur by a radical mechanism, but only ionic additions are discussed here. Thiols add to olefins, forming sulphides [$Me_2C{=}CHMe + RS \rightarrow Me_2C(SR)CH_2Me$]; with acetylenes in step-wise additions first an unsaturated sulphide and then a saturated disulphide is formed: $HC{\equiv}CH \xrightarrow{RSH} CH_2{=}CH(SR) \xrightarrow{RSH} RSCH_2CH_2SR$. By addition to nitriles, iminothiothers are formed: $RC{\equiv}N \rightarrow R(R'S)C{=}NH.HCl$. With carbonyl groups, hemithioacetals and thioacetals are obtained. Conjugated systems, $C{=}C-C{=}X$ (X = C, N, O), usually give addition of thiols across the C=C bond: $MeCH{=}CHCHO \xrightarrow{RSH} MeCH(SR)CH_2CHO$. Three-membered rings containing O or S react through ring opening, producing hydroxy sulphides or thiol sulphides. Finally, thiols also add across C=N bonds in cyclic compounds (e.g. thiazolium salts).

310 references up to 1972.
Complementary: none.
Relevant: none.

18, *17: Oxidation of thiols* by *G. Capozzi and G. Modena*

I. Introduction (785); **II**. Electrochemical oxidation (787); **III**. Chemical oxidation (789); **IV**. Oxidation by molecular oxygen (806); **V**. Photo-oxidation (832); **VI**. References.

Thiols are oxidized by a variety of reagents to disulphides and higher oxidation products.

Section II describes electrochemical oxidation in various conditions (Table 1). Section III discusses chemical methods of oxidation, including the use of H_2O_2, hydroperoxides and peroxy acids, halogens, sulphoxides (Tables 2–4), diethyl azodicarboxylate (Table 5), nitroso, nitro, iodoso and sulphoxonium compounds, metal ions and oxides [Fe^{3+} (Table 6), Ce^{4+}, Co^{3+}, V^{5+} etc.].

Section IV deals with oxidation by O_2 and considers the catalysis of this process by strong bases (Tables 7, 9 and 11) and the effect of solvents (Tables 8, 10, 12). The disulphides produced may also undergo base-catalysed oxidation to sulphonic acids (Table 13). The thiol → disulphide oxidation is also catalysed by amines (Table 14) and by metal ions (Tables 15–19), when the rate of oxidation is strongly influenced by the presence of ligands (Table 20) and also by complex biological ligands such as porphyrins (Table 21). Catalysis by organic redox systems (e.g. quinone–hydroquinone) and co-oxidation of thiols with other organic compounds (alkenes, acetylenes) is also described. Finally, in Section V photo-oxidation is briefly treated.

197 references up to 1973.
Complementary: none.
Relevant: 27, 8, 12, 22.

18, *18: The synthesis and uses of isotopically labelled thiols* by *A. Lapidot and C. S. Irving*

I. Introduction (841); **II.** Motional processes (842); **III.** Cleavage of the S—H bond (846; **IV.** Tracing ^{35}S-labelled thiols in biological systems (856); **V.** Applications of ^{35}S-tracer studies to agricultural science and industry (865); **VI.** Isotope labelling and counting in practice (866); **VII.** References (880).

The chapter treats thiols (RSH) in which the H atom is replaced by ^{2}H or ^{3}H or the S atom by ^{35}S, and also labelling in R.

Section II reviews measurements of translational, rotational and vibrational motions of thiols. Isotopic labelling overcomes the problem that loss of, e.g., an H atom from different sites of a group (C— or S—bonded H in a —CH_2SH group) gives fragments of identical mass in MS and isotopically labelled molecules can be used to obtain sufficient molecular parameters in microwave and in infrared spectroscopy.

Section III discusses S—H bond cleavage, the primary isotope effect in this process and the nature of the transition state. As a tracer, ^{2}H is indispensable in following the fate of the thiol hydrogen in the reaction mixture and ^{35}S is used similarly to trace the fate of the sulphur atom in both chemical and enzymatic reactions.

Section IV describes the use of ^{35}S-labelled thiols in macromolecular systems (involving, e.g., transfer-RNA, proteins and nucleic acids in viruses), and also in whole-body systems, where lipid-soluble thiols may pass freely in and out of cellular structures and are used in enzymatic studies, drug design, distribution of drugs in the body, uptake of amino acids by proteins, the study of the neurosecretory system, etc.

Section V reviews very briefly the uses of ^{35}S tracer studies in agriculture and industry.

Section VI describes technical problems associated with isotope labelling experiments. Thus, methods of synthesis are described and tabulated (Table 1), followed by a review of the counting methods involving ^{35}S and the methods used in preparing the samples for counting (wet ashing, oxygen flask combustion, etc.), and finally some methodological considerations relevant to body tracer studies are given.

168 references up to 1971.
Complementary: 27, 10.
Relevant: 5, 10; **12**, 13; **13**, 15; **17**, 12; **19**, 17; **22**, 14; **23**, 16; **24**, 20; **26**, 11; **28**, 8; **29**, 28; **33**, 18; **36**, 12.

The chemistry of the hydrazo, azo and azoxy groups (1975)

19, *1: Electronic structures of the azo, azoxy and hydrazo groups* by *M. B. Robin*

I. Introduction (1); **II**. The azo group (2); **III**. The azoxy group (13); **IV**. The hydrazo group (15); **V**. References (21).

The chapter discusses the relevant semi-empirical and *ab initio* calculations, compared predicted structures with the results of spectroscopic experiments and compared with the three title groups between themselves.

Section II describes the azo group, including theoretical calculations for diimide (HN=NH) (Table 1) and diazirine (Table 2), ionization potentials (Table 3) and photoelectron spectra and optical spectra of diazirines 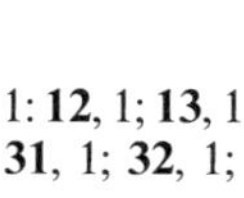 and azoalkanes (RN=NR′).

Section III treats briefly the azoxy group and Section IV considers theoretical calculations for the hydrazo group (Table 4) and photoelectron (Table 5) and optical spectra.

41 references up to 1973.
Complementary: none.
Relevant: 1, 1; **2**, 1; **3**, 1; **4**, 1; **5**, 1; **6**, 1; **8**, 1; **11**, 1: **12**, 1; **13**, 1; **14**, 1; **15**, 1; **16**, 1; **17**, 1; **18**, 1; **20**, 1; **22**, 1; **23**, 1; **24**, 1; **25**, 1; **26**, 1; **27**, 9; **28**, 1; **31**, 1; **32**, 1; **33**, 1; **36**, 1.

19, *2: Structural chemistry* by *R. Allmann*

I. Introduction (23); **II**. General remarks about bond lengths (24); **III**. Structural examples (36); **IV**. References (51).

Section II discusses the determination of bond lengths by various methods and their accuracy (Table 1) and the standard bond length for various relevant atom pairs (N—N, N—C, C—C, C—O, N—O) (Table 2), and gives structural details for a number of molecules containing N—N, N=N and related groups.

Section III describes the structures of hydrazine derivatives and other compounds with N—N single bonds, azo compounds, including simple ones such as HN=NH and FN=NF, and alkyl and aryl derivatives, and finally considers azoxy compounds and nitroso dimers (e.g. azobenzene dioxide 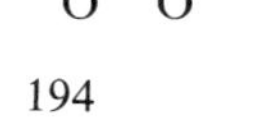).

41 references up to 1972.
Complementary: none.
Relevant: **16**, 2; **17**, 2; **18**, 2; **20**, 2; **21**, 1; **22**, 2; **23**, 3; **24**, 2; **26**, 2; **27**, 4; **28**, 2; **29**, 1; **30**, 1; **31**, 24; **32**, 24; **33**, 12; **34**, 10; **36**, 3.

19, *3: Thermochemistry of hydrazo, azo and azoxy groups* by *R. Shaw*

I. Introduction (53); **II**. Estimation by group additivity (54); **III**. Kinetics and thermochemistry (55); **IV**. Hydrazine and substituted hydrazines (59); **V**. Azo compounds (64); **VI**. Acknowledgements (68); **VII**. References (68).

Section II discusses the estimation of thermochemical data by group additivity, and Section III deals with kinetics and thermochemistry in general and presents the relevant equations.

Section IV describes hydrazine and substituted hydrazines (Tables 1 and 2) and the N—N homolysis, and presents the thermochemical properties of amino and substituted amino radicals (Table 3) and the relevant group values for their calculation (Table 4). Subsequently, four-centre concerted loss of ammonia and of molecular hydrogen is described.

Section V deals with azo compounds (Table 6–9) and their decomposition in the gas phase.

17 references up to 1972.
Complementary: **21**, 3.
Relevant: **16**, 16; **18**, 3; **20**, 11; **22**, 6; **23**, 4; **24**, 3; **25**, 2; **26**, 4; **27**, 9; **28**, 4; **29**, 24; **30**, 2; **31**, 2; **33**, 3; **36**, 4.

19, *4: Preparation procedures* by *J. W. Timberlake and J. C. Stowell*

I. Preparation of hydrazines (69); **II**. Preparation of acyclic azoalkanes (83); **III**. Preparation of azoxy compounds (96); **IV**. References (102).

Section I reviews the preparative procedures for obtaining hydrazines. These include alkylations of hydrazine and its derivatives or of *N*, *N'*-diacylhydrazines by halides, sulphates, epoxides, aziridines or active olefins. Other routes involve addition of RLi to azo compounds or addition of RMgX to hydrazones, cycloadditions between azo compounds and dienes to yield cyclic azo compounds, reductions of *N*-nitroso compounds, azines, hydrazones, azo compounds and acylhydrazines and *N*-aminations by displacement of good leaving groups from ammonia derivatives [e.g. $Et_2NCH_2CH_2NHPh + NH_2Cl \rightarrow Et_2NCh_2CH_2N(Ph)NH_2$]. Diaziridines on hydrolysis yield ketones and hydrazines, and ureas and sulphamides on treatment with NaOH and NaOCl also yield hydrazines:

$$Et_2NSO_2NH_2 \xrightarrow[(2)HCl(3)NaHCO_3]{(1)NaOH + NaOCl} Et_2NNH_2$$

Section II describes preparations of acyclic azoalkanes by oxidation of 1,2-dialkylhydrazines ($R_2CHNHNHCHR_2 \xrightarrow{HgO} R_2CHN{=}NCHR_2$), by oxidation of amines (oxidative coupling by IF_5 for *tert*-alkylamines: $Me_3CNH_2 \xrightarrow{IF_5} Me_3CN{=}NCMe_3$), by oxidation of ureas with *t*-BuOCl and base, from hydrazodicarboxylates in a complicated multi-step procedure and finally by oxidation of *N*, *N'*-dialkylsulphamides:

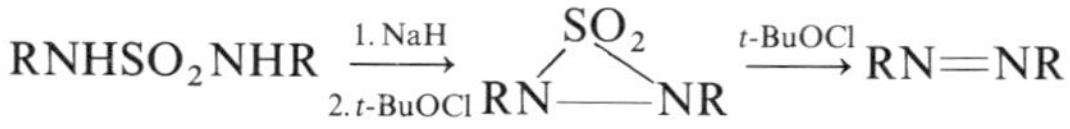

Section III deals with the preparation of azoxy compounds by oxidation of azo compounds with peracetic or chloroperbenzoic acid or with H_2O_2, by reduction of azodioxy compounds (nitroso dimers), by condensation of nitroso compounds with hydroxylamines [$RNO + R'NHOH \rightarrow RN(O){=}NR'$], by alkylation of diazotates:

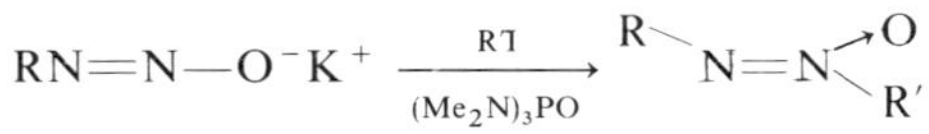

by Grignard reagents from diimide *N*-oxides containing F or TsO leaving groups [$ArN(O){=}NF \xrightarrow{RMgX} ArN(O){=}NR$] and finally by photochemical rearrangements of various azoxy compounds to produce new azoxy derivatives.

176 references up to 1973.
Complementary: none.
Relevant: none.

19, *5: Mass spectra of hydrazo, azo and azoxy compounds* by *S. W. Tam*

I. Introduction (109); **II**. Simple fragmentations (110); **III**. Fragmentations involving rearrangements (116); **IV**. Conclusions (125); **V**. References (126).

Section II describes simple fragmentations which take place without rearrangement, emphasizing processes which are characteristic of the presence of the functional groups discussed. The compounds treated are azobenzenes, azomethanes, aromatic azoxy compounds, polyfluorinated azoxymethanes, substituted hydrazines, aliphatic hydrazones and aromatic azines.

Section III discusses fragmentations involving rearrangements (either skeletal or H migration or both) of azobenzenes, azomethanes, aromatic azines, azoxy compounds and substituted hydrazines.

30 references up to 1972.
Complementary: none.
Relevant: **9**, 7; **10**, 5; **13**, 9; **16**, 4; **17**, 5; **18**, 6; **22**, 7; **24**, 6; **25**, 4; **26**, 6; **27**, 7; **29**, 3; **30**, 22; **31**, 3; **32**, 3; **33**, 4; **36**, 7.

19, *6: Chiroptical properties of the azo and the azoxy chromophores* by *G. Snatzke and F. Snatzke*

I. The azo chromophore (129); **II**. The azoxy chromophore (147); **III**. Acknowledgement (150); **IV**. References (150).

Section I discusses the azo chromophore, considering the geometry and electron absorption spectra of azo compounds on the basis of simple MO treatment, photoelectron spectroscopy and transitions of *trans*- and *cis*-azo compounds (Tables 1 and 2). Subsequently, CD and ORD of azoalkanes, phenylazo compounds with chiral first sphere, azo-containing biopolymers and induced Cotton effects (including magnetic ORD and induced CD spectra in complexes with biopolymers) are considered.

Section II describes briefly the MO treatment and chiroptical properties of the azoxy chromophore.

59 references up to 1973.
Complementary: none.
Relevant: **8**, 4; **18**, 7; **20**, 2; **22**, 4; **25**, 3; **26**, 3; **27**, 6; **28**, 10; **29**, 23; **31**, 3; **36**, 17.

19, *7: Basicity hydrogen bonding and complex formation involving hydrazo, azo and azoxy groups* by *P. J. Krueger*

I. Introduction (154); **II**. Structural considerations (155); **III**. Basicities (159); **IV**. Hydrogen bond formation (175); **V**. Complex formation (212); **VI**. References.

Section II deals with conformations of hydrazines and with geometric isomerism in azomethine, azo and azoxy compounds and considers also triazenes and their tautomerism (Ar—NH—N=N—Ar′⇌Ar—N=N—NHAr′) and also triazene oxide–hydroxytriazene tautomerism [R^1—N=N—N(OH)R^2⇌R^1NH—H=$\overset{+}{N}$—($\overset{-}{O}$)R^2].

Section III describes the basicity of hydrazines (Tables 1 and 2) and carboxylic acid hydrazides ($RCONHNH_2$) (Table 3), azomethines (e.g. hydrazones, Table 4; oxines, Table 5), azobenzenes with various substituents in different solvents (Tables 6–8), azoxy compounds (Tables 9 and 10), and triazenes and their oxides (Tables 11 and 12).

Section IV considers hydrogen bond formation of hydrazines with various OH-containing molecules (Table 13), hydrogen bond lengths in hydrazine derivatives in the crystalline state (Table 14), intermolecular hydrogen bonds between *p*-nitrophenol and substituted benzalanilines (Table 15) and also polarographic, IR, UV, NMR, X-ray, association (Table 16) and tautomerization studies of various azomethines. Azo compounds accept protons from alcohols and phenols (Table 17) and also from OH groups intramolecularly. In azoxy compounds probably the O atom is the electron donor site. Hydrogen bonding is also important in triazenes (NH···N bonds in crystals and in dimers as well as intromolecularly), and finally triazene oxides give both intramolecular hydrogen-bonded structures and also intermolecular associations.

Section V discusses metal–ion complexes of the title compounds which often involve hydrogen-bonds and chelate rings and finally deals with addition products with Lewis acids (Tables 18 and 19).

305 references up to 1973.
Complementary: none.
Relevant: **2**, 9; **3**, 6; **4**, 4; **5**, 6; **7**, 6; **8**, 5; **11**, 3; **12**, 3, 10; **13**, 7; **15**, 4; **16**, 5; **17**, 6; **18**, 8; **20**, 12; **22**, 8; **23**, 6, 7; **24**, 4; **25**, 5, 6; **29**, 17; **31**, 17, 21; **32**, 21; **33**, 5.

19, *8: Directing and activating effects* by *D. A. R. Happer and J. Vaughan*

I. Introduction (226); **II**. Directing and activating effects of the azo group (226); **III**. Directing and activating effects of the azoxy group (244); **IV**. Directing and activating effects of the hydrazo group (254); **V**. References (255).

Section II discusses effects exerted by the azo group on reactions of aromatic side-chains. The phenylazo substituent PhN=N— is considered as a unit influencing side-chain reactivity in groups attached to the second aryl group in PhN=N—C_6H_4X, where X may be OH, COOH, COOR, etc. Various substituent constants (σ^0, σ^-, σ^+) are discussed.

Arylazo groups are treated next, i.e. $YC_6H_4N{=}N{-}$ groups with Y substituents at different positions and of different natures and the side-chains in the second aryl group again may be OH, COOR, etc. Transmission of resonance effects through the azo group is discussed in relation to structural, kinetic and NMR data. Electrophilic substitutions in azo compounds are reviewed, when azobenzenes are rather unreactive and in many cases the N=N linkage or side-chains are attacked preferentially. However, the phenylazo groups activate both nucleophilic and free radical aromatic substitution. Protonated azo (azonium) groups have a powerful $-R$ effect.

Section III deals similarly with the azoxy group. Here, two isomers exist, $Ar{-}N{=}\overset{O}{\overset{\uparrow}{N}}{-}$ (aryl-NNO-azoxy) and $Ar\overset{O}{\overset{\uparrow}{N}}{=}N{-}$ (aryl-NNO-azoxy). The basicities of p-$Me_2NC_6H_4N{=}NC_6H_5$ have been studied, in addition to electrophilic aromatic substitutions in both types ($-N\overset{O}{\overset{\uparrow}{N}}O-$ and $-ONN-$) of azoxy compounds and also in alkylazoxybenzenes, $ArCH_2N{=}NPh$ and $ArCH_2N{=}\overset{O}{\overset{\uparrow}{N}}Ph$. The reactions are complicated by the possibility of acid-catalysed migration of oxygen in the azoxy group. Only qualitative data exist as to the activating power of the two isomeric arylazoxy groups, but both have a strong $+R$ effect. They also activate nucleophilic aromatic substitutions. NMR chemical shifts are given for aliphatic azoxy and azo (Table 1) and for aromatic azoxy (Table 2) compounds.

Finally (Section IV), directing and activating effects of the hydrazo groups are briefly treated, since under most reaction conditions their rearrangements and/or oxidations are rapid.

78 references up to 1971.
Complementary: none.
Relevant: **3**, 3; **4**, 5; **5**, 16; **6**, 8; **7**, 2, 3; **11**, 12; **12**, 5; **13**, 8; **14**, 4; **15**, 5; **16**, 6; **18**, 9; **23**, 11; **28**, 9; **29**, 16; **31**, 8; **36**, 15.

19, *9: The transition metal chemistry of azo compounds* by *M. I. Bruce and B. L. Goodall*

Section II describes complexes of Mo, Ir, Ni and Pt with azo compounds which involve no interaction between the metal and C atoms, e.g.

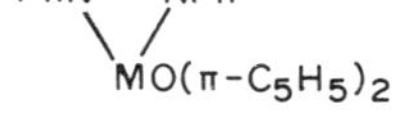

Section III deals with complexes of iron carbonyls which contain a cyclic Fe_2N_2 system, e.g.

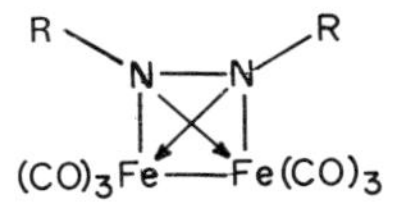

Similar complexes are also formed with heterocycles containing N=N bonds.

Section IV discusses the syntheses of complexes from azobenzene, $C_6F_5N{=}NC_6H_5$ and $C_6F_5N{=}NC_6F_5$, with Mo, Mn, Re, Fe, Ru, Co, Rh, Ir, Ni, Pt and Pd. In these the *ortho*-H of the azo compound is eliminated (or F, in the case of C_6F_5). The structures of these compounds are described in detail, together with mechanisms of the complex formation and the ^{19}F chemical shifts of complexes formed from m-$FC_6H_4N{=}NPh$.

Section V treats Mo, Fe, Ru, Co and Rh complexes with *o*-semidine (*o*-aminodiphenylamine) and the mechanism of the metal-promoted *o*-semidine rearrangement.

Section VI discusses processes in which azobenzene complexes undergo various synthetically useful reactions. The simplest case is the addition of one-carbon units:

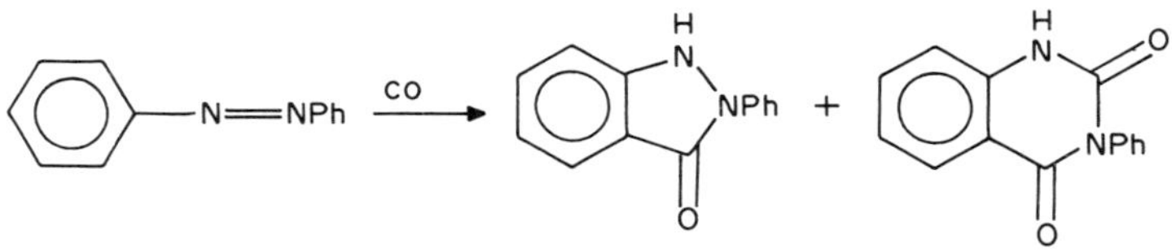

Similar reactions are also observed with two- and three-carbon units, with H ($LiAlH_4$) and with halogens.

Section VII reviews the reactions of diazonium salts with transition metal complexes (usually containing hydride ligands), which yield complexes with ArN=N or ArN=NH groups. Thus, $[(\pi\text{-}C_5H_5)Mo(CO)_3]^-$ reacts with ArN_2^+ BF_4^- and yields $(\pi\text{-}C_5H_5)Mo(CO)_2N{=}NAr$.

Related complexes are also formed from derivatives containing W, Re, Fe, Ru, Os, Co, Rh, Ir and Pt. The structures of some of these complexes have been studied. Finally (Section VIII), the survey is summarized and examples of ligands for which *ortho*-metallation or related reactions have been reported are tabulated (Table 2). The chapter contains an Appendix added in December 1973.

145 references up to 1974.
Complementary: none.
Relevant: **9**, 5; **21**, 12; **25**, 14; **30**, 4–7; **32**, 26; **34**, 5–8; **35**, 3, 6.

19, *10: Radiation chemistry of hydrazo and azo compounds* by *A. Horowitz and L. A. Rajbenbach*

Section II discusses the radiation chemistry of hydrazine and its methyl derivatives (Table 1), the radiochemical synthesis of hydrazine from NH_3, aqueous solutions of hydrazine and derivatives, including rate constants with OH and with e_{aq}^- (Table 2), the use of 1, 1-diphenyl-2-picrylhydrazyl in radiolytic studies in order to evaluate the yield of radiolytically formed radicals and ESR studies of γ-irradiated hydrazo compounds.

Section III describes the radiation chemistry of azomethane in the vapour phase, of azo esters and of azo dyes in aqueous and in organic solutions.

48 references up to 1973.
Complementary: none.
Relevant: **9**, 8; **10**, 6; **11**, 6; **13**, 17; **15**, 9; **16**, 12; **17**, 10; **18**, 11; **22**, 11; **25**, 12; **27**, 23, 24; **29**, 7; **31**, 6; **32**, 9; **33**, 21; **36**, 8.

19, *11: Formation and fragmentation of cyclic azo compounds* by *K. Mackenzie*

I. Introduction (330); **II**. Monocyclic azo compounds (330); **III**. Polycyclic azo compounds (394); **IV**. Appendix (432); **V**. Acknowledgement (424); **VI**. References (434).

The chapter deals only with azo functions contained in alicyclic structures.

Section II describes the preparation and general properties of diazirines and their thermal and photochemical fragmentation, yielding alicyclic and olefinic products in the thermal route and products derived from triplet methylene in the photochemical route. Very few studies deal with diazetines.

Several routes for the synthesis of pyrazolines are described, including the relevant mechanism of cycloadditions leading to these compounds. Pyrazolines undergo easy thermal fragmentation, producing N_2 and cyclopropanes (Table 1) or olefins (Table 2). In the photolysis of substituted pyrazolines the products are substituted olefins and cyclopropanes. The rates and mechanisms of these processes are described in detail. Other five-membered rings are discussed, such as methylene, pyrazolines, pyrazolenines and vinylpyrazolines and also miscellaneous pyrazoline rearrangements. Six- and seven-membered rings and larger rings are treated, in addition to large rings containing two azo groups.

Section III deals with bicyclic pyrazolines and other non-bridged systems and also with diazabicyclo[2.2.1]heptenes and related bridged bicyclic azo compounds, bridged tri-, tetra- and pentacyclic azo compounds and related caged structures (diazabicycloheptenes, diazabicyclooctenes, bridged polycyclic azo compounds derived from triazines and tetrazines and other caged structures containing the azo function). In each section the preparation, reactions, thermochemistry and photochemistry of the compounds are discussed in detail. The section ends with a description of the use of 4-phenyl-1, 2, 4-triazoline-3, 5-dione:

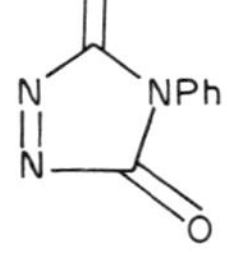

(easily accessible through phenyl isocyanate) as a powerful dienophile reacting easily and in excellent yields with a wide variety of dienes, thus providing a facile route to many polycyclic azo compounds.

The Appendix (Section IV) discusses the theory of the addition of diazomethanes to olefins to form pyrazolines.

230 references up to 1974.
Complementary: none.
Relevant: none.

19, *12: The electrochemistry of azoxy, azo and hydrazo compounds* by *F. G. Thomas and K. G. Boto*

I. Introduction (443); **II**. Electrolytic preparation of azoxy, azo and hydrazo compounds (445); **III**. Electrochemical reactions (451); **IV**. References (490).

Section II describes the electrolytic preparation of azoxy compounds by the reduction of nitrobenzene. Higher current densities and more concentrated alkaline solutions lead to the formation of azo compounds, and further to hydrazo compounds (Tables 1 and 2).

Section III deals with the electrochemistry of azoxy compounds in protolytic solvents, compares unsubstituted and substituted azoxy compounds and discusses conditions (pH, solvent, electrolyte) and the position, basicity and electroreactivity of substituents, especially NO_2 groups (Table 3). Azo compounds are reduced to hydrazo compounds and the reaction is complex, depending on the substituents, concentration, solvent composition, presence of surfactants and pH (Table 4). Finally, the electrochemical behaviour of the three title compounds in aprotic solvents is described and extensive data are given (Table 5) for the polarographic behaviour of azo and hydrazo compounds. Mechanistic details are presented.

82 references up to 1973.
Complementary: none.
Relevant: **5**, 2; **8**, 11; **13**, 5; **16**, 15; **17**, 14; **20**, 5; **21**, 5; **22**, 9; **23**, 10; **24**, 17; **25**, 13; **26**, 12; **27**, 8; **29**, 8, 9; **31**, 7; **32**, 6; **34**, 1.

19, *13: Biological formation and reactions of hydrazo, azo and azoxy groups* by *T. Miyadera*

I. Introduction (496); **II**. Natural occurrence of hydrazo, azo and azoxy compounds (499); **III**. Metabolic reactions of hydrazo, azo and azoxy compounds (501); **IV**. Oxidation of thiols to disulphides by azo compounds and the resulting biological effects (515); **V**. Antifungal action of hydrazo and azo compounds (522); **VI**. Enzymic formation of azobenzenes from anilines in soils (526); **VII**. Biological nitrogen fixation possibly involving diimide and hydrazine intermediates (529); **VIII**. Introduction of phenylazo group into proteins and amino acids (530); **IX**. References (532).

The title compounds are involved in inhibition of protein synthesis, carcinogenesis, thiol–disulphide interchange, azo reduction, monoamine oxidase inhibition and nitrogen fixation.

Section II describes the natural occurrence of azoxy and hydrazo compounds. Section III deals with metabolic reactions, such as acetylation of hydrazides and hydrazines to hydrazo compounds, azo reduction of various azo dyes and drugs [e.g. prontosil, 2, 4-$(NH_2)C_6H_3N{=}NC_6H_4SO_2NH_2$-4′), oxidative and hydrolytic cleavage of N—C bonds of antidepressive and monoamine oxidase inhibitory drugs such as iproniazid and carcinogenic alkylation of aliphatic compounds, involving cycasin or its aglycone, methylazoxymethanol [$CH_3N(\rightarrow O){=}NCH_2OH$, 'MAM']. Various hydrazo,

azo and azoxyalkanes are potent carcinogens, in many cases owing to the methylating action of MAM; other carcinogens include 1-methyl-2-benzylhydrazine, aryldimethytriazenes and various azobenzene derivatives.

Section IV deals with the oxidation of thiols to disulphides by azo compounds (RN=NR′ + R″SH → RNHNHR′ + R″SSR″). Thus, glutathione is oxidized to the corresponding disulphide by the azo ester PhN=NCOOCH$_3$ (Table 1) or by the diamide Me$_2$NCON=NCONMe$_2$. The effect of these two azo compounds and some others on bacterial growth, protein synthesis and some other processes is described.

Section V treats the antifungal action of hydrazo and azo compounds, their mode of action and structure–activity relationships.

Some azobenzenes (e.g. 3, 3′4, 4′-tetrachloroazobenzene) are formed in soils enzymatically from the corresponding anilines. The reaction can also be carried out *in vitro* by a mixture of partially purified peroxidase and aniline oxidase, and also photochemically (Section VI).

Section VII deals briefly with biological N$_2$ fixation and the mechanisms proposed for this, involving diimide (HN=NH) and hydrazine intermediates.

Section VIII describes the possibilities of introducing phenylazo groups into proteins, e.g. by reacting —HgCl substituents on the azo compound with —SH groups of proteins or amino acids, or by coupling phenyl or imidazole rings in proteins (or amino acids) with aromatic diazonium ions.

181 references up to 1972.
Complementary: none.
Relevant: **2**, 7; **3**, 5; **4**, 9; **5**, 18; **7**, 4; **11**, 11; **12**, 12; **13**, 14; **15**, 10; **16**, 13; **17**, 13; **18**, 13; **20**, 6; **22**, 20; **24**, 14; **26**, 13; **28**, 17; **31**, 9.

19, *14: Oxidation and synthetic uses of hydrazo, azo and azoxy compounds* by *B. R. Newbold*

Section II describes the oxidation of aliphatic hydrazines (both symmetrical and unsymmetrical) to azo compounds, using HgO (preferred reagent), H_2O_2, HNO_3, NBS, Br_2, etc. Cyclic hydrazines behave similarly, although in some cases N_2 is lost and unsaturated hydrocarbons are formed. Cyclic acylhydrazines are easily oxidized to the corresponding diones of azo compounds and bicyclo- and polycyclohydrazines undergo facile oxidation to azo compounds by using $CuCl_2$, air in the presence of CuCl, HgO, etc.

Section III discusses the oxidation of aliphatic symmetrical, unsymmetrical (Table 1) and cyclic azo compounds to the corresponding azoxy derivatives, using various peracids (usually in aprotic solvents) or H_2O_2–KOH.

Section IV deals with the oxidation of arylhydrazines to azo derivatives (Tables 2 and 3), electrochemically or using $Pb(OAc)_4$, H_2O_2 or NBS, with the oxidation of various hydrazobenzenes to azobenzenes by I_2, $Pb(OAc)_4$, air, $K_2Cr_2O_7$, NaOCl and other reagents and with the oxidation of hydrazobenzenes to azoxybenzenes with H_2O_2 (Table 4).

Section V treats the oxidation of azobenzenes to variously substituted azoxybenzenes (Tables 5–7) with H_2O, peracetic acid or perbenzoic acid and with the similar oxidation of azo- to azoxynaphthalenes.

140 references up to 1973.
Complementary: none.
Relevant: 29, 25.

19, *15: Reduction and synthetic uses of hydrazo, azo and azoxy compounds* by *B. R. Newbold*

Section II describes the reduction of aliphatic azo to hydrazo compounds by N_2H_4 or H_2/PtO or $H_2/Pd–C$, of aliphatic azoxy to azo compounds (by $LiAlH_4$ or Mg–MeOH) and of aliphatic azoxy to hydrazo compounds (by $SnCl_2$–HCl, catalytically or electrochemically). Section III discusses the reduction of azobenzene and its derivatives (Table 1) to hydrazo compounds, using diimide, hydrazine, $LiAlH_4$, diborane, H_2–catalyst or electrochemical techniques. Azoxybenzenes are reduced to azobenzenes by $LiAlH_4$, Na and ethylene glycol, Co catalysts, electrochemically and by various other methods (Table 2). The Wallach rearrangement of azoxy- to hydroxyazobenzenes is also relevant, since the reaction may be conducted so as to produce azobenzene almost quantitatively, either by using PCl_3 or conc. H_2SO_4 (Table 3). Azoxybenzenes may also be reduced directly to hydrazobenzenes by Na_2S_2, $LiAlH_4$, Zn–KOH and electrochemically.

Section IV deals with reductive cleavages of the title compounds, leading to the corresponding amines by using HI, H_2–Ni, Zn–EtOH, Sn–HCl and other reducing agents.

100 references up to 1973.
Complementary: none.
Relevant: none.

19, *16: Ionic reactions involving hydrazo, azo and azoxy groups* by *A. F. Hegarty*

Hydrazines are strong bases (Table 1). Alkylations of hydrazines and of hydrazine anions occurs with alkyl halides or by expoxides or aziridines, e.g. $RNHNH_2 +$  $\rightarrow RN(NH_2)CH_2CH_2OH$. The enhanced nucleophilic reactivity ('α-effect') is shown by hydrazines and discussed here. Subsequently, the site of acylation, use of protecting groups, nature of acylating agents, diacylation and reactions of hydrazides are discussed, followed by reactions of hydrazines with activated alkenes, alkynes, haloarenes, aldehydes and ketones, and with bifunctional reagents (e.g. diketones, yielding cyclic products). Arylhydrazones give characteristically electrophilic substitution, and reactions with halogens, hydroxymethylation, alkylation and acylation are discussed. The section continues by treatment of the nitrosation of hydrazines, their reactions with aryldiazonium ions [yielding tetrazenes: $ArNHNH_2 \xrightarrow{PhN_2^+} ArN(NH_2)N{=}NPh$] and intramolecular condensations of *o*-nitrophenylhydrazines to give benzotriazole oxides or other cyclic products. Finally, reactions of diaziridines are reviewed and it is found that they react very similarly to their cyclic analogues, except for the easy ring cleavage of the strained small ring (Section II).

Section III describes the basicity and tautomerism of azo compounds and its main reactions, e.g. alkylation (by powerful agents such as $MeOSO_2CF_3$), oxidation by peracids to azoxy compounds, electrophilic and nucleophilic aromatic substitutions on the ring atoms (activated by the N=N group by the latter case and strongly deactivated in the former, necessitating the use of indirect or catalytic methods). This is followed by a description of intramolecular cyclization reactions with *o*-nitroso, *o*-nitro, *o*-CON_3, etc., groups next to the azo moiety, and with some reactions in which chloride adjacent to azo groups is labile and can be replaced by various anions (CH_3COO^-, N_3^-, CN^-, etc). Finally, azo compounds containing strongly electron-withdrawing groups (CN, CO_2R, COR) attached directly to the azo linkage are considered. They are highly reactive towards nucleophilic reagents, e.g. will attack activated aromatic nuclei:

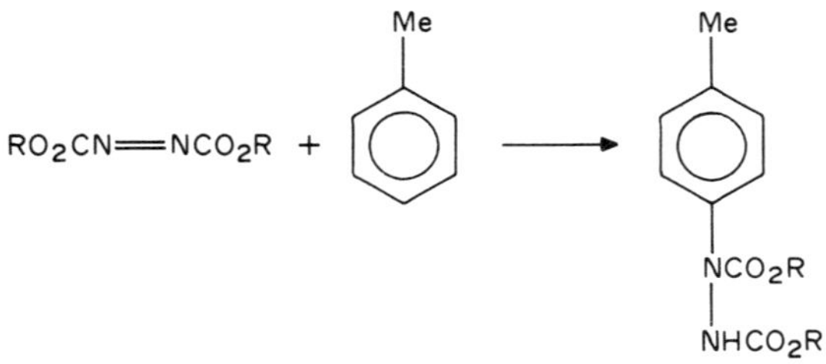

and will undergo similar electrophilic reactions towards activated alkenes, thioureas, etc.

Section IV deals with acidity and basicity of azoxy compounds and with their reactions in acids and bases, with electrophilic substitution (on the ring remote from the NO group) and with alkylation.

284 references up to 1973.
Complementary: none.
Relevant: none.

19, *17: Syntheses and uses of isotopically labelled hydrazo, azo and azoxy compounds* by *A. J. Dolenko and E. Buncel*

Section II describes the syntheses of azo compounds containing labelled N, C, H or O atoms and the similarly labelled hydrazo and azoxy compounds.

Section III discusses studies with the labelled title compounds. Thus, labelled azo compounds are employed in the study of the rearrangement of diazonium salts in the course of their conversion to azo compounds, in the azo–hydrazone tautomerism of hydroxy derivatives of azoarenes, in the thermolysis of azoalkanes (Tables 1–3), in polymerization studies involving the azobisisobutyronitrile catalyst and in some cage effect, photolytic and biological studies. Labelled hydrazo compounds were employed in the study of the benzidine rearrangement (Table 4). Labelled azoxy compounds were used in the study of the mechanism of the nitrosobenzene–phenylhydroxylamine condensation, of the Wallach rearrangement (Table 5), in photochemical rearrangements of azoxybenzenes and their reaction with arenesulphonyl chlorides (Table 6) and in the NMR identification of α- and β-azoxy isomers (Table 7).

97 references up to 1973.
Complementary: none.
Relevant: **5**, 10; **12**, 13; **13**, 15; **17**, 12; **18**, 18; **22**, 14; **23**, 16; **24**, 20; **26**, 11; **27**, 10; **28**, 8; **29**, 28; **33**, 18; **36**, 12.

19, *18: Rearrangement of hydrazo, azoxy and azo compounds* by *R. A. Cox and E. Buncel*

I. General introduction (776); **II**. Rearrangements of hydrazo compounds (777); **III**. Rearrangements of azoxy compounds (808); **IV**. Rearrangements of azo compounds (838); **V**. Acknowledgements (851); **VI**. References (852).

Section II describes the acid-catalysed benzidine rearrangement, its reaction products (Tables 1–3), kinetics (Table 4), mechanism, intermediates and various theories proposed. Next, the products and mechanism of the thermal benzidine rearrangement and the photochemical rearrangement are discussed. Hydrazo compounds with two heterocyclic rings (2, 2′-hydrazopyridine and related compounds) do not rearrange even under forcing conditions. The conversion of phenylhydrazones to indoles (in the Fischer indole synthesis) and some examples of the Bucherer reaction (Schemes 3 and 4) are related to the benzidine rearrangement.

Section III deals with the Wallach rearrangement of azoxy compounds:

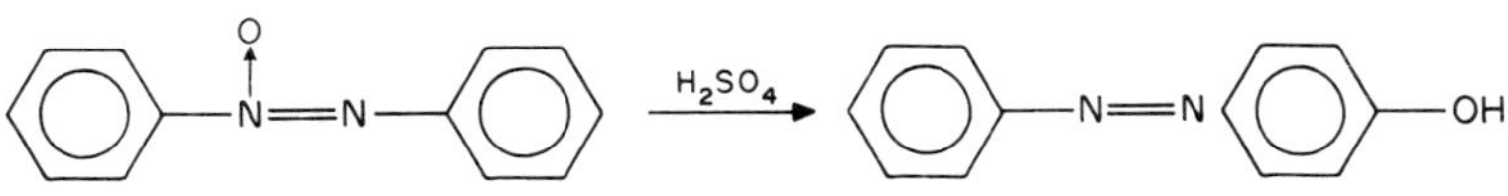

with its products (Table 5), catalysts, isotope studies, intermediates, kinetics and theories of mechanisms (involving *N, N*-oxide, dicationic or quinonoid intermediates). The photochemical rearrangement of azoxybenzenes yields *ortho*-hydroxy products by an intramolecular route. A unique aromatic rearrangement of 4-nitro-2-(benzeneazoxy)benzylidene dichloride leading through a carbene and cyclization route to a cyclic product is mentioned. Finally, aliphatic azoxy compounds rearrange through a wide variety of routes involving unknown mechanisms (Table 6).

Section IV describes rearrangements of azo compounds. Among these are the azo–hydrazone tautomerism of aromatic azo compounds, the effect of various substituents on the phenyl group of 4-phenylazo-1-naphthol on the amount of azo tautomer present (Table 7) in the equilibrium

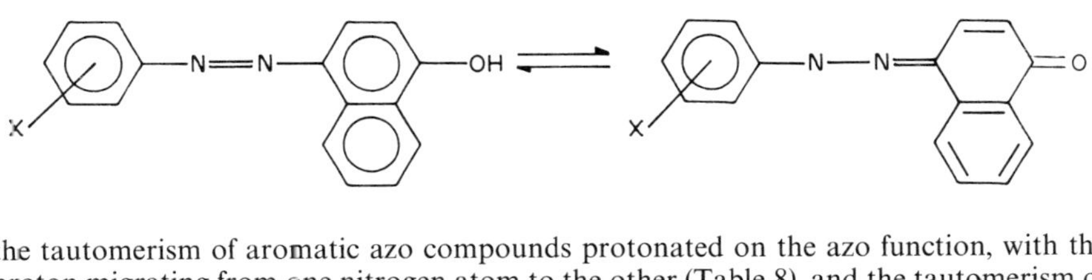

the tautomerism of aromatic azo compounds protonated on the azo function, with the proton migrating from one nitrogen atom to the other (Table 8), and the tautomerism of aliphatic azo compounds, e.g.

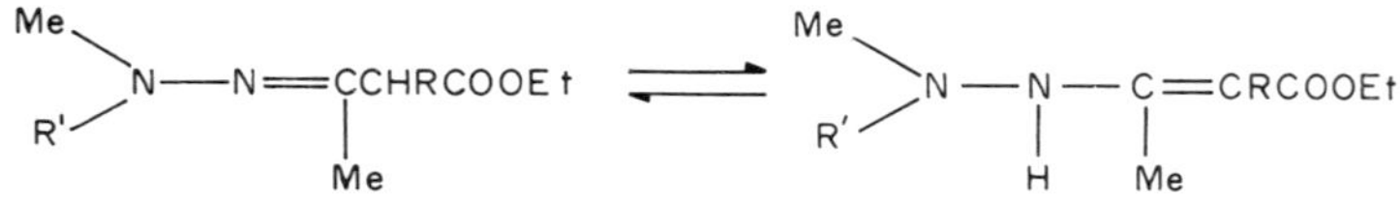

and finally some rare skeletal rearrangements of cyclic azo compounds, allylic diazenes and diazoaminoaromatics.

232 references up to 1974.
Complementary: 4, 10.
Relevant: 1, 7, **2**, 15; **3**, 14; **4**, 10; **5**, 8; **8**, 9; **9**, 3; **10**, 3; **11**, 4; **12**, 16; **13**, 13; **14**, 7; **15**, 8; **16**, 17; **17**, 8; **18**, 15; **20**, 4; **23**, 13; **24**, 10; **26**, 15; **29**, 4.

19, *19: Radical reactions of azo, hydrazo and azoxy compounds* by *G. Koga, N. Koga and J.-P. Anselme*

The chapter deals with reactions of the title compounds with radicals or with processes which involve radical intermediates.

Section II describes the thermal decomposition of azoalkanes into molecular nitrogen and an alkane (RN=NR → N_2 + RR) or, in the case of cyclic azo compounds, into N_2 and a cycloalkane or by a carbene route into an olefin. Photolytic processes are also discussed.

The free-radical thermal decomposition of azocarbonyl compounds gives a variety of products, including N_2, CO and also (in some cases dominantly) heterocyclic compounds containing two N atoms and one O atom:

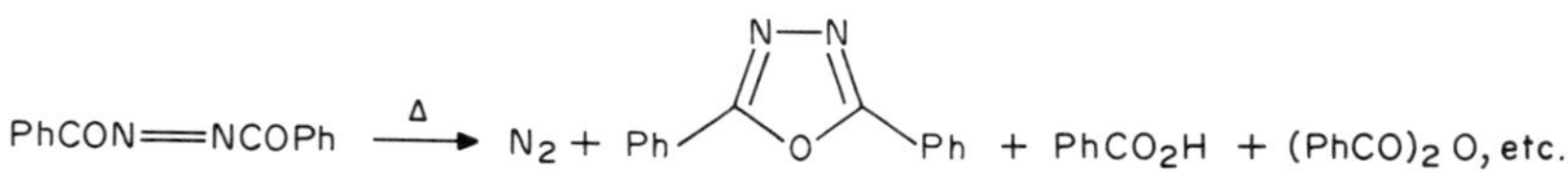

Azosulphones are also treated:

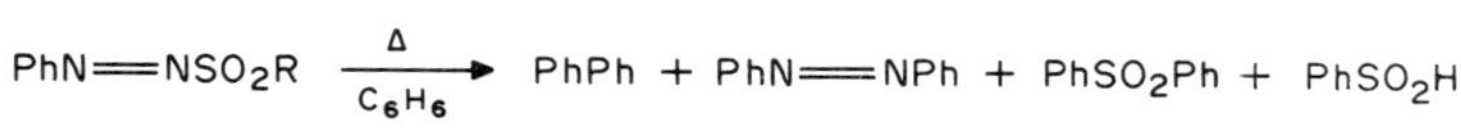

The reactions of azocarbonyl compounds with olefins also involve a wide variety of products, although the route is almost always a radical one. Similarly poorly defined products are obtained from azodiformate esters ($RO_2CN=NCO_2R$) or from azodiacyls (RCON=NCOR) with alkylbenzenes, esters, ketones, amines, etc. A valuable synthetic use of azodiformate esters is oxidation of thiols to disulphides, alcohols to carbonyl compounds and hydrazines to azo compounds, etc.

Azobenzene itself is very stable to heat but $PhN=NCPh_3$ decomposes to $Ph_3C^{\bullet}$ and $Ph^{\bullet}$; azobenzene itself undergoes facile homolytic arylations on the phenyl ring and is also easily reduced homolytically by H-donors to hydrazobenzene, so that it can be used as a preparative oxidizing agent.

Section III describes (presumably) free-radical oxidations of hydrazo to azo compounds, often accompanied by N_2 loss. Arguments are presented for the possibility of a free-radical mechanism in the benzidine rearrangement, which seems to be valid in some special cases (e.g. with tetraphenylhydrazine). Finally, some uncatalysed thermal reactions are treated briefly, such as disproportionation of hydrazobenzenes (ArNHNHAr → ArN=NAr → 2H$^{\bullet}$ → 2ArNH_2), involving also rearrangements to *o*- and *p*-semidines.

Section IV discusses relatively little known radical reactions of azoxy compounds. These include in rare cases loss of N_2O ($CH_3N{=}N(O)CH_3 \rightarrow CH_3CH_3 + CH_4 + N_2O + N_2$) and in others formation of oxaziridines, oxygen migration and isomerization.

182 references up to 1973.
Complementary: 4, 10.
Relevant: 1, 9; **16**, 8; **31**, 10; **33**, 8; **34**, 3; **33**, 8; **34**, 3.

19, *20: The photochemistry of the hydrazo, azo and azoxy groups* by *R. J. Drewer*

I. The hydrazo group (936); **II**. Azo compounds (942); **III**. Azoxy compounds (1006); **IV**. References (1010).

Section I describes the photochemistry of aliphatic and aromatic hydrazo compounds (including photolysis with N—N bond fission, photochemical disproportionation and the photobenzidine rearrangement).

Section II first discusses acyclic aliphatic azo compounds, gas-phase kinetics of *trans*-azoalkanes (Tables 1 and 2), photolysis in condensed phases (including quantum yields and cage effects), the nature of the dissociating states, photoisomerization (Table 3), the photolysis of some sterically bulky azo compounds, the problem of concerted or stepwise decomposition in acyclic azoalkanes, evidence for hot radicals from short-wavelength photolysis and some secondary reactions (Table 4) and unusual products.

Next cyclic aliphatic azo compounds are considered, including the mechanism and the primary processes in gas-phase and condensed-phase photolysis of 1-pyrazolines and bicyclic azo compounds. The photolysis of 1-pyrazolines and 3*H*-pyrazolines, the formation of the trimethylenemethyl radical by photolysis of 4-alkylidene-1-pyrazolines and bicyclic and polycyclic systems are considered.

General photolyses of three-, six- (Table 5), seven- and eight-membered cyclic azo compounds are discussed, usually occurring with extrusion of N_2.

Among mixed aliphatic–aromatic azo compounds, decomposition via thermolysis of unstable *cis* isomers of compounds of the type ArN=NR and the photolysis of (*tert*-butylazo)benzene (PhN=NBu-*t*) are described, followed by the photoisomerization and cyclodehydrogenation of aromatic azo compounds and by photoadditions to the azo group (e.g. photoreduction by H-abstraction from donors, or addition on irradiation:

$$\mathrm{PhN{=}NPh} \xrightarrow[\mathrm{CH_3COCl}]{h\nu} \mathrm{PhN(Ac)N(Ac)C_6H_4Cl\text{-}}p$$

Finally, Section III discusses recent advances in the photochemistry of aliphatic azoxy compounds (mainly

$$\mathrm{Me\overset{+}{N}(O^-){=}NMe} \longrightarrow 2\mathrm{Me}^{\bullet} + \mathrm{N_2O}$$

but also elimination of N_2 and oxaziridine formation) and of aromatic azoxy compounds mainly undergoing *cis–trans* photoisomerization and also migration of the azoxy oxygen atom to the *ortho*-position of the more distant aromatic ring.

205 references up to 1973.
Complementary: none.
Relevant: **2**, 16; **3**,8; **6**, 4; **8**, 12; **9**, 6; **11**, 5; **13**, 16; **14**, 8; **15**, 9; **16**, 11; **17**, 9; **18**, 10; **22**, 10; **23**, 9; **24**, 12; **25**, 11; **27**, 21, 22; **28**, 5; **29**, 5, 6; **32**, 29; **33**, 21.

19, *21: Conformational analysis of hydrazines* by *Y. Shvo*

I. Introduction (1017); **II**. Conformational processes in bipyramidal hydrazines (1040); **III**. Pyramidal–planar hydrazines (1075); **IV**. Biplanar hydrazines (1079); **V**. Conclusion (1091); **VI**. Acknowledgement (1092); **VII**. References (1093).

Section I reviews the general conformational aspects of hydrazines, the concepts on which conformational analysis is based [thermodynamic and kinetic properties, relative stabilities, populations of various conformers, torsional functions (Table 1), use of IR, NMR and PES data for the study of conformations, quantum chemical calculations and inversion barriers in hydrazine systems (Table 2) compared with compounds with a single N atom (Table 3)].

Sectional II discusses the two main conformational processes occurring in hydrazines (which are regarded as bipyramidal). These are the inversion of the N pyramids (Scheme 2) and the rotations about the N—N bond (Scheme 3). These conformational interconversions are summarized for differently substituted hydrazines in Tables 4–6. The analysis shows the occurrence in both inversion and rotation of 'low-energy' and 'high-energy' processes in each case. Cyclic bipyramidal hydrazines (including ring-substituted compounds) are considered in Schemes 6 and 7 and in Table 7. Again, a combination of spectroscopic techniques (IR, NMR, PES) is used for the detailed analysis of the processes.

Section III deals with hydrazines in which mesomeric electron-withdrawing groups substituted on one of the N atoms deform the pyramidal configuration so as to render it nearly planar, owing to conjugation of the lone pair with the adjacent π-system. Activation data on such hydrazines are given in Table 8. Here the rate of inversion is too fast for detection by NMR and the rotational process is the rate-determining step.

Section IV describes biplanar hydrazines (Table 9) in which each N atom carries at least one mesomerically electron-withdrawing group so that near planarity is achieved and the only process which may possibly be detected by NMR is rotation [e.g. $PhCH_2N(Ph)N(Ph)CH_2Ph$].

Planar structures have been revealed by X-ray studies for various cyclic biplanar hydrazines and activation data for them are given in Table 10. In these cases, not only the NCO_2R moiety, but the *whole* molecule is planar:

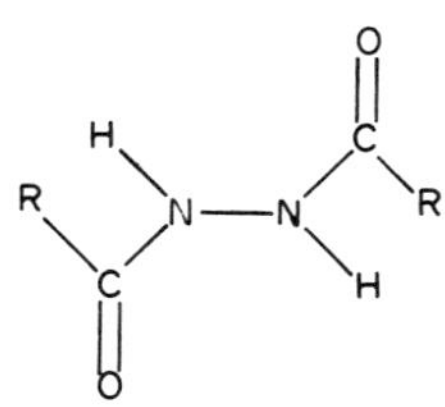

93 references up to 1973.
Complementary: none.
Relevant: **20**, 2; **21**, 1; **22**, 3; **27**, 5; **28**, 10; **29**, 2; **31**, 19; **32**, 1; **33**, 2.

The chemistry of amidines and imidates (1975)

20, *1: General and theoretical aspects of amidines and imidic acid derivatives* by *G. Häfelinger*

I. Introduction (2); **II.** Physico-chemical properties (4); **III.** Theoretical considerations (18); **IV.** Electronic spectra (37); **V.** Infrared and Raman spectra (43); **VI.** Nuclear magnetic resonance spectra (57); **VII.** Mass spectra (75); **VIII.** References (79).

Section II describes the molecular structure (bond lengths, angles) of amidines and amidinium salts (Table 1), the dipole moments of amidines (Table 2) and of imidates (Table 3) and the basicities of the two groups (Tables 4–6). Section III discusses the Hückel MO treatment of amidinium cations, amidines, imidic acid derivatives (Table 7), amidrazones ($RN{=}CR^1NR^2NR^3R^4$) (Table 8), HMO bond orders and bond lengths (Table 9) and the effect of phenyl substitution on amidines (Table 10). Empirical parameters for Pariser–Parr–Pople calculations (Table 11) and the results of these calculations (Table 12), including dipole moments (Table 13), are presented.

Section IV deals with the electronic spectra of amidines and amidinium salts and gives PPP transition energies (Table 14) and UV absorption maxima of amidines and amidinium salts (Table 15) and imidates (Table 16). Section V treats the IR and Raman spectra of acetamidine (Table 17) and of the acetamidinium cation (Table 18), and the force constants for the latter (Table 19). C=N valence vibrations are given in Tables 20 and 21. Next, the tautomerism of amidines $RC({=}NH)NHR' \rightleftarrows RC(NH_2){=}NR'$ and also their molecular association by hydrogen bonding are discussed.

Section VI describes NMR spectra (Tables 22–24), the *cis–trans* and rotational isomerism of amidines and their derivatives, the study of *N, N'*-disubstituted acetamidinium cations and the equilibrium constants for the *cis–trans* isomerization between the *E, E* and *E, Z* isomers in the latter. Rotational barriers are large enough to allow observation of NMR coalescence temperatures and to determine activation parameters for the rotation around the C—N single bond in $RC({=}NH)NHMe_2$ and in $RC({=}NR')NMe_2$ to be determined (Tables 26–28). ^{13}C NMR spectra are presented in Table 29 and ^{15}N–H coupling constants are also discussed.

Finally mass spectra and fragmentation schemes are presented for a few amidines (Section VII).

156 references up to 1974.
Complementary: none.
Relevant: 1, 1; **2**, 1; **3**, 1; **4**, 1; **5**, 1; **6**, 1; **8**, 1; **11**, 1; **12**, 1; **13**, 1; **14**, 1; **15**, 1; **16**, 1; **17**, 1; **18**, 1; **19**, 1; **22**, 1; **23**, 1; **24**, 1; **25**, 1; **26**, 1; **27**, 9; **28**, 1; **31**, 27; **32**, 1; **33**, 1; **36**, 2.

20, *2: Constitution, configurational and conformational aspects and chiroptical properties of imidic acid derivatives* by *G. Fodor and B. A. Phillips*

Section II discusses the constitution of imidic acid derivatives and primarily the question of whether amides can also exist in the tautomeric imidic acid form (as their derivatives the imidates certainly do):

$$RC(=O)NH_2 \rightleftharpoons RC(OH)=NH$$

The section deals with the tautomerism of alkyl and aryl imidates, amidines, *N*-trisubstituted amidinium salts, aryl sulphonylamidines [$ArSO_2N{=}C(Ph)NH_2 \rightleftharpoons ArSO_2NHC(Ph){=}NH$], amidrazones, *N*-halogenoamidines and imidoyl halides, which tautomerize by migration of both protons and halogen atoms:

$$PhC(=NCH_2Ar)Cl \rightleftharpoons PhCH_2N{=}C(Ar)Cl$$

$$Ar = p-NO_2C_6H_4$$

Section III describes the configuration of amidines, including *syn–anti* geometric isomers detected by NMR. Geometric isomerism in imidates was similarly proved (Tables 1–3) and activation parameters (Table 4) and isomer ratios (Table 5) have been determined. Imidoyl halides were less studied but *syn–anti* isomers were detected in several cases at room temperature by NMR (Table 6) and by dipole moments (Table 7).

Section IV deals with the existence of conformational isomers due to restricted rotation in imidic acid derivatives. Thus, restricted rotation occurs in benzamidines (Tables 8–10), formamidines (Tables 11–13), amidinium salts (Tables 14–20) and amidoximes (Table 21). Conformations of imidic esters are shown (Tables 22 and 23) and the relationship between conformation and reactivity of imidates is discussed, including the effect of bulky groups (Table 29).

Section V deals with chiroptical properties of amidines and presents ORD data (Tables 25 and 27) and UV absorption data (Table 26) on amidinium chlorides and ORD data on Cu complexes of amidinium chlorides (Table 28) and cyclic α-hydroxyamidines (Table 29).

94 references up to 1973.
Complementary: none.
Relevant: 19, 21; **21**, 1; **22**, 3; **27**, 5; **28**, 10; **29**, 2; **31**, 19; **32**, 1; **33**, 2; (also ORD/CD chapters): **8**, 4; **18**, 7; **19**, 6; **22**, 4; **25**, 3; **26**, 3; **27**, 6; **28**, 10; **29**, 23; **31**, 1; **36**, 15.

20, *3: Detection and determination of imidic acid derivatives* by *W. H. Pritchard*

Section II describes the IR (Table 1), Raman, UV–visible, NMR and mass spectra and chromatography of amidines, while Section III treats the IR (Table 2), NMR and ESR spectra of amidoximes.

Amidrazones [RC(=NR^1)NHNR^{2}R^3] and their IR and NMR spectra are discussed in Section IV. Section V gives pK, IR, Raman, UV–visible, NMR and chromatographic data on formazans (ArN=NCR=NNHAr′) and Section VI the very sparse material on hydrazidines (R$_2$NN=CR′NHNR$_2$″).

Section VII deals with IR, Raman UV–visible and NMR spectra of imidates [RC(=NH)OR′] and their chromatography, dipole moments and X-ray crystallography.

Section VIII discusses briefly various spectroscopic features of imidoyl halides [RC(Hal)=NR′] and their NMR spectra and dipole moments.

91 references up to 1972.
Complementary: none.
Relevant: **1**, 5; **2**, 8; **3**, 15; **4**, 3; **5**, 17; **8**, 3; **12**, 4; **13**, 6; **14**, 3; **15**, 3; **16**, 3; **17**, 4; **18**, 5; **22**, 5; **23**, 5; **24**, 5; **25**, 10; **26**, 5; **28**, 3; **29**, 21; **30**, 14–22; **36**, 5.

20, *4: Rearrangements involving imidic acid derivatives* by *C. G. McCarty and L. A. Garner*

I. Introduction (189); **II**. Rearrangements of imidates and related compounds (190); **III**. Rearrangements of amidines and related compounds (230); **IV**. Some related rearrangements (234); **V**. References (236).

Section II discusses the thermal uncatalysed rearrangement of aryl imidates to N-aroyldiarylamines [Chapman rearrangement: Ar^{1}C(=NAr^2) → Ar^{1}CONAr^{2}Ar^3]. The preparative importance (for obtaining diarylamines by hydrolysis of the primarily formed amide (Ar^{1}CONAr^{2}Ar^3 → Ar^{2}NHAr^3) and the mechanism of the reaction are discussed in detail (Tables 1–3), including kinetics and migrating aptitudes together with some related O to N migrations of aryl groups. Next, thermal and catalysed (by alkyl halides, acids, various Hg, Sn, Cu and other metal salts) rearrangements of alkyl and allyl imidates are described and also some other related O to N migrations of alkyl groups. Finally, rearrangements of acyclic and cyclic acyl imidates (iosoimides) are treated. In these the acyl group migrates from O to N:

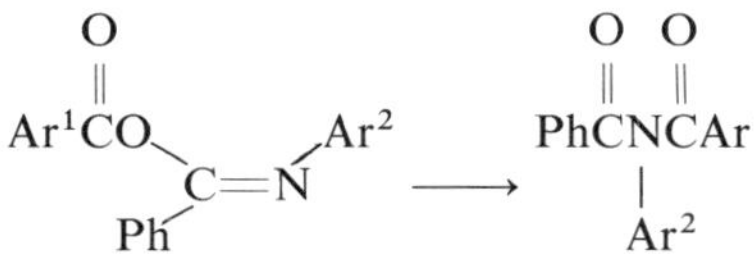

and the reaction may be either thermal or base (CH$_3$COONa, Et$_3\overset{+}{N}$H$\overset{-}{O}$OCCH$_3$, aziridine) catalysed. Again, some related O to N migrations of acyl groups are described. Thioimidates [ArC(SAr)=NAr] rearrange thermally to the corresponding thioamide [ArC(=S)NAr$_2$], but the equilibrium is almost completely at the side of the thioimidate.

Section III deals with some N to N migrations of aryl or alkyl groups in amidines:

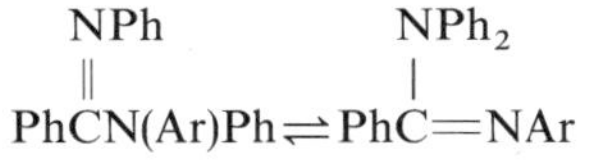

A similar N to N migration of acyl groups in acylamidines is more facile and related processes were observed with chloro- and hydroxyamidines.

Finally, Section IV presents, for the sake of comparison, some O to N rearrangements of imidocarbonates thioimidocarbonates and isoureas (MeNC=NCN) and related compounds.

130 references up to 1973.
Complementary: none.
Relevant: 1, 7; **2**, 15; **3**, 14; **4**, 10; **5**, 8; **8**, 9; **9**, 3; **10**, 3; **11**, 4; **12**, 16; **13**, 13; **14**, 7; **15**, 8; **16**, 17; **17**, 8; **18**, 15; **19**, 18; **23**, 13; **24**, 10; **26**, 15; **29**, 4.

20, *5: The electrochemistry of imidic esters and amidines* by *H. Lund*

I. Introduction (241); **II**. Electrochemical preparation of imidic acid derivatives (242); **III**. Electrochemical reactions of imidic acid derivatives (246); **IV**. References (252).

Section II discusses the electrochemical preparation of imidic esters at controlled cathode potentials. Amidines can be obtained by anodic oxidation [$CH_3CH_2OH \rightarrow CH_3CHO \xrightarrow{NH_3} CH_3CH(OH)NH_2 \rightarrow CH_3C({=}NH)NH_2$] or by reduction of compounds $RC(NR'_2){=}NY$, where Y is a heteroatom (O, N), e.g. $ArC(NH_2){=}NOH \longrightarrow ArC(NH_2){=}N^+H_2$. Similarly, amidrazones [$RC(NR'_2){=}NNHR''$] may be electrosynthesized by reduction of oxime hydrazones and hydrazidines from tetrazolium chlorides.

Section III describes the electrochemical reactions of imidic esters (yielding amines, Table 1), of amidines (Table 2), yielding under different conditions alkanes, cyanamide, alkylamines, etc., and of amidrazones, hydrazidines, amidoximes, hydroxyamidoximes and hydrazooximes, all of which undergo either reductions or oxidations, usually without a cleavage of a C—N or N—N bond. Finally, imidic acid halides [RC(Hal)=NR′] are easily reduced and may yield diimines [R′N=C(R)—C(R)=NR′] or aldimines (RCH=NR′).

31 references up to 1973.
Complementary: 8, 11; **21**, 5.
Relevant: 5, 2; **13**, 5; **16**, 15; **17**, 14; **19**, 12; **22**, 9; **23**, 10; **24**, 17; **25**, 13; **26**, 12; **27**, 8; **28**, 7; **29**, 8, 9; **31**, 7; **32**, 6; **34**, 1.

20, *6: Biological reactions and pharmaceutical uses of imidic acid derivatives* by *R. J. Grout*

I. Introduction (256); **II**. Biological formation of the amidine group (256); **III**. Biological activity of imidate derivatives (262); **IV**. References (277).

Section II describes the involvement of amidines during the biosynthesis of purines and histidine and the catabolism of the latter, in which a one-carbon unit is returned to the 'one carbon pool' via a formamidine intermediate. Folic acid derivatives are also formed by enzymatic ring closure of a formamidine (5-formimino-5, 6, 7, 8-tetrahydrofolate).

Section III discusses the antiviral activity of naturally occurring amidines (amidinomycin, netropsin, distamycin, etc.), synthetic antibacterial, antifungal and antiprotozoal drugs, many of which contain two amidine groups, cancer chemotherapeutic agents (especially terephthalanilide derivatives containing two amido and two amidino groups), protease inhibitors (potentially useful in the chemotherapy of cardiovascular disease and organ transplants), anthelmintic drugs, antihypertensive agents, drugs for the control of cardiac arrythmias, tranquillizers (of the diazepine family) and anti-inflamatory and antipyretic agents obtained by the substitution of a carboxylic acid function by a hydroxamic acid residue.

127 references up to 1973.
Complementary: none.
Relevant: 2, 7; **3**, 5; **4**, 9; **5**, 18; **7**, 4; **11**, 11; **12**, 12; **13**, 14; **15**, 10; **16**, 13; **17**, 13; **18**, 13; **19**, 13; **22**, 20; **24**, 14; **26**, 13; **28**, 17; **31**, 9.

20, *7: Preparation and synthetic uses of amidines* by *J.-A. Gautier, M. Miocque and C. C. Farnoux*

I. Introduction (284); **II**. Preparation of amidines (285); **III**. Synthetic uses of amidines (312); **IV**. References (338).

Section II describes the methods of preparation of amidines from nitriles. The three main routes are addition of alkali metal amides [$RC{\equiv}N + KNH_2 \rightarrow RC({=}NK)NH_2 \rightarrow \rightarrow RC({=}NH)NH_2$], addition of NH_3 or amines or their salts [$RCN + R'NH_2 \rightarrow RC({=}NR')NH_2$] and the Pinner synthesis involving imido ester intermediates [$RCN \xrightarrow{R'OH} RC({=}NH)OR' \xrightarrow{R''NH_2} \rightarrow RC({=}NH)NHR''$]. Another synthetic pathways starts from amides or thioamides with amines in the presence of halogenating agents [$RC({=}O)NH_2 \rightarrow RC(Cl){=}NH \rightarrow RC({=}NH)NHR$] and through modifications of this method. Other special cases are the use of amides with Ti complexes and amides with isocyanates [$RNHCOR' + R''N{=}C{=}O \rightarrow R''C({=}NR')NHR$]. Thioamides with NH_3 or amines yield H_2 salts of the amidine. Other less general routes use as starting materials Schiff bases, hydrazones, amidoximes, carbodiimides, halides, ortho esters, amines and amidines (by ammonolysis or alkylation).

Section III discusses reactions in which amidines are starting materials for the preparation of, e.g. thioamides [$RC({=}NH)NR'_2 \xrightarrow{H_2S} RC({=}S)NH_2 + HNR'_2$], amidrazones, amidoximes, aldehydes (by reduction with Na–EtOH), condensation products with active methylene groups [$CH({=}NR)HNR + CH_2XY \rightarrow RNH_2 + RHNCH{=}CXY$] and for the synthesis of various three-, four-, five- and six-membered rings.

381 references up to 1973.
Complementary: **21**, 13.
Relevant: none.

20, *8: Kinetics and mechanisms of reactions of amidines* by *R. H. De Wolfe*

I. Introduction (349); **II**. Hydrolysis reactions of amidines (350); **III**. Amidines as nucleophiles; kinetics of reactions of amidines and amidoximes with esters, aryl halides and acid halides (373); **IV**. Conformational isomerizations of amidines and amidinium salts (380); **V**. Pyrolysis and thermal isomerization reactions of amidines (381); **VI**. References (382).

Only in the few cases when kinetics have been studied does solid evidence exist regarding the mechanism of amidine reactions, although proposals abound.

Section II describes the kinetics of the hydrolysis of acyclic and cyclic amidines (imidazolines, imidazolinium ions) and also *N*, *N'*-dihydroguanidines. The mechanisms of these reactions are discussed in detail.

Section III deals with reactions in which amidinic N atoms are nucleophiles and are attacked by electrophiles such as esters, aryl halides and acyl halides. Competition experiments show that amidines (although only slightly more basic) react many thousands of times faster than amines. Kinetic and mechanistic details are given about these amidinolysis reactions.

Section IV describes the rotation of trivalent N atoms of *N*,*N*-dimethylamidines and their ions about acyl C—N bonds, and Section V some pyrolysis and thermal isomerization reactions [PhC(=NAr)NArAr′ $\rightleftharpoons$ PhC(=NAr′)NAr_2] (Table 1).

69 references up to 1972.
Complementary: none.
Relevant: none.

20, *9: Imidates including cyclic imidates* by *D. G. Neilson*

Sections I and II describe the nomenclature, scope and tautomerism of the subject compounds, Section III discusses the Pinner synthesis [RCN + ROH + HCl → RC(=NH)OR.HCl), it scope and limitations, base-catalysed reactions (RCN + ROH) (Table 1), reactions of imidoyl halides with alkoxides and phenoxides [RC(=NR′)Cl + NaOR″ → RC(=NR′)OR″], alkylation of amides and thioamides [RC(=S)NH_2 + R′I → RC(=NH)SR′], reactions of amines with ortho esters [$HC(OR)_3$ + $ArNH_2$ → → HC(=NAr)NHAr], transesterification of imidates, *N*-alkylation of unsubstituted imidates, syntheses starting from unsaturated systems [e.g. R_2C=C=NR′ + MeOH → RCH(=NR′)OMe], from isonitrile and nitrile complexes, from organometallic compounds, from boron–imidate derivatives, thione esters with amines, cyanic esters with carbanions [ArOCN + $\bar{C}HR_2$ $\xrightarrow{\text{base}}$ R_2CHC(=NH)OAr] and by photochemical reactions of, e.g., thiols with isonitriles.

Section IV discusses general properties, *syn–anti* isomerism, conformations and spectra. Next the thermal decomposition of unsubstituted and *N*-substituted imidate salts and bases is treated with special reference to the Chapman rearrangement [$Ar^1C(=NAr^2)OAr^3 \rightarrow Ar^1C(=O)NAr^2Ar^3$], the reactions of imidates with H_2O, H_2S, ROH, NH_3 and primary, secondary and tertiary amines (yielding amidines), unsubstituted, mono- and disubstituted hydrazines (yielding amidrazones or, with the latter under more vigorous conditions, dihydroformazanes), with NH_2OH, amino acids and their esters and with low MW proteins (e.g. insulin).

The next few subsections present various reactions in which heterocyclic compounds are formed (imidazoles, oxazoles, thiazoles, isothiazoles, oxadiazoles, triazoles, tetrazoles, pyridines, pyrimidines, pyridazines, triazines, tetrazines, oxazines and azepins) through a wide variety of condensation reactions with different synthones.

Next, Grignard reactions [HC(=NPh)OR + ArMgBr → → ArCH=NPh → ArCHO] and oxidations and reductions (to amines) are treated, followed by the preparation of acyl and sulphonyl derivatives of imidates and the properties of *N*-haloimidates and of P and Sb derivatives of imidates. A few photoreactions are treated briefly and also iminoanhydrides (containing the normally unstable —N=$\overset{|}{C}$—O—$\overset{|}{C}$=O grouping). Finally, some industrial, bactericidal, herbicidal and pesticidal uses of imidates are mentioned.

611 references up to 1973.
Complementary: none.
Relevant: none.

20, *10: The chemistry of amidrazones* by *K. M. Watson and D. G. Neilson*

Amidrazones, $RC(=NNR^1R^2)NR_3R_4$, are hydrazides of the hypothetical imidic acids [RC(=NH)OH].

Section III describes preparations of amidrazones by addition of hydrazines to nitriles [$RCN + NH_2NH_2 \rightarrow RC(=NNH_2)NH_2$], nitrile complexes, ketimines, carbodiimides or *s*-triazine [e.g. $Ph_2C{=}C{=}NAr \xrightarrow{NH_2NH_2} Ph_2CHC(=NNH_2)NHAr$].

Other routes are substitutions by hydrazines on a variety of synthones such as imidates, thioimidates, imidoyl halides [$PhC(=NPh)Cl \xrightarrow{PhNHNH_2} PhC(=NNHPh)NHPh + PhC(=NPh)NPhNH_2$], amidines, amides and thioamides. Amidrazones are also obtained from the reaction of NH_3 or amines on hydrazonate esters and thioesters or on other hydrazonyl derivatives.

Ring opening by nucleophilic attack involving hydrolysis on oxadiazoles, triazolium salts or *s*-tetrazines also leads to amidrazones:

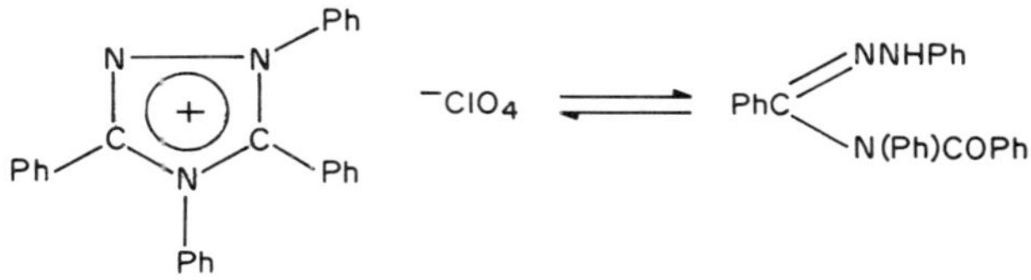

Section IV discusses the tautomerism of amidrazones:

their physical properties (basicity, spectra), their alkylation, silylation and nucleophilic substitution (hydrolysis, aminolysis), their reactions with acid derivatives (acid chlorides, anhydrides, esters, acids, dicarboxylic acids, ortho esters, imidates) when usually the N^1-nitrogen is attacked:

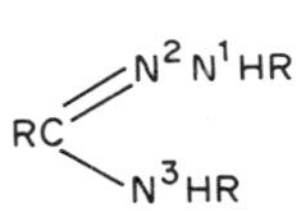

e.g.,

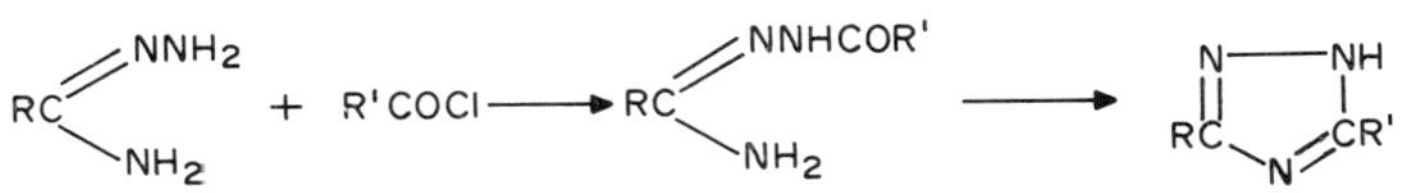

Amidrazones also condense with aldehydic and ketonic CO groups, with dicarbonyl compounds, with α,β-unsaturated carbonyl compounds, with compounds of the type

XC(Z)=Y where X and Z are good leaving groups (e.g. $COCl_2$, $CSCl_2$) and with compounds of the type X=C=Y (CS_2, ketenes, isocyanates, etc.), with which 1, 2, 4-triazoles or 1, 2, 4-triazines are formed. Other cyclic systems are obtained with HNO_2, P compounds, boronic acid derivatives, etc. Finally, the oxidation, metal complexes (colours with transition metal ions) and the industrial and medicinal applications of amidrazones are mentioned.

161 references up to 1972.
Complementary: none.
Relevant: none.

20, *11: Estimation of the thermochemistry of imidic acid derivatives* by *R. Shaw*

I. Introduction (547); **II.** Estimation by group additivity (548); **III.** Assumptions, derivation of group values and some examples of their use (550); **IV.** Acknowledgements (564); **V.** References (564).

At the time of writing, no experimental thermochemical data existed on imidic acid derivatives.

Section II reviews the method of estimating chemical thermodynamic properties of molecules by group additivity (i.e. from the contributions of individual groups that make up the molecule).

Section III gives the assumptions and the derivation of group values, and some examples of the use of the method for imines, imidic acids, amidines, omidrazones, amidoximes and imidoyl halides.

14 references up to 1975.
Complementary: none.
Relevant: 16, 16; **18**, 3; **19**, 3; **21**, 3; **22**, 6; **23**, 4; **24**, 3; **25**, 2; **26**, 4; **27**, 9; **28**, 4; **29**, 24; **30**, 2; **31**, 2; **33**, 3; **36**, 4.

20, *12: Complex formation, H-bonding and basicity of imidic acid derivatives* by *J. Sevcik and F. Grambal*

I. Introduction (566); **II.** Complex formation (567); **III.** Hydrogen bonding (590); **IV.** Basicity (600); **V.** References (612).

Section II describes Cu, Ag, Hg and other metal complexes of α-hydroxyamidines (Tables 1 and 2), other amidines (Table 3), amidinourea derivatives (Table 4) and amidinothiourea. Boron complexes of amidines are obtained in the boronation of carbodiimides and also on heating mixtures of benzamidine, benzonitrile and a trialkylboron (Table 5). Amidoxime complexes of metals have widespread applications in analytical chemistry (Table 6). Their structures were determined by X-ray crystallography.

Section III discusses inter- and intramolecular hydrogen bonds of amidines, the strong intramolecular hydrogen bond in sulphonylamidines (Table 7) and hydrogen bonds in amidoximes and imidates. Section IV deals with the basicity of mono- (Table 8), di- (Table 9) and trisubstituted (Table 10) benzamidines and their S- and P-analogues, with diacidic benzamidines (Tables 11 and 12), which have an additional tertiary amino group, and with other variously substituted amidines (Tables 13 and 14). Next, the basicity of amidoximes (Tables 15 and 16), amidoximehydrazides of malonic acid and finally 3-aminoamidoximes are considered.

180 references up to 1972.
Complementary: none.
Relevant: **2**, 9; **3**, 6; **4**, 4; **5**, 6; **7**, 6; **8**, 5; **11**, 3; **12**, 3, 10; **13**, 7; **15**, 4; **16**, 5; **17**, 6; **18**, 8; **19**, 7; **22**, 8; **23**, 6, 7; **24**, 4; **25**, 5, 6; **29**, 17; **31**, 17, 21; **32**, 21; **33**, 5.

Supplement A: The chemistry of double-bonded functional groups (1977)

21, *1: Dipole moments, configurations and conformations of molecules containing X=Y groups* by *O. Exner*

Section II deals with stereochemistry, classical and partial double bonds and the concept of conformation and the abundance of rotamers (Fig. 1).

Section III describes some specific features of the dipole moment approach to structural and stereochemical problems and its contribution with other physical methods (e.g. molar Kerr constant, NMR, IR).

Section IV discusses configurations on X=Y bonds, considering C=C bonds, C=N bonds in imines, imidates (Table 1), hydroxylamine and hydrazine derivatives and C=N compounds with two substituents on N (e.g. *N*-oxides); N=N bonds in azo, azoxy and azodioxy compounds and briefly also C=S, S=N and B=N bonds.

Section V deals with conformation on partial double bonds, such as occur in conjugated systems (dienes, α, β-unsaturated aldehydes, ketones, esters, etc.). Conformations are considered in the case of simple carboxylic esters (prevailingly planar, *Z* conformation), α-substituted esters and analogues and derivatives with two axes of rotation (e.g. dialkyl carbonates and carboxylic anhydrides are discussed). Amides and related compounds are treated next, and finally the conformations of aromatic derivatives.

360 references up to 1975.
Complementary: none.
Relevant: 19, 21; **20**, 2; **22**, 3; **28**, 10; **29**, 2; **31**, 19; **33**, 2.

21, *2: Liquid crystals with X=Y groups* by *J. P. Van Meter*

Liquid crystals are capable of forming a liquid crystalline or mesomorphic state, which is intermediate between the crystalline solid and the liquid phase, being less ordered than the former but more than the latter. The compounds forming such phases are usually long, flat and rigid about the long axis. Section II explains the terms smectic, nematic and cholesteric and considers the theory of the liquid crystalline state. Section III describes the detection and identification of liquid crystals.

Section IV presents methods of preparation of liquid crystals, containing imine (Table 1), nitrone, imidoyl cyanide, acetylene, ethylene, azo, azoxy (Table 2), ester (Tables 3 and 4) and biphenyl groups.

After some comments on mesophase thermal stability (Section V), Section VI describes electrooptical applications, spectroscopic studies, temperature and solvent sensing and the use of liquid crystals as stationary phases in chromatographic separations.

178 references up to 1975.
Complementary: none.
Relevant: none.

21, *3: Thermochemistry of X=Y groups* by *R. Shaw*

I. Introduction (131); **II.** Electrostatics and thermochemistry (132); **III.** Latest group values for compounds containing X=Y (138); **IV.** Some recent experimental work (138); **V.** Acknowledgements (146); **VI.** References (147).

Section II considers the simple electrostatic models that account for the heat of formation of alkanes and alkenes (Tables 1 and 2 compare values calculated by the electrostatic method with observed values).

New or revised group values are given for X=Y compounds (Table 3, Section III).

Section IV discusses some recent experimental work, such as π-bond strengths (Table 4), stabilization energies of the acetonyl radical and olefin isomerizations.

17 references up to 1975.
Complementary: 19, 3.
Relevant: 16, 16; **18**, 3; **20**, 11; **22**, 6; **23**, 4; **24**, 3; **25**, 2; **26**, 4; **27**, 9; **28**, 4; **29**, 24; **30**, 2; **31**, 2; **33**, 3; **36**, 4.

21, *4: Mechanisms of elimination and addition reactions involving the X=Y group* by *A. F. Cockerill and R. G. Harrison*

I. Scope (152); **II.** Alkene-forming eliminations (153); **III.** The reactions of nucleophiles with vinyl halides and esters (221); **IV.** Addition and elimination reactions involving the carbonyl group (223); **V.** Addition and elimination reactions involving C=N (288); **VI.** Thiocarbonyl additions and eliminations (306); **VII.** Acknowledgement (310); **VIII.** References.

Section II describes alkene-forming eliminations and their terminology. The *E*1cB mechanisms including several variants [$(E1)_{anion}$ $(E1cB)_R$, $(E1cB)_{irr}$, $(E1cB)_{ip}$] are discussed in detail, considering rates, mechanisms, structural, solvent and isotope effects (Tables 1–4) and the controversy of concerted versus stepwise eliminations. The variable TS for concerted eliminations in protic media is presented, including its terminology and the effects of leaving groups, stereochemistry, base, solvent and substituents at C_α and C_β (Tables 5–8). The theory, methodology and results regarding the stereochemistry of concerted eliminations in alicyclic and acylic systems in protic media are treated (Table 9). Next the orientation of elimination in protic media (effects of leaving group, base and solvent and of alkyl structure) (Table 10) are dealt with, followed by eliminations in aprotic media, mechanisms with carbonium ion character (including pyrolytic eliminations and dehydrations) and finally dehalogenations and related eliminations.

Section III comments briefly on nucleophilic additions to vinyl halides and esters.

Section IV discusses additions and eliminations involving the C=O group. First, condensations are treated, including the aldol, Knoevenagel, Darzens, Wittig and related reactions, considering in each case kinetics, mechanisms, stereochemistry, solvent and other effects. Next, among organometallic additions to carbonyl compounds, reactions with Grignard reagents and with organolithium, -aluminium, -zinc and -cadmium reagents are described, including the Reformatsky reaction. Reductions of CO groups with metal hydrides (including asymmetric reductions) and the Meerwein–Ponndorf–Verley reduction are treated. This is followed by descriptions of the hydration of the C=O group, the addition of alcohols, thiols and H_2S and the Baeyer–Villiger peracid oxidation of ketones to lactones or esters.

Various elimination mechanisms for alcohol oxidations are discussed (cyclic and non-cyclic proton transfers, cyclic and non-cyclic hydride transfer, radical mechanisms, etc.).

The *A*1, *A*2 and A–S_E2 mechanism for acetal hydrolysis in biological systems, intramolecular general acid catalysis and hydrolysis of ortho esters and vinyl ethers are considered in detail.

Section V deals with additives and eliminations involving C=N groups. The mechanisms of carbonyl–amine condensations yielding aldimines, ketimines, Schiff bases, semicarbazones, hydrazones, thiosemicarbazones and oximes are described in detail. This is followed by a discussion of the main addition reaction of C=N, including hydrolysis, addition of alcohols, thiols, phenols (to Schiff bases), HCN, Grignard reagents (to oximes), reduction and peracid oxidation.

Section VI describes elimination reactions yielding thiocarbonyl groups, and the addition reactions of the latter.

641 references up to 1974.
Complementary: 1, 2, 3, 4, 8, 9; **2**, 2, 3, 11, 12, 13; **4**, 7, 8; **8**, 2, 6, 7, 10; **9**, 4; **10**, 1; **13**, 12; **16**, 9; **21**, 9; **27**, 11, 16; **32**, 5, 23.
Relevant: 1, 11; **21**, 6.

21, *5: The electrochemistry of X=Y groups* by *A. J. Fry and R. G. Reed*

Section II reviews the general aspects of the electrochemical behaviour of double-bonded groups.

Section III describes the reduction and oxidation of olefins. Oxidation leads to the formation of various cations or radical cations, which in turn supply products of substitution, dimerization or cyclization.

Section IV discusses the mechanisms of the reduction of the carbonyl group in diaryl ketones, aryl alkyl ketones, aryl aldehydes, diketones and carboxylic acid derivatives. The stereochemistry (Table 1) and energy correlations of the reduction (Table 2) are treated, in addition to anodic oxidation of C=O groups which leads to C—C bond cleavages similar to MS fragmentation (Table 3).

Section V deals with the C=N linkage, the reduction of which is very similar to that of C=O groups. The stereochemistry of the reduction of some imines (Table 4) and of some bicyclic oximes (Table 5) is presented.

Finally, the N=N linkage is treated. Under different experimental conditions azobenzene can yield hydrazobenzene, aniline or benzidine, and the reaction has synthetic applications.

130 references up to 1974.
Complementary: 8, 11; **17**, 14; **19**, 22; **26**, 12.
Relevant: 5, 2; **13**, 5; **16**, 15; **20**, 5; **22**, 9; **23**, 10; **24**, 17; **25**, 13; **27**, 8; **28**, 7; **29**, 8; **31**, 7; **32**, 6; **34**, 1.

21, *6: 1,3-Dipolar cycloadditions involving X=Y groups* by *G. Bianchi, C. De Micheli and R. Gandolfi*

I. Introduction (371); **II**. General (371); **III**. Cycloadditions of allyl and heteroallyl anion systems (406); **IV**. Cycloadditions of 1,3-dipoles with an orthogonal double bonds (471); **VI**. Acknowledgements (518); **VII**. References (518).

Section II reviews the general aspects of 1,3-dipoles (Tables 1–3), dipolarophiles, stereospecificity, solvent effects (Table 4 and 5), activation parameters (Table 6), transition states, 'allowedness,' reactivity, regioselectivity, periselectivity, *syn–anti* and *endo–exo* isomerism and 1,1-additions of 1,3-β-dipoles.

Section III deals with allyl and heteroallyl anions reacting with double bonds to give $3^- + 2 \rightarrow 5^-$ cycloadditions.

Section IV discusses synthetic approaches, reactivity and regioselectivity of the cycloadditions of nitrile ylides with C=O, C=N, C=S, N=N and N=O double bonds as well as with cumulated double bonds and with olefins (Table 7). Subsequently, 1,3-dipolar cycloadditions of nitrile imines and nitrile oxides are described with the same (and some additional) double-bonded systems (Tables 8 and 9). Diazo compounds (Table 10) undergo analogous cycloaddition with C=C, C=N, C=O and C=S linkages, and so do azides with alkenes, enamines, vinyl ethers, azomethines and conjugated electron-poor olefins. Reactivity and regiochemistry are presented (Tables 12 and 13) and also the reactions of azides with α-ester and α-keto phosphorus ylides, isocyanates, isothiocyanates and active methylene compounds.

Section V describes the preparation, reactions and reactivity of azomethine ylides, azomethine imines, nitrones (azomethine oxides, including treatment of their structure and cycloadditions with C=C, —NCO, —NCS, C=S, C=P, N=P and N=S), carbonyl ylides and carbonyl oxides obtained as intermediates in the decomposition of the primary ozonides of C=C compounds (Table 14). Ozone reacts with aliphatic and

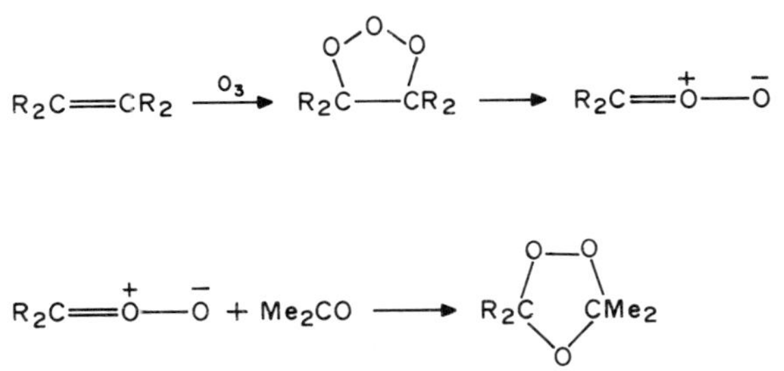

aromatic double bonds (Table 15) and also with C=N and C=S groups.

Finally, the structure, preparation and cycloadditions of thiocarbonyl ylides and thiocarbonyl imines are presented.

493 references up to 1975.
Complementary: 1, 11; **8**, 7; **31**, 8.
Relevant: 14, 6; **24**, 11.

21, *7: Reactions of carbenes with X=Y groups* by *A. P. Marchand*

I. Introduction (534); **II**. Reactions with alkenes (C=C) (534); **III**. Reactions with C=X (607); **IV**. Reactions with X=Y (619); **V**. Miscellaneous reactions with X=Y (622); **VI**. Addendum (629); **VII**. Acknowledgements (635); **VIII**. References (635).

Section II discusses the relation between the spin state (singlet or triplet) of the carbene and the stereochemistry of its addition to C=C (Table 1). Carbenoids (i.e. metal-complexed carbenes) are considered separately (Table 2). Reactivity studies are reviewed, including electronic effects in both the carbene and the olefin, LFER correlations (Table 3), steric effects in both reactants, stereoselectivity involving effects in both reactants, attempted intramolecular reactions in which the completion of the methylenation process is sterically impossible ('foiled reactions'), reactions of nucleophilic carbenes (in which the C bears a negative charge, e.g. in competition between carbene addition and ylide formation in

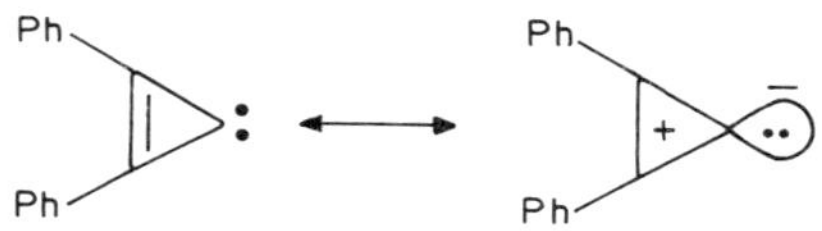

reactions with olefins bearing heteroatomic substituents (Table 4), reactions of unsaturated carbenes (alkylidene carbenes, $R_2C{=}C{:}$, or vinylidenecarbenes, $R_2C{=}C{=}C{:}$) with olefins (Table 5) and finally reactions with olefins of elemental carbon (C_1–C_5) and carbynes ($R\dot{C}{:}$), which bear a formal resemblance to carbenes (Table 7).

Section III describes reactions of carbenes and carbenoids with C=O, C=S, C=N and C=P linkages, usually producing the corresponding three-membered heterocyclic rings, which in turn may undergo various fragmentation and rearrangement reactions.

Section IV deals with carbene and carbenoid additions to N=N, N=O and N=S groups.

Finally, Section V discusses 1, 3-dipolar cycloadditions of carbenes to X=Y dipolarophiles under special conditions [e.g. in the case of carboalkoxycarbenes RO—C(O)—$\ddot{C}$H] and also 1,4-additions of carbenes ot conjugated systems.

The Addendum (Section VI) presents new material which was published between April and November 1974.

350 references up to 1974.
Complementary: 1, 9; **30**, 4, 5.
Relevant: none.

21, *8: The formation of unsaturated groups by heterolytic fragmentation* by *K. B. Becker and C. A. Grob*

I. Introduction (653); **II**. Mechanism and stereochemistry (659); **III**. The scope of heterolytic fragmentation (681); **IV**. Outlook (716); **V**. References (717).

Section I defines the fragmentation process a—b—c—d—X → a—b + c = d + X: where a—b is an electrofugal group, c and d two atoms which are able to form double bonds (C,

N, O) and X is a nucleofuge which leaves with its bonding electron pair (halogen, RSO_3^-, —N=N; $—OR_2$, $—NR_3$, SR_2) (Table 1, a–c). As a rule, fragmentation must lead to three fragments, although exceptions exist.

Section II discusses the reaction paths, the basic mechanisms, the two-step C^+ ion mechanism (i.e., S_N1, $E1$), the concerted mechanism and its stereoelectronic and conformational requirements (necessitating as a rule rigid, cyclic starting materials), the competition between concerted fragmentation, elimination and cyclization with open-chain, freely rotating molecules (Table 2), the frangomeric effect (i.e. the increase in ionization rate due to a concerted mechanism) and the extremely rare two-step carbanion mechanism in which an electrofugal ion is released first, yielding an intermediate carbanion.

Section III presents the scope of heterolytic fragmentations, classified according to the unsaturated fragment c=d. Examples are given to illustrate the scope and general principles involved and they are fragmentations forming olefins, imines, cyanates (e.g.

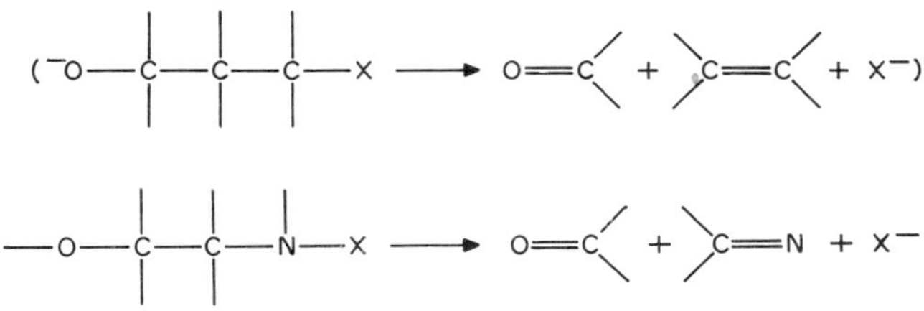

Hoffmann, Curtius, Lossen degradations of amides), carbonyl compounds, oxides of B, S,

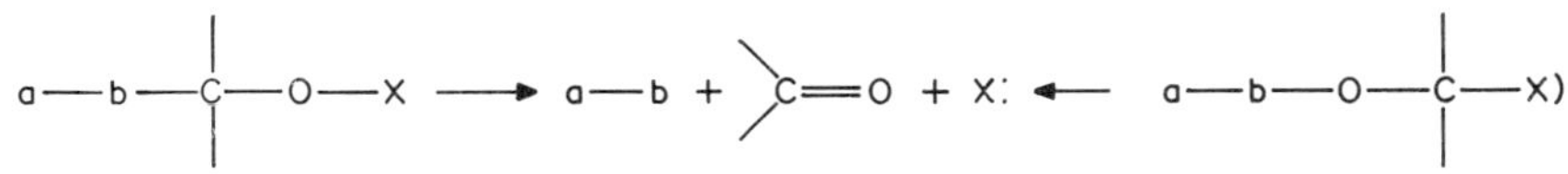

P ($ROH + BCl_3 \rightarrow ROBCl \rightarrow R^+Cl^- + O{=}BCl$), nitrogen [e.g., heterolytic fragmentation of azo compounds ($R^1N{=}NR^2$) in which R^1 is electrofugal and R^2 is nucleofugal ($a—b—N{=}N—X \rightarrow a—b + N_2 + X{:}$)], diimides and azo compounds:

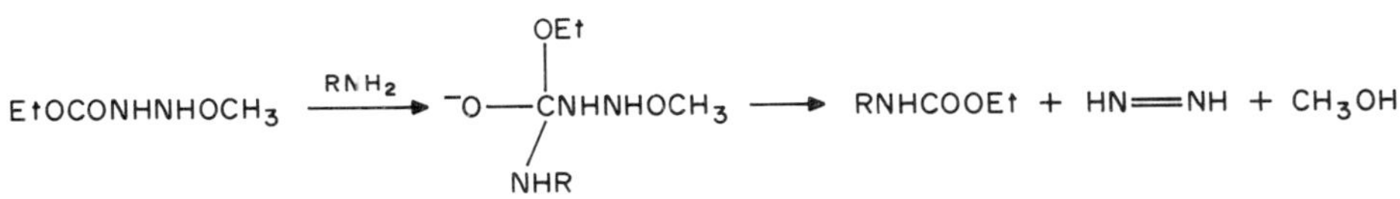

Finally, fragmentations occur in displacement reactions on atom b: Y:b—c—d—X → Yb + c=d + X:. and in the α-fragmentation on the electrofugal group a—b and a nucleofugal group X can be released from same c atom, when the latter has two available valency states: a—b—c—X → a—b + :c + X:

(e.g., ROCCOOH → ROCCOX → ROC^+ + CO + X:).

149 references up to 1974.
Complementary: none.
Relevant: 1, 2.

21, *9: Electrophilic additions to carbon—carbon double bonds* by *G. H. Schmid and D. G. Garratt*

Section I reviews the mechanisms, orientations and stereochemistry of the main electrophilic routes of addition to C=C groups.

Section II discusses the addition of hydrogen halides (when solvent-incorporated and rearranged products can also be formed) and the acid-catalysed additions of water (Table 1), alcohols and carboxylic acids. Section III describes additions of F_2, Cl_2 (Table 2), Br_2 to alkenes (Tables 3–7) and to cycloalkenes (Table 8) and I_2.

Section IV deals with the interhalogens and considers the addition of BrF, IF (Tables 9 and 10), chlorine, bromine and iodine perfluorides, ClF, BrCl (Tables 11 and 12), ICl (Table 13) and IBr (Table 14). Product distributions and configurations are described.

Section V treats electrophilic reactions of pseudohalogens, including kinetic data on INCO (Table 15), NCSCl (Table 16), ISCN (Table 17), halogen nitrates (e.g. INO_3) (Tables 18–20) and halogen azides (Table 21).

Section VI reviews the electrophilic epoxidation of alkenes by organic peroxy acids and the support for the three-membered transition state (Tables 22 and 23) and deals also with diepoxidation and with the epoxidation of allenes. The addition of sulphenyl halides is treated next, presenting specific rate constants. Hammett σ values (Tables 24–26), data showing the effect of configuration on the rate of addition (Table 27) and the formation of kinetically and thermodynamically controlled products (Tables 28–32) are presented. This is followed by addition of divalent selenium compounds, such as selenenyl halides (Tables 33–35), areneselenenyl hexafluorophosphates, hexafluoroantimonates and acetates. Among tetravalent Se and Te compounds, the additions of arylselenium trichlorides, aryltellurium trichlorides, alkylselenium trichlorides, $SeCl_4$ and $TeCl_4$ are treated.

Section VII discusses the addition of some metal salts to alkenes. Thus oxymercuration (Table 37) and methoxymercuration (Table 38) are considered, and also the effect of the alkene structure on the stereochemistry (Table 39) and the effect of solvents (Table 40). Finally, the thallation reaction is described (Table 41) and is compared with mercuration (Table 42).

Mechanisms for the whole chapter are summarized and classified (Section VIII, Table 43) and various correlations are presented (Tables 44 and 45).

540 references up to 1974.
Complementary: none.
Relevant: 24, 8; **34**, 7.

21, *10: The olefin metathesis reaction* by *N. Calderon*

Olefin metathesis is a bond reorganization reaction between two olefins, in which the total number and type of bonds remain unchanged. A short review on bonding and interchange of olefins on transition metals is given.

Section II discusses the evolution of the metathesis concept, product distributions of α-

olefins disproportionated over an $Mo(CO)_6$–alumina catalyst (Table 1) and variation of polymer structures over different catalyst systems (Table 2).

Section III deals with metathesis catalysis, first with some typical heterogeneous catalysts (Table 3), with views regarding the rate-determining step and with the enhancement of catalytic activity by using a variety of experimental techniques and catalyst modifications (Table 4). Next, homogeneous catalysts (W or Mo salts and a wide variety of organometallic compounds of Li, Mg, Al, Zn, Sn and Hg) and their contributions are discussed, together with procedures for catalyst preparation and the use of oxygenated modifiers which may strongly activate the homogeneous metathesis catalysts).

Section IV discusses the redistribution by transalkylidenation (Table 6) and the metathesis of acyclic vinylenic (Table 7) and cycloolefinic substrates (Table 8), followed by geometric isomerization in metathesis reactions (*cis–trans* equilibration, Table 9) and structural selectivity in cross-metathesis reactions [e.g. addition of pent-2-ene to pent-1-ene not only results in cross-metathesis between the two, but pent-1-ene also suppresses the self-metathesis of pent-2-ene (Table 10)].

Section V deals with mechanistic aspects, with the proposed four-centred transition state and with the tetramethylene–metal transition state, the metallocycle transition state and the mechanism involving a metal–carbene complex. At present, no clear preference exists for any of the four schemes suggested. Finally, the exchange step (i.e. accommodation of the incoming olefinic ligand within the coordination sphere of the metal and disengagement of the leaving ligand) is considered.

Section VI describes synthetic applications, especially to polymer chemistry.

91 references up to 1974.
Complementary: none.
Relevant: none.

21, *11: Oxidation of C=C and C=N groups* by *P. M. Henry and G. L. Lange*

I. Introduction (966); **II**. Oxidation of C=C bonds (968); **III**. Oxidation of C=N bonds (1067); **IV**. References (1082).

Section II first describes the oxidation of C=C bonds by non-transition metal ion oxidants, among which the two-electron oxidants Hg^{II}, Tl^{III} and Pb^{IV} are the most important, e.g. oxymercuration [$Hg^{2+} + CH_2{=}CHCH_3 + ROH \rightarrow {}^{+}HgCH_2CH(OR)CH_3 + H^{+}$] and oxidation of alkenes by Tl^{III} to glycols or carbonyl compounds (Table 1). The olefin oxidation products with $Pb(OAc)_4$ are very dependent on the conditions and may be 1,2-diacetates, hydroxyacetates, glyoxylic acid derivatives, rearrangement products, lactones, products of allylic oxidation, etc., and the routes of reaction may be heterolytic or homolytic. Bismuth(V) and cerium(IV) salts are mentioned briefly.

Transition metal oxidants are treated next. The most widely used process involves Pd^{II} as a catalyst (Wacker process: Pd-catalysed oxidation of ethylene to acetaldehyde and related olefin → carbonyl oxidations). The scope, conditions, mechanism and uses of this reaction are considered in detail. Similar processes occur with Pt^{II}, Rh^{III}, Ru^{III}, Os^{III}, Ir^{III} and Cu^{II}. OsO_4, RuO_4 and Re_2O_7 are used for the hydroxylation of olefins, often in conjunction with H_2O_2. $KMnO_4$ with olefins yields glycols, hydroxy ketones, diketones, epoxides or cleavage products. Chromic acid gives primarily epoxides, which may then be further oxidized with or without rearrangement.

Transition metals are also used as catalysts for oxidations by other reagents. These include H_2O_2 or peroxides (usually yielding *trans*-diols), hydroperoxides (giving epoxides) or oxygen (yielding allylic oxidation products, epoxides or cleavage products). A special case is the use of I_2 with silver carboxylate (Prevost and Woodward reactions).

Non-metallic oxidants are treated next. These include peroxy acids (to epoxides), H_2O_2 (epoxidation in the presence of nucleophilic catalysts), alkyl hydroperoxides, hypohalites and related oxidants (NBS, CH_3COOCl, etc.), ozone (considering the mechanism, stages and applications of the ozonolysis reaction), molecular oxygen, especially in its singlet state generated photochemically or chemically, forming allylic hydroperoxides or 1,2- or 1,4-cycloaddition products, and finally SeO_2, giving allylic oxidation [$CH_3C(CH_3){=}CHCH_3 \rightarrow AcOCH_2C(CH_3){=}CHCH_3$].

Section III discusses the oxidation of C=N bonds in imines, oximes and hydrazones by both non-transition and transition metal oxidants and also be non-metal oxidants (peroxy acids, ozone, dioxyne, NO_2 and HNO_2). Depending on the oxidant and the substrate, a wide variety of products are obtained (aldehydes, ketones, dimers, *gem*-nitrosoacetals, etc.).

566 references up to 1974.
Complementary: none.
Relevant: 31, 13; **33**, 6, 7, 8, 11, 14, 15, 17.

21, *12: Transition metal catalysed carbonylation of olefins* by *J. K. Stille and D. E. James*

I. Introduction (1100); **II**. Mechanism of homogeneous catalysis. The nature of the reactive intermediates (1101); **III**. Solvocarbonylation (1111); **IV**. Hydrocarbonylation (1132); **V**. Olefin coupling–carbonylation reactions (1148); **VI**. References (1158).

Section II reviews complexation in metal–olefin (Table 1) and π-allyl complexes, the formation of σ-bonded intermediates and the stereochemistry of the nucleophilic addition in various olefin metallations (Table 2), CO insertions, oxidative addition–reductive elimination and the mechanism of olefin carbonylation.

Section III discusses solvometallation of diolefins, carbonylation of σ-bonded enyl and diolefin complexes and the carbonylation of diolefins without prior complex formation. Catalysts, conditions, stereochemistry and mechanism are considered. Next, the transmetallation–carbonylation and solvometallation–carbonylation of monoolefins (Tables 3 and 4) are treated similarly.

Section IV describes the 'oxo' or hydroformylation reaction (olefin + CO + H_2 $\rightarrow$ aldehyde). The mechanism, catalysts and scope of the reaction are treated, and also the use of the reaction for the synthesis of carbonyl compounds other than aldehydes (ketones, lactones, etc.). The stereochemistry of the addition of H and CO across the olefinic double bond is considered, together with the use of the technique to produce optically active products in the presence of optically active catalysts.

In protic solvents carbonylation of olefins occurs with involvement of water or alcohols and the product is an aliphatic acid from olefins and an acrylic acid from acetylenes. This 'hydrocarbonylation' reaction can also be carried out so as the afford optically active products.

Section V deals with transition metal-catalysed coupling reactions of olefins (such as cyclodimerization, oligomerization, olefin metathesis of polymerization), which in the presence of CO take place with incorporation of a carbonyl function into the carbon skeleton. The catalysts, mechanism, scope and practical uses of the reaction are described.

289 references up to 1974.
Complementary: 30, 7; **35**, 6, 7, 10.
Relevant: 1, 6; **9**, 5.

21, *13: Imidines and diamidides (1,3,5 triazapentadienes)* by *J. A. Elvidge and N. R. Barot*

Imidines are analogues of cyclic anhydrides of dicarboxylic acids, in which all three oxygen atoms are replaced by N.

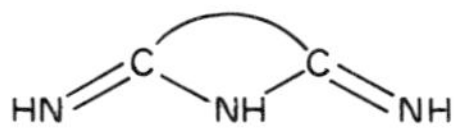

Open-chain compounds containing the same functional grouping are called diamidides (or by their systematic name, 1, 3, 5-triazapentadienes).

Section II describes syntheses of imidines by cyclization of diamidine salts,

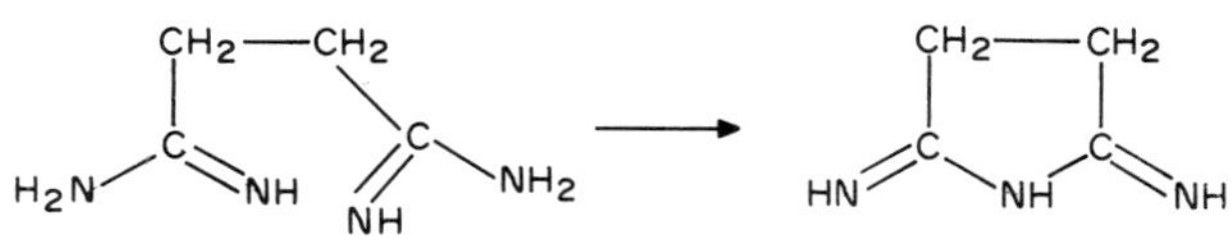

by the reaction of diimidoates with amines, by the addition of HN_3 or amines to dinitriles (uncatalysed, alkoxide-catalysed, $NaHN_2$- or amine salt-catalysed), by the reaction of halogen-, alkoxy- or thio-substituted precursors with NH_3 or amines or by treatment of a trinitrile with dry HBr.

Section III deals with the following reactions which lead to the formation of diamidides: benzamidines with benzimidochlorides:

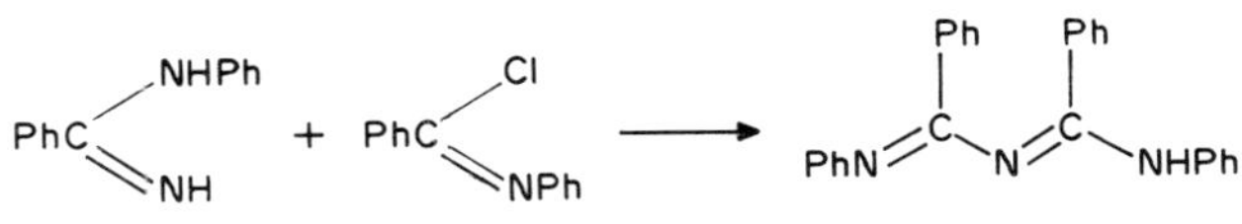

N-benzoyl-*N*-phenylbenzamidine and amines, imidosulphonates and amidines, thiobenzoylbenzamidines and amines and, finally arylamidines and aryl aldehydes.

Section IV discusses physical characteristics (m.p., Table 1; pK_a values, Table 2) of imidines and their reactions with water (hydrolysis of one or more =NH groups to C=O), NH_2OH (yielding imidine oximes) and amines (condensations, substitutions, adduct formation, formation of cross-conjugated macrocycles, dehydrogenation, reduction and ring-expansion). Next, alkylation, arylation self- and mixed condensations, reactions with active methylene compounds, open-chain and macrocyclic chelates with various metal ions, phthalogen compounds (used for printing fabrics) and the hydrogenolysis and dehydrogenation of glutarimidines are considered.

Section V deals with general characteristics of conjugated and non-conjugated diamidides and a triamidide. Section VI treats UV (Tables 3 and 4), IR (Table 5) and NMR data relevant to imidines.

86 references up to 1972.
Complementary: none.
Relevant: none.

The chemistry of cyanates and their thio derivatives (1977)

22, *1: Electronic structure of the cyanato and thiocyanato groups—ground state and excited states* by *K. Wittel, J. L. Meeks and S. P. McGlynn*

I. Introduction (2); **II.** Ground state properties (3); **III.** Concerning the difficulties inherent to a discussion of electronically excited states (23); **IV.** Excited state properties (25); **V.** Conclusion (65); **VI.** References (66).

Section II discusses first anions of triatomic anionic species (NCO^- and NCS^-), their molecular orbitals, geometry, shapes and energies of valence orbitals, X-ray emission spectra, charge distributions and ESCA chemical shifts (Tables 1 and 2). Next the geometry, energies, PES and X-ray spectra (Table 3), valence electron distribution and ESCA data (Tables 4 and 5) of covalent compounds are given and the main results and their limitations are discussed.

Section III considers the experimental and theoretical problems inherent to a discussion of electronically excited states.

Section IV deals with MO calculations and with spin and symmetry adaptations, tabulates different eigenfunction results (Tables 6, 7, 8 and 8a) and treats intra- and inter-configurational splitting, spin–orbit coupling, bending of excited states and intensities of transitions. The results of various computations are shown in Figs 13–16 for the NCO^- anion and for HNCO, CH_3NCO and PhNCO. Next, singlet–triplet absorption and emission (Tables 10 and 11 and Figs 17–19), singlet excited states (Table 13), Rydberg transitions (Tables 14 and 15) and charge-transfer to solvent trasitions are considered and summarized.

89 references up to 1975.
Complementary: none.
Relevant: **1**, 1; **2**, 1; **3**, 1; **4**, 1; **5**, 1; **6**, 1; **8**, 1; **11**, 1; **12**, 1; **13**, 1; **14**, 1; **15**, 1; **16**, 1; **17**, 1; **18**, 1; **19**, 1; **20**, 1; **23**, 1; **24**, 1; **25**, 1; **26**, 1; **27**, 9; **28**, 1; **31**, 27; **32**, 1; **33**, 1; **36**, 2.

22, *2: Structural chemistry of the cyanates and their thio derivatives* by *I. Hargittai and I. C. Paul*

I. Introduction (69); **II.** Experimental information on the geometry of molecules in the gas phase (74); **III.** Structural variations in the isocyanate and isothiocyanate groups in the gas phase (78); **IV.** Structural variations involving the atom adjacent to the isocyanate (or isothiocyanate) group (81); **V.** Conformations of molecules in the gas phase (86); **VI.** Geometrical variations in the rest of the molecule as observed in the gas phase (94);

VII. Two thiocyanates in the gas phase (96); **VIII**. Crystal structure data on covalent cyanates and isocyanates (98); **IX**. Crystal structure data on covalent thiocyanates and isothiocyanates (101); **X**. Crystal structure data on selenocyanates and related compounds (113); **XI**. Crystal structure data on metal complexes containing cyanates, thiocyanates and related ligands (119); **XII**. Conclusions (124); **XIII**. Acknowledgements (124); **XIV**. References (124).

The chapter surveys the geometry and conformation of ROCN, RNCO, RSCN and RNCS molecules (Table 1), mainly by X-ray diffraction studies and by microwave spectroscopy (Sections I and II). Sections III and IV deal with variations in bond distances in RNCO and RNCS (Tables 2 and 3) and in the X—N=C bond angle when X is the atom adjacent to N in RNCO or RNCS (Tables 4–6). Gas-phase conformations (Section V) and the influence of the functional groups on structural variations in the rest of the molecule (Tables 7–9) are presented. Microwave spectroscopic results for bond lengths and angles in MeSCN and EtSCN are given (Table 10, Section VII).

Crystal structure data are given for covalent ROCN and RNCO (Table 11, Section VIII), covalent RSCN and RNSC (Tables 12–14, Section IX, including many stereoscopic drawings), selenocyanates and related compounds (Tables 15 and 16, Section X) and metal complexes containing the title groups (Tables 17 and 18, Section XI).

127 references up to 1975.
Complementary: 32, 24.
Relevant: 16, 2; **17**, 2; **18**, 2; **19**, 2; **20**, 2; **23**, 3; **24**, 2; **26**, 2; **27**, 4; **28**, 2; **29**, 1; **30**, 1; **31**, 24; **33**, 12; **34**, 10; **36**, 3.

22, *3: Stereochemical and conformational aspects of cyanates and related groups* by *G. C. Corfield*

I. Introduction (131); **II**. Conformational and configurational analysis (135); **III**. References (150).

Section I surveys the importance of geometric, optical and conformational isomerism on the reactions and properties of molecules containing the title groups.

Section II deals with conformations, rotational barriers, microwave spectra and dipole moments of acyclic compounds. This is followed by a discussion of the same factors in mono-, di- and trisubstituted cyclohexane derivatives of the title compounds, considering axial and equatorial conformations, NMR results and conformational free energies.

82 references up to 1973.
Complementary: none.
Relevant: 19, 21; **20**, 2; **21**, 1; **27**, 5; **28**, 10; **29**, 2; **31**, 19; **32**, 1; **33**, 2.

22, *4: The optical rotatory dispersion and circular dichroism of the cyanates and related groups* by *C. Toniolo*

I. Introduction (153); **II**. Chirospectroscopic properties of cyanates and isocyanates (154); **III**. Chirospectroscopic properties of thiocyanates (161); **IV**. Chirospectroscopic properties of isothiocyanates (167); **V**. Chirospectroscopic properties of cyanamides and azides (170); **VI**. Chirospectroscopic properties of the reaction products of cyanates and related groups (175); **VII**. References (187).

Section II describes the ORD and CD spectra of isocyanates in various solvents, including some which are optically active and/or polymeric. No data are available on cyanates. Section III discusses thiocyanates, their octant rule and CD data for monochromophoric steroidal thiocyanates (Table 1). Isothiocyanates are similarly treated (Section IV, Table 2), followed by chirospectroscopic properties of cyanamides and azides (Table 3), including their Cotton effect and octant rule. Section VI deals with reaction products of the title compounds, e.g. those obtained from CH_3NCS with amino acids and peptides (Table 4) and from cyanates and related groups with alcohols or thiols.

121 references up to 1976.
Complementary: none.
Relevant: **8**, 4; **18**, 7; **19**, 6; **20**, 2; **25**, 3; **26**, 3; **27**, 6; **28**, 10; **29**, 23; **31**, 1; **36**, 17.

22, *5: Detection and determination of cyanates, isocyanates, isothiocyanates and thiocyanates* by *D. A. Ben-Efraim*

Section I deals with chemical, chromatographic, IR, Raman, UV (Table 1), NMR and MS methods for the analysis of cyanates.

Section II describes chemical methods for the qualitative analysis of isocyanates, their determination by amine addition, by reaction with peroxides, derivatization as ureas or urethanes, by polarography and gas chromatography, followed by IR, Raman, UV (Table 2), fluorescence, NMR and mass spectroscopy. Section III discusses the analogous methods as applied to isothiocyanates [including UV (Table 3) and MS (Table 4) and also Cotton effects].

Section IV treates the qualitative analysis of thiocyanates, their quantitative determination by reduction or by cleavage by various reagents [$Na_2Pb(OH)_4$, NH_3–$AgNO_3$, NaOH–EtOH, NaOEt, Na_2S), polarography, chromatography, IR, Raman, UV (Table 5), fluorescence, NMR (Table 6) and MS (Table 7).

185 references up to 1974.
Complementary: none.
Relevant: **1**, 5; **2**, 8; **3**, 15; **4**, 3; **5**, 17; **8**, 3; **12**, 4; **13**, 6; **14**, 3; **15**, 3; **16**, 3; **17**, 4; **18**, 5; **20**, 3; **23**, 5; **24**, 5; **25**, 10; **26**, 5; **28**, 3; **29**, 21; **30**, 14–22; **36**, 5.

22, *6: Thermochemistry of cyanates, isocyanates and their thio derivatives* by *R. Shaw*

Owing to the lack of experimental data, most of the chapter deals with estimation.

Section II gives the estimated thermochemical data for HOCN (Table 1), CH_3OCN and C_2H_5OCN (Table 2), $CH_2{=}CHOCN$ and PhOCN (Table 3). Section III presents data for the five analogues isocyanate derivatives, RNCO (R = H, Table 4; R = Me, Table 5; R = Et, Table 6; R = $CH_2{=}CH$ and R = Ph, Table 7). Section IV deals in the same manner with thiocyanates (Tables 8–11) and Section V with isothiocyanates (Tables 12–15).

Section VI and Table 16 give a summary of the measured and estimated molar heats of formation, entropies and heat capacities of the title compounds, while Section VIII and Table 17 present group values for the title compounds, which can be used for the estimation of thermochemical properties by group additivity.

26 references up to 1975.
Complementary: none.
Relevant: 16, 16: **18**, 3; **19**, 3; **20**, 11; **21**, 3; **23**, 4; **24**, 3; **25**, 2; **26**, 4; **27**, 9; **28**, 4; **29**, 24; **30**, 2; **31**, 2; **33**, 3; **36**, 4.

22, *7: Mass spectra of cyanates, isocyanates and related compounds* by *K. A. Jensen and G. Schroll*

Section II describes the MS of alkyl cyanates and phenyl cyanate, their fragmentation schemes and main peaks (Figs. 1–3 present the MS of ROCN, RNCO, RSCN, RNCS, RSeCN and RNCSe, with both aliphatic and aromatic R groups).

Section III deals with alkyl (Tables 2 and 3) and phenyl isocyanates, Section IV with thiocyanates, Section VI with selenocyanates and Section VII with isoselenocyanates.

Section VIII describes the limitation of the method and especially the differences in MS obtained with different instruments (Table 4), when the variations due to instrumentation may be larger than those due to the study of different compounds. Prominent peaks, molecular ions, eliminations and fragmentations in the title compounds are compared.

21 references up to 1972.
Complementary: none.
Relevant: 9, 7; **10**, 5; **13**, 19; **16**, 4; **17**, 5; **18**, 6; **19**, 5; **24**, 6; **25**, 4; **26**, 6; **27**, 7; **29**, 3; **30**, 22; **31**, 3; **32**, 3; **33**, 4; **36**, 7.

22, *8: Hydrogen bonding and complex formation* by *D. Hadži and S. Milićev*

Section II describes hydrogen bonding, including relative proton-accepting propensities between phenyl pseudohalides and phenol as H-donor (Table 1), and presents frequency shifts and thermodynamic quantities of association between phenol and some of the title compounds (Table 2). The site of bonding and the influence of the structure of the alkyl or aryl group in the title compounds and of the proton donors are discussed.

Section III deals with CT complexes between I_2 and some of the title compounds (Table 5). Section IV discusses complexes between the title compounds and Lewis acids, identified by IR, NMR and MS. Section V treats the formation of complexes between isocyanates and various catalysts (organometallic compounds, metal salts, chelates of transition metals, etc.) used mainly in the process of formation of urethanes. Section VI describes salts and addition compounds formed between the title compounds and HCl, HBr, tertiary phosphines, etc.

Finally, Section VII discusses transition metal complexes (Tables 3, 4 and 6) and their spectra, structure, optical isomerism and stability.

89 references up to 1974.
Complementary: none.
Relevant: **2**, 9; **3**, 6; **4**, 4; **5**, 6; **7**, 6; **8**, 5; **11**, 3; **12**, 3, 10; **13**, 7; **15**, 4; **16**, 5; **17**, 6; **18**, 8; **19**, 7; **20**, 12; **23**, 6, 7; **24**, 4; **25**, 5, 6; **29**, 17; **31**, 17, 21; **32**, 21; **32**, 5.

22, *9: The electrochemistry of cyanates and related compounds* by *O. Hammerich and V. D. Parker*

I. Introduction (321); **II**. Thiocyanates (331); **III**. Isocyanates (336); **IV**. Isothiocyanates (339); **V**. References (341).

Half-wave potentials are tabulated for isocyanates (Table 1), thiocyanates (Table 2) and isothiocyanates (Table 3).

Section II deals with polarographic studies of the reduction of thiocyanates ($RSCN + 6e + 6H_2O \rightarrow RSH + CH_3HN_2 + 6OH^-$). Phenacyl thiocyanates yield ketones ($ArCOCH_2SCN \rightarrow ArCOCH_3 + SCN^-$) and benzyl thiocyanates give the corresponding toluene and bibenzyl derivatives.

Section III discusses isocyanates, which can give a crystalline stereoregular polyphenyl isocyanate by cathodic reduction. Under different conditions, polyurethanes can be obtained.

Section IV treats isothiocyanates (RNCS). When R = Ph, electrochemical reduction yields thioformanilide ($PhNCS + 2e + 2H_2O \rightarrow PhNHCHS$), and the latter can be further reduced to *N*-mercaptomethylanilide ($PhNHCH_2SH$) or alternatively to $PhNH_2$, CH_2O and H_2S.

32 references up to 1973.
Complementary: none.
Relevant: **5**, 2; **8**, 11; **13**, 5; **16**, 15; **17**, 14; **19**, 12; **20**, 5; **21**, 5; **23**, 10; **24**, 17; **25**, 13; **26**, 12; **27**, 8; **28**, 7; **29**, 8, 9; **31**, 7; **32**, 6; **32**, 1.

22, *10: Photochemistry of cyanates and related groups* by *R. Jahn and U. Schmidt*

I. Introduction (343); **II**. Spectra (344); **III**. Isocyanates (346); **IV**. Photolysis of cyanates (350); **V**. Thiocyanates and isothiocyanates—photoiosomerization (352); **VI**. Thiocyanation (353); **VII**. Elimination of atomic sulphur from isothiocyanates (353); **VIII**. References (355).

Section I reviews the main photochemical routes available with the title compounds. Section II deals with UV spectra of alkyl isothiocyanates (Table 1) and related compounds.

Section III describes the photosynthesis of isocyanates from β-ketoamides or acyl azides and also the formation of nitrenes by photolysis of HNCO, RNCO and acyl, aryl and styryl isocyanates and the photoaddition of alcohols to isocyanates to yield urethanes.

Section IV describes free-radical photolyses of cyanates, leading to O—C fission ($ArOCN + h\nu \rightarrow ArO\cdot + \cdot CN$) or to rearrangements (e.g. $ROCN \rightarrow RNCO$).

The photoisomerization of RSCN and RNCS is discussed in Section V. In addition to the equilibration between the two isomers, a variety of other primary (e.g. $RSCN \rightarrow RS\cdot + \cdot CN$) and secondary (dehydrogenation of solvent) reactions occur.

Section VI deals with photochemical thiocyanation by $(SCN)_2$ or by KSCN and finally Section VII describes the RNCS $\xrightarrow{h\nu}$ RNC + $S(^3P)$ reaction.

40 references up to 1974.
Complementary: none.
Relevant: 2, 16; **3**, 8; **6**, 4; **8**, 12; **9**, 6; **11**, 5; **13**, 16: **14**, 8; **15**, 9; **16**, 11; **17**, 9; **18**, 10; **19**, 20; **23**, 9; **24**, 12; **25**, 11; **27**, 21, 22; **28**, 5; **29**, 5, 6; **32**, 29; **33**, 21.

22, *11: Radiation chemistry of organic compounds containing OCN, CNO, SCN and CNS groups* by *A. Horowitz*

Section II reviews the radiation chemistry of the title compounds in the pure form and in mixtures.

Section III deals with aliphatic and aromatic thiocyanates studied mainly by ESR, with the yields of various products (radicals, H), with RSCN → RNCS isomerization and radiation protection of alkanes by RSCN. Section IV describes similarly ESR studies of irradiated aliphatic and aromatic isothiocyanates and Section V those of isocyanates.

Section VI discusses the use of free-radical chain reactions involving high-energy ionizing radiation for the syntheses of compounds containing one of the title functional groups, from simple starting materials carrying OCN and related groups and olefins or haloolefins. The radiation-induced synthesis of aliphatic thiocyanates from ethanol or acetone and KSCN or NH_4SCN and some other reactions is also described, dealing finally (Section VII) with polymerization of isocyanates and copolymerizations of isocyanates with ketene or formaldehyde.

75 references up to 1974.
Complementary: none.
Relevant: 9, 8; **10**, 6; **11**, 6; **13**, 17; **15**, 9; **16**, 12; **17**, 10; **18**, 11; **19**, 10; **25**, 12; **27**, 23, 24; **29**, 7; **31**, 6; **32**, 9; **33**, 21; **36**, 8.

22, *12: Kinetics and mechanisms of reactions of cyanates and related compounds D. E. Giles*

The Introduction defines the scope of the chapter and especially the terms basicity and nucleophilicity (the former being used for equilibrium and the latter for kinetic properties).

Section II deals with the —SCN/—NCS isomerization of alkyl, benzyl, cycloalkyl, acyl, aryl (Table 2) and allyl thiocyanates and the mechanisms of the reactions, including ionization and return to a carbonium ion, intramolecular group transfer, direct S_N2 displacement and addition–elimination involving carbonyl or aryl carbon. Organic thiocyanates are considered as trifunctional electrophiles (the sites being the C atom of the

R group bonded to SCN, the S atom and the cyanide C atom). Products of reactions with various bases with three aromatic thiocyanates are given in Table 3 and rate constants in Table 4. The competition between cyanide and aryl carbon in ArSCN is illustrated by Table 5. Subsequently, reactions of RSCN with $(RO)_3P$ (yielding nitriles and thiophosphate) are described and also additions to thiocyanates and to isothiocyanates (acids, alcohols, amines, dipolar cycloadditions).

Section III discusses the thermal ROCN$\rightleftharpoons$RNCO isomerization and its photoinduced counterpart (Table 6) and considers cyanates as trifunctional electrophiles, with the O atom being much less reactive than the S in thiocyanates and the competition arising mainly between the alkyl and cyanide carbon atoms (Table 7). Next, addition reactions of cyanates are treated, e.g.

$$\mathrm{ArOC{\equiv}N} \xrightarrow{\mathrm{ROH}} \mathrm{ArO}\overset{\overset{\displaystyle\mathrm{OR}}{|}}{\mathrm{C}}\mathrm{{=}NH} \xrightarrow{\mathrm{ROH}} \mathrm{RO}\overset{\overset{\displaystyle\mathrm{OR}}{|}}{\mathrm{C}}\mathrm{{=}NH + ArOH}$$

with the dialkoxyimine reacting further to give triazines.

Isocyanates are the most reactive of the title compounds towards additions: they add alcohols, amines, water and phenols, e.g. $\mathrm{RNCO + R'OH \rightarrow RNHCOOR'}$ (i.e. urethane formation, the kinetics and mechanism of which are considered in detail, including the use of organometallic catalysts). 1, 2-Cycloadditions to RNCO favour the C=N bond, although when R is electron-withdrawing the C=O bond becomes favoured.

Section IV considers the polar substituent effects of the four title groups. —OCN and —SCN are electron-withdrawing and so are (although less effectively) the —NCS and —NCO groups.

244 references up to 1974.
Complementary: none.
Relevant: none.

22, *13: Acyl and thioacyl derivatives of isocyanates, thiocyanates and isothiocyanates* by *O. Tsuge*

I. Introduction (446); **II.** Methods of preparation (446); **III.** Reactions with active hydrogen compounds (453); **IV.** Di- and trimerization reactions (465); **V.** Cycloadditions and related reactions (468); **VI.** Miscellaneous reactions (493); **VII.** References (501).

Section II reviews the syntheses of acyl isocyanates, thioacyl isocyanates (Table 1) and acyl and thioacyl thiocyanates and isothiocyanates, mainly by reactions of the corresponding acyl halides with the anions of the title groups.

Section III describes the reactions of the title compounds with amines, alcohols, hydrazines, hydrazobenzenes, hydrazones, amidines and CH-acidic compounds. The reactions are complex owing to competition between addition to the NCO or NCS system and substitution on the carbonyl carbon, e.g. $\mathrm{RCONCS + R'NH_2 \rightarrow RCONHR' + RCONHCSNHR'}$.

Section IV discusses the di- and trimerization reactions of benzoyl and acetyl isocyanates to yield six-membered cyclic compounds in the presence of appropriate catalysts. Section V considers cycloadditions of the title compounds to olefins, enamines, enol ethers and related compounds, and to C=C, C=N, and C=O bonds, to cumulated compounds such as allenes, ketenes, diazo compounds (included here by virtue of the contribution of the structure $\mathrm{RR'\bar{C}{-}\overset{+}{N}{\equiv}N}$) and carbodiimides and to isonitriles. Next, reactions with O-, S- and N-containing three-membered heterocycles are treated (yielding various five-membered heterocyclic rings) and reactions with nitrones and with nitrosobenzenes (again yielding five-membered heterocycles) are mentioned.

Section VI deals with the following reactions: chlorination, affording dichlorides ($RCONCO \xrightarrow{PCl_5} RCCl_2NCO$ and $RCONCS + Cl_2 \rightarrow RCON{=}CCl_2$), reactions with organometallics containing Mg, Si, Ge and Sn, reactions with ylides, pyrrole, dioxaphospholene and nitrosonium salts, cyclizations with $AlCl_3$ (benzoyl isothiocyanates and phenylacetyl isothiocyanates yielding monothiophthalimides and monothiohomophthalimides, respectively) and reaction with S_8 yielding five-membered heterocycles with one or two sulphur atoms.

194 references up to 1974.
Complementary: none.
Relevant: none.

22, *14: Syntheses and uses of isotopically labelled cyanates and related groups* by *A. Ceccon and U. Miotti*

I. Introduction (508); **II**. Labelled inorganic ions (508); **III**. Syntheses of labelled organic cyanates, isocyanates, thiocyanates and isothiocyanates (510); **IV**. Uses of labelled isocyanates, isothiocyanates and thiocyanates (531); **V**. Acknowledgements (545); **VI**. References (545).

Section II describes the preparation of the alkali metal cyanates labelled with ^{14}C, ^{13}C, ^{15}N or ^{18}O and of alkali metal thiocyanates labelled with ^{14}C or ^{35}S.

Section III deals with the synthesis of labelled aryl and alkyl cyanates, isocyanates (by Curtius rearrangement of acyl azides, from amines with $COCl_2$, by exchange with $PhN^{14}CO$ or by isomerization of nitrile oxides) (Table 1), thiocyanates by substitutions with labelled —SCN (Tables 2 and 3), forming mixtures of RSCN and RNCS, or by replacement with labelled CN ion, and isothiocyanates by exchange of the S atom with labelled S, using elemental S or H_2S or by exchange of the whole group by ^{35}SCN— (Tables 4 and 5). Isothiocyanates can also be prepared from amines with CS_2 or $CSCl_2$ using appropriately labelled compounds and also in some cases by heating, e.g. 2-naphthylamine with labelled NH_4NCS, by isomerizations of thiocyanates to (the thermodynamically more stable) isothiocyanates and also from halides by replacement with SCN^-. Table 6 summarizes the methods, isotope sources and positions of labelling in the products for isothiocyanates.

Section IV discusses studies in which the labelled title compounds have been used. These include the conversion of isocyanates into carbamates (Table 7) and ureas (Table 8), which have important industrial applications; the conversion of isothiocyanates into thiourea derivatives (Table 9), much used in physiological studies, and some other uses of labelled inorganic cyanates (Table 10) and thiocyanates in mechanistic and spectroscopic studies.

105 references up to 1973.
Complementary: none.
Relevant: **5**, 10; **12**, 13; **13**, 15; **17**, 12; **18**, 18; **19**, 17; **23**, 16; **24**, 20; **26**, 11; **27**, 10; **28**, 8; **29**, 28; **33**, 18; **36**, 12.

22, *15: Pyrolytic reactions of cyanates and related compounds* by *N. Barroeta*

I. Introduction (549); **II**. Cyanic and thiocyanic acids (550); **III**. Kinetics and mechanism of high-temperature pyrolysis (552); **IV**. Thermal synthesis and other reactions (560); **V**. Thermodynamics of the elimination (565); **IV**. References (566).

Section I and Table 1 deal with relative reactivities of various RX compounds (X = AcO, Cl, NCS, SCN) in their gas-phase pyrolysis to yield olefins.

Section II describes in brief the characteristics of the parent acids of the four title compounds, of which only the 'iso' forms (HNCO and HNCS) have been detected in the gas phase.

Section III discusses the general stoichiometry of the pyrolysis, the distribution of products in the pyrolysis of isopropyl isothiocyanate (Table 2) and proportions of olefinic products from the decomposition of *sec*-butyl compounds including the title compounds (Table 3) and Arrhenius parameters for olefin-forming eliminations in various thiocyanates, isothiocyanates and isocyanates (Table 4). Next the mechanisms of the decomposition of thiocyanates (Table 5), isothiocyanates and of isocyanates are discussed.

Section IV treats thermal syntheses and some other reactions of the title compounds, while Section V considers the thermodynamics of the elimination.

75 references up to 1975.
Complementary: **1**, 3; **16**, 20.
Relevant: **3**, 4; **25**, 15; **27**, 11; **32**, 7; **33**, 20.

22, *16: Syntheses and preparative applications of cyanates (esters of cyanic acid)* by *K. A. Jensen and A. Holm*

I. Introduction (570); **II**. Syntheses of cyanates (571); **III**. Physical properties (576); **IV**. Chemical properties: reaction mechanisms (579); **V**. Chemical properties: reactions of cyanates (585); **VI**. Conclusion (614); **VII**. References (615).

Section II describes the syntheses of cyanates by six main routes (Table 1): by thermolysis of thiatriazoles

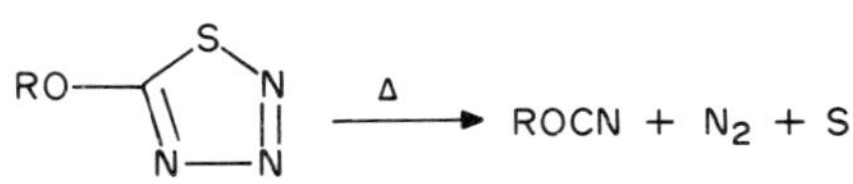

by reaction of alcohols or phenols with cyanogen halides (ROH + ClCN → ROCN), from *O*-alkyl thiocarbamates with heavy metal oxides (HgO, Ag_2O), from *O*-alkyl-*N*-hydroxythiocarbonates, from thiocyanates by treatment with HgO and also from diazo compounds or halides by reaction with cyanic acid or inorganic cyanates (when a mixture of cyanates and isocyanates is formed).

Section III deals with general properties, atomic distances, IR, Raman, NMR and UV spectra, with salts, complexes and associates and very briefly with the photochemistry of cyanates.

Section IV discusses the scope and mechanism of cyanates in isomerizations (ROCN → RNCO), in nucleophilic substitutions (ROCN $\xrightarrow{Cl^-}$ RCl) and in nucleophilic additions to the triple bond {including MeOH, ArOCN + MeOH → ArOC(OMe)=NH and the analogous additions of phenols and *N*-methylanilines and of Grignard reagents resulting in the formation of a complex, $ROCN + RMgX + Et_2O \rightarrow [RCN.ROMgX\frac{1}{2}Et_2O]\}$.

Section V treats cyanates as synthones. The reactions described are isomerizations, trimerizations (to cyanurates), nucleophilic substitutions, eliminations and nucleophilic additions to the C=N bond (including H_2O, alcohols, phenols, Me_2SO, H_2S, H_2Se, RSH, RCOOH, NH_3, RNH_2, RR′NH, RR′R″N, silylamines, hydrazines, hydroxylamines, oximes, amides, ureas, hydrazides, organometallic compounds, active CH_2 compounds, enamines, pyrroles, ylides and allenetetramines). Next, nucleophilic additions to the O atom are mentioned, followed by electrophilic additions of proton acids [$ArOCN + 2HCl + SbCl_5 \rightarrow ArOC(Cl){=}NH_2^+ SbCl_6^-$), Lewis acids, acyl halides and anhydrides and by *N*-alkylation ($ArOCN + Ph_3C^+ \xrightarrow{Me_2SO} ArOCONHCPh_3$).Finally, cycloadditions with azides, diazo compounds, nitrones and nitrile oxides (yielding the expected 1, 3-dipolar cycloaddition products) are mentioned.

124 references up to 1974.
Complementary: none.
Relevant: none.

22, *17: Syntheses and preparative applications of isocyanates* by *R. Richter and H. Ulrich*

Section II describes the synthesis of alkyl and aryl isocyanates by phosgenation of amines, imines, carbamates and ureas (Table 1), via nitrene intermediates, by thermal decomposition of carbamates (Table 2), ureas, polymers of isocyanates and some heterocycles, from organic halides or sulphates with salts of cyanic acid, from olefins, aldehydes and imines with isocyanic acid and by some other less general methods.

Next the synthesis of acyl, thioacyl and imidoyl isocyanates is treated, followed by the preparation of isocyanates, where the group is attached to B, Si, Ge, Sn, Pb, N, P, As, Sb, O, S or halogen atoms and not directly to carbon.

Section III discusses the reactions of isocyanates across their C=N bond, such as dimerization to four-, five- and six-membered ring dimers, trimerization to 1,3,5-trisubstituted hexahydro-*s*-triazenetriones (isocyanurates) and polymerization to linear polymers, $RN{=}C{=}O \rightarrow [NR{-}CO]_n$, but also to products of different structures. Next, [2 + 2] cycloadditions with olefins, cycloalkenes, dienes, allenes, ketenes, aldehydes, ketones, amides and other compounds are described, and [2 + 4] cycloadditions with dienes, heterodienes, cumulenes, carbonyl compounds and imines, followed by [2 + 3] cycloadditions with various 1, 3-dipolar compounds (*N*-oxides, nitrile oxides, azomethine imines, azides, epoxides, aziridines, etc.), which yield a wide variety of heterocyclic compounds. Cycloadducts are also obtained with acetylenes and with carbenes and by some less general routes.

Insertion reactions of isocyanates occur with O—H and S—H bonds, primary and secondary amines, hydroxylamines and related compounds ($RNCO + R_2NH \rightarrow RNHCONR_2$). Insertion also takes place into the C—H bond of arenes ($ArH + RNCO \rightarrow ArCONHR$), alkanes and olefins, into COC bonds, producing carbamates, and into C—Hal bonds, affording carbamoyl halides. Organometallic compounds containing Li, Mg, Hg, B, Al, Sn, Si, Pb, P, etc., react with isocyanates to yield amides in the simplest cases ($R'Li + RNCO \rightarrow RNHCOR'$) and a variety of complicated adducts in others.

Finally, reactions of isocyanates across their C=O bond are treated. The reagents are usually compounds of P, As and Sb, such as phosphonium ylides, phosphinimines and triphenylarsine oxide.

883 references up to 1976.
Complementary: none.
Relevant: none.

22, *18: Syntheses and preparative applications of thiocyanates* by *R. G. Guy*

Since isothiocyanates RNCS are thermodynamically more stable, they are often a major side-product in the preparations of thiocyanates (RSCN).

Section II describes methods of synthesis of thiocyanates from halides ($RX + {}^{-}SCN \rightarrow RSCN$; Tables 1 and 2), sulphonate or benzoate esters ($ArCO_2R + {}^{-}SCN \rightarrow RSCN + ArCO_2^-$), diazonium salts, organothallium compounds, epoxides, aziridines or oxaziridines:

by electrophilic addition of HNCS to alkenes or to the C=C bond of α, β-unsaturated carbonyl compounds and by some other less general methods.

Section III deals with syntheses involving $(SCN)_2$ with aromatic compounds (Ar_{El} substitution), heterocyclic compounds and metal complexes (e.g., porphyrins) and by addition to alkenes:

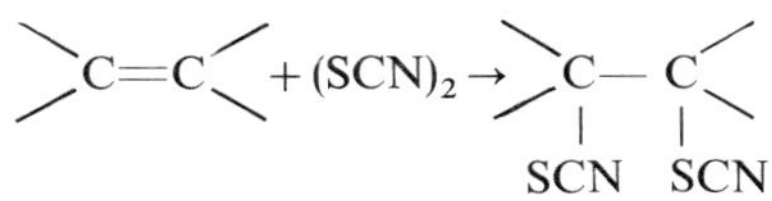

(Table 3) or to cycloalkenes (Table 4), to alkynes and to phosphoranes. $(SCN)_2$ also reacts homolytically (Table 5) and it attacks side-chains of aralkanes, alkenes (yielding both α, β-dithiocyanates and allylic thiocyanates), allenes and alkynes, either under the influence of UV light or in the presence of peroxide initiators. Thiocyanogen halides give similar heterolytic and homolytic reactions to $(SCN)_2$, and so do aryl sulphonyl and sulphenyl thiocyanates.

Section IV discusses the synthesis of thiocyanates by cyanation of sulphur derivatives, by CN^- (e.g. $ArSCl + CN^- \rightarrow ArSCN$) or by cyanogen halides. Some syntheses of more limited scope are presented in Section V.

Section VI considers reactions in which the SCN group behaves as a pseudohalide, undergoing substitution or isomerization ($RSCN \rightarrow RNCS$) by a variety of carbonium or radical or (in the case of allylic thiocyanates) sigmatropic routes. Thiocyanates also undergo, 1,2-elimination (to yield alkenes) and additions and oxidations.

Section VII treats reactions in which the thiocyanates show a sulphenyl cyanide (RS—CN) structure and react by S—C fission or by C=N addition.

S—C fission takes place in oxidations, reductions, nucleophilic displacements of the CN group (e.g. conversion of thiocyanates into disulphides), homolytic fission and base-catalysed 1,2-eliminations of HCN to yield thioketones ($Ar_2CHSCN \rightarrow Ar_2C{=}S + HCN$). The C=N bond undergoes additions of H_2O, ROH, alkenes, H_2S, PCl_5, etc., and also cyclizations to yield a wide variety of five- and six-membered heterocycles containing various combinations of N, S and O atoms, e.g.

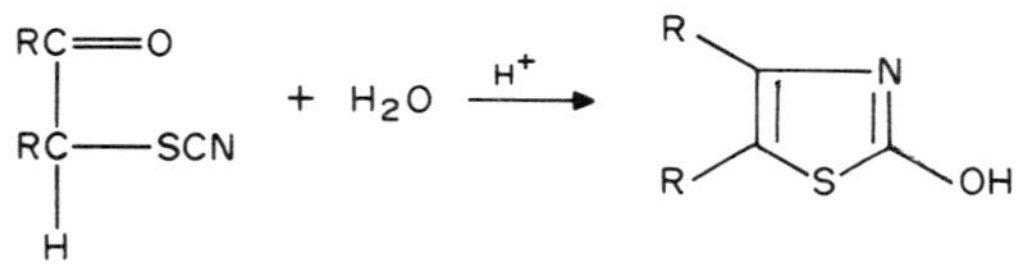

250 references up to 1975.
Complementary: none.
Relevant: none.

22, *19: Selenocyanates and related compounds* by *E. Bulka*

Section II deals with general properties of HSeCN, and Section III deals with the preparation of esters of selenocyanic acid (RSeCN) from halogens, diazonium salts, from selenenyl derivatives ($ArSeCl + AgCN \rightarrow ArSeCN$), by direct selenocyanation of arenes and by substitution of various other groups by $^{-}SeCN$. Next, the physical properties and reactions are described, including reduction (to RSeH or RSeSeR), hydrolysis (yielding RSeSeR), oxidation, halogenation, cyclization and substitutions (e.g. $ArSeCN + RSH \rightarrow ArSeSR + HCN$). Subsequently, selenenyl selenocyanates (ArSeSeCN), sulphenyl selenocyanates (ArSSeCN) are discussed, followed by treatment of the preparation, physical properties and reactivity of isoselenocyanates ('seleno mustard oils,' RNCSe).

Finally, Section IV considers briefly the much less known tellurocyanates (RTeCN).

147 references up to 1975.
Complementary: none.
Relevant: none.

22, *20: Biological formation and reaction of cyanates* by *S. Cohen and E. Oppenheimer*

The chapter deals with compounds containing —OCN, —NCO and —SCN groups (natural product chemistry and biochemistry of RNCS are the subject of Chapter 22,

Sections II and VIII). The term 'cyanate' is used to denote ROCN, RNCO and RSCN.

Section II discusses the biological formation of cyanates in living cells by rhodanase (Table 1) and other enzymes.

Section III describes reactions of the title compounds with amino groups (Table 2), thiols (Table 3), hydroxyl groups and imidazoles. The significance of these processes in biological systems is discussed.

Section IV considers the reactions of the title compounds with haemoglobin (Table 4), chymotrypsin, elastase, other peptidohydrolases, papain, ribonuclease, phosphatase, β-galactosidase, hormones (Table 5), plasma and immunoproteins, and in Section V with nucleic acids and their constituents. Section VI treats the action of cyanates at the cellular level (Table 6) and finally Section VII deals with the toxicity of cyanates (Table 7) and with their pharmacology.

141 references up to 1973.
Complementary: **22**, II, VIII.
Relevant: **2**, 7; **3**, 5; **4**, 9; **5**, 18; **7**, 4; **11**, 11; **12**, 12; **13**, 14; **15**, 10; **16**, 13; **17**, 13; **18**, 13; **19**, 13; **20**, 6; **24**, 14; **26**, 13; **28**, 17; **31**, 9.

22, *21: Syntheses and reactions of isocyanide dihalides* by *H. Ulrich and R. Richter*

Section II describes the preparation of isocyanide dihalides ($RN{=}CHal_2$) by addition of halogen to isocyanides ($RNC + Cl_2 \rightarrow RN{=}CCl_2$), by halogenation of isothiocyanates ($RNCS \xrightarrow{Cl_2} RN{=}CCl_2$), isocyanates, carbamoyl chlorides, imidoyl chlorides ($[ArN^+H{=}CHCl]Cl^- \xrightarrow{SO_2Cl_2} ArN{=}CCl_2$) or amines ($ArNMe_2 \rightarrow ArN{=}CCl_2$, with many ring-halogenated by-products), by addition of $:CCl_2$ to aliphatic carbodiimides, by the use of cyanogen chloride and nitrogen trichloride and some other minor methods. Next, syntheses of acyl and aroyl isocyanide dihalides, sulphenyl isocyanate dihalides, sulphonyl and phosphoryl isocyanide dihalides and of dichloroformaldoxime ($Cl_2C{=}NOH$) derivatives are dealt with.

Section III discusses nucleophilic displacement reactions of isocyanide dichlorides Table 1), formation of four-, five- and six-membered heterocycles (Tables 2 and 3); electrophilic reactions (e.g. $RN{=}CCl_2 + C_6H_6 \xrightarrow{AlCl_3} RNHCOC_6H_5$), additions to the C=N bond and reactions with phosphorus compounds and with alkali metal fluorides ($R_2CClN{=}CCl_2 + NaF \rightarrow R_2C{=}NCF_3$).

118 references up to 1974.
Complementary: none.
Relevant: none.

22, *22: The chemistry of the —NCS group* by *Ľ. Drobnica, P. Kristián and J. Augustin*

VII. Reactivity of isothiocyanates (1160); **VIII**. Biochemically important reactions of isothiocyanates (1160); **IX**. References (1199).

The chapter presents a comprehensive description of isothiocyanates (RNCS), including their occurrence, preparation, structure, reactions and biological interactions.

Section II describes the occurrence and origin of RNCS in nature, especially in glucosinolates ('mustard oil glucosides'), and the biosynthesis (Table 1), synthesis and biological importance of the latter.

Section III deals with the methods of preparation of RNCS from alkyl halides (RX + —NCS → RNCS) or by addition of HSCN or KSCN to C=C bonds. The classification and data on the preparation and properties of carbonyl, thiocarbonyl, azacarbonyl and phosphoryl isothiocyanates are summarized in Tables 2 and 3. Other methods of preparation are from amines with $CSCl_2$ ($RNH_2 + CSCl_2 \rightarrow RN{=}C{=}S + 2HCl$) or with thiocarbamoyl derivatives [e.g. ClC(=S) NEt_2]. RNCS are also obtained by decomposition of dithiocarbamic acid derivatives by heavy metal salts [$RNHCSS^- \ {}^+NH_4 + Pb(NO_3)_2 \rightarrow RNCS$], chlorine compounds ($COCl_2$, NaOCl), H_2O_2, carbodiimides, organosilicon compounds, etc., and also by acid-catalysed decomposition of diarylthioureas. Next, the preparation of isothiocyanates containing multiple bonds in the molecule are described, and of *N*-isothiocyanato-amines and -imines, and also some less general methods of synthesis of RNCS.

Section III.B treats methods of synthesis of isotopically labelled RNCS (including ^2H, ^3H, ^{14}C and ^{35}S) by chemical and biological methods and considers the uses of these labelled compounds and (Section III.C) also of macromolecular polyisothiocyanates (Table 4).

Section IV discusses compounds which on hydrolysis may yield an —NCS group by spontaneous decomposition in aqueous or organic solutions at room temperature (termed 'synthetic producers of isothiocyanates' or 'mustard oil formers'). These have various biological effects (antimicrobials, cytotoxic and toxic effects, etc.). The group includes *N*-monosubstituted di- and monothiocarbamates and their thione and thiol esters, *N*-substituted thioamides, rhodanines and thiohydantoins and 3,5-disubstituted tetrahydro-1, 3,5-thiadizine-2-thiones. Kinetic and mechanistic aspects of these decomposition are considered.

Section V describes spectral and MW studies relevant to the structure of the —NCS group, its polar character and Hammett constants (Table 5), electron-transfer effects of substituents through different conjugated system of isothiocyanates (Tables 6 and 7), dipole moments and electronic interactions (Tables 6–13), theoretical calculations (Table 14) and bond distances and angles from X-ray and MW data (Table 15).

Section VI treats nucleophilic additions of the type $RNCS + HX \rightarrow RNHC(X){=}S$, where X is a nucleophile and HX represents a nitrogen base (RNH_2 etc.), ROH, RSH, RCOOH, RCOSH or an active methylene compound.

In Section VI.B cycloadditions are considered. These include [2 + 2] additions on the C=S or on the C=N bond of RNCS leading primarily to four-membered heterocycles, [3 + 2] cycloadditions of various 1,3-dipoles leading to five-membered heterocycles, [4 + 2] cycloadditions (and dimerizations) of isothiocyanates containing C=O, C=S or C=NR groups with electron-rich multiple (C=C or C=N) bonds leading to six-membered heterocycles, including the cases of isothiocyanates as dienophiles, and finally [4 + 1] cycloadditions of carbonyl isothiocyanates to isonitriles (RNC), producing five-membered heterocycles.

Section VII discusses quantitative studies relating to the reactivity of RNCS. These include nucleophilic attacks by OH^- and RNH_2, presenting rate constants (Tables 16 and 17) and Hammett constants (Table 18), rates of reactions with amino acids (Tables 19 and 20) and their Hammett values (Table 21). Polycyclic isothiocyanates are usually more reactive than PhNCS (Table 22) and arylalkyl isothiocyanates less so (Table 23). Rate constants and Hammett values for attacks by HS^- and RS^- ions are given in Tables 24–27 and the addition of alcohols to the N=C bonds isothiocyanates is treated next. Section VII.B deals with addition–cyclization reactions (Table 28) and the hydrolysis of the products so formed (Tables 29 and 30). Table 31 compares the reactivity of different nucleophiles with respect to PhNCS and summarizes mechanistic proposals.

Section VIII considers the use of isothiocyanates in the determination of the primary structure of proteins and in their labelling and also in the modification of enzymes by using isothiocyanates as specific potent and reversible inhibitors by blocking SH groups. Isothiocyanates as metabolic inhibitors are treated next, and their biological activity and mode of action (Tables 32 and 33) are discussed. Finally, structure–activity relationships of isothiocyanates are presented.

806 references up to 1975.
Complementary: none.
Relevant: none.

The chemistry of diazonium and diazo groups (1978)

23, *1: General and theoretical aspects of the diazonium and diazo groups* by *J. B. Moffat*

I. Introduction (2); **II**. Diazo compounds (4); **III**. Diimides (diazenes) (30); **IV**. Diazonium compounds (58); **V**. Acknowledgement (64); **VI**. References (64).

The introduction reviews some energy and geometry parameters (Tables 1–4) of relevant groups, ions and molecules.

Section II discusses diazo compounds, such as the prototype CNN (Table 5), diazomethane and its theory (Tables 6 and 7), thermochemistry (Table 8) and its isomers, nitrilimine, diazirine (Tables 9–16) and perfluorodiazirine F_2CN_2 (Tables 17–20). In all cases spectra, structures and the kinetics of thermal decomposition (Table 21) are considered.

Section III discusses the results of theoretical and spectral studies of diimide and its mono- and disubstituted derivatives. These include the prototypes N_2H^+ (Table 22) and N_2H, diimide (or diazene) itself HN=NH, its spectroscopy and theoretical studies, energy and geometry (Tables 23–28), disubstituted diimides, especially N_2F_2 (Tables 29–36) and tetrazene (N_4H_4) and monosubstituted diimides (RN=NH) (Tables 37–40).

Section IV considers structure (Tables 41), spectroscopic data (Tables 42 and 43) and theoretical studies on diazonium compounds and also includes treatment of PES charge distribution results on diazocyclopentadiene.

241 references up to 1976.
Complementary: 19, 1.
Relevant: 1, 1; **2**, 1; **3**, 1; **4**, 1; **5**, 1; **6**, 1; **8**, 1; **11**, 1; **12**, 1; **13**, 1; **14**, 1; **15**, 1; **16**, 1; **17**, 1; **18**, 1; **20**, 1; **22**, 1; **24**, 1; **25**, 1; **26**, 1; **28**, 1; **31**, 27; **32**, 1; **33**, 1; **36**, 2.

23, *2: Diazonium–diazo equilibrium* by *V. Sterba*

I. Introduction (71); **II**. Diazonium–diazo equilibrium of aliphatic diazo compounds (72); **III**. Arenediazonium ion⇌diazohydroxide⇌diazotate equilibria (72); **IV**. Diazonium–diazo equilibria of heterocyclic diazonium ions (84); **V**. Reactions of diazonium ions with other nucleophiles (86); **VI**. References (92).

The diazo⇌diazonium⇌diazotate equilibrium in alkyl systems cannot be studied owing to the instability of the alkanediazonium ion (Section II).

In aryl systems (Table 1) the equilibria $ArN_2^+ + OH^- \rightarrow ArN_2OH$ and $ArN_2OH + OH^- \rightarrow ArN_2O^- + H_2O$ can be measured conveniently; further equilibria exist between the *syn* and *anti* forms of ArN_2OH and ArN_2O^-. The kinetics and mechanism of all the above processes are discussed (Tables 2–4), together with their dependence on pH and on substituents (Section III).

Section IV considers the analogous equilibria involving heterocyclic diazonium ions. In these cases, nitrosamines (and the isomeric diazohydroxides) are much more stable than in homocyclic cases.

Arenediazonium ions react with a variety of nucleophiles to give diazo compounds. Thus alkoxide ions yield diazo ethers, secondary amines yield arenediazoamino compounds ($ArN{=}NNR^1R^2$), sulphite ions give arenediazosulphonates (Table 5), sulphinate anions yield arenediazosulphones, thiols give diazo thioethers (Table 6) and cyanide ion gives diazocyanides, ArN=NCN (Table 7).

89 references up to 1975.
Complementary: none.
Relevant: none.

23, *3: Structural chemistry* by *S. Sorriso*

I. Introduction (96); **II**. Diazonium salts (97); **III**. Diazotates (105); **IV**. Diazoalkanes (106); **V**. Aromatic diazocyanides (109); **VI**. Diazoketones (113); **VII**. Diazo oxides (126); **VIII**. References (131).

The Introduction reviews the limitations of X-ray data. Section II deals with structural data of outer diazonium salts, giving bond length, angles, bond orders and electronic structures based on X-ray and spectral data. Inner diazonium salts (i.e. in which the anion is also a substituent on the ring and the compound has zwitterionic structure) are also treated.

Section III considers isomerism in diazotates, and especially the structure of potassium *syn*-methyl diazotate.

Section IV discusses the scarce data available on diazoalkanes, bromodiazofluorene and phenyl(triphenylsilyl)diazomethane.

Section V describes the synthesis and structures of four aromatic diazocyanides and their *syn–anti* isomerism (Tables 1 and 2). Section VI treats the isomerism, UV–visible, IR and NMR spectra of diazoketones, (Table 3), their electric dipole moments, MO calculations, configurations and structures. Finally, Section VII discusses diazo oxides (obtained by addition of alkali to *o*- and *p*-hydroxy-substituted diazonium salts, their IR spectra (Table 4), electronic structures, dipole moments (Table 5) and X-ray data.

206 references up to 1974.
Complementary: none.
Relevant: **16**, 2; **17**, 2; **18**, 2; **19**, 2; **20**, 2; **21**, 1; **22**, 2; **24**, 2; **26**, 2; **27**, 4; **28**, 2; **29**, 1; **30**, 1; **31**, 24; **32**, 24; **33**, 12; **34**, 10; **36**, 3.

23, *4: Thermochemistry of diazo compounds and organic azides* by *R. Shaw*

I. Introduction (137); **II**. Diazo compounds (138); **III**. Organic azides (141); **IV**. Summary of thermochemical properties of diazo compounds and organic azides (145); **V**. Values for estimating thermochemical properties by group additivity (146); **IV**. Acknowledgements (146); **VII**. References (146).

Owing to a lack of experimental data, the emphasis of the chapter is on estimation by additivity techniques.

Section II deals with the experimental values for CH_2N_2 and the estimation of data for other diazo compounds (Table 1).

Section III treats organic azides (Tables 2–5) and describes the techniques for the estimation of the data.

Section IV compares measured and estimated thermochemical properties of diazo compounds and of organic azides (Table 6) and Section V tabulates group values for estimating thermochemical properties by group additivity (Table 7).

21 references up to 1977.
Complementary: none.
Relevant: **16**, 16; **18**, 3; **19**, 3; **20**, 11; **22**, 6; **24**, 3; **25**, 2; **26**, 4; **27**, 9; **28**, 4; **29**, 24; **30**, 2; **31**, 2; **33**, 3; **36**, 4.

23, *5: Detection and determination of diazo and diazonium groups* by *D. A. Ben-Efraim*

Section I describes chemical methods of detection of diazoalkanes, α-diazo ketones, α-diazo esters, arenediazonium salts and triazenes, and Section II deals with the quantitative determination of the same compounds by chemical methods (gravimetric, colorimetric, gasometric, reductometric, coupling, etc.).

Section III discusses the polarography and Section IV the chromatography of the title compounds.

Section V deals with IR and Raman spectroscopy, including also diazophenols, diazotates, diazosulphonates, diazo ethers, diazo cyanides and diazo isocyanides.

The electronic spectra of the compounds are treated in Section VI, including also fluorescence and phosphorescence; Section VII deals with NMR, Section VIII with MS, Section IX with ESR and CIDNP, Section X with PES and ESCA and Section XI with dipole moments.

192 references up to 1973.
Complementary: none.
Relevant: **1**, 5; **2**, 8; **3**, 15; **4**, 3; **5**, 17; **8**, 3; **12**, 4; **13**, 6; **14**, 3; **15**, 3; **16**, 3; **17**, 4; **18**, 5; **20**, 3; **22**, 5; **24**, 5; **25**, 10; **26**, 5; **28**, 3; **29**, 21; **30**, 14–22; **36**, 5

23, *6: Basicity, acidity and hydrogen bonding* by *J. F. McGarrity*

Section II discusses the gas-phase, solution and thermodynamic basicity of diazoalkanes, acid-catalysed H–D exchange, base-catalysed decomposition of *N*-nitrosourethanes and ureas, quaternization of onium salts by diazoalkanes (Table 2), alkylation of acidic species in general and the kinetic basicity of aliphatic diazo compounds, giving kinetic parameters (Table 3) and Hammett ρ values (Table 4) for diazoalkane protonations.

Section III deals with basicity scales of diazoalkanes towards Lewis acids, with the kinetics of electrophilic addition of Lewis acids to Ph_2CN_2 (Table 6) and with reactions of diazoalkanes with electrophilic carbon centres (esters, aliphatic diazonium ions, aromatic diazonium ions and carbenes (Table 7).

Section IV describes the proton acidity of diazoalkanes and various reactions which occur through primary proton abstraction. Section V deals with the Lewis acidity of the title groups, considers coupling of aromatic diazonium ions with nucleophilic carbon (e.g. $Ar\overset{+}{N}{\equiv}N + {}^{-}CHR_2 \rightarrow ArNHN{=}CR_2$) and with heteronucleophiles ($ArN_2 + MeNH_2 \rightarrow ArN{=}NNHMe$), followed by coupling of aliphatic diazonium ions, in cases when these could be trapped, e.g. $R\overset{+}{N}{\equiv}N \xrightarrow{EtNH_2} RN{=}NNHEt$, where R = cyclopropyl. The section ends with a description of coupling reactions of aliphatic diazo compounds, e.g. with phosphines: $R_2C^{-}{-}N{=}N^{+} + PPh_3 \rightarrow R_2C^{-}{-}N{=}N{-}{}^{+}PPh_3 \leftrightarrow R_2C{=}N{-}N{=}PPh_3$.

Section VI considers diazoalkanes as both proton acceptors and proton donors in hydrogen bonding (e.g. $ROH + {}^{-}CH_2N_2^{+} \rightarrow RO{-}H \cdots CH_2N_2$ in first case and the action of α-diazo sulphones, α-diazo ketones and other diazo compounds containing electron-attracting groups in the second).

Section VII deals with acidity and basicity of diazohydroxydes ($ArN{=}NOH$), with the dissociation of *syn*- and *anti*-diazohydroxides and with their reactions with acids (to yield diazonium ions). Finally, acid–base reactions of diazophenols and of triazenes are mentioned briefly.

288 references up to 1975.
Complementary: 23, 7.
Relevant: 2, 9; **3**, 6; **4**, 4; **5**, 6; **8**, 5; **11**, 3; **12**, 3, 10; **13**, 7; **15**, 4; **16**, 5; **17**, 6; **18**, 8; **19**, 7; **20**, 12; **22**, 8; **24**, 4; **25**, 5, 6; **29**, 17; **31**, 17, 21; **32**, 21; **33**, 5.

23, *7: Complex formation* by *H. M. Niemeyer*

Section II describes complexes of diazoalkanes with metals in which the coordination between the two moieties is either through the diazo carbon atom or through one of the diazo N atoms. A third variety is the 'side-on' coordination which is known in alkyne–metal complexes.

Section III deals with the structure, synthesis and IR (Table 1) and NMR (Table 2) spectroscopy and the chemical properties of complexes formed between diazonium ions and transition metals. Almost all of these contain arene–diazonium ligands and almost none alkane–diazonium ligands.

105 references up to 1976.
Complementary: 23, 6.
Relevant: 2, 9; **3**, 6; **4**, 4; **5**, 6; **6**, 6; **8**, 5; **11**, 3; **12**, 3, 10, **13**, 7; **15**, 4; **16**, 5; **17**, 6; **18**, 8; **19**, 7; **20**, 12; **22**, 8; **24**, 4; **25**, 5, 6; **29**, 17; **31**, 17, 21; **32**, 21; **33**, 5.

23, *8: Synthetic applications of diazonium ions* by *D. S. Wulfman*

Section II deals with aryl diazonium ions, classified according to Saunders (see ref. 1103 and the Table of Contents of this chapter). First, reactions are discussed in which the changes occur in the aryl group of the aryldiazonium (ArN_2^+) ion, including substitution, solvolysis and anion exchanges between the ring and the N_2 function. Next, reactions are treated in which the diazo group is involved, but remains functionally intact. Here, the preparation and use of diazotates, diazo anhydrides, diazo ethers, diazo cyanides, diazo carboxylates, diazo sulphates, diazo perhalides and diazo sulphides as synthones is discussed. Next, reactions are described in which the diazo function disappears as such, but the two N atoms remain in the product. This class includes formations of triazenes (ArN=NNHAr′), aryl azides and especially formation of azo compounds by coupling with phenols, anilines, active methylene compounds, esters, acids, nitriles, sulphones, ketones, tropolones and also the Japp–Klingemann reaction and complex formation with transition metals.

Subsequently, reactions are described in which the diazo nitrogens are retained with the formation of a variety of heterocyclic rings, such as cinnolines, indazoles, benztriazoles, benztriazenes and benzothiodiazoles. On the other hand, in many reactions of the diazo compounds the diazo nitrogen is lost, and the product is formed in a route which resembles the Pschorr phenanthrene synthesis.

Possibly the largest class of diazonium reactions is that in which the N_2 group is replaced by another atom or group. This new group may be an H atom (Griess reaction), any of the halogens (including astatine), NO_2, N_3, OH, OR, OAr or CN. Other routes lead to the formation of biaryls, quinones, etc., and to the attachment of new groups through S, Se or Te, As, Sb, or Bi and a variety of heavy metals, e.g. in —HgCl or —$SnCl_2$ groups.

Section III describes the synthetic applications of alkyldiazonium ions, often carried out without separating the latter and obtaining only final products of decomposition or rearrangement.

Section IV deals with the preparation of benzynes and hetarynes through routes involving diazonium compounds and Section V mentions briefly a similar route to cycloalkynes.

1370 references up to 1975.
Complementary: 32, 15, 16.
Relevant: 23, 13, 18.

23, *9: Photochemistry of the diazonium and diazo groups* by *W. Ando*

The Introduction reviews photochemical energy and excitation potential energy curves and the Jablonsky diagram.

Section II deals with photolytic decomposition of diazonium salts to yield arenes and ethers derived from the solvent (Table 1), or in special cases phenols and other products (Table 2). BF_4^- and PF_6^- salts on photolysis yield fluorides $ArN_2^+BF_4^- \xrightarrow{h\nu} ArF$,

Table 3) while diazonium chlorides yield usually aryl chlorides. The spectra which identify the excited states are presented (Tables 4 and 5) and the quantum yields, charge densities and transition energies of diazonium cations are tabulated (Table 6). Next, various reaction intermediates in photodecompositions are given (Tables 7–9) together with detailed reaction schemes. *Cis–trans* photoisomerization of diazo compounds and dye-sensitized photolysis of diazonium salts are discussed.

Section III treats the photolysis of aliphatic diazonium salts of bornyl, pinyl, terpinyl and fenchyl derivatives.

The bulk of the chapter (Section IV) describes the photolysis of diazo compounds, especially to yield carbenes ($R_2C{=}N_2 \rightarrow R_2C\colon + N_2$). It deals with the structure of carbenes (Table 10), with CIDNP studies, with spin multiplicity and reactivity of CH_2 (Table 11). Next it describes the photolysis in olefins of phenyl carbene (Table 12), of $Ph_2C\colon$ (Tables 13 and 14), of fluorenylidene (Tables 15 and 16) and of cyclic carbenes with 3, 5 or 7 carbons in the ring (Table 17). Subsequently, the photolysis (again yielding carbenes) of diazocarbonyl compounds such as N_2CHCO_2R, $N_2C(COOMe)_2$ and related compounds in olefins and intermolecular additions of the two synthons are discussed (Tables 18–23), followed by intramolecular photocyclization of diazoketones and by the adduct formation between diazo-derived carbenes and arenes.

Next, insertions of carbenes into C—H bonds of saturated hydrocarbons

$$\left(\begin{matrix} | \\ -\mathrm{C}- \\ | \end{matrix}\mathrm{H} + {:}\mathrm{CH_2} \rightarrow \begin{matrix} | \\ -\mathrm{C}- \\ | \end{matrix}\mathrm{CH_3}\right)$$

(Tables 24–26) and of ethers (Table 27) are described. Intramolecular reactions (mainly self-insertion into β- and γ-C—H bonds give olefins or cyclopropanes:

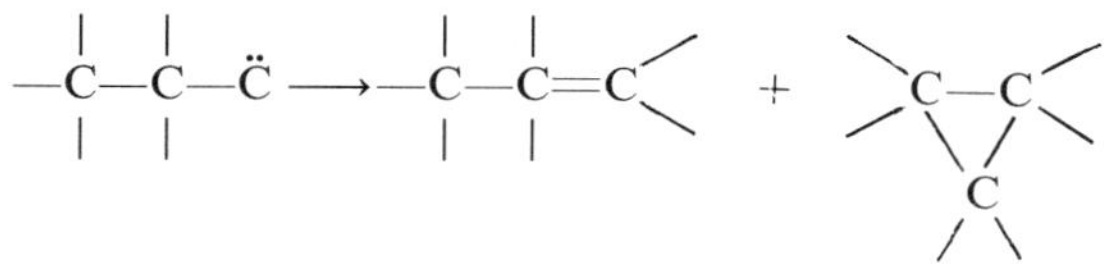

in some cases accompanied by rearrangements.

Photolyses to carbenes and attacks by the latter at C—X bonds are considered, where X is a substituent having a lone pair of electrons. Thus methylene reacts with polyhalomethanes, e.g. CCl_4 produces $C(CH_2Cl)_4$ as the major product. Similar reactions occur with other carbenes derived from diazo esters and related compounds (Tables 29–33). Carbene reactions with N compounds are complicated and take many routes. Reactions with O_2 (Table 34), alcohols (Table 35), esters (Table 36), allyl ethers (Table 37), thiols (Table 38), sulphonium ylides (Table 39) and sulphides (Tables 40–43) are treated and the variety of possible routes and mechanisms are described in detail, considering different diazo compounds and different heteroatom-containing compounds, competitive reactions between singlet and triplet carbenes (Table 32), solvent effects (Tables 33 and 42), sensitizers (Table 48), competition between two substrates (Table 40), various 1, 2-alkyl shifts and other rearrangements, etc.

The Wolff rearrangement of diazocarbonyl compounds leads to ketenes or to esters:

$$RC({=}N_2)C({=}O)R' \xrightarrow{h\nu} RR'C{=}C{=}O \xrightarrow{R''OH} RR'CHCO_2R''.$$

The synthetic utility of the reaction including the relative migration tendencies of various R and R′ groups are discussed (Table 44) and its mechanism is considered both in the gas phase and in solution, and in the presence of sensitizers and with irradiations by various wavelengths. Finally, bisdiazo compounds

$$N_2{=}\overset{|}{C}{-}(C)_n{-}\overset{|}{C}{=}N_2$$

and their uses for the syntheses of cyclic compounds are mentioned briefly.

445 references up to 1974.
Complementary: 31, 15.
Relevant: 2, 16; **3**, 8; **6**, 4; **8**, 12; **9**, 6; **11**, 5; **13**, 16; **14**, 8; **15**, 9; **16**, 11; **17**, 9; **18**, 10; **19**, 20; **22**, 10; **24**, 12; **25**, 11; **27**, 21, 22; **28**, 5; **29**, 5, 6; **32**, 29; **33**, 21.

23, *10: The electrochemistry of the diazo and diazonium groups* by *A. J. Fry*

I. Introduction (489); **II**. The diazo group (490); **III**. The diazonium group (496); **IV**. References (498).

Section II deals first with the electrochemical reduction of diazo compounds. The process, e.g. with diazoacetophenone, involves three stages: $PhCOCHN_2 \xrightarrow{6e} PhCOCH_2NH_2 \xrightarrow{2e} PHCOCH_3 \xrightarrow{2e} PhCHOHCH_3$. In various reaction conditions, diazo compounds may yield different products, e.g. $R_2CN_2 \rightarrow R_2C{=}NNH_2$ or $R_2CN_2 \rightarrow RCHNH_2$ or $R_2CN_2 \rightarrow R_2CH_2$.

Anodic oxidation of diazo compounds occurs through a complex free-radical chain reaction, involving the radical cation and often a dimer as the principal product ($Ph_2CN_2 \rightarrow Ph_2C{=}CPh_2$). Section III treats the electrochemistry of diazonium salts.

33 references up to 1974.
Complementary: none.
Relevant: 5, 2; **8**, 11; **13**, 5; **16**, 15; **17**, 14; **19**, 12; **20**, 5; **21**, 5; **22**, 9; **24**, 17; **25**, 13; **26**, 12; **27**, 8; **28**, 7; **29**, 8, 9; **31**, 7; **32**, 6; **34**, 1.

23, *11: The influence of the diazo and diazonium groups* by *E. S. Lewis*

I. Introduction (499); **II**. Influence of the diazo group (500); **III**. The diazonium ion group (504); **IV**. Conclusions (508); **V**. References (509).

Diazo and diazonium groups are expected to have very strong influences on the rest of the molecule; however, owing to their extraordinary reactivity it is very difficult to carry out reactions in other parts of the molecule without destroying these functions.

Section II considers the diazo group, taking as models either the carbonyl or preferentially the carbanion group, justified by protonation and by reactions with electrophiles. Diazo compounds are bases, but can also act as acids (e.g. they are lithiated by CH_3Li). The effect of the diazo group on the NMR spectra shows that they have a significant partial negative charge.

Section III deals with the N_2^+ group, with Hammett substituent constants obtained for it (Table 1), with electrophilic aromatic substitutions (when it is *meta*-directing and deactivating), with its activating effect on Ar_N reactions, and finally with some reactions of aromatic bisdiazonium salts which allow the study of the effect of one of the N_2^+ groups on the reactions of the second.

66 references up to 1975.
Complementary: 23, 12.
Relevant: 3, 3; **4**, 5; **5**, 16; **6**, 8; 7, 2, 3; **11**, 12; **12**, 5; **13**, 8; **14**, 4; **15**, 5; **16**, 6; **18**, 9; **19**, 8; **28**, 9; **29**, 16; **31**, 8; **36**, 15.

23, *12: Kinetics and mechanisms of reactions involving diazonium and diazo groups* by *A. F. Hegarty*

Section II deals with the scope, kinetics and mechanism of the diazotization of alkyl- and arylamines at low, intermediate, moderate and high acidities, also in the presence of halide ions as catalysts and in non-aqueous solvents.

Section III discusses the dediazonization reaction, first the replacement of nitrogen by nucleophiles through the aryl cation pathway (Table 1) considering isotope effects, selectivity and structure of Ar^+, substituent effects, the Schiemann reaction ($ArN_2^+BF_4^- \rightarrow ArF$), bimolecular mechanisms, reversal of the position of the two N atoms in the course of hydrolysis and benzyne formations.

Next, reactions of nucleophiles at the terminal N are treated, yielding azo adducts and more stable products by reactions of the latter. The reagents include oxygen nucleophiles (OH^-, RO^-, ArO^-, $RCOO^-$) (Tables 2–5), S nucleophiles (e.g. $ArN_2^+ + ArS^- \rightarrow ArN{=}NSAr$), carbon nucleophiles ($CN^-$, ketones, carbanions, activated aromatic substances yielding azoarenes) (Tables 7–9) and N nucleophiles (amines and NH_3 yielding triazenes). Next, Ar_N reactions activated by the diazonium group are treated, followed by metal-catalysed reactions (Sandmeyer, biaryl and azoarene formation, Meerwein reactions), Gomberg–Bachman (free-radical) arylations, arylations by aryl cations and intramolecular ring closures involving displacement of the N_2^+ group (Pschorr reaction). Finally, reductions (to form as a rule arenes) are described.

Section IV deals with the preparation of diazoalkanes by nitrosation of primary amines, from *N*-nitrosamides, from alkanediazotates and by some other methods. Next, Section V describes the reactions of diazoalkanes with electrophilic species (protic acids to yield carbonium ion rearrangement product, alkenes, dimers, etc., alcohols to yield ethers, aldehydes, ketones, esters, etc., to yield a variety of alkylation products). Other reactions of diazoalkanes with arenediazonium ions, halogens, carbonium ions and Lewis acids are also treated. Finally, 3 + 2 cycloadditions of diazoalkanes to unsaturated centres (C=C, C≡C, C=N, C≡N, N=N, etc.) are discussed, including orientation and substituent effects in this reactions.

398 references up to 1975.
Complementary: 23, 11.
Relevant: 29, 22.

23, *13: Rearrangements involving the diazo and diazonium groups* by *D. Whittaker*

Section II reviews the isomers of CH_2N_2 and of other diazo and diazonium compounds.

Section III deals with rearrangements involving 1,3-dipolar addition (e.g. with alkenes, primary formation of Δ^1-pyrazolines and their rearrangement to the Δ^2-isomer) or nucleophilic addition with carbonyl compounds, to yield epoxides and also ring expansion products.

Section IV deals with the reactions of carbenes and carbenoids obtained by decomposition (heat or $h\nu$) of diazo compounds. These include a wide variety of rearrangements in the course of the reactions which may be insertions to C—H bonds, olefin-forming insertions, reactions with nucleophiles, addition to multiple bonds, fragmentations, carbene to carbene rearrangements, rearrangements of carbenes to nitrenes and to other reactive intermediates, Wolff rearrangements and finally the so-called 'carbenoid' rearrangements, which involve similar, but not true divalent carbon species.

Section V discusses the formation of aliphatic diazonium ions and their decomposition, which produces a variety of rearranged products (derived from high-energy or 'hot' carbonium ions). Next, the importance of external stabilization of carbonium ions by neutral species or by counter ions (in ion pairs) is considered, together with the nature of the high-energy carbonium ion. Among rearrangements accompanying diazonium ion decomposition, the following are discussed: semi-pinacol rearrangements, ring expansion reactions, ring contraction reactions, cyclization and ring opening, transannular interactions, delocalized (non-classical) ions and allylic rearrangements. Finally, rearrangements of α-ketodiazonium ions and some (rare) rearrangements of aromatic diazonium ions are treated.

218 references up to 1975.
Complementary: none.
Relevant: **1**, 7; **2**, 15; **3**, 14; **4**, 10; **5**, 8; **8**, 9; **9**, 3; **10**, 3; **11**, 4; **12**, 16; **13**, 13; **14**, 7; **15**, 8; **16**, 17; **17**, 8; **18**, 15; **19**, 18; **20**, 4; **24**, 10; **26**, 15; **29**, 4.

23, *14: Preparation of diazonium groups* by *K. Schank*

I. Introduction (645); **II**. Formation of aromatic diazonium salts from the corresponding amines by diazotization (647); **III**. Formation of aromatic diazonium salts starting from aromatic diamines (648); **IV**. Diazotization of heteroaromatic amines (649); **V**. Formation of arenediazonium compounds from other arenediazonium compounds (649); **VI**. Formation of aromatic diazonium salts by transdiazotization (651); **VII**. Formation of aromatic diazonium salts by azo decoupling (651); **VIII**. Formation of aromatic diazonium compounds from nitroso acyl amines (652); **IX**. Formation of aromatic diazonium salts by nitrosation of aromatic imines (652); **X**. Formation of aromatic diazonium salts by nitrosation of aromatic nitroso compounds and related species (653); **XI**. Formation of aromatic diazonium ions by aldol-like condensations of nitroso compounds and of nitrobenzene (654); **XII**. Formation of aromatic diazonium salts from arylhydrazines and their derivatives (654); **XIII**. Miscellaneous (654); **XIV**. References (655).

Section II describes preparation procedures of diazonium salts from amines in dilute or concentrated acids or in organic solvents. Syntheses starting from aromatic diamines (Section III), from heteroaromatic amines (Section IV) and from other diazonium compounds (Sections V and VI) are treated next. Section VII deals with azo decoupling of triazenes, VIII with syntheses from nitrosoacyl amines, IX with syntheses by nitrosation of aromatic imines and X by nitrosation of aromatic nitroso compounds. Sections XI–XIII treat syntheses by aldol-like condensations of nitroso and nitro compounds, from arylhydrazines and by some other routes.

Descriptions are very brief and only leading references are given.

71 references up to 1975.
Complementary: **23**, 15.
Relevant: **23**, 2.

23, *15: Synthesis of diazoalkanes* by *M. Regitz*

Section II describes first diazotization methods for the synthesis of diazoalkanes, diazocycloalkadienes and diazoheteroaromatics. Ring-carbon atoms of some pyrrole and pyrazole derivatives are transformed into $\rangle C{=}N_2$ groups by $NaNO_2 + HCl$; alkyl and aryldiazomethanes can be prepared by the Forster reaction ($R^1R^2C{=}NOH + NH_2Cl \rightarrow R^1R^2C{=}N_2$). Dehydrogenation of hydrazones by HgO, AgO, MnO_2, $Pb(OAc)_4$ or I_2 leads to diazo groups (Tables 1 and 2) ($R_2C{=}NNH_2 \xrightarrow{-2H} R_2C{=}N_2$). Next, the Bamford–Stevens reaction, i.e. the alkaline cleavage of tosylhydrazones to yield a variety of substituted diazoalkanes (including silyl, germyl and α,β-unsaturated), is described: $R^1R^2C{=}NNHSO_2C_6H_4CH_3 \rightarrow R^1R^2C{=}N_2 + {}^-O_2SC_6H_4CH_3$.

This is followed by syntheses through cleavage of β-(*N*-alkyl-*N*-nitrosoamino) ketones and sulphones, and by consideration of the acyl cleavage of *N*-alkyl-*N*-nitrosamides by alkali [$RCON(NO)CH_2R \rightarrow N_2{=}CHR$]. The mechanism of the reaction is discussed, together with the routes starting from *N*-alkyl-*N*-nitrosourethanes (Table 3), *N*-alkyl-*N*-nitrosoamides, *N*-alkyl-*N*-nitrosoureas (Table 4), *N*-alkyl-*N′*-nitro-*N*-nitrosoguanidines and *N*-alkyl-*N*-nitroso-*p*-toluenesulphonamides.

Next the diazo group transfer (i.e. transfer of a complete N_2 group from a donor (azide, diazoalkane) to an acceptor (CH-acid, double-or triple-bond system) is treated:

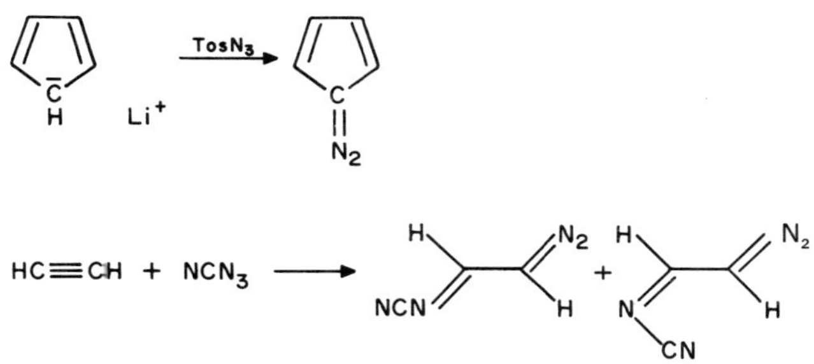

The acceptors dealt with are cyclopentadienes, cyclohexadienes, enamines, enol ethers and acetylenes.

Finally, the preparation of new diazo compounds by substitution in molecules which already contain a diazo group is described. These substitutions include halogenation ($H_2C{=}N_2 \rightarrow HalCH{=}N_2$), metallation ($H_2C{=}N_2 \rightarrow LiCH{=}N_2$), aldol additions and cleavage reactions. Some relevant special methods are also mentioned.

356 references up to 1975.
Complementary: **23**, 14.
Relevant: 23, 2.

23, *16: Preparation and uses of isotopically labelled diazonium and diazo compounds* by *P. J. Smith and K. C. Westaway*

Section I describes the syntheses of diazonium salts and of diazo compounds labelled with deuterium at various sites of the molecules, and also of diazonium salts labelled with ^{15}N at either the α-or the β-N (Table 1).

Section II deals with the use of ^{13}C labelling to determine the nature of bonding in diazoalkanes (Table 2) by NMR spectra.

Section III discusses the use of ^{15}N as tracer in reactions of diazonium ions with N_3^-, amines and OH^-, in the study of N_α–N_β exchange in ArN_2^+ decomposition (Tables 3–6) and in mechanistic studies involving alkanediazonium ions. Next, deuterium tracers studies in the reactions of diazo compounds (Tables 7 and 8), in alkylamine diazotization (Table 9), in the alkaline decomposition of ArN_2^+ ions (Table 10) and reactions involving arynes generated from diazonium salts are presented (Table 11), together with the use of ^{18}O as a tracer in diazo ester decomposition.

Section IV treats the theory of kinetic isotope effects in general, followed by discussion of N kinetic isotope effects in diazonium salt reactions (Tables 12–14) and secondary β-deuterium kinetic isotope effects in the same.

94 references up to 1975.
Complementary: none.
Relevant: **5**, 10; **12**, 13; **13**, 15; **17**, 12; **18**, 18; **19**, 17; **22**, 14; **24**, 20; **26**, 11; **27**, 10; **28**, 8; **29**, 28; **33**, 18; **36**, 12.

23, *17: Carbonyl, phosphoryl and sulphonyl diazo compounds* by *M. Regitz*

I. Introduction **II**. Synthetic methods (752); **III**. References (812).

The chapter deals with the preparation of diazoalkanes bearing electron-attracting substituents.

Carbonyl diazo compounds are obtained by diazotization of NH_2 groups (Table 1), e.g. $EtO_2CCH_2NH_2 \rightarrow EtO_2CCH{=}N_2$, by the Forster reaction involving oximation and subsequent reaction with NH_2Cl:

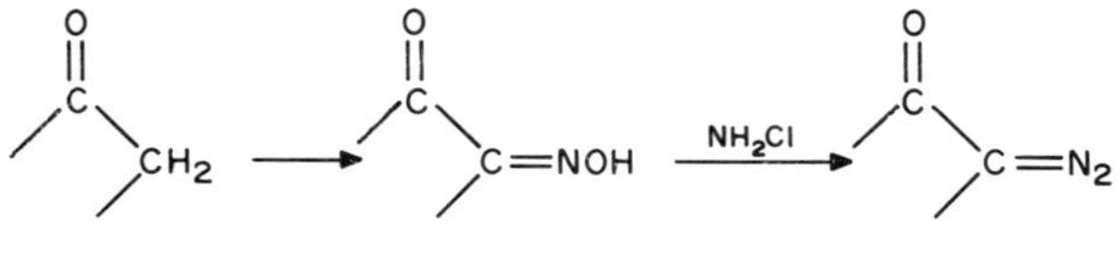

by dehydrogenation of hydrazones (Table 2), by the Bamford–Stevens reaction starting from diketones (Table 3), by cleavage of *N*-alkyl-*N*-nitrosamides and by diazo group transfer of a complete diazo group from a donor to an acceptor (Table 4), which can be active methylene compounds, acyl aldehydes, alkenes, alkynes, cyclopropenes etc.

Since the diazo group is usually stable, new diazo compounds can be prepared by a wide variety of substitutions without affecting the N_2 group. The reactions may be nitrations, acylations (Table 5), metallation (and metallations followed by substitution of the metal atom by other groups), additions and acyl cleavages in basic media

$$[R^1COC({=}N_2)COR^2 \xrightarrow{H_2O} R^1COOH + N_2{=}CHCOR^2].$$

Section IIB describes the preparation of phosphoryl diazo compounds $[R^1R^2P({=}O)CH{=}N_2]$ by the same methods as reviewed in Section IIA, including the Bamford–

Stevens reaction (Table 6), diazo group transfer (Table 7), substitutions and also the formation of diazoaldol compounds from 1,2-di- and 1,2,3-tricarbonyl compounds with diazomethylphosphoryl compounds (Table 8).

Section IIC reviews the syntheses of sulphonyl diazo compounds [$R^1SO_2C(=N_2)R^2$] by the Forster reaction, by dehydrogenation of hydrazones, by cleavage of *N*-alkyl-*N*-nitrosourethanes, by diazo group transfer (Table 9) and by substitutions which are, however, much rarer than in the other two groups considered in the chapter.

393 references up to 1975.
Complementary: none.
Relevant: **23**, 14, 15.

23, *18: Synthetic applications of diazoalkanes, diazocyclopentadienes and diazoazacyclopentadienes* by *D. S. Wulfman, G. Linstrumelle and C. F. Cooper*

I. Introduction (823); **II**. Rearrangements (906); **III**. The formation of dimers and telomers from diazoalkanes (912); **IV**. Cyclopropanation reactions (918); **V**. Insertions into X—Y bonds (945); **VI**. Acknowledgements (956); **VII**. References (956).

Section I deals first in general with 1,3-dipolar additions of diazoalkanes to olefins (Table 1), cycloalkenes (Table 2) and styrenes (Table 3) and considers the effect of pressure on the reactions (Table 4). Various X=Y systems are reviewed in detail, e.g. the addition of diazophosphonates with norbornadiene and norbornene (Table 5), additions of substituted diazoalkanes to substituted olefins (Table 6), of diazoacetates to alkylidene cyanoacetates (Table 7), to activated double bonds (Table 8) and to benzylidene cyanoacetates (Table 9). Decompositions of Δ^1-pyrazolines (Tables 11 and 12) and other reactions of diazo compounds with double bonds (Tables 13–16) are considered, presenting reaction conditions, mechanistic schemes, etc., in great detail. Next, reactions between diazoalkanes and imines and the reversibility of these processes are reviewed (Tables 18–20), followed by 1,3-additions with isonitriles, carbonyl and thiocarbonyl groups, —N=N— and —N=O systems and with O_2.

Next, 1,3-dipolar additions of the title compounds to X=Y=Z cumulene systems are described. The cumulenes include allenes (Tables 22 and 23), ketenes, sulphines (to yield episulphones; Table 24), which in turn can be decomposed to olefins (Tables 25 and 26), isocyanates and isothiocyanates, azasulphines (RN=S=O; Table 27) and SO (Table 28). Subsequently, additions to acetylenes, enynes, acetylenic and diyne ketones (Tables 29–31) and nitriles (Table 32) are presented. Cycloadditions, presumably involving carbonyl ylides (Tables 33 and 34), as and ketocarbenes (obtained by loss N_2 from diazocarbonyl compounds) are treated. Finally, 1,4-cycloadditions of arenediazonium ions and of diazocyclopentadiene and its aza-analogues are discussed.

Section II describes mainly the thermal, photolytic and catalytic version of the Wolff rearrangement and some other rearrangements which compete with it.

Section III discusses reactions in which diazoalkanes yield products containing two or more fragments of the starting material, e.g. $2XYCN_2 \rightarrow XYC{=}CXY$ (*cis* and *trans*) or $2XYCN_2 \rightarrow XYC{=}N{-}N{=}CXY$. The processes may be photochemical, thermal or metal salt catalysed and the products in various solvents may be azines, olefins, glycols or glycol esters (Tables 35–41).

Section IV treats first the various problems arising in choosing the optimal reaction conditions for cyclopropanations involving diazo compounds. Table 42 compares the selectivity of bis (methoxycarbonyl) carbene, $(MeOOC)_2C$:, generated by several methods; Tables 43–47 present data on the effect of various catalysts on yields, on the effect of additives and on the distribution of different products in the same reaction as a function of catalyst concentration. Next, sensitized vs non-sensitized photochemical methods are compared, followed by studies of various copper-containing catalysts in the presence and absence of peroxides, common-ion and solvent effects (Tables 48–52) and differences in the cyclopropanation of *cis*-and *trans*-octenes in photolytic and catalytic reactions, varying also the amounts of reactants, solvents, reaction times and catalysts (Tables 54 and 55). After discussing in detail reaction routes and mechanisms, the cyclopropanation of arenes by various diazo compounds is considered, followed by similar reactions of olefins, employing in both cases diazoalkanes, diazo esters and diazo ketones.

Section V treats insertion reactions into X—Y bonds, in which as a rule the N_2 group is lost, e.g.

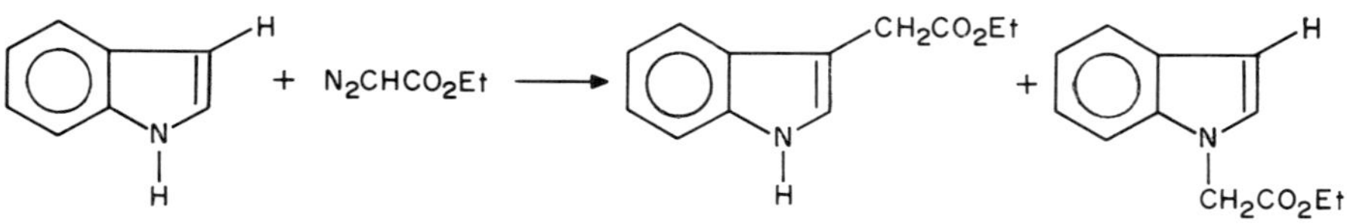

Insertions are described involving C—H, C—B, C—C, C—N, C—O, C—S, C—Se, C—Te, C—Hal and C—metal bonds, and also into O—H bonds (alcohols, enols, acids), S—H, Se—H, Te—H, N—H, H—Hal, N—Hal, S—Hal, P—Hal, O—Hal and S—S bonds.

981 references up to 1975. Numbered 1400–2381; numbers 1–1400 occur in Chapter **23**, 8.
Complementary: 31, 15, 16.
Relevant: 23, 8, 13.

The chemistry of the carbon–carbon triple bond (1978)

24, *1: General and theoretical aspects of the acetylenic compounds* by *M. Simonetta and A. Gavezzotti*

I. Introduction (2); **II**. General structural features (3); **III**. Energetics (18); **IV**. IR, UV and NMR, data (27); **V**. Interaction with transition metal atoms (36); **VI**. Quantum-mechanical calculations (49); **VII**. References (50).

Section II outlines the valence MO of acetylenes, considering extended Hückel theory (Tables 1 and 2), X-ray and electron diffraction data for simple alkynes, enynes, poly-ynes, (Tables 3–10), acetylene–allene equilibria (Table 11) and MW data on bond lengths and dipole moments (Table 12).

Section III contains an account of the molecular mechanics method and deals with the conformational analysis by the force field method (Table 13), heats of formation (Table 14), bond dissociation energies (Table 15), rotational isomerism and barriers (Tables 16–18), MO energies and ionization potentials (Tables 19–21).

Section IV presents IR data and molecular vibrations (Tables 22–24), force constants (Table 25), electronic spectra (Table 26), red-shift (Table 27) and NMR data (Tables 29, 30) and a discussion of solid-state properties (Table 31) of some acetylenes.

Section V opens with X-ray structural data of complexes with transition metals (Tables 32–36) and considers acetylene–metal bonding theories (especially the EHT–MO view) and also chemisorption. Section VI mentions quantum-mechanical calculations (Table 37).

255 references up to 1977.
Complementary: 31, 27.
Relevant: 1, 1; **2**, 1; **3**, 1; **4**, 1; **5**, 1; **6**, 1; **8**, 1; **11**, 1; **12**, 1; **13**, 1; **14**, 1; **15**, 1; **16**, 1; **17**, 1; **18**, 1; **19**, 1; **20**, 1; **22**, 1; **23**, 1; **25**, 1; **26**, 1; **27**, 9; **28**, 1; **31**, 27; **32**, 1; **33**, 1; **36**, 2.

24, *2: The structural chemistry of the C≡C bond* by *J. L. Hencher*

I. Introduction (57); **II**. The C≡C bond (58); **III**. The C_a—C bond (60); **IV**. Conjugated C≡C and C—C bonds (62); **V**. Bonds between C_a and atoms other than carbon (63); **VI**. Summary and conclusions (64); **VII**. References (66).

Section II describes the effect of substitution on C≡C bond lengths (Table 1) and Section III deals with the single bond between the acetylenic carbon and an adjacent carbon atom (Table 2). Section IV treats the effect of conjugation on bond lengths in systems incorporating a C≡C bond (Table 3). Section V discusses the bond between an acetylenic carbon atom and an atom other than carbon (C—H, C—Hal, C—Si, C—Ge, C—O, C—S and C—B) (Table 4).

68 references up to 1976.
Complementary: **31**, 24.
Relevant: **16**, 2; **17**, 2; **18**, 2; **19**, 2; **20**, 2; **21**, 1; **22**, 2; **23**, 3; **26**, 2; **27**, 4; **28**, 2; **29**, 1; **30**, 1; **32**, 24; **32**, 29; **33**, 12; **34**, 10; **36**, 3.

24, *3: Thermochemistry of acetylenes* by *R. Shaw*

I. Introduction (69); **II**. The strength of carbon—hydrogen bonds in ethyne (69); **III**. Estimation of the thermochemical properties of acetylenes (70); **IV**. Acknowledgements (73); **V**. References (73).

Section II describes C—H bond strengths in ethyne and in the ethynyl radical obtained by classical and by MS methods. Section III discusses the available methods (including the additivity principle) for the estimation of the thermochemical properties of acetylenes. A comparison of observed values with values obtained by group additivity is presented in Table 1.

14 references up to 1977.
Complementary: none.
Relevant: **16**, 16; **18**, 3; **19**, 3; **20**, 11; **21**, 3; **22**, 6; **23**, 4; **25**, 2; **26**, 4; **27**, 9; **28**, 4; **29**, 24; **30**, 2; **31**, 2; **33**, 3; **36**, 4.

24, *4: Acidity, hydrogen bonding and complex formation* by *A. C. Hopkinson*

I. Introduction (76); **II**. Proton transfer (78); **III**. Hydrogen bonding (88); **IV**. Metal complexes (97); **V**. References (127).

Section II discusses equilibrium acidities (Tables 1 and 2), kinetic acidities (Tables 3 and 4), gas-phase acidities (Tables 5 and 6) and acidities computed by MO calculations (Table 7) and the allene–propyne interconversion occurring by removal of a methyl proton to form a mesomeric anion:

$$CH_3C{\equiv}CR \xrightarrow{-H^+} {}^-CH_2C{\equiv}CR \leftrightarrow CH_2{=}C{=}\bar{C}R \xrightarrow{H^+} CH_2{=}C{=}CHR$$

Section III deals with alk-1-ynes as proton donors in hydrogen bonding, including IR, equilibria and NMR studies (Tables 8–11), and also with alkynes as proton acceptors (Tables 12a and b), with intermolecular association of alk-1-ynes and briefly with heterosubstituted alkynes and with the theory of hydrogen bonding.

Section IV describes the structure of metal–alkyne complexes (Table 13) and the formation and stabilities of transition metal–acetylene complexes (involving Ti, Zr, V, Nb, Ta, Cr, Mo, W, Mn, Re, Fe, Ru, Os, Co, Rh, Ir, Ni, Pd, Pt, Cu, Ag and Au).

Oligomerization of acetylenes occurs through metallic complexes, often involving 3-, 5- and 7-membered rings with one metal atom. Other reactions take place through carbonyl complexes of transition metals, through vinyl complexes, by reaction of macrocyclic diynes with metal carbonyls or from alkoxycarbenes.

427 references up to 1976.
Complementary: none.
Relevant: **2**, 9; **3**, 6; **4**, 4; **5**, 6; **7**, 6; **8**, 5; **11**, 3; **12**, 3, 10; **13**, 7; **15**, 4; **16**, 5; **17**, 6; **18**, 8; **19**, 7; **20**, 12; **22**, 8; **23**, 5, 6; **25**, 5, 6; **29**, 17; **31**, 17, 21; **32**, 21; **33**, 5.

24, *5: Detection and determination of alkynes* by *K. A. Connors*

I. Introduction (137); **II**. Detection and identification (138); **III**. Quantitative analysis (148); **IV**. References (153).

Section II deals with detection and identification of alkynes, including chemical methods (Table 1), UV (Tables 2–5), IR (Table 6) and NMR spectra (Tables 7–11) and chromatography.

Section III describes quantitative methods such as reaction with Cu, Ag (Table 12) and Hg salts, and hydration to form carbonyl compounds; spectroscopic methods and gas chromatography are also discussed.

120 references up to 1974.
Complementary: **31**, 25.
Relevant: **1**, 5; **2**, 8; **3**, 15; **4**, 3; **5**, 17; **8**, 3; **12**, 4; **13**, 6; **14**, 3; **15**, 3; **16**, 3; **17**, 4; **18**, 5; **20**, 3; **22**, 5; **23**, 5; **25**, 10; **26**, 5; **28**, 3; **29**, 21; **30**, 14–22; **36**, 5.

24, *6: Mass spectrometry of acetylenes* by *C. Lifshitz and A. Mandelbaum*

I. Mass spectrometry of acetylenes (157); **II**. Ionization methods (158); **III**. Appearance potentials and thermochemical measurements (165); **IV**. Theoretical treatments and fundamental aspects of fragmentation processes of simple acetylenes (167); **V**. Location of the triple bond by mass spectrometry (170); **VI**. Decomposition of acetylenes under electron impact (171); **VII**. Ion–molecule reactions (186); **VIII**. References (193).

Section II reviews gas-phase ionization methods which have been employed in the study of acetylenes. These include photoionization, electron impact (Table 1), charge exchange (Tables 2 and 3) and chemical ionization and chemi-ionization (interaction of electronically excited neutral particles with ground-state (species).

Section III describes ionization potentials (Table 4) and heats of formation (Table 5). Section IV discusses fragmentation processes and within this subject H–D isotope effects, 'metastable' transitions, the quasi-equilibrium theory, predissociation by electronic transition and fragmentation in the series of isomeric $C_4H_6^+$ ions. Section V considers the application of MS to structure determination and especially to the location of triple bonds.

Section VI describes C—H and C—C bond cleavages, rearrangements involving H-migrations ('McLafferty rearrangements') with or without triple bond migration, skeletal rearrangements and also some alkoxy and alkyl migrations, eliminations of H_2O, and migrations of Me_2SiH and Me_3Si groups.

Section VII treats reactions between positive ions and molecules (Tables 6 and 7), their reaction rates and mechanisms in pure acetylene or in its mixtures with other gases (H_2, CH_4, C_2H_6, SiH_4). Finally, negative ion–molecule reactions are considered, e.g. the reaction of acetylene with O^- (Table 8) and the reactions of negative ions derived from acetylene (C^-, C_2^-, C_2H^-) with various neutral molecules.

178 references up to 1977.
Complementary: none.
Relevant: **9**, 7; **10**, 5;, **13**, 19; **16**, 4; **17**, 5; **18**, 6; **19**, 5; **22**, 7; **25**, 4; **26**, 6; **27**, 7; **29**, 3; **30**, 22; **31**, 3; **32**, 3; **33**, 4; **36**, 7.

24, *7: Applications of acetylenes in organic synthesis* by *P. F. Hudrlik and A. M. Hudrlik*

I. Introduction (200); **II**. Hydrometalation and carbometalation (202); **III**. Addition reactions of H^+Z^- to acetylenes (239); **IV**. Friedel–Crafts type acylations and alkylations of acetylenes (247); **V**. Anionic and organometallic cyclizations involving acetylenes (254); **VI**. Carbon-chain extension reactions (256); **VII**. Acknowledgements (258); **VIII**. References (259).

Section I reviews the most useful synthetic uses of acetylenes. Section II discusses the additions of metal hydride derivatives to the C≡C bond. These include hydroboration (Table 1) and related reactions (in general $RC{\equiv}CR + R'_2BH \rightarrow RCH{=}CRBR'_2$) and the synthetic uses of alkenylboranes, alkeneboronic esters and related compounds so obtained (e.g. protonolysis to form alkenes, oxidation to carbonyls, halogenation, rearrangements). Dihydroboration generates 1, 1-diboroalkanes [$RC{\equiv}CH + 2R'_2BH \rightarrow RCH_2CH(BR'_2)_2$] and the products are again important synthones. Alkynylborates ($RC{\equiv}CBR'_3$), on being attacked by various electrophiles, provide a wide variety of routes (Table 2) to obtain variously substituted olefins or ketones.

Section IIC deals with hydroalumination by R_2AlH, carboalumination by R_3Al, reactions with $LiAlH_4$ and the synthetic uses of alkenylaluminium compounds by their reactions with electrophiles (Table 3), and also with the dihydroalumination of alkynes. Section IID treats the hydrosilylation of alkynes and other methods for the preparation of alkenylsilanes ($RCH{=}CHSiR_3$) and with electrophilic substitutions of the latter. Next, the preparations and uses of alkenylstannanes ($RCH{=}CHSnR_3$) and of alkenylzirconium compounds are discussed. Section IIG considers the preparation of alkenylcopper intermediates (Table 4), using various organocopper reagents (RCu, R_2CuLi, R_2CuMgX, $RCuMgX_2$, etc.). The intermediates are not isolated, but can be converted to a variety of specifically substituted olefins (Table 5) or also to allenes (starting from propargylic acetates, halides, ethers, etc.). Finally in Section IIH the carbonylation of acetylenes, using various transition metal catalysts and yielding a variety of products (acids, lactones, etc.), is briefly mentioned.

Section III discusses the addition of HZ acidic compounds (H_2O, ROH, RNH_2, $RCONH_2$, RSH, HX, etc.) to the C≡C bond. Among these, hydration of acetylenes yields aldehydes or ketones (usually catalysed by Hg salts) whereas addition of alcohols yields ketals or enol ethers and carboxylic acids yield enol esters [$RC{\equiv}CR + AcOH \rightarrow RC(OAc){=}CHR$]. Hydrogen halides can add both by ionic and radical routes.

Section IV describes Friedel–Crafts-type acylations, yielding β-halovinyl ketones. Intramolecular alkylations are useful for the synthesis of 3–8-membered cyclic products, and can occur on either carbon of the C≡C group. Finally, carbanionic or organometallic cyclizations (Section V) and chain-extension reactions (involving metal acetylides or metalated propargyl derivatives) are mentioned briefly (Section VI).

643 references up to 1976.
Complementary: **24**, 8, 11, 19; **31**, 10, 23.
Relevant: none.

24, *8: Electrophilic additions to carbon—carbon triple bonds* by *G. H. Schmid*

Section I reviews the scope, mechanisms (single-step Ad_E2 and Ad_E3 and also multi-step Ad_E2 and Ad_E3 mechanisms), orientation rules, stereochemistry and product formation.

Addition of HX, ROC_2H and H_2O occurs with formation of the corresponding vinyl compounds ($HC{\equiv}CH + HX \rightarrow HCX{=}CH_2$). In some cases diadducts (HCX_2CH_3) and rearranged products can be obtained. Mechanisms, rates, stereochemistry and product compositions are presented in Tables 1–4.

Section III describes reactions with boranes when, depending on the nature of the alkyne, trivinylboranes (from internal alkynes) or dihydroboration products (from terminal alkynes) or monohydroboration products (e.g. from alkynes with dialkylborenes) can be obtained (Table 5).

Section IV treates the hydroalumination of alkynes [e.g. $RC{\equiv}CR + R'_2AlH \rightarrow RCH{=}CR(AlR'_2)$] which, in the case of terminal alkynes, may also be a substitution as well as an addition reaction: $RC{\equiv}CH + R'_2AlH \rightarrow RCH{=}CH(AlR'_2) + RC{\equiv}CAlR'_2$ (Tables 6–8). Section V describes reactions of alkyl halides, alcohols and acyl halides with alkynes, under conditions in which carbonium ions are formed (concentrated H_2SO_4, $AlCl_3$, etc.). The stereochemistry of the products depends on the nature of both reactants (Tables 9 and 10). The oxidation of alkynes with peracids (Section VI) is a complex reaction as shown, e.g., by $PhC{\equiv}CH + PhCO_3H \xrightarrow{CHCl_3} PhCH_2CO_2C_2H_5 + PhCH_2CO_2CH_3 + PhCHO + PhCO_2H + PhCO_2CH_3$. The reaction involves important solvent effects, oxirene and dioxabicyclo intermediates and possibly also oxocarbenes ($—\overset{O}{C{=}C}— \rightleftharpoons —\ddot{C}—\overset{O}{\overset{\|}{C}}—$).

Sulphenyl (and especially arenesulphenyl) halides form β-halovinyl sulphides with alkynes: $RSX + —C{\equiv}C— \rightarrow —CX{=}C(SR)—$. The rates are strongly dependent on the alkyne structure (Table 11) and on the substituents in the arenesulphenyl ring (Table 12), whereas the regiochemistry of the reaction depends greatly on the structure of the alkyne (Table 14) and on the solvent (Table 15) (Section VII).

Section VIII deals with the addition of divalent (RSeX; Tables 16 and 17) and quadrivalent ($RSeCl_3$; Table 18) selenium compounds, and Section IX treats compounds containing mercury and thallium.

Section X discusses the additions of F_2, Cl_2 (Tables 19–22), Br_2 (Tables 23–25) and I_2 (Table 26), Section XI deals with interhalogens (ClF, ICl, IBr) and with pseudohalogens (—NCO, —SCN, —N_3) and finally Section XII compares the electrophilic additions of alkenes and of alkynes (Tables 27–29), emphasizing differences between the two.

221 references up to 1976.
Complementary: none.
Relevant: 21, 9.

24, *9: Propargylic metalation* by *J. Klein*

(365); **V**. Rearrangements (371); **VI**. Some synthetic applications (375); **VII**. References (377).

Propargylic metalation (e.g. $Ph_3SiC{\equiv}CCH_3 \rightarrow Ph_3SiC{\equiv}CCH_2Li$) has been known only since mid-1960s. Section II describes the scope of the reaction, its mechanism and product composition and presents PMR data (Table 1 and Figs 6–10) and UV spectra (Figs 3–5).

Section III deals with the structure of the products obtained by propargylic metalation, with the involvement of allenic structures, with quantum-mechanical calculations, IR spectra (Table 2), rates and ρ and σ values. Next (Section IV), the mechanism of the reaction is discussed in detail, including the equilibrium between allenyl and propargyl compounds ($RCM{=}C{=}CH_2 \rightleftharpoons RC{\equiv}CCH_2M$), various electrophilic (including cyclic) mechanisms, the steric course and the regioselectivity of the reaction.

Section V treats the possibility of rearrangements in metalated derivatives, 1,3-migrations of protons and 1,3-sigmatropic proton shifts involving simultaneous migration of Li and H (a 'dyotropic' reaction).

Section VI presents some synthetically useful routes involving metalated alkynes leading to alkylation and ketone formation, mentioning other routes not yet fully exploited.

121 references up to 1977.
Complementary: none.
Relevant: 26, 10.

24, *10: Rearrangements involving acetylenes* by *F. Théron, M. Verny and R. Vessière*

I. Introduction (381); **II**. Prototropic rearrangements (381); **III**. Rearrangements involving organometallic intermediates (398); **IV**. Aniontropic rearrangements (401); **V**. Triple bond participation (417); **VI**. Sigmatropic rearrangements (421); **VII**. References (437).

Section II describes prototropic rearrangements (in general $R'C{\equiv}CCH_2R \rightleftharpoons R'CH{=}C{=}CHR \rightleftharpoons R'CH_2C{\equiv}CR$) of unsubstituted and substituted acetylenes (and allenes) and the mechanism of these (usually base-catalysed) rearrangements; rearrangements of ene-ynes and related compounds; diacetylenes, e.g. $HC{\equiv}CC{\equiv}CCH_2R \rightleftharpoons CH_3C{\equiv}CC{\equiv}CR$, and in appropriate cases yielding also benzene derivatives by base-catalysed routes involving acetylene–allene rearrangements before the cyclization step; and finally acetylene–allene rearrangements in acidic media.

During the formation of organometallic intermediates derived from propargylic or from allenic halides, acetylene–allene rearrangements occur (Section III) and various other rearrangements take place in subsequent reactions of the organometallic derivatives.

Section IV discusses anionotropic rearrangements including acid-catalysed rearrangements of acetylenic (α- and β-) alcohols to ketones or aldehydes, Ag^I or Cu^I salt catalysed isomerizations of propargylic compounds to allenes, conversion of α-acetylenic alcohols to haloallenes by various reagents (HX, $SOCl_2$, PX_5) and various substitution processes (solvolytic S_N1, carbenoid, S_N2, S_N2' organometallic alkylations and reductions by $LiAlH_4$ and related reagents).

Section V deals with triple bond participation in homopropargylic rearrangement and in the cases of remote triple bonds, especially in cyclizations of 6-substituted alkynes.

Section VI discusses the Claisen and the thio-Claisen rearrangements and some related [3,3] sigmatropic rearrangements of propargyl (or allenyl) derivatives, among them Cope and oxy-Cope rearrangements. Finally [2,3] sigmatropic rearrangements of propargyl derivatives are presented, including propargyl phosphite, phosphinite, sulphinate, sulphenate and sulphite rearrangements, and also propargylic anion, acetylenic sulphonium ylide and acetylenic ammonium ylide rearrangements.

378 references up to 1975.
Complementary: **1**, 13.
Relevant: **1**, 7; **2**, 15; **3**, 14; **4**, 10; **5**, 8; **8**, 9; **9**, 3; **10**, 3; **11**, 4; **12**, 16; **13**, 13; **14**, 7; **15**, 8; **16**, 17; **17**, 8; **18**, 15; **19**, 18; **20**, 4; **23**, 13; **26**, 15; **29**, 4.

24, *11: Cycloadditions and cyclizations involving triple bonds* by *J. Bastide and O. Henri-Rousseau*

I. Introduction (448); **II**. Cycloaddition reactions (449); **III**. Theoretical study of cycloadditions (482); **IV**. Cyclization reactions (504); **V**. Acknowledgements (517); **VI**. References (517).

Section II reviews the different types of cycloadditions (Table 1). The classification in the chapter uses the number of electrons participating in the TS and originating from both reactants. First, $[\pi^2 + \pi^2]$ cycloadditions of carbenes with alkynes leading generally to cyclopropenes are considered, then $[2+2] \rightarrow 4$ between alkynes and double or triple bonds yielding cyclobutenes or cyclobutadienes, respectively. Next, 1,3-dipolar cycloadditions of alkynes are treated, with the 1,3-dipole being a diazoalkane (giving a pyrazole), an azide (giving triazoles), a nitrile oxide (giving isoxazoles), an azomethine oxide (giving 4-isoxazolines), an azomethine ylide (giving 3-pyrrolines), an azomethine imine (giving 3-pyrazolines) or carbonyl ylides (giving furans), and reactions of alkynes with azimines, *N*-epxides, nitrile sulphides, oxathioazolines, thiazolines and other S-containing compounds are also described.

Diels–Alder reactions are discussed, involving both fluorinated alkynes (e.g. $CF_3C{\equiv}CCF_3$) and fluorinated dienes, and reactions yielding furan, pyridine and benzene derivatives are also described.

Section II.C deals with $[\pi^6 + \pi^2]$, $[\pi^8 + \pi^2]$, $[\pi^{10} + \pi^2]$, $[\pi^{12} + \pi^2]$ and $[\pi^2 + \pi^2 + \pi^2]$ and with intramolecular cycloadditions, yielding a variety of polycyclic ring systems with different heteroatoms.

Section III discusses the theoretical problems involved in cycloadditions. Thus, the mechanism of the reaction, its experimental criteria (intermediates, stereochemistry, entropy and enthalpy of activation, activation volume and solvent effects) and theoretical criteria (Woodward–Hoffmann rules, TS calculations) are treated, as relevant to $[2+2]$ cycloadditions to triple bonds, to Diels–Alder reactions and to 1,3-dipolar cycloadditions. Next, the conclusions as to the reactivity in cycloadditions, obtained from the perturbation molecular orbital theory, are presented (Tables 2–4 and Figs 12 and 13), and the perturbation method is again used to predict regioselectivity, especially in 1,3-dipolar cycloadditions (Tables 5–10) and in Diels–Alder reactions.

Section IV describes some recent results on intramolecular cyclizations involving OH, NR and CH groups with C≡C bonds in the same molecule. This is followed by a description of reactions in which in the first step one or two molecules add to the C≡C bond (or to another functional group in its vicinity) and the reaction is completed by cyclization of the intermediate.

255 references up to 1976.
Complementary: **1**, 11; **21**, 6; **31**, 23.
Relevant: **8**, 7.

24, *12: Photochemistry of the C=C bond* by *J. D. Coyle*

I. Introduction (523); **II**. Photofragmentation (524); **III**. Photorearrangement (526); **IV**. Photoaddition (528); **V**. Photocycloaddition (533); **VI**. References (548).

Section II describes homolytic bond cleavages occurring on (vacuum) UV irradiation of alkynes, when usually either a bond adjacent to C≡C or a 'propargylic' bond (the term is used analogously to 'allylic' in the case of alkenes) breaks. In Section III photolytic rearrangements are discussed, occurring, e.g., by propargylic cleavage followed by recombination or possibly by two consecutive [1, 3] sigmatropic shifts or in some cases by di-π-methane routes. Photoadditions (Section IV) of saturated hydrocarbons, alcohols, amines, amides or H_2O to C≡C bonds may occur, involving alkyne excited states, while other additions may occur through the formation of free radicals by photocleavage of the addend (e.g. $CH_3C{\equiv}CH + HBr \rightarrow \rightarrow CH_3CH{=}CHBr + CH_3CHBrCH_2Br$), and some of these (e.g. the addition of formamide) require sensitizers.

Section V deals with photocycloaddition of alkynes to alkenes (forming cyclobutenes) and treats also trimerizations to benzene derivatives or in the presence of H_2S to thiophenes. Next, photocycloaddition to aromatic compounds (forming, e.g., with benzene first a bicyclooctatriene and then yielding by thermal electrocyclic ring a cyclooctatetraene), and finally photocycloadditions to C=O, C=S and NO_2 groups are considered.

113 references up to 1976.
Complementary: none.
Relevant: 2, 16; **3**, 8; **6**, 4; **8**, 12; **9**, 6; **11**, 5; **13**, 16; **14**, 8; **15**, 9; **16**, 11; **17**, 9; **18**, 10; **19**, 20; **22**, 10; **23**, 9; **25**, 11; **27**, 21, 22; **28**, 5; **29**, 5, 6; **32**, 29; **33**, 21.

24, *13: Synthetic acyclic polyacetylenes* by *W. D. Huntsman*

I. Introduction (554); **II**. Properties (554); **III**. Synthesis (565); **IV**. Reactions (577); **V**. References (614).

Section II deals with the stability, geometry, electronic effects (Tables 1 and 2), UV spectra (Tables 3–6), NMR (Tables 8 and 9), MS and PES (Table 10) of polymers $R(C{\equiv}C)_nR$. Section III describes syntheses by oxidative coupling of Cu and other organometallic derivatives [$RC{\equiv}CH \xrightarrow{Cu^{II}} R(C{\equiv}C)_2R$], by coupling of terminal alkynes with 1-bromoalkynes [$RC{\equiv}CH + BrC{\equiv}CR' + B \xrightarrow{Cu^+} R(C{\equiv}C)_2R'$] or with propargyl, vinyl and allenyl halides. Alkynyl Grignard reagents undergo self-coupling or coupling with propargyl halides. Other routes involve various eliminations of HX or H_2O and pyrolysis of hydrocarbons (CH_4, HC≡CH at 1600–2400 K).

Section IV discusses nucleophilic additions, many of which fail with simple alkynes [e.g. CN^-, RLi, $CH_2(CO_2R)_2$ and $LiAlH_4$ add readily to conjugated tetraynes and pentaynes]. Other addends considered are oxygen nucleophiles (ROH, ArOH), sulphur nucleophiles (RSH, etc.) and nitrogen nucleophiles (RNH_2, R_2NH). Nucleophilic attacks may lead to the formation of heterocycles (e.g. thiophenes with H_2S; Table 11). Next, electrophilic addends are treated, including halogens (usually giving polyhalogenated products), hydrogen halides and water, the rate decreasing with the number of C≡C bonds in the molecule, falling to zero with 2, 4, 6, 8-decatetrayne). Subsequently, reduction to polynes and enynes, cycloadditions, 1, 3-dipolar cycloadditions, dimerizations and solid-state (thermal, UV or X-ray catalysed) polymerization are described. Electrophilic substitution (e.g. cleavage of alkynyl—metal bonds in organometallic derivatives by H^+,

Ag^+, RCO^+, etc.) are dealt with, together with the diyne reaction [two acetylene groups less than 3.4 Å from each other in a molecule reacting with a transition metal compound to produce a five-membered (4C and one transition metal) ring], cyclotrimerization, prototropic and thermal rearrangements and finally some photochemical reactions.

284 references up to 1976.
Complementary: 31, 22.
Relevant: 24, 14, 15.

24, *14: Natural acetylenes* by *Sir E. R. H. Jones and V. Thaller*

I. Introduction (621); **II**. Classification of natural acetylenes (622); **III**. Polyacetylenes (623); **IV**. References (632).

After a review of the classification of natural acetylenes (Section II), the distribution and detection of natural polyacetylenes are described, together with their structures in different sources (Table 1), methods for their structural elucidation and synthesis, their biogenesis and physiological properties and finally taxonomic implications.

58 references up to 1977.
Complementary: none.
Relevant: 2, 7; **3**, 5; **4**, 9; **5**, 18; **7**, 4; **11**, 11; **12**, 12; **13**, 14; **15**, 10; **16**, 13; **17**, 13; **19**, 13; **18**, 13; **20**, 6; **22**, 20; **26**, 13; **28**, 17; **31**, 9.

24, *15: Cyclic acetylenes* by *M. Nakagawa*

I. Introduction (636); **II**. Shortlife cyclic acetylenes (637); **III**. Medium-ring acetylenes (643); **IV**. Large-ring acetylenes (654); **V**. Dehydroannulenes (663); **VI**. References (707).

Section II deals with cyclic acetylenes smaller than cycloheptyne, for which there is evidence as intermediates in various reactions and which can be trapped as adducts with diphenylbenzofuran or with tetracyclone (tetraphenylfuranone) (Tables 1 and 2). These intermediates may be prepared from 1, 2-dibromocycloalkenes:

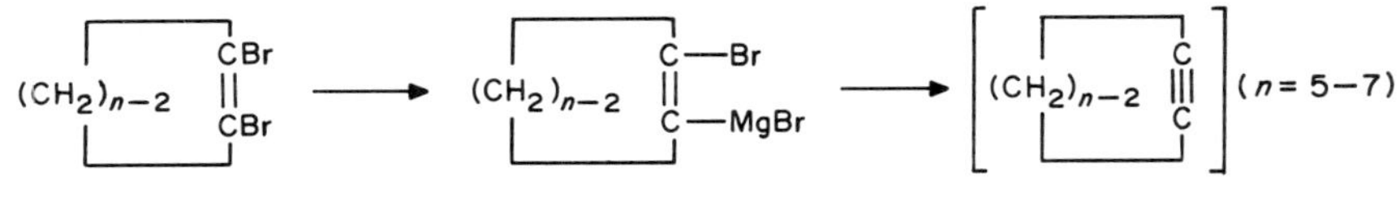

by oxidation of bishydrazones of cycloalkane-1, 2-diones, by base elimination from bromoethylene derivatives and by oxidation or photolysis of aminotriazole derivatives. Intermediacy of cyclopentyne was firmly proved, but no evidence exists for the intermediacy of cyclobutyne in any reaction. The C_5–C_7 cycloalkynes undergo polar additions, cycloadditions, isomerization and oligomerization (possibly through primary cyclobutadiene products).

Section III describes medium-ring (8–11-membered) acetylenes, which can be isolated. Their enthalpies of hydrogenation are given in Table 3 and the cycloalkyne–cycloallene equilibrium composition in Table 4. Although the parent cycloheptyne cannot be isolated, its tetramethyl derivatives can, in addition to 3, 3, 6, 6-tetramethylthiacycloheptyne, and syntheses, physical and chemical properties of these are described. Next, eight-membered rings with one or two C≡C bonds and their benzo derivatives, and also ten-membered rings with one, two or three C≡C groups are discussed.

Section IV deals with the synthesis of macrocyclic (12 and over) alkynes, with their conformation and with the strain, molecular geometry and electronic spectra of cyclic diphenyldiacetylenes and cyclic tolans (Table 5).

Section V treats the synthesis of dehydroannulenes (cyclic polyenepolyenes, fully conjugated), the problem of aromaticity and anti-aromaticity studied mainly by NMR (Table 6), dehydroannulenones, dehydroheteroannulenes containing one or more heteroatoms (Table 7) and dehydroannulene anions and cations, some of which are aromatic, planar and stable. 'Acetylene–cumulene' dehydroannulenes, i.e. cyclic compounds containing both C≡C and cumulative double bonds, their synthesis, structure and properties are considered (Tables 8–12). Finally, dehydroannulenes annelated with 6π ring systems (Table 13) and dehydroannulenoannulenes (i.e. condensed annulenes in which two systems have two common carbon atoms, as in naphthalene) are discussed.

259 references up to 1976.
Complementary: none.
Relevant: none.

24, *16: Proximity interactions of acetylenes* by *S. Misumi and T. Kaneda*

I. Introduction (713); **II**. Proximity interactions in spectral behaviour (713); **III**. Proximity interactions in chemical reactions (723); **IV**. References (735).

Interactions between two C≡C groups are reminiscent of the effect of hydrogen bonds. The chapter deals mainly with π–π interactions between two C≡C bonds or between C≡C and another unsaturated system.

Section II describes the effects of such proximity interactions in electronic absorption spectra, in charge-transfer spectra (Tables 1 and 2), in NMR and in PES.

Chemical reactions also show the proximity interactions (Section III). Thus, C≡C/C≡C interactions affect favourably intramolecular cyclizations by thermal, photochemical or transition metal catalysed reactions. In cyclic polyacetylenes, transannular C—C bond formation is favoured by proximity interactions, often resulting in one-step syntheses of complicated polycyclic systems. Interactions between C≡C bonds and aromatic rings are important in various cycloaddition modes of, e.g., [3.3]paracyclophadiyne with tetracyanoethylene.

71 references up to 1976.
Complementary: none.
Relevant: none.

24, *17: The electrochemistry of the carbon—carbon triple bond* by *J. H. P. Utley and R. Lines*

I. General introduction (739); **II**. Cathodic reduction (743); **III**. Anodic oxidation (751); **IV**. References (752).

Section I summarizes data from polarographic (Table 1) and cyclic voltammetric (Table 2) experiments and reviews the mechanism of electroreduction and the ease of oxidation (Table 3).

Section II deals with direct reduction of acetylenes (Table 4) in aprotic solvents at Hg or Pb cathodes, and with indirect reduction (Table 5) involving the generation of a reagent which reacts with acetylene (H_2, solvated electron, etc.) at low overvoltage cathodes (Pt, Ni, Co).

Section III discusses the anodic oxidation of acetylenes, especially at a gold anode in aqueous solution, producting CO_2 and polymers, and a variety of carbonyl compounds from diphenylacetylene.

37 references up to 1975.
Complementary: none.
Relevant: 5, 2; **8**, 11; **13**, 5; **16**, 15; **17**, 14; **19**, 12; **20**, 5; **21**, 5; **22**, 9; **23**, 10; **25**, 13; **26**, 12; **27**, 8; **28**, 7; **29**, 8, 9; **31**, 7; **32**, 6; **34**, 1.

24, *18: The preparation of acetylenes and their protection* by *D. A. Ben-Efraim*

I. Introduction (756); **II**. Acetylenes by elimination reactions (757); **III**. Acetylenes by substitution reaction (790); **IV**. Miscellaneous methods (800); **V**. Protection of the C—H and C—C bonds of acetylenes (800); **VI**. References (805).

Section IIA describes the preparation of acetylenes by dehydrohalogenations from a wide variety of starting materials and bases, including alkali metal carbonates, hydroxides, alkoxides and also $NaNH_2$, RLi and other organometallic compounds, metal hydrides, amines and quaternary ammonium hydroxides and fluorides. Vinyl halides undergo photochemical dehydrohalogenation and the Fritsch–Buttenberg–Wiechell rearrangement ($Ar_2C{=}CHX \xrightarrow{\text{base}} ArC{\equiv}Ar$) occurs under the influence of KOH, KOR, $NaNH_2$ and other bases.

Section IIB deals with routes involving the elimination of alcohols, thiols, sulphides, sulphonic and phosphoric acids and trialkyltin hydrides. Next, Hoffmann eliminations and dehalogenations (from tetrahaloalkanes and dihaloalkenes), deoxygenations of α-diketones [$RCOCOR + (EtO)_3P \rightarrow RC{\equiv}CR$] and eliminations of Ph_3PO from enol phosphonium salts and of CO, SO_2 and related species are considered, then C≡C-forming eliminations of N_2 from mono- and dihydrazones [$RC({=}NNH_2)C({=}NNH_2) + O_2 \rightarrow\rightarrow RC{\equiv}CR + 2N_2 + 2H_2O$], from 3-nitroso-2-oxazolidones, from 2-pyrazolin-5-ones and from some other related compounds are described.

Section III deals with alkylations of acetylenes ($RC{\equiv}CM + R'X \rightarrow RC{\equiv}CR'$). The synthons discussed are alkali metal acetylides, Grignard reagents (which do not react with primary alkyl halides, but do so with allylic, propargylic and benzylic halides, e.g. $HC{\equiv}CCH_2Br + BrMgC{\equiv}Ph \rightarrow HC{\equiv}CCH_2C{\equiv}CPh$) and copper acetylides (alkylated by alkyl, allyl, propargyl, vinyl and aryl halides). Terminal acetylenes are alkylated in the presence of Pd complexes and 1-alkynyltrialkyl borates rearrange on reaction with I_2 and produce new alkynes: $RC{\equiv}CH \rightarrow RC{\equiv}CLi \xrightarrow{R'_3B} Li^+[R'_3BC{\equiv}CR] \xrightarrow{I_2} R'C{\equiv}CR + R'_2BI + LiI$.

Section V treats the protection of CH groups in acetylenes by groups which can be removed at a later stage. The protecting groups include MgX, $SiMe_3$ and $SiEt_3$. Finally, protection of the C≡C group is mentioned. This can be carried out with dicobalt octacarbonyl, which forms stable complexes with alkynes.

262 references up to 1976.
Complementary: 31, 28.
Relevant: none.

24, *19: Nucleophilic attacks on acetylenes* by *J. I. Dickstein and S. I. Miller*

I. Introduction (814); **II**. Reactivity and orientation (816); **III**. Nucleophilic additions (843); **IV**. Nucleophilic substitutions (911); **V**. References (938).

Section I defines the scope of the chapter and Section II reviews the subjects of reactivity and orientation in nucleophilic attacks on acetylenes, including a comparison of alkynes and alkenes (in general, nucleophiles react faster with alkynes), *anti* vs *syn* selectivity in additions (Table 1), regioselectivity and the effects operating in it [substitution effects in terminal acetylenes (Table 2) and on disubstituted acetylenes (Tables 3 and 4), solvent effects (Table 5) and effects of substituents, nucleophiles and solvents in the especially important attacks on 1-haloalk-1-ynes (Table 6)]. Next, the influence of the size of the ring formed in internal nucleophilic cyclizations is discussed.

Section III deals with electrochemical reduction (which is considered to be a nucleophilic attack by an electron on an alkyne) (Tables 7 and 8), conventional (non-electrochemical) electron transfer to alkynes by various metals and hydrides (Table 9), hydride transfer to alkynes from $LiAlH_4$, R_3SnH and related compounds to yield alkenes or allenes (Tables 10 and 11), halide attacks on activated alkynes to yield the corresponding haloalkene [$RC{\equiv}CR + HalE \rightarrow RC(Hal){\equiv}CRE$], additions of oxygen nucleophiles such as HO^-, RO^-, ArO^- and AcO^- (Tables 13 and 14) and the addition of oxygen nucleophiles to form cyclic products (in at least two steps) such as furans, pyrans and heterocycles with various combinations of O, S and N atoms (Tables 15 and 16). Sulphur nucleophiles (CNS^-, RS^- ArS^-, etc.) (Table 17) are treated next, followed by N nucleophiles (NH_3, RNH_2, R_2NH, anilines, *N*-heterocyclics, etc.) forming either acyclic (Tables 18 and 19) or cyclic (Table 20) products. Carbon nucleophiles (anions of carbon acids, organometallics, ylids, enamines, enol ethers, etc.) usually add to alkynes only in the presence of some activating substituent, reaction-facilitating solvent, coordination site or catalyst (Table 22). In many cases, cyclic products are formed, e.g.

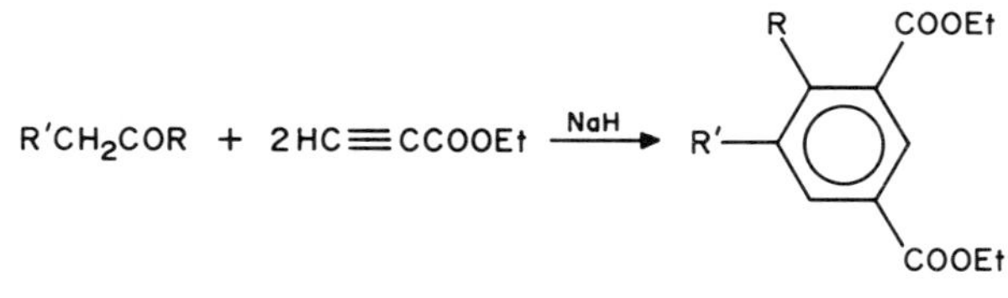

A number of less general nucleophilic additions are collected in Table 23, including nucleophiles such as R_3GeX, Na_2Se, Na_2Te, Ar_3P, $(RO)_3P$ and Et_3SnNa.

Section IV describes nucleophilic substitutions on haloalkynes, which usually, compared with other halides, are very sluggish, although in exceptional cases (e.g. with p-$CH_3C_6H_4S^-$) haloalkynes are more reactive than haloalkanes. The kinetics and mechanisms of the reactions of haloalkynes are discussed in detail, and rate constants, enthalpies and entropies of activation of haloalkynes are tabulated and compared with those of other organic halides (Table 24). The element effect (i.e. the rates of the same reaction system differing only in the nature of the halogen) (Tables 25 and 27) and Hammett ρ values (Table 26) are used as diagnostic probes for the mechanisms proposed (Scheme 6, p. 916), each of which are presented in detail.

Finally, synthetic applications of the process $RC{\equiv}CX + Nu^- \rightarrow RC{\equiv}CNu + X^-$ are given in Section IV.C and Table 28. The latter presents starting materials, reaction conditions, products and yields, while the text describes the scope and limitations of the reactions.

743 references up to 1976.
Complementary: none.
Relevant: **1**, 8; **8**, 6, 8; **16**, 7; **24**, 7.

24, *20: Synthesis and uses of isotopically labelled acetylenes* by *J. C. Lavalley and J. Saussey*

I. Introduction (958); **II**. Preparation (958); **III**. Isotope purity determination and isotope effects (964); **IV**. Examples of uses of labelled acetylenes (968); **V**. Acknowledgements (974); **VI**. References (974).

The preparation of acetylenes labelled with D or T is described in Section II. This is accomplished in terminal acetylenes by hydrolysis of alkali metal salts of alk-1-ynes by D_2O or by T_2O or by isotopic exchange of the acidic C≡CH hydrogen atom. D-containing alkynes can be prepared from carbides and D_2O, from alkaline acetylides with labelled alkyl or alkynyl halides and by various rearrangements (e.g. alk-2-yne → alk-1-yne, $RC{\equiv}CCD_3 \rightarrow RCD_2C{\equiv}CH$ and alk-1-yne → alk-2-yne isomerizations).

^{13}C-or ^{14}C-labelled acetylenes are usually prepared by hydrolysis of labelled acetylides ($^{13}CO_2 + Li \rightarrow Li_2{}^{13}C \xrightarrow{H_2O} H^{13}C{\equiv}{}^{13}CH$). Preferred methods of preparing some simple labelled acetylenes are presented in Table 1).

Section III discusses the determination of isotopic purity and the effects of the presence of isotopes by gas chromatography, MS, IR, ^{1}H NMR (Table 2) and ^{13}C NMR.

Section IV describes the uses of labelled acetylenes in spectroscopic studies, mechanistic studies, isomerizations, metathesis of acetylenes and their cyclotrimerization to benzene derivatives by transition metal catalysts, kinetic isotope effects involved in the reaction of but-2-yne with labelled and unlabelled arene–Cr complexes and also in the Ni complex-catalysed trimerization of acetylene. Finally, adsorbed species formed on the surface of catalysts can be studied by spectroscopy, and in these cases the shifts of the characteristic bonds due to H → D substitutions are of great assistance (Table 3).

94 references up to 1976.
Complementary: none.
Relevant: **5**, 10; **12**, 13; **13**, 15; **17**, 12; **18**, 18; **19**, 17; **22**, 14; **23**, 16; **26**, 11; **27**, 10; **28**, 8; **29**, 28; **33**, 18; **36**, 12.

Supplement B: The chemistry of acid derivatives (1979)

25, *1: Recent advances in the theoretical treatment of acid derivatives* by *I. G. Csizmadia, M. R. Peterson, C. Kozmuta and M. A. Robb*

Section II reviews the main theoretical concepts to be used in the chapter, including many-electron wave functions, the variation theorem, closed-shell and open-shell Hartree–Fock problems, the correlation of theoretical results with experimental observations and the limitations molecular size imposes or rigorous calculations.

Section III discusses general theoretical considerations, the search for optimal geometries by theoretical methods (Table 1) and calculations on acyl halides, such as for excitation energies (Table 2), dipole moments (Table 3), geometries of HCOF (Table 4) and HCOOH (Tables 5 and 6), electronic transition energies (Table 7) and ionization potentials (Table 8) of HCOOH and conformational energy differences and barriers to rotation for $HCOOCH_3$. Next, geometries (Table 10), total energy and dipole moments (Table 11) and electronic transition energies of $HCONH_2$ are presented, followed by considerations of higher amides, amino acids and some biological molecules.

Section IV treats protonation, deprotonation and proton transfer in acid derivatives, presents experimental acidities and basicities of acids (Tables 13 and 14) and deals with hydrogen bonding, dimerization and solvation and with the site of hydration of formamide and *N*-methylacetamide. Finally, the hydrolysis of amides and esters is considered, together with *ab initio* calculations, and the mechanisms corresponding to the theoretical results are discussed with the aid of energy profiles, computed topologies of conformational energy surfaces and pseudo-three-dimensional representations of the conformational energy surfaces of the intermediates (Figs 21–29).

124 references up to 1977.
Complementary: **6**, 1; **11**, 1; **15**, 1.
Relevant: 1, 1; **2**, 1; **3**, 1; **4**, 1; **5**, 1; **6**, 1; **8**, 1; **11**, 1; **12**, 1; **13**, 1; **14**, 1; **15**, 1; **16**, 1; **17**, 1; **18**, 1; **19**, 1; **20**, 1; **22**, 1; **23**, 1; **24**, 1; **26**, 1; **27**, 9; **28**, 1; **31**, 27; **32**, 1; **33**, 1; **36**, 2.

25, *2: Thermochemistry of acid derivatives* by *R. Shaw*

The chapter is concerned with the standard molar heats of formation (ΔH_f^0) of the title compounds.

Section II deals with aliphatic, alicyclic and aromatic acids (Table 1), Section III with esters and halogenated esters, Section IV with amides, including observed and estimated values (Table 2), and Section V presents observed ΔH_f^0 data and derived group values for estimating heats of formation of acyl halides (Table 3).

48 references up to 1977.
Complementary: none.
Relevant: **16**, 16; **18**, 3; **19**, 3; **20**, 11; **21**, 3; **22**, 6; **23**, 4; **24**, 3; **26**, 4; **27**, 9; **28**, 4; **29**, 24; **30**, 2; **31**, 2; **33**, 3; **36**, 4.

25, *3: Chiroptical properties of acid derivatives* by *R. Håkansson*

The chapter considers Cotton effects (CE), optical rotatory dispersion (ORD) and circular dichroism (CD) of the title compounds.

Section II reviews the electronic properties of the COOR group, theoretical aspects and experimental facts (Tables 1 and 2) which support an energy level scheme presented for the ester chromophore.

Section III considers various rules which have been formulated for the correlation of the geometries of lactones and lactams with their chiroptical properties, including sector and ring chirality rules and also theoretical studies for saturated lactones and lactams (Section III.A) and for unsaturated lactones (Section III.B).

Section IV deals with the conformations of α-substituted acids, esters and amides, presents CD data for α-hydroxy and α-amino acids (Table 3), for esters (Table 4) and for the absorption and sign of CE of some *N*-substituted derivatives of amino acids (Table 5), and treats also metal complexes of amino and hydroxy acids. Next, acetamides of steroids and carbohydrates, acetates of carbohydrates, glycerides, thiolacetates and thionamides, some natural complex acids and esters and some unsaturated and allenic acids are considered.

Finally, in Section IV.B acids containing aromatic rings are discussed, including phenylacetic acids and related compounds and the corresponding acyl chlorides and aromatic acids and derivatives containing thiophene and benzene rings.

318 references up to 1977.
Complementary: **20**, 2.
Relevant: **8**, 4; **18**, 7; **19**, 6; **22**, 4; **26**, 3; **27**, 6; **28**, 10; **29**, 23; **31**, 1; **36**, 17.

25, *4: Mass spectra of acid derivatives* by *S. W. Tam*

Section II deals with aliphatic, aromatic, dicarboxylic, unsaturated, halo and thio acids, describing in each case simple fragmentations, fragmentations with rearrangements and other relevant subjects, and also includes a sub-section (Section II.E) on negative-ion mass spectra.

Section III treats analogously aliphatic, aromatic, dicarboxylic, unsaturated, hydroxy, methoxy, keto and thio esters and briefly also their negative-ion and chemical ionization mass spectra.

Section IV on amides discusses α- and γ-cleavage and fragmentation involving H-migration in aliphatic amides and treats next aromatic, unsaturated, thio- and trifluoroacetamides. Brief sections describe the MS of acyl halides (Section V), of anhydrides and imides (VI) and finally of lactones and lactams (VII).

226 references up to 1976.
Complementary: none.
Relevant: **9**, 7; **10**, 5; **13**, 19; **16**, 4; **17**, 5; **18**, 6; **19**, 5; **22**, 7; **24**, 6; **26**, 6; **27**, 7; **29**, 3; **30**, 22; **31**, 3; **32**, 3; **33**, 4; **36**, 7.

25, *5: Complexes of acid anhydrides* by *R. Foster*

The chapter concerns electron donor–acceptor (EDA) (also described as charge-transfer, CT) complexes of acid anhydrides (Table 1).

Section II reviews the stabilities of EDA complexes in solution with anhydrides being acceptors and arenes, ethers, amines, halides, esters etc. being donors. Table 2 gives equilibrium constants and λ_{max} and ε_{max} of the CT bond.

Section III deals with the crystal structures of solid complexes (obtained by X-ray diffraction) and especially with those of pyromellitic dianhydride (PMDA) with various polycyclic aromatic compounds. Section IV discusses absorption spectra, presents frequencies of maxima of intermolecular CT bonds (Table 3), electron affinity values of some anhydrides (Table 4), fluorescence maxima of some EDA complexes and their phosphorescence maxima and lifetimes (Table 5). Section V describes some thermal and photochemical reactions involving EDA complexes in which new products are obtained by bond formation between the constituents of the complex.

155 references up to 1976.
Complementary: **25**, 6.
Relevant: **2**, 9; **3**, 6; **4**, 4; **5**, 6; **7**, 6; **8**, 5; **11**, 3; **12**, 3, 10; **13**, 7; **15**, 4; **16**, 5; **17**, 6; **18**, 8; **19**, 7; **20**, 12; **22**, 8; **23**, 6, 7; **24**, 4; **29**, 17; **31**, 17, 21; **32**, 21; **33**, 5.

25, *6: Hydrogen bonding in carboxylic acids and derivatives* by *D. Hadži and S. Detoni*

Section I reviews the structural aspects and IR, Raman, NMR and UV spectra of hydrogen bonding and the electronic theory of the hydrogen bond and its dynamics. Section II deals with the structural aspects of hydrogen bonds in acids such as association between neutral acids and between ions, with molecular complexes involving bases and with aqueous systems. Next it describes the vibrational, UV (CT), NMR and NQR spectra of carboxylic acids, their salts and complexes, discusses quantum-chemical calculations relevant to hydrogen bonding (especially in dimers) and also the thermodynamics and kinetics of acid dimers and hydrates and other molecular complexes, together with intramolecular hydrogen bonds in carboxylic acids.

Section III treats esters, lactones, anhydrides and acyl halides as H acceptors, and Section IV discusses amides, including theoretical aspects of the hydrogen bonds, association in solid and in dissolved amides and deals with amides both as proton donors and as acceptors and with intramolecular hydrogen bonds in amides. Finally, Section V describes hydrogen bonds in thio derivatives of acids and amides.

481 references up to 1977.
Complementary: 25, 5.
Relevant: 2, 9; **3**, 6; **4**, 4; **5**, 6; **7**, 6; **8**, 5; **11**, 3; **12**, 3, 10; **13**, 7; **15**, 4; **16**, 5; **17**, 6; **18**, 8; **19**, 7; **20**, 12; **22**, 8; **23**, 6, 7; **24**, 4; **29**, 17; **31**, 17, 21; **32**, 21; **33**, 5.

25, *7: The synthesis of carboxylic acids and esters and their derivatives* by *M. A. Ogliaruso and J. F. Wolfe*

I. Introduction (270); **II**. Synthesis of carboxylic acids (270); **III**. Synthesis of esters (411); **IV**. Synthesis of acid anhydrides (437); **V**. Synthesis of acyl halides (438); **VI**. Synthesis of amides (442); **VII**. Synthesis of imides (458); **VIII**. Acknowledgements (461); **IX**. References.

The chapter attempts to present the most widely applicable and well tested methods of synthesis of the title compounds.

Section II describes syntheses of carboxylic acids by hydrolysis of esters, nitriles, amides, acyl halides and anhydrides, trihalides and dihalides. Next, condensation reactions (Perkin, Doebner, Stobbe, Darzens, Wittig, acetoacetic ester and malonic ester syntheses, Michael and Claisen reactions, coupling and other condensation reactions) in which carbanions are key intermediates are presented. This is followed by syntheses involving free-radical processes, such as radical additions to unsaturated systems, carboxylalkylations ($ArH + \dot{C}H_2COOH \rightarrow ArCH_2COOH$); hydrocarboxylation [by use of CO plus catalysts, e.g. $HC{\equiv}CH + CO + Ni(CO)_4 \xrightarrow{H_2O} H_2C{=}CHCOOH$), carbonation of organometallic reagents (Grignard, RLi, etc.) and electrophilic substitution of arenes by dicarboxylic anhydrides (ArH + 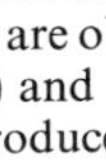 + $AlCl_3 \rightarrow ArCOCH_2CH_2COOH$). Next (Section II.G) a variety of oxidations are presented. These include oxidations of alcohols, (Tables 1–3), aldehydes (Tables 4–7), arenes and alkylarenes (Tables 8–12), double and triple bonds (Tables 13–20), ethers, acetals and ketals (Tables 21–26) and ketones (Tables 27–39). In each case the most useful oxidants are considered (including hypohalites, metal salts, air, O_2, peroxides, peracids, ozone, nitrogen oxides, periodates and photochemical and electrochemical methods).

Reactions in which carboxylic acids are obtained by cleavage of other functional groups include cleavages of ethers (Table 40) and of ketones (Table 41) (Section II.H).

A wide variety of rearrangements produce acids and esters. Among these are the Arndt–Eisert, Wolff (from diazoketones), Beckmann (to obtain the next lower homologue), benzilic acid (diketones to α-hydroxy acids), Claisen (allyl aryl ethers to *o*-allylphenols), Favorskii (α-halo ketones to esters of acids), Henkel (or Raecke process, thermal rearrangement of, e.g., phthalic to terephthalic acid), oxidative ring contraction (cyclic ketones to cyclic acids containing one C atom less in the cycle) and the Pummerer, Stevens and some other reactions. In each case, the scope, mechanism and stereochemistry of the rearrangements are discussed (Section II.H).

Section III treats the preparation of esters by direct esterification of acids, alkylation of carboxylate salts (Table 42), alcoholysis of acyl halides, anhydrides, ketenes, nitriles and amides and transesterification. Next the useful ester-producing condensations (Knoevenagel, Darzens, Wittig, Reformatsky, acetoacetic and malonic ester syntheses, use of α-anions, Michael and other conjugate additions and Claisen condensations) are discussed (Section II.B). Free-radical addition and substitution processes, acyloxylations and anodic dimerizations are dealt with and also some other processes using organoboranes with halo esters, acetylenes in multi-step processes involving HgO-catalysed hydration of an ethoxythynyl group to an ester and introduction of CHCOOR groups by the use of diazo esters.

Section IV considers anhydride syntheses involving dehydrative coupling of carboxylic acids and acylation of carboxylate salts.

Section V treats the preparation of acyl halides from carboxylic acids and anhydrides (by $SOCl_2$, PCl_3, PCl_3, $COCl_2$ and related reagents), from esters and lactones by cleavage with triphenylphosphine dihalides, from trihalides ($ArCX_3 \rightarrow ArCOX$, using SO_2 or SO_3) and by carbonylation of benzylic olefinic and allylic halides in the presence of various Pt, Pd or Rh-containing catalysts.

Amides (Section VI) are obtained by acylation of NH_3 and amines with carboxylic acids, acyl halides, esters, ketenes and isocyanates and also by transamidation ($R^1CONHR^2R^3 + HNR^4R^5 \xrightarrow{H_2O} R'CONR^4R^5 + R^2R^3NH$). Other methods involve hydrolysis of nitriles, cleavage of ketones with $NaNH_2$, various rearrangements (Beckmann, Favorskii, Willgerodt, Chapman, etc.), oxidations, acylaminations (Ritter addition of nitrile to a carbonium ion: $R^1R^2R^3C^+ + R^4C{\equiv}N \rightarrow R^1R^2R^3\overset{+}{N}{\equiv}CR_4 \xrightarrow{H_2O} R^1R^2R^3NHCOR^4$), carboxamidation (direct electrophilic introduction of a $CONH_2$ group by urea or formamide ($ArH + H_2NCONH_2 \xrightarrow{AlCl_3} ArCONH_2$) and condensation reactions of various electrophiles attacking α-carbanions derived from amides. Finally (Section VII), the synthesis of imides by acylations, oxidations and alkylations is described briefly.

1420 references up to 1976.
Complementary: **5**, 3, 4, 5, 7, 11; **15**, 2; **25**, 16.
Relevant: **25**, 19.

25, *8: The chemistry of lactones and lactams* by *G. V. Boyd*

I. Introduction (492); **II**. Lactones (493); **III**. Lactams (516); **IV**. References (526).

Section II describes the general methods of synthesis of lactones (Table 1) and their general physical and chemical properties, followed by treatment separately of α-lactones, β-lactones, saturated and unsaturated γ- and δ-lactones and some higher lactones, dealing in each class with methods of synthesis and properties and with some special cases, such as β,γ- and α,β-butenolides, lactones containing exocyclic double bonds and the macrolide group of antibiotics which also contain a lactone grouping.

Section III discusses the general methods of synthesis and physical and chemical properties of lactams and, considers separately and in more detail α-, β-, γ- and higher lactams.

255 references up to 1977.
Complementary: none.
Relevant: none.

25, *9: The chemistry of orthoamides of carboxylic acids and carbonic acid* by *W. Kantlehner*

I. Introduction (534); **II**. Preparation of orthoamides (535); **III**. Physical characteristics and structure of orthoamides (553); **IV**. Reactions of orthoamides (554); **V**. Acknowledgements (586); **VI**. References (586); **VII**. Addendum (594).

The chapter deals with N-containing derivatives of orthoacids [$RC(OH)_3$]. Section II describes the preparation of the following compounds: amide acetals,

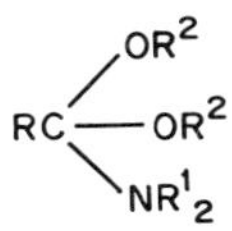
; amide thioacetals,
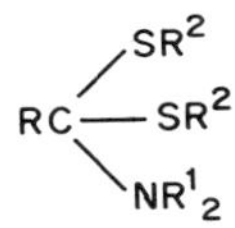
; lactam acetals,
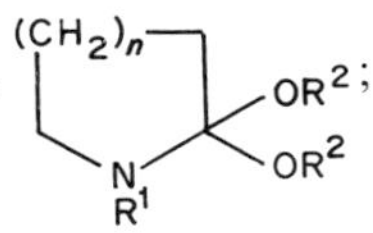
;

α-dialkylamino-α-alkoxynitriles,
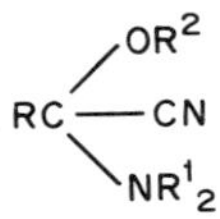
; aminal esters,

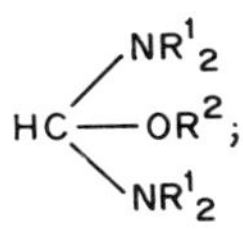
; triaminoalkanes, $RC(NR^1_2)_3$; α, α-bis(dialkylamino)

nitriles,
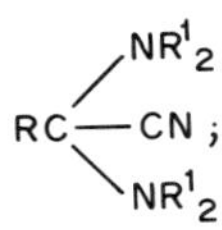
; orthocarbamic acid esters $(EtO)_3CNR_2$; urea acetals, $(R_2N)_2C(OR^1)_2$;

tris(dialkylamino)alkoxymethanes, $(R_2N)_3COR^1$; and tetrakis(dialkylamino)methanes, $(R_2N)_4C$. In each class, several methods of synthesis are presented and the scope, general applicability and, if known, the mechanisms of the various routes are discussed.

Section III treats the physical characteristics and structure of the title compounds.

Section IV considers the reaction of the title compounds. These include nucleophilic attack [by H_2O, ROH, ArOH, Ar(R)COOH, with compounds containing acidic SH, NH_2, NH or PH groups, with organometallic compounds, with carbon acids and by cyclizations and condensations). These reactions may lead to another orthoamide or a product derived from it, e.g.

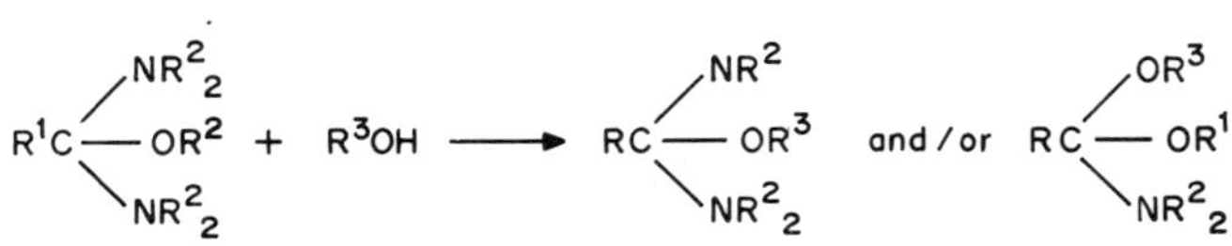

or the orthoamide function may be destroyed:

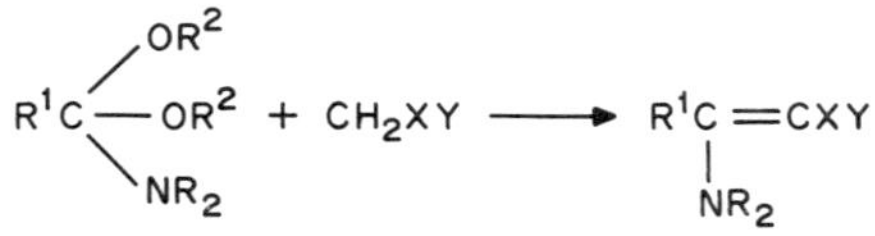

Next, reactions with electrophilic reagents are discussed (acids, aldehydes, alkylating and acylating agents, isocyanates, etc.). Usually in these reactions the orthoamide function is converted to a different functional group, e.g. substituted triaminomethanes with alkylating agents yield formamidinium salts:

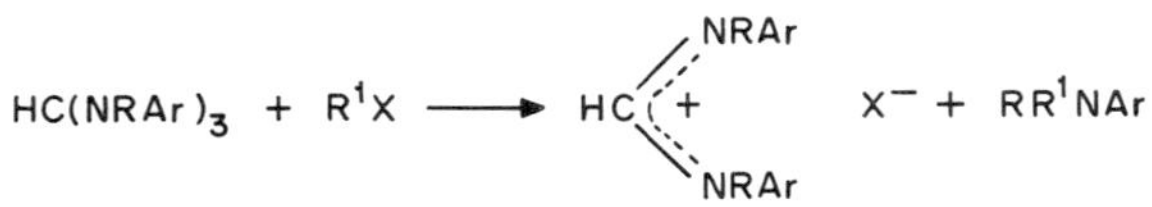

Reactions of α-CH_2 groups of amides and lactam acetals involve alkylation, acylation, condensations and allylic rearrangements. However, these occur due to the equilibrium with ketene *O*, *N*-acetals:

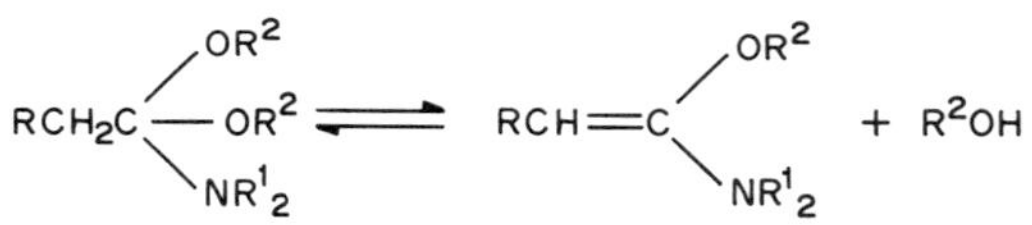

and hence are not orthoamide reactions but reactions of electron-rich olefins.

398 references up to 1977.
Complementary: none.
Relevant: **5**, 13.

25, *10: Detection and determination of acid derivatives* by *W. H. Prichard*

Section I describes analytical methods relating to amides and lactams, including IR and Raman spectra of primary, secondary and tertiary amides and lactams, NMR, ORD and CD, UV and visible and mass spectra, chromatography and polarography and also some titration methods.

Section II deals with IR and Raman spectra of δ-, γ-, β- and α-lactones, with their NMR, ORD and CD, UV and visible and mass spectra and with their chromatography.

Section III discusses the same experimental methods relating to carboxylic anhydrides.

Acids and acyl halides are not reviewed since no significant developments have taken place since the appearance of Chapters **5**, 17 and **15**, 3 respectively.

219 references up to 1976.
Complementary: **5**, 17; **15**, 3.
Relevant: **1**, 5; **2**, 8; **3**, 15; **4**, 3; **8**, 3; **12**, 4; **13**, 6; **14**, 3; **16**, 3; **17**, 4; **18**, 5; **20**, 3; **22**, 5; **23**, 5; **24**, 5; **26**, 5; **28**, 3; **29**, 21; **30**, 14–22; **36**, 5.

25, *11: The photochemistry of organic acids, esters, anhydrides, lactones and imides* by *R. S. Givens and N. Levi*

The chapter reviews only processes in which the title functional groups are directly involved in the bond-making or -breaking processes occurring from the excited reactants.

Section II discusses excitation, excited-state decay and reactivities, the determination of excited state multiplicities (Tables 1 and 2) and excited state reactions.

Section III deals with the photodecarbonylation of esters and lactones (Table 3), anhydrides (Table 4) and lactams and imides, and especially with the mechanisms of these reactions. Section IV describes the photodecarboxylation of acids (Tables 5–7), esters, lactones (Tables 8 and 9) and anhydrides and deals with fluorescence quantum yields for singlet processes of α- and β-naphthyl esters (Table 10). Again, detailed mechanistic schemes are presented.

Section V reviews hydrogen-abstraction processes by the carbonyl group of esters (Tables 11 and 12), which may be either inter- or intramolecular, and the latter process can again be subdivided into type II photofragmentations and photodeconjugation reactions. Section V.B deals with other H-abstraction processes in which carbonyl groups of acids, unsaturated acids, lactones and imides are involved (Table 13).

Section VI discusses photorearrangements of the title compounds, i.e. bond reorganizations which occur without fragmentation and produce isomers. These include migrations of aryl, carboxyl, carbonyl and acyl migrations and also skeletal rearrangements. Finally, photosolvolysis of some of the title compounds is discussed and a tentative mechanism is proposed. The Addendum (Section VIII) presents some more recent material, mainly from 1977–78.

235 references up to 1978.
Complementary: 11, 5; **15**, 9.
Relevant: 2, 16; **3**, 8; **6**, 4; **8**, 12; **9**, 6; **13**, 16; **14**, 8; **16**, 11; **17**, 9; **18**, 10; **19**, 20; **22**, 10; **23**, 9; **24**, 12; **27**, 21, 22; **28**, 5; **29**, 5, 6; **32**, 29; **33**, 21.

25, *12: Radiation chemistry of acids, esters, anhydrides, lactones and lactams* by *O. I. Mićić and O. S. Gal*

Section II describes radiolytic reactions of acids initiated by ionization, followed by other processes involving ions or neutral radicals. In all acids, CO_2 is liberated on irradiation, and often also H_2, H_2O, dimers and other stable products. Radiolysis of acetic, oxalic and oleic acid is described in detail. Rate constants for the reaction of various carboxylic acids with H, e_{aq}^- and OH are presented in Table 1 and optical characteristics and pK values of some radicals formed in the radiolysis of acids in aqueous systems in Table 2, and the nature of adducts with OH, H and e_{aq}^- are discussed.

Section III deals with the mechanism of the decomposition of esters, with the stable radiolytic products, with the oxidation of methyl oleate and with studies is aqueous solutions. Anhydrides (Section IV), lactones (Section V), including especially ascorbic acid (Table 3), and lactams (Section VI) are treated more briefly.

101 references up to 1976.
Complementary: **11**, 6.
Relevant: **9**, 8; **10**, 6; **11**, 6; **13**, 17; **15**, 9; **16**, 12; **17**, 10; **18**, 11; **19**, 10; **22**, 11; **27**, 23, 24; **29**, 7; **31**, 6; **32**, 9; **33**, 21; **36**, 8.

25, *13: The electrochemistry of carboxylic acids and derivatives; cathodic reductions* by *J. P. Coleman*

I. Introduction (782); **II**. Cathodic reduction reactions (782); **III**. Acknowledgements (821); **IV**. References (821).

The chapter describes cathodic reduction of aromatic carboxylic acids to aldehydes (Tables 1 and 2), alcohols (Table 3) and other, e.g., ring-hydrogenated, products. Heterocyclic acids are treated similarly (Table 4), again considering aldehydic, alcoholic and other reduced (hydrogenated) products, and so are, more briefly, aliphatic acids (Section II.A).

Section II.B discusses reductions of esters, lactones and anhydrides, anion-radical intermediates (Table 5), formation of aldehydes, alcohols, ethers and some rearranged products and coupling of radicals to yield hydrocarbons. Next, the electrohydrodimerization (EDH) reaction of unsaturated compounds is treated [$2CH_2{=}CHX \rightarrow X(CH_2)_4X$; X = CN, CO_2R, $CONR_2$, etc.] and products and yields from esters are given (Table 6). Cathodic carboxylation of α,β-unsaturated esters in the presence of CO_2 is considered, together with cleavages in which carboxylate ion is a leaving group or in which it activates the cleavage of other groups (Table 7). Finally (Section II.C), reductions of amides (Tables 8 and 9), imides, and lactams are dealt with.

172 references up to 1976.
Complementary: **5**, 2.
Relevant: **8**, 11; **13**, 5; **16**, 15; **17**, 14; **19**, 12; **20**, 5; **21**, 5; **22**, 9; **23**, 10; **24**, 17; **26**, 12; **27**, 8; **28**, 7; **29**, 8, 9; **31**, 7; **32**, 6; **34**, 1.

25, *14: Decarbonylation reactions of acid halides and aldehydes by chlorotris(triphenylphosphine)rhodium(I)* by *M. C. Baird*

I. Introduction (825); **II**. The preparation, structure and chemistry of $RhCl(PPh_3)_3$ (**1**) (826); **III**. Reactions of **1** with acyl halides (827); **IV**. Reactions of **1** with aldehydes (843); **V**. References (855).

The most useful and most studied decarbonylations of acid halides and aldehydes involve $RhCl(PPh_3)_3$ (**1**), the preparation, structure and chemistry of which are described in Section II.

Section III discusses decarbonylation with **1** of aroyl halides and cyanides ($ArCOX \rightarrow ArX$; X = F, Cl, Br, I, Cn) (Table 1) and also of aliphatic and benzylic acyl halides (Table 2). The mechanism involves an oxidative addition of **1** with the acyl halide, and this intermediate has been studied by NMR (Tables 3 and 4) and its equilibrium constant for the rearrangement reaction $RhCl_2\ (COR)\ (PPh_3)_2 \rightleftharpoons RhRCl_2CO(PPh_3)_2$ has been determined with a variety of R groups (Table 5).

Section IV considers the analogous decarbonylations of aldehydes $RCHO \xrightarrow{1} RH$ (Table 6) and allylic alcohols (probably after isomerization by **1** to the corresponding aldehyde tautomer $RCH{=}CHCH_2OH \xrightarrow{1} [RCH_2CH{=}CHOH] \rightarrow RCH_2CH_2CHO$) (Table 7). Olefinic aldehydes are usually not decarbonylated by **1**, but cyclized to the corresponding ketone or to similar compounds (Table 8). Intermediates and mechanisms of aldehyde reactions with **1** are discussed.

97 references up to 1976.
Complementary: **2**, 14.
Relevant: none.

25, *15: Pyrolysis of acids and their derivatives* by *R. Taylor*

I. Introduction (860); **II**. Pyrolysis of carboxylic acids (860); **III**. Pyrolysis of acid halides (867); **IV**. Pyrolysis of acid amides (867); **V**. Pyrolysis of acid anhydrides (871); **VI**. Pyrolysis of esters (874); **VII**. Pyrolysis of lactones (908); **VIII**. Pyrolysis of lactams (909); **IX**. References (909).

The chapter attempts to correlate the mechanisms of the pyrolyses of the title compounds, considering cyclic processes which are non-synchronous (i.e. in the transition state some polarization is produced within the cyclic structure, which is often six-membered).

Section II deals with the pyrolysis of unsaturated acids, yielding CO_2 and an alkene (Tables 1 and 2), and of saturated acids, in which the route is either decarboxylation or dehydration (Table 3). Next, acid halides (Section III) and acid amides (Section IV) are considered, including primary amides and higher amides lacking β-H in the *N*-alkyl group, which produce nitriles, ketenes, isocyanates, etc., and amides with *N*-alkyl groups containing β-H atoms, which are able (like esters) to eliminate alkenes and yield a new amide $RCON(R')CH_2R'' \rightarrow RCONHR' + CH_2{=}CHR''$.

Acid anhydrides (Section V) usually yield a carboxylic acid and a ketene, although in special cases different products are obtained. Anhydrides which lack β-H atoms or are α,β-unsaturated or are cyclic behave differently and may eliminate CO and CO_2 and a variety of other products such as aldehydes, acids and even benzyne (from the pyrolysis of phthalic anhydride).

Section VI deals in detail with the scope, mechanism and the products of the pyrolysis of esters, which behave differently in cases when they do not have β-H atoms in the alkyl group, or when they are vinylic or allylic. In the latter case, the major product is often a conjugated diene (Table 4).

If the ester contains non-vinylic β-H atoms (Section VI.D), usually an alkene and a carboxylic acid are formed by a cyclic concerted β-elimination. The mechanism is discussed in detail, and product distributions of aliphatic acetates (Table 5) and of cycloalkyl acetates (Table 6) are given. Statistical effects, thermodynamic stability of the products and electronic effects are discussed (Tables 7–12), and also O and H isotope effects, neighbouring group effects and rearrangements.

The ester elimination as studied in the pyrolysis of 1-arylethyl acetates can be used as a model for the reactivities of various homoaromatic and heteroaromatic compounds in the course of electrophilic substitutions (Table 13). The most significant implications of this method are discussed briefly.

Finally, pyrolyses of lactones (Section VII) and of lactams (Section VIII) are described.

243 references up to 1977.
Complementary: none.
Relevant: **1**, 3; **3**, 4; **5**, 12; **16**, 10; **22**, 15; **27**, 11; **32**, 7; **33**, 20.

25, *16: Transcarboxylation reactions of salts of aromatic carboxylic acids* by *J. Ratuský*

I. Introduction (915); **II**. The intermolecular ionic mechanism of transcarboxylation reactions (915); **III**. Mixed transcarboxylations of salts of aromatic carboxylic acids (940); **IV**. Conclusion (942); **V**. References (942).

The title reaction has been proved to be intermolecular; it occurs by an ionic mechanism and involves various benzenecarboxylates as intermediates:

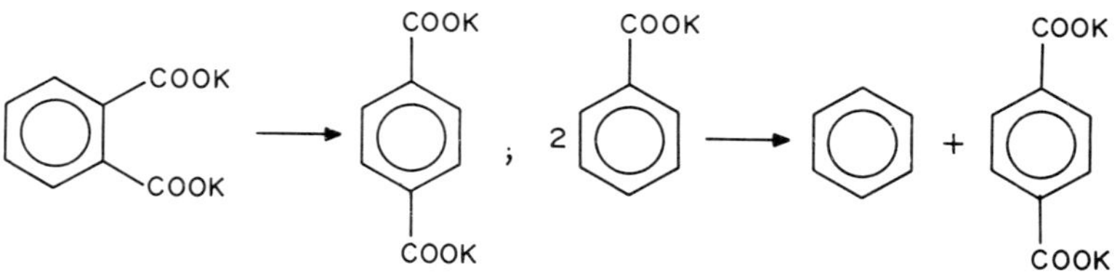

Section II discusses the course, mechanism and time dependence of the product composition in the transcarboxylation reactions of the potassium salts of benzoic acid, benzenedicarboxylates and benzenepolycarboxylates. The reactions were also studied in fused inert salts such as KOCN or KSCN. In an atmosphere of $^{14}CO_2$, incorporation of ^{14}C into the carboxylate groups was observed. Also H and D atoms in mixtures of benzencarboxylates were exchanged faster than the transcarboxylation itself. The effect of catalysts and of various cations on the reaction is considered. Next, reactions of COOH derivatives of naphthalene, biphenyl, heteroarenes, etc., are described (Table 1) and finally (Section III) mixed transcarboxylations of the salts of two different aromatic acids.

74 references up to 1975.
Complementary: none.
Relevant: none.

25, *17: Micellar effects upon deacylation* by *C. A. Bunton and L. S. Romsted*

I. Symbols used in text (945); **II**. Introduction (947); **III**. Carboxylic esters as probes for micellar catalysed reactions (948); **IV**. Micellar structure (951); **V**. The simple kinetic model (957); **VI**. Kinetic models for micellar catalysed reactions (960); **VII**. Miscellaneous (984); **VIII**. Experimental problems (986); **IX**. Preparative aspects (988); **X**. Deacylations in non-functional micelles (989); **XI**. Functional micelles and comicelles (995); **XII**. The question of bifunctional catalysis (1011); **XIII**. Conclusions, connections and conjectures (1013); **XIV**. Acknowledgements (1015); **XV**. References.

Micelles are submicroscopic aggregates and may catalyse or inhibit reactions on their surface. Current models to explain this and the explanatory and predictive power of models are considered in this chapter.

Section III describes micellar effects on rates of ester saponification for various types of surfactants and substrates (Table 1) and justifies the use of the deacylation of carboxylic esters as a probe for micellar catalysis. Section IV discusses the structural features of micelle and of micellar solutions, of various types of surfactants (anionic, cationic, zwitterionic, non-ionic, etc.) (Table 2) and the distribution of hydrophilic ions in micellar solutions.

Section V considers micellar catalysis and inhibition (especially in enzyme-like models) and the effect of micelles on unimolecular and bimolecular reactions. Section VI presents the quantitative treatment of micellar catalysed bimolecular reactions (Table 3), of second-order reactions between hydrophobic substrates (Tables 4 and 5) and of second-order reactions between hydrophilic ions and organic substrates, and also considers the as yet unsolved problem of the effect of buffers on reaction rates. Next, theory and experiment are compared qualitatively and various kinetic problems involving third-order reactions, micellar inhibition of bimolecular reactions, the effect of added electrolytes (Table 6), the effect of changing the concentration of the hydrophilic nucleophile and other problems for future research are considered.

Section VII deals with a variety of questions, involving non-ionic micelles, the effects of non-ionic additives, zwitterionic surfactants, self-aggregating systems and 'reversed' micelles (in which the apolar groups are at the exterior of the micelle and the ionic or polar head groups in the interior, which also contains water while the solvents are aprotic and non-polar such as benzene or hexane.

Section VIII treats various experimental problems and Section IX the preparative aspects of work with micelles.

Section X describes deacylations of esters, amides and anhydrides in non-functional micelles such as hexadecyltrimethylammonium bromide, and Section XI deals with functional micelles in which a nucleophile or basic group is covalently bound to the surfactant. Various factors are discussed in micelles which contain thiol, amine, imidazole, oxime, hydroxamate or hydroxyalkyl groups in the surfactants and which are powerful catalysts for deacylation and other reactions.

Finally, bifunctional catalysis at micellar surfaces is considered (Section XII) and some conclusions, connections and conjectures are presented on the subject of the chapter (Section XIII).

203 references up to 1977.
Complementary: 5, 15; **11**, 8; **15**, 11.
Relevant: none.

25, *18: The chemistry of thio acid derivatives* by *J. Voss*

I. Structural and physical properties (1022); **II.** Syntheses (1036); **III.** Chemical properties of thiocarboxylic acid derivatives (1044); **IV.** References (1053).

Section I describes structure and physical properties of the title compounds, including bond lengths and angles (Table 1), dipole moments (Tables 2 and 3), UV spectra of monothio- and dithiocarboxylic acid derivatives (Tables 4–6), IR bands (Tables 7–10), NMR (Tables 11 and 12), ESR (Table 13) and MS.

Section II discusses the syntheses of thio- and dithiocarboxylic acids and their salts, thiolo esters, thiono esters, dithio esters and thioacyl halides and anhydrides.

Section III considers prototropic behaviour and presents some pK_a values (Table 14) and discusses electrochemical reductions of thio esters (Table 15), reactions with carbanions and solvolysis. Next, oxidation, alkylation and acylation of some of the title compounds are treated, followed by some non-polar (photochemical) reactions and processes that form S-containing heterocycles.

415 references up to 1978.
Complementary: 5, 15; **11**, 8; **15**, 11.
Relevant: 2, 17; **3**, 13.

25, *19: The synthesis of lactones and lactams* by *J. F. Wolfe and M. A. Ogliaruso*

I. Introduction (1064); **II**. Synthesis of lactones (1065); **III**. Synthesis of lactams (1194); **IV**. Acknowledgements (1315); **V**. References (1315).

Section II describes the preparation of lactones by a wide variety of methods. These include intramolecular cyclization of hydroxy acids and related compounds (Tables 1–3) and of unsaturated acids and esters (Tables 4–7), including acid-catalysed, photochemical, electrochemical and Diels–Alder methods, condensations including acetoacetic and cyanoacetic esters, the von Pechman reaction (Table 8), malonic acid and ester (Table 9), Perkin, Stobbe, Grignard (Table 10), Reformatsky (Table 11) and Wittig reactions. Next, routes involving α-anions (or dianions) or carboxylic acids (Tables 12 and 13), lithium salts of 2-alkyl-2-oxazolines (Table 14) and direct functionalization of pre-formed lactones [e.g. alkylations (Table 15) and Reformatsky and Wittig reactions (Table 16)] are considered. Lactones are also obtained by reaction of ketenes with carbonyl compounds (Table 17) or with diazoketones (Table 18), by reduction of acid anhydrides (Table 19) or of acids and esters containing ketonic or aldehydic groups (Table 20), by oxidation of diols (Table 21) of ketones (Table 22), of ethers and of olefins (Tables 23 and 24) and also by carbonylation of unsaturated esters with CO, H_2 and $Co_2(CO)_8$ (Table 25) and similarly of unsaturated alcohols (Table 26), and by the silver-induced addition of nitrones to olefins.

Lactone-producing rearrangements include the Claisen method (Table 27), carbonium ion and photochemical rearrangements and lactone interconversions. Other routes which do not fall into the preceding categories but nevertheless have general importance are the Barton reaction, and the photolysis of α-diazo esters and amides, of 2-alkoxyethanes and of 1, 2-dioxolane-3, 5-diones. Some less often used routes are also mentioned (Section II.S, 5–9).

Section III discusses the preparation of lactams first by ring closures of amino acids and related compounds (Table 18), from various substituted amides (e.g. by base-catalysed intramolecular alkylation of α- and β-haloamides) (Table 29) and by intramolecular photocyclizations of amides, unsaturated amides and their derivatives (Table 30).

Cycloadditions are considered next, including addition of isocyanates to olefins (Table 31), reactions of imines with ketenes (Table 32) or with Reformatsky reagents (Table 34) and also reactions of ketenes with azo compounds (Table 33) and of ketenes with nitrones and nitroso compounds.

This is followed (Section III.D) by the treatment of various rearrangement routes, such as ring contractions and ring expansions, namely the Beckmann rearrangement (Table 35), the Schmidt rearrangement (Table 36) and Claisen-type rearrangements (Table 37). Direct functionalization of a pre-formed lactam (e.g. by electrophilic substitution on a carbanion centre adjacent to the lactam carbonyl) provides new lactams.

Section III.F deals with the preparation of lactams by oxidative routes, using halogens, CrO_3, OsO_4, MnO_2, $KMnO_4$, PtO_2 and RuO_2 in the presence of $NaIO_4$, by sensitized and unsensitized photooxidation and by some less important oxidants such as $K_3Fe(CN)_6$, *m*-$ClC_6H_4CO_3H$, Me_3CCO_3H and also Hg^{II} salts.

A few specialized synthetic procedures are also mentioned (Section III.G).

723 references up to 1977.
Complementary: 25, 8.
Relevant: 25, 7.

The chemistry of ketenes, allenes and related compounds (1980)

26, *1: Theoretical methods and their application to ketenes and allenes* by *C. E. Dykstra and H. F. Schaefer, III*

I. Introduction (2); **II**. The Hartree–Fock method and its reliability (3); **III**. The problem of electron correlation (10); **IV**. The electronic structure of ketene (18); **V**. The electronic structure of allene (34); **VI**. References (40).

The aim of the chapter is to review developments in *ab inito* molecular electronic structure theory since 1962, when this was summarized by Coulson and Stewart in the first chapter of the first volume of this series (**1**, 1).

Section II deals with the Hartree–Fock method, defines the expressions used and discusses basis sets for the carbon atom (Tables 1 and 2) and gives theoretical predictions for the structures of methylenecyclopropane (Table 3), for NH_2F and PH_2F (Table 4) and for HCNO (Table 5), and compares in each case theory and experiment. Energy predictions are also discussed.

Section III deals with the concept of electron correlation, correlation energy, configuration interaction and structural predictions for HOCN (Table 6), HCN (Table 7) and the bond distances in first- and second-row diatomic hydrides (Table 8), and also with energy predictions for simple carbenes (Table 9).

Section IV describes the ground-state structure of ketene (Table 10), ground-state orbital energies and Mulliken populations for ketene (Table 11), ketene excited states (Tables 12–14) and considers the oxirene HC=CH (bridged by O) and ethynol HC≡COH isomers of ketene (Tables 15 and 16) and the analogous isomers of thioketene (Table 17).

Section V discusses the electronic structure of allene in the ground state (Table 18), allene excitation energies (Table 19), allene SCF orbital energies at optimum geometries (Table 20), the predicted structures of planar allene states (Table 21) and their energies (Table 22) and finally the energies of the C_3H_4 isomers, namely allene, propyne and cyclopropene (Table 23).

149 references up to 1977.
Complementary: 1, 1.
Relevant: 2, 1; **3**, 1; **4**, 1; **6**, 1; **8**, 1; **11**, 1; **12**, 1; **13**, 1; **14**, 1; **15**, 1; **16**, 1; **17**, 1; **18**, 1; **19**, 1; **20**, 1; **22**, 1; **23**, 1; **24**, 1; **25**, 1; **28**, 1; **31**, 27; **32**, 1; **33**, 1.

26, *2: Structural chemistry* by *W. Runge*

I. Introduction (45); **II**. Bond lengths and bond angles in cumulenes (48); **III**. Isomerism of cumulenes and barriers to isomerization (75); **IV**. Acknowledgement (93); **V**. References (93).

Section I reviews general considerations and techniques for the determination of structural features and the structural implications of the valence bond theory of cumulenes.

Section II discusses differences between structures determined experimentally by different methods and warns against the interpretation of small structural variations. Bond lengths and angles are given for allenes (Table 1), ketenes, ketene imines, carbodiimides and thioketenes (Table 2), higher cumulenes (Table 3) and for some special systems containing P=C, Cr=C and Ir=C bonds. Bond lengths and angles obtained from theoretical calculations are presented (Table 5).

Section III deals with isomerism in cumulenes and with barriers separating stereoisomers, tabulates calculated rotational barriers for cumulenes (Table 6), experimental characteristics for rotational barriers (Table 7), theoretical barriers for stereoisomerization in diazacumulenes ($RN{=}(C)_n{=}NR$) (Table 8) and also experimental barriers for the stereoisomerization of ketene imines and of carbodiimides (Table 9). Next, magnetic nonequivalences in allenes and pentataetrenes of ^{1}H and ^{13}C nuclei are discussed, followed by treatment of isomerism resulting from rotational around single bonds including dihedral angles, relative rotamer populations, molecular conformations and barriers to internal rotations around single bonds (Table 11). Conformational preferences with various substituents are considered and finally calculated and experimental values for five-bond proton–proton NMR coupling constants are presented (Table 12).

176 references up to 1978.
Complementary: none.
Relevant: 16, 2; **17**, 2; **18**, 2; **19**, 2, 21; **20**, 2; **21**, 1; **22**, 2, 3; **23**, 3; **24**, 2; **27**, 4, 5; **28**, 2, 10; **29**, 1, 2; **30**, 1; **31**, 19, 24; **32**, 24; **33**, 12; **34**, 10; **36**, 3, 7.

26, *3: Chirality and chiroptical properties* by *W. Runge*

I. Introduction (99); **II**. Theoretical (101); **III**. Elements of chirality and optical activity (104); **IV**. Semiempirical calculation of optical rotations of acyclic allenes (119); **V**. Spectroscopic and stereochemical aspects of circular dichroism of allenes (131); **VI**. Acknowledgement (152); **VII**. References (152).

Theoretical aspects of chiroptical properties are reviewed in Section II. Section III discusses chirality and chirality functions in general, the 'axial chirality' of allenes, various skeletal symmetries and symmetry operations, chirality functions for D_{2d} molecules (which include allenes), elements which induce optical activity in cumulenes (due to a chiral environment) (Table 2), and finally considers chira! cumulenes other than allenes (e.g. ketenes and azacumulenes).

Section IV describes semiempirical calculations of molar rotations of acyclic allenes and compares experimental and calculated rotations (Tables 3–5), tabulates some calculated values relating to various hetero substituents (Table 6) and considers conformational chirality effects and finally deals with the ORD of allenes.

Section V treats excited states in allenes and in alkylallenes and presents CD data for them (Table 7), in addition to calculated energies and reduced rotatory strengths (Tables 8 and 9). Allenic carboxylic acids and ketones are considered next, and UV and CD data are given (Tables 10 and 11). Similar data for phenylallene carboxylic acids are given in Table 12, and Table 13 presents experimental and calculated rotatory strengths for (*S*)-phenylallene carboxylic acid. UV and CD characteristics of (*S*)-1, 3-diphenylallene are given in Table 14.

80 references up to 1979.
Complementary: none.
Relevant: **8**, 4; **18**, 7; **19**, 6; **20**, 2; **22**, 4; **25**, 3; **27**, 6; **28**, 10; **29**, 23; **31**, 1; **36**, 17.

26, *4: The thermodynamics of allenes, ketenes and related compounds* by *R. L. Deming and C. A. Wulff*

I. Introduction (155); **II**. Allenes (157); **III**. Ketenes (162); **IV**. References (164).

Section I reviews the thermochemical data which became available since the publication of chapters on cumulenes and ketenes in the first volume of this series (**1**, 13, 14) and tabulates the enthalpies of hydrogenation of multiply unsaturated compounds (Table 1).

Section II presents thermochemical data for allenes and gives the six types of carbon centres =C= from which the enthalpies of allenes can be constructed (Table 2), uses these values for the calculation of ΔH_f^0 and compares the calculated and observed values (Table 3) and presents also ΔH_f^0 and rotational barriers (V_0) from MINDO/2 calculations (Table 4). The available experimental data for the thermodynamic properties of allenes are summarized in Table 5.

Section III discusses similarly ketenes and again summarizes the thermodynamic properties (ΔH_f^0, S^0 and ΔG_f^0) in Table 6.

27 references up to 1977.
Complementary: none.
Relevant: **16**, 16; **18**, 3; **19**, 3; **20**, 11; **21**, 3; **22**, 6; **23**, 4; **24**, 3; **25**, 2; **27**, 9; **28**, 4; **29**, 4; **30**, 2; **31**, 2; **33**, 3; **36**, 4.

26, *5: Detection, determination and identification of allenes and ketenes* by *J. W. Munson*

I. Introduction (165); **II**. Chemical methods (166); **III**. Physical methods (169); **IV**. References (186).

Most of the chapter is devoted to the two parent molecules and their simple derivatives. MS (**26**, 6) and chiroptical properties (**26**, 3) are treated in other chapters in this volume.

Section II describes the chemical methods available for analysis of allenes (Br_2 addition, oxidation, etc.) and of ketenes (hydrolysis to acids, formation of amides, hydroxamic acids, etc.).

Section III discusses UV–visible spectroscopy of allenes (Tables 1–5) and ketenes, IR spectroscopy of allenes (Tables 6–9) and of ketenes (Tables 10 and 11), ^{1}H NMR (Tables 13–15) and ^{13}C NMR data (Tables 16–19) for allenes and for ketenes (Table 20), including their Ge, Si and Sn derivatives (Table 21). Next, gas chromatography is treated and retention indices (Table 22) and retention times (Table 23) are presented for allenes; finally, gas chromatographic methods applied to ketenes are considered.

84 references up to 1975.
Complementary: **26**, 3, 6.
Relevant: **1**, 5; **2**, 8; **3**, 15; **4**, 3; **5**, 17; **8**, 3; **12**, 4; **13**, 6; **14**, 3; **15**, 3; **16**, 3; **17**, 4; **18**, 5; **20**, 3; **22**, 22; **23**, 5; **24**, 5; **25**, 10; **28**, 3; **29**, 21; **30**, 14–22; **36**, 5.

26, *6: The generation of neutral and ionized allenes, cumulenes and heterocumulenes by electron impact* by *H. Schwarz and C. Köppel*

The chapter deals with the MS and fragmentation pathways and mechanism of the title compounds.

Section II discusses electron impact induced reactions of allenes and cumulenes (and of the relevant acetylene compounds) and heterocumulenes containing one or more heteroatoms (ketenes, ketene imines, isocyanates, isothiocyanates). Section III considers the formation of neutrals with cumulated double bonds during the fragmentation of the title compounds.

Section IV deals with the formation of ionized title compounds and treats the formation of ketene-like structures from 1,2-disubstituted arenes by interaction of the vicinal substituents ('*ortho* effect'). Next, cycloreversions of four-, five- and six-membered ring systems (e.g. diketene to ionized ketenes) are presented, together with some complex multi-step rearrangements which have been analysed in detail. Detailed mechanistic schemes are given for all processes discussed.

122 references up to 1979.
Complementary: 22, 7.
Relevant: 9, 7; **10**, 5; **13**, 19; **16**, 4; **17**, 5; **18**, 6; **19**, 15; **22**, 7; **24**, 6; **25**, 4; **27**, 7; **29**, 3; **30**, 22; **31**, 3; **32**, 3; **33**, 4; **36**, 7.

26, *7: The preparation of ketenes* by *R. S. Ward*

The chapter deals with methods through which ketenes can be prepared either as stable compounds or as intermediates to be used *in situ* and also as transient reactive intermediates, if the formation of these is sufficiently documented.

Section II describes eight methods and procedures (Fig. 1). These are the Wolff rearrangement of α-diazocarbonyl compounds, the dehalogenation of 2-haloacyl halides ($RCHBrCOBr \xrightarrow{Zn} RCH{=}C{=}O$), the dehydrohalogenation of acyl halides ($Me_2CHCOCl \xrightarrow{Et_3N} Me_2C{=}C{=}O$), thermal decomposition of ketones, carboxylic acids or anhydrides, decomposition of ketone or ketene acylals, decomposition of ketene dimers, thermal cleavage of alkoxyalkynes and decomposition of 2,5-diazidoquinones.

Section III considers separately the preparation of various types of ketenes and thioketenes. These include alkyl-, aryl-, halo-, alkoxy-, thioalkyl-, amino- silyl-, germyl-, stannyl-, acyl-, thioacyl-, imino-, cyano-, azido-, alkoxycarbonyl- and vinylketenes. Finally, the synthesis of bisketenes, phosphoranylideneketenes and thioketenes are considered.

235 references up to 1977.
Complementary: 1, 14; **26**, 14, 16, 17, 19.
Relevant: none.

26, *8: Synthetic uses of ketenes and allenes* by *W. T. Brady*

I. Introduction (279); **II**. Synthetic uses of ketenes (280); **III**. Synthetic uses of allenes (298); **IV**. References (304).

Section II deals with the uses of ketenes for the preparation of a wide variety of compounds by different routes. Thus, cycloaddition of ketenes with olefins provides cyclobutanones; adducts of dichloroketene with cyclopentadiene yield tropolone by hydrolysis:

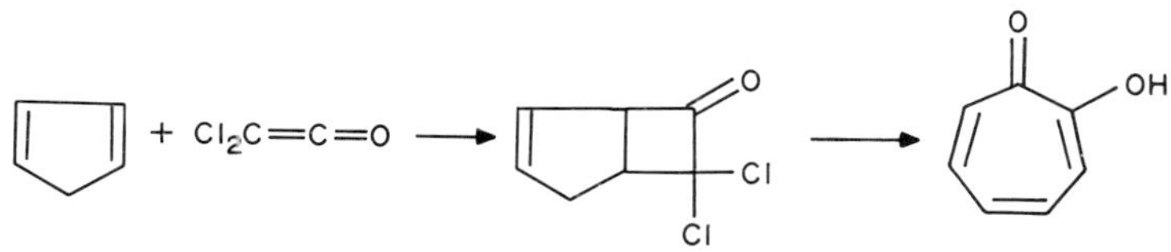

and other tropolones and tropones can be prepared by related routes; halocyclobutanones prepared from haloketenes and olefins yield cyclopropanes by ring contraction; ketenes add to allenes to yield alkylidenecyclobutanones and to alkoxyalkynes to yield cyclobutenones, while their cycloaddition to methylenecycloalkanes yields spiroketones. Dialkylketenes dimerize to 1,3-cyclobutanediones:

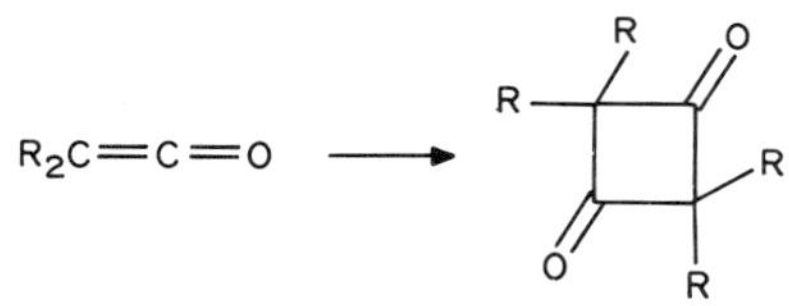

and the cycloaddition of ketenes with carbonyl compounds yields 2-oxetanones. Other routes lead to the formation of 2-azetidinones and other heterocyclic compounds and 2-oxetanones which can easily be prepared from ketenes yield allenes by decarboxylation:

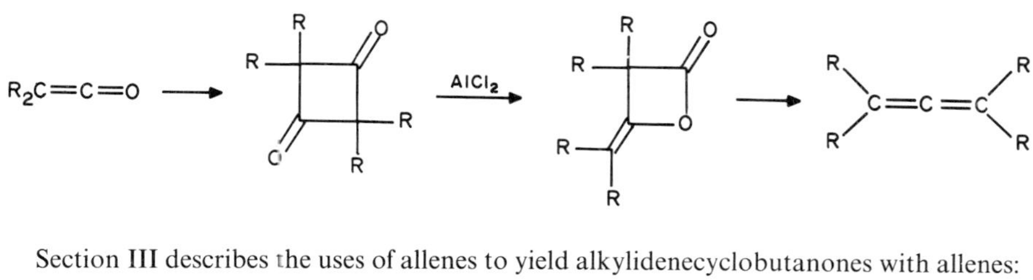

Section III describes the uses of allenes to yield alkylidenecyclobutanones with allenes:

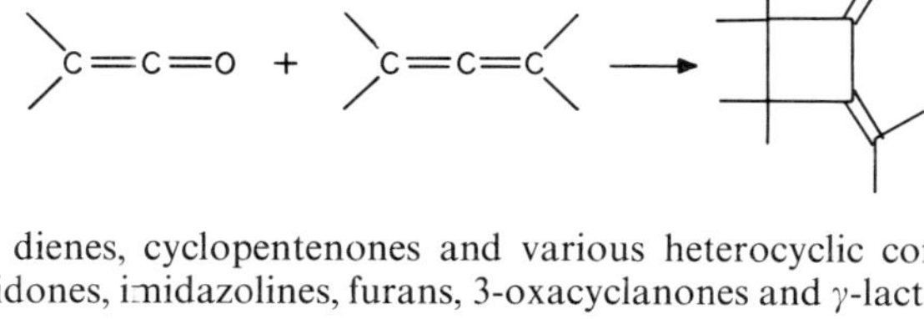

and conjugated dienes, cyclopentenones and various heterocyclic compounds (2-pyridones, 3-pyrrolidones, imidazolines, furans, 3-oxacyclanones and γ-lactones) by different addition, cycloaddition and oxidation reactions. Finally, some less general synthetic methods involving ketenes and organometallic derivatives are mentioned.

202 references up to 1977.
Complementary: **1**, 13, 14; **26**, 10, 14, 16, 17, 19.
Relevant: none.

26, *9: Kinetics and mechanisms (excepting cycloadditions)* by *P. Blake*

The main reactive centre of ketenes is the electrophilic carbonyl carbon on which a positive charge appears in two of the canonical forms $R_2C{=}C{=}O \leftrightarrow R_2C{=}\bar{C}{-}\bar{O}$ $\leftrightarrow R_2\bar{C}{-}\overset{+}{C}{=}O$. The most frequent reactions of ketenes are reviewed generally and briefly (Section I).

Section II deals with the kinetics and mechanism of the hydrolysis of ketenes to carboxylic acids, and with the effect of *p*-aryl substituents in phenylketenes (Table 1). In the uncatalysed addition of alcohols to ketenes the kinetics of the reaction, the effect of the ketene structure (Tables 2 and 3) and solvent effects are discussed. Catalysis of ROH addition by acids (Table 4), amines (Table 5) and amides (Table 6) is considered. Next, reactions with thiols are discussed, followed by addition of carboxylic acids in solution and in the gas phase (Table 7). Reactions with other acids, with amines (Table 8) and with H atoms are treated, in addition to some other additions [e.g. acid halides ($R^1_2C{=}C{=}O + R^2COX \rightarrow R^2COCR^1_2COX$), anhydrides, BH_3, R_3SiPEt_2).

Section III is concerned with the thermal decomposition of ketenes and Section IV with their oxidation by free radical routes.

Section V discusses the polymerization of ketenes to yield cyclic dimers and the effect of solvents, structures and catalysts (Tables 9–11), trimers and higher polymers and also the reactions of ketene dimers [including hydrolysis (Table 12), pyrolysis, reactions with Grignard reagents and other organometallic compounds].

Section VI deals with hydrolysis, rearrangements and polymerization of ketene acetals and thioacetals, with carbodiimides (Table 13) and with ketene imines.

Allenes are introduced in Section VII and Section VIII deals with their isomerization (especially to alkynes) and thermal decomposition. Section IX describes free-radical additions of H atoms, alkyl radicals (Table 14), halogens and O- and S-containing radicals to allenes.

Section X treats electrophilic additions of halogens (Table 15), sulphenyl halides, oxymercury compounds and boron compounds (Table 16) to allenes.

Section XI discusses gas-phase and solution oxidation of allenes and Section XII mentions briefly some other reactions (transition metal catalysed hydrogenation, etc.).

203 references up to 1977.
Complementary: **1**, 13, 14.
Relevant: none.

26, *10: Organometallic derivatives of allenes and ketenes* by *J.-L. Moreau*

The chapter is limited to organometallics in which the metal is directly attached to the allenic or ketenic C atom.

Section II describes the preparation of compounds containing Mg, Zn and Al and their stability, spectra (Tables 1 and 2), NMR parameters (Table 3), and reactions with H_2O, amines, alkylating and halogenating agents and with carbonyl compounds, especially in Reformatsky-type reactions (Tables 4–6) to produce allenic alcohols, including the mechanism, stereochemistry (Tables 7 and 8) and reversibility of these processes. Next, reactions with α-unsaturated ketones and esters, Schiff's bases, esters, amides, orthoesters, CO_2 and some other synthones are considered.

Addition of RLi to enynes produces allenic lithium compounds ($R^1C{\equiv}CCH{=}CH_2 + R^2Li \rightarrow R^1CLi{=}C{=}CHCH_2R^2$). In many reactions of the latter, the allenic structure is preserved; their hydrolysis gives allenic hydrocarbons; carbonyl compounds yield allenic alcohols (Section III). RLi compounds yield metalated derivatives with allenes and acetylenes ($H_2C{=}C{=}CH_2 + RLi \rightarrow H_2C{=}C{=}CHLi$) and in the presence of excess of RLi polymetalated allenes (and acetylenes) are formed (Section II.B). Acetylenic and allenic ethers, thio esters and silylated ethers are very easily metalated and may exchange one, two or even three hydrogens with Li. The corresponding mono- and dianions have preparative importance and some of them can be regarded as 'masked' acyl anions, e.g. $H_2C{=}C{=}C\begin{matrix}\nearrow OCH_3 \\ \searrow Li\end{matrix}$ is equivalent to the acyl anion of acrolein, $CH_2{=}CH{-}\bar{C}{=}O$.

Section IV deals with the preparation and reactions of allenic and propargylic boron derivatives and Section V treats similarly derivatives of Group IVb elements (Si, Ge, Sn, Pb).

Section VI considers organometallic derivatives of ketenes. The preparation of compounds of the type $R_3MCH{=}C{=}O$ (where M may be any of the Group IVb elements) is described; their structure is stated to be ketenic with no equilibrium formation of the oxyacetylene tautomer, i.e. $R_3NCH{=}C{=}O \not\leftrightarrow HC{\equiv}COMR$. Some addition reactions with H_2O, alcohols, amines, etc., leading to acids, esters and amides, respectively, are mentioned, and also reactions with Br_2, RLi and Ph_2PH, cycloadditions, etc. Finally, metal ketenides, $M_2C{=}C{=}O$ (where M = Ag, Cu, Au), which can act as powerful oxidation catalysts, are mentioned.

368 references up to 1978.
Complementary: 24, 9.
Relevant: 30, 7.

26, *11: Preparation and uses of isotopically labelled allenes* by *M. D. Schiavelli*

Section II describes the use of isotopically labelled allenes for the elucidation of the mechanism of [2 + 2], [2 + 3] and [2 + 4] cycloadditions, including secondary deuterium isotope effects (Table 1). Section III deals with the solvolysis of haloallenes and the utilization of isotope effects (H/D, Table 2) in their study. H/D secondary isotope effects were also used in the study of addition reactions (HCl, HBr, ArSCl, $HgSO_4$/AcOH) to allenes (Table 3) and in the investigation of propargyl–allenyl rearrangements (Section V) and of base-catalysed isomerization of acetylenes (Section VI) employing primary H/D isotope effects. Section VII deals with some reactions which are also convenient for the preparation of allenes and isotopically labelled allenes (e.g. $Me_2CClC{\equiv}CH \xrightarrow{LiAlD_4} Me_2C{=}C{=}CHD$), and also deuterated allenic ketones.

51 references up to 1977.
Complementary: 24, 20.
Relevant: 5, 10; **12**, 13; **13**, 15; **17**, 12; **18**, 18; **19**, 17; **22**, 14; **23**, 16; **24**, 20; **27**, 10; **28**, 8; **29**, 28; **33**, 18; **36**, 12.

26, *12: Electrochemistry of allenes and cumulenes* by *D. G. Peters, W. F. Carroll, Jr, D. M. La Perrière and B. C. Willett*

I. Introduction (431); **II**. Allenes (432); **III**. Cumulenes (447); **IV**. References (450).

Section II describes the electrochemical reduction of allenes. Allene itself is not reducible, except under destructive conditions, when it is nearly quantitatively converted into CO_2($H_2C{=}C{=}CH_2 + 6H_2O \rightarrow 3CO_2 + 16H^+ + 16e^-$). Reduction is facile in the presence of one or more phenyl groups or other activating substituents. Thus tetraphenylallene reacts ($Ph_2C{=}C{=}CPh_2 \rightarrow \rightarrow Ph_2CHCH_2CHPh_2$) and 1,1-diphenylallene yields 1,1-diphenylpropane (via 1,1-diphenylprop-1-ene). Allenes substituted on C-1 by Ph and on C-3 by halogen yield either organomercurials (on an Hg cathode) or the corresponding allenes in which the halogen is exchanged for a hydrogen atom. These arylallenes may then be further reduced in turn, as stated above. Next, electrochemical studies relating to allenic ethers, ketones and sulphones are discussed, and also the electrolytically induced isomerization of alkynes to allenes. Section III treats the electrochemistry of cumulenes and their stepwise reduction until the fully reduced corresponding hydrocarbon is obtained.

21 references up to 1977.
Complementary: 24, 17.
Relevant: 5, 2; **8**, 11; **13**, 5; **16**, 15; **17**, 14; **19**, 12; **20**, 5; **21**, 5; **22**, 9; **23**, 10; **25**, 13; **27**, 8; **28**, 7; **29**, 8, 9; **31**, 6; **32**, 6; **34**, 1.

26, *13: Biological formation and reactions* by *C. H. Robinson and D. F. Dovey*

I. Introduction (452); **II**. Allenic natural products (452); **III**. Allenes, ketenes and ketene imines as enzyme-generated inactivators (462); **IV**. Allenes which affect insect steroid transformations (481); **V**. Acknowledgements (482); **VI**. References (482).

The chapter deals with allenes, ketenes and ketene imines. Section II discusses fungal allenes and fungal allenic polyacetylenes (Table 1) and their biosynthetic pathways (Scheme 1). Next, unconjugated allenes and allenic carotenoids (Table 2) from higher plants are described, followed by allenes from insects and allenic alkaloids from neotropical frogs.

Section III treats enzyme-generated irreversible inhibitors (inactivators, 'suicide substrates'), which are able, under the influence of an enzyme, to develop a carbanionic site, which in turn may rearrange to an allene and the latter (being susceptible to Michael addition) may react with an enzyme nucleophile. The systems treated are allenic substances interacting with isomerases (Table 3), with a variety of model nucleophiles (Table 4), with pyridoxal-linked and flavin-linked enzymes (α-hydroxy acid oxidases and dehydrogenases, α-amino acid oxidases and amino oxidases) and with the non-flavin linked plasma monoamine oxidase. Finally, ketenes and ketene imines as inhibitors and allenes which affect insect steroid transformations are considered.

143 references up to 1978.
Complementary: **1**, 13 (pp. 1027–1029); **24**, 14.
Relevant: **2**, 7; **3**, 5; **4**, 9; **5**, 18; **7**, 4; **11**, 11; **12**, 12; **13**, 14; **15**, 10; **16**, 13; **17**, 13; **18**, 13; **19**, 13; **20**, 6; **22**, 20; **28**, 17; **31**, 9.

26, *14: Ketene O, O-acetals* by *P. Brassard*

I. Introduction (488); **II**. Preparation of ketene acetals (488); **III**. Electrophilic additions (492); **IV**. Reactions with carbonyl compounds and their sulphur analogues (497); **V**. Cycloadditions (503); **VI**. Claisen rearrangements (511); **VII**. Bisketene and vinyl ketene acetals (513); **VIII**. References (515).

The pronounced dipolar character of ketene acetals:

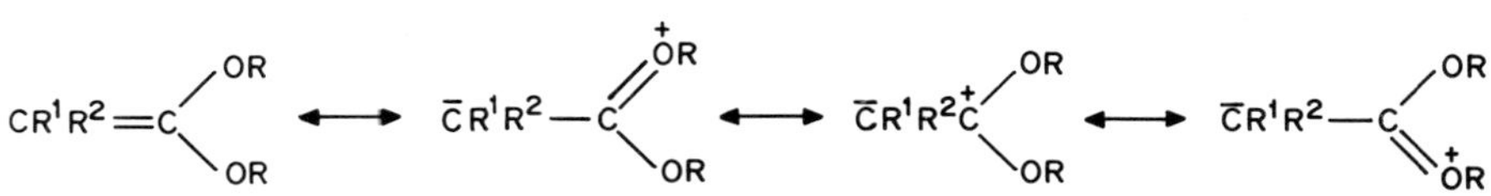

confers on them great reactivity and synthetic versatility. Section II describes the preparation of ketene dialkyl acetals, alkyl trialkysilyl acetals, bis(trialkylsilyl) acetals and some other related compounds, such as $(RO)_2C{=}C(OR)_2$ and $(RO)_2C{=}C{=}C(OR)_2$.

Section III deals with electrophilic additions to the electron-rich C=C bond of the title compounds. The addends include compounds containing labile H, such as amines (reacting through complex addition–elimination sequences), H_2O, ROH, ArOH and RCOOH. Halogens usually yield α-halo esters [$H_2C{=}C(OR)_2 + Br_2 \rightarrow CH_2BrCOOR + RBr$]; oxidations lead to carbonates, carbonyl compounds and other products through pathways involving fragmentation; carbenes yield cyclopropanes [$R^1R^2C{=}C(OR)_2 + \ddot{C}H_2 \rightarrow R^1R^2C\overset{CH_2}{\text{——}}C(OR)_2$]; reactive (allyl, benzyl, benzhydryl, etc.) halides react with ketene acetals at high temperatures [$R^1R^2C{=}C(OR)_2 + ArCH_2Br \rightarrow ArCH_2CR^1R^2CO_2R + RBr$]. Some reactions with diazonium and cyclopropenylium ions are also mentioned.

Section IV treats reactions with aldehydes and ketones resulting in various addition products, e.g. $R'CHO + H_2C{=}C(OR)_2 \rightarrow R'CH{=}CHCO_2R$. Additions to cyclic enediones (e.g. maleic anhydride) yield a variety of different products through complex and competing processes, although in some cases the reaction is synthetically useful. The same is the case in the reactions with acid chlorides, esters and lactones.

Section V discusses first thermal [2 + 2] cycloadditions to olefins and acetylenes yielding derivatives of cyclobutane and cyclobutene; ketenes with ketene acetals probably yield a zwitterionic initial product [$R^1R^2C{=}C(OR)_2 + R^3R^4C{=}CO \rightarrow R^3R^4CCOCR^1R^2C(OR)_2$], which in turn may give a variety of products by cyclization, isomerization and other processes depending on the nature of the R^1, R^2, R^3 and R^4 groups.

Next photochemical [2 + 2] cycloadditions to double bonds, 1,3-dipolar cycloadditions and Diels–Alder reactions are described.

Section VI treats reactions related to the Claisen rearrangement involving vinyl benzyl ether, allyl vinyl ethers and other related compounds, in each case yielding specific products depending on the structure of the starting material.

Finally, Section VII deals with bisketene acetals [$(RO)_2C{=}CR'{-}CR'{=}C(OR)_2$] and with vinyl ketene acetals $CH_2{=}C(R)CH{=}C(OR)_2$, which have recently been shown to possess great potential as synthons.

207 references up to 1978.
Complementary: none.
Relevant: 26, 16.

26, *15: Rearrangements involving allenes* by *W. D. Huntsman*

I. Introduction (522); **II**. Propargylic and retropropargylic rearrangements (522); **III**. Allene–diene rearrangements involving α-allenic halides, alcohols, etc. (559); **IV**. Acid-catalysed rearrangements (561); **V**. Homoallenic participation (564); **VI**. Oxidative cyclization (576); **VII**. Rotation about the allene axis (579); **VIII**. Pericyclic reactions (582); **IX**. Rearrangement of alkenylidenecyclopropanes (640); **X**. Rearrangement of cyclopropylallenes (643); **XI**. Rearrangements involving carbene intermediates (644); **XII**. Photochemical rearrangements (650); **XIII**. References (657).

Section II describes the prototropic propargylic and retropropargylic rearrangements of hydrocarbons and their functionally substituted derivatives and also rearrangements at gas–solid interfaces and the thermal interconversion of propyne and allene. Free-radical rearrangements are treated, followed by anionotropic routes involved in the replacement of OH by halogen, e.g.

$$HC{\equiv}CCH_2OH \xrightarrow{(PhO)_3\overset{+}{P}Me\overset{-}{I}} H_2C{=}C{=}CHI + HC{\equiv}CCH_2I$$

in the hydrolysis of allenic halides forming propargyl alcohols in reactions of organometallics with propargylic compounds (containing a leaving group L at the propargylic position:

$$R^4M + R^1C{\equiv}CCLR^2R^2 \longrightarrow R^4R^1C{=}C{=}CR^2R^3 + ML$$

(M = metal) and in the reduction of either propargylic or allenic halides, alcohols and ethers to the corresponding rearranged hydrocarbons.

Propargylic and allenic halides give identical Grignard reagents:

$$XCH_2C{\equiv}CH \xrightarrow{Mg} H_2C{=}C{=}CHMgX \xleftarrow{Mg} H_2C{=}C{=}CHX$$

The latter may undergo electrophilic substitutions either with retention of the original halide structure or with rearrangement, yielding from either starting material a mixture of both isomeric products in proportions depending on the structure of the synthone and on the reaction conditions.

Section III discusses the rearrangement of allenic substrates to conjugated dienes and the reversal of this process.

Section IV deals with acid-catalysed rearrangements of allenes to conjugated dienes:

$$R_2C{=}C{=}CR(OCOAr) \xrightarrow{H^+} H_2C{=}CRCH{=}CR(OCOAr)$$
$$+ Me_2C{=}CHC(OCOAr){=}CH_2 \ (Ar = 4\text{-}NO_2C_6H_4).$$

Section V discusses the problem of homoallenic participation in solvolysis reactions, often yielding cyclopropyl or cyclobutyl products in the solvolysis of β-allenic substrates (Table 1). Mechanistic proposals and their evidence are presented in detail.

Section VI treats oxidative cyclizations by H_2O_2 and by peracids, yielding in various cases allenic epoxides or cyclopentenones depending on the substituents present in the allenes.

Section VII considers the rotation barriers for the interconversion of enantiomeric forms of 1, 3-disubstituted allenes and of cumulenes (Table 2).

Section VIII considers pericyclic reactions, including electrocyclization, internal cycloaddition and related cyclizations (e.g. 1, 2, 4, 5-hexatetraene yields 3, 4-bimethylenecyclobutane), and sigmatropic rearrangements of various types such as [1, 5] and [1, 7] H-shifts, silapropynylic rearrangements, and [2, 3] sigmatropic rearrangements. Cope-type rearrangements of various enzymes and polyenes (Table 3), including thermal cyclization of alka-1, 2, 5-trienes to 3- and 4-methylenecyclopentenes (Table 4), Claisen-type rearrangements of aryl propargyl, propargyl vinyl and allenyl vinyl ethers and including thio- and amino-Claisen rearrangements, propargyl ester–allenyl ester rearrangements, dienol–benzene and dienone–phenol rearrangements and also ene- and retro-ene reactions.

Sections IX and X deal with the rearrangements of alkenylidenecyclopropanes and cyclopropylallenes, respectively.

Section XI considers rearrangements which involve alkenylidenecarbenes ($R_2C{=}C{=}C{:}\leftrightarrow R_2\overset{+}{C}{-}C{\equiv}\overset{-}{C}{:}$) or cyclopropylidenes as intermediates and reactions leading to fulvenallene (vinylidenecyclopentadiene) by a multiplicity of thermal routes (Scheme 53, p. 646).

Section XII describes photochemical rearrangements. These include sigmatropic rearrangements, e.g. to obtain allenes from conjugated trienes, intramolecular cycloaddition, electrocyclizations and cycloreversions (e.g. the interconversion of 1-methyl-3-methylenecyclobutene with 4-methyl penta-1, 2, 4-triene) and photorearrangements involving carbene intermediates and also metallic catalysts.

502 references up to 1978.
Complementary: 24, 10.
Relevant: 1, 7, 13; **2**, 15; **3**, 14; **4**, 10; **5**, 8; **8**, 9; **9**, 3; **10**, 3; **11**, 4; **12**, 16; **13**, 13; **14**, 7; **15**, 8; **16**, 17; **17**, 8; **18**, 15; **19**, 18; **20**, 4; **23**, 13; **29**, 4.

26, *16; Ketene thioacetals* by *M. Kolb*

Section I describes the preparation of the title compounds and the scope of the main routes available. These include the Peterson olefination $[(RS)_2\bar{C}{-}SiMe_3 + R^1R^2C{=}O \rightarrow (RS)_2C{=}CR^1R^2]$ (Table 1), the Horner–Emmons olefination $[(RS)_2\bar{C}PO(OR^3)_2 + R^1R^2C{=}O \rightarrow (RS)_2C{=}CR^1R^2]$ (Table 2) and preparations from esters with $R_2AlS(CH_2)_nSAlR_2$ (Table 3). The synthesis of ketene thioacetal monosulphoxides,

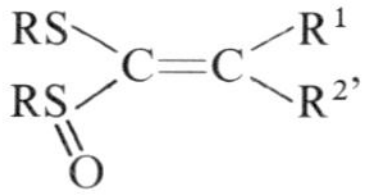, (Table 4) is also discussed.

Section III deals with physical properties such as UV (Table 5), IR, NMR, and MS.

Section IV considers reactions of the title compounds. These include various 'umpolung' reactions (see explanation in text and in Fig. 5, p. 680) such as reactions of ketenes with nucleophiles (at the C-2, C-4, C-6, etc. atoms, Table 6), the analogous reactions with ketene thioacetal sulphoxides (Table 7) and the monosulphonium salts (Table 8), reactions of metalated ketene thioacetals with electrophiles (reacting on the C-1 or C-3 atom, Tables 9–10), reactions of ketene thioacetals with electrophiles at the C-2 atom, hydrolysis of ketene thioacetals to yield acid derivatives (Table 11), cycloadditions and a selection of miscellaneous reactions with nucleophiles (amines, etc.) to produce *N*, *S*- and *N*, *N*-acetals and other products (Table 12).

206 references up to 1977.
Complementary: 26, 14.
Relevant: 3, 7.

26, *17: Ketene imines* by *M. W. Barker and W. E. McHenry*

I. Introduction (702); **II**. Structure (702); **III**. Preparation (703); **IV**. Spectroscopy (705); **V**. Reactions (706); **VI**. References (718).

Section II discusses the structure of ketene imines:

$$R_2C{=}C{=}NR \leftrightarrow R_2\overset{-}{C}{-}C{\equiv}\overset{+}{N}R \leftrightarrow R_2C{=}\overset{+}{C}{-}\overset{-}{N}R$$

Section III deals with methods of synthesis by dehydration of amides, dehydrohalogenation of imino chlorides, from the reaction of amides, Ph_3PBr_2 and Et_3N; from the condensation of diphenyl ketene with phosphinimines, by the alkylation of nitriles (e.g. by CH_2N_2) and from various heterocyclic intermediates.

Section IV deals with IR and NMR spectroscopy. Section V describes nucleophilic additions (Scheme 1, p. 706) of halogens, HX, H_2O, ROH, RNH_2, etc., of carbanions (Schemes 4 and 5) and of carboxylic acids. It also deals with alkylation of aromatic rings with ketene imines, addition of carbenes to the title compounds, cycloadditions to produce four-, five- and six-membered rings and with various oxidations.

58 references up to 1977.
Complementary: none.
Relevant: none.

26, *18: Carbodiimides* by *Y. Wolman*

I. Introduction (722); **II**. Structure and physical properties (722); **III**. Preparation (724); **V**. Reactions (731); **VI**. Application of carbodiimides in organic synthesis (742); **VI**. References (752).

Section II discusses the structure and physical properties of carbodiimides ($R^1N{=}C{=}NR^2$). Section III describes methods of preparation from thioureas, ureas, isocyanates, isothiocyanates, tetrazoles and some other compounds.

Section IV deals with the addition to the C=N double bond of various nucleophiles and electrophiles (Scheme 1, p. 731) and with isomerization, oxidation and cycloaddition to produce four-, five- and six-membered rings, concentrating mainly on the scope and mechanism of the reactions, while Section V describes the main synthetic application of the title compounds, including intermolecular and intramolecular dehydration, oxidation (by Me_2SO + carbodiimide in acidic media), formation of heterocyclic compounds by cycloaddition and some other less general applications.

172 references up to 1977.
Complementary: one.
Relevant none.

26, *19: Methyleneketenes* by *R. F. C. Brown and F. W. Eastwood*

Section I reviews general aspects of the unstable title compounds ($H_2C{=}C{=}C{=}O$). Section II deals with methods of synthesis by photochemical routes, by pyrolysis of acrylic and propiolic acid derivatives, e.g. $H_2C{=}CHCOOCOCH{=}CH_2 \rightarrow H_2C{=}C{=}C{=}O + CH_2{=}CHCO_2H$, and by thermal decomposition of various methylenemalonic acid derivatives. Several reactions are discussed in which the title compounds were postulated to be intermediates.

Section III gives some IR and dipole moment data. Section IV describes reactions with nucleophiles (when the title compounds behave as typical ketenes, e.g., $R_2C{=}C{=}C{=}O + HNuc \rightarrow R_2C{=}CHCONuc \rightarrow$ other products); dimerization:

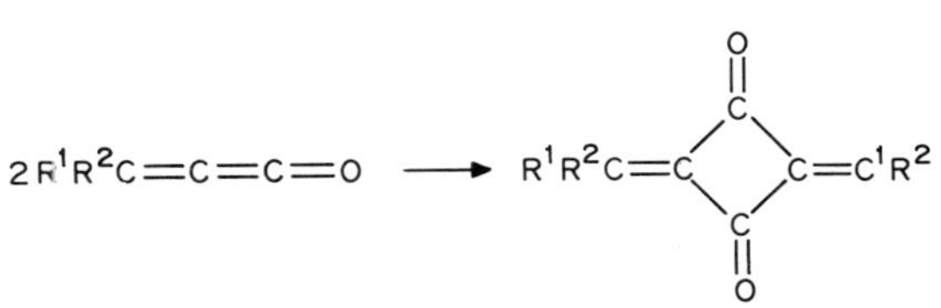

decarbonylation by flash vacuum pyrolysis to yield methylenecarbenes ($R_2C{=}C$:) and reactions of the latter and also intramolecular rearrangements.

40 references up to 1978.
Complementary: none.
Relevant: none.

26, *20: The preparation of allenes and cumulenes* by *H. Hopf*

Section II surveys the general methods available for the synthesis of allenes, including eliminations from olefins, Witting reactions, 1,4-additions to enynes (e.g. $R^1C{\equiv}CCR^2{=}CH_2 + R^3Li \rightarrow R^1CLi{=}C{=}CR^2CH_2R^3$); ring opening and elimination from 1,1-dihalogenocyclopropanes, cycloreversions and fragmentations and also conversions of one allene to another.

Specific methods for preparing alkyl-substituted allenes and also allenes carrying unsaturated groups (vinyl-, cyclopropyl-, ethinyl- and allyl allenes and diallenes) are presented, followed by the preparation of arylallenes, allenic alcohols, aldehydes and ketones and their derivatives and also allenic carboxylic acids, esters, amides and nitriles. In many cases the yields obtained by various methods with synthones carrying different substituents are given.

Section II. C deals with the preparation of endocyclic allenes containing six to eleven (and more) C atoms in the ring, and also of exocyclic allenes (vinylidenecyclopropanes, -cyclobutanes, -cyclopentanes, -cyclohexanes and higher cycloalkanes) and of bicyclic allenes. Section II. D treats heterosubstituted allenes including derivatives of Li, Hg, Al, B, Si, Ge, Sn, Pb, N, P, O, S, Se, Te, F, Cl, Br and I. Again, in most methods of preparation the scope and the yields with various substituents are given.

Section III starts with a short survey of the general methods used to prepare cumulenes (eliminations, Wittig reactions, propargylic rearrangements, etc.). Next, specific methods for preparing acyclic cumulenes with three to five (and more) cumulated double bonds and carrying various alkyl and aryl groups and other functions are described, followed by methods relating to cyclic cumulenes (including endocyclic, exocyclic and bicyclic cumulenes). Finally, heterosubstituted cumulenes including ethers, thioethers, sulphonium salts and various halogenocumulenes are dealt with.

953 references up to 1978.
Complementary: 1, 13.
Relevant: one.

Supplement E: The chemistry of ethers, crown ethers, hydroxyl groups and their sulphur analogues (1980)

27, *1: Synthesis of crown ethers and analogues* by *D. A. Laidler and L. F. Stoddart*

Section I gives a brief history of crown ethers since the synthesis of the first macrocyclic polyethers by Lüttringhaus (Ref. 7). Section II discusses the template effect in the synthesis of crown ethers by using cations of a certain diameter for the production of macrocycles with a corresponding cavity diameter (Tables 1–3). The *gauche* effects stems from the fact that the C—C bonds in the —OCH_2CH_2—O— units prefer to adopt the *gauche* conformation (Tables 4–6). Section III reviews the design of different crown ethers so that they should be able to act as receptor molecules for appropriate purposes.

Section IV describes the syntheses of monocyclic multidentate ligands containing O, N and S and their various combinations in the macrocycle, and also crown compounds incorporating benzene, furan, pyridine or thiophene rings. Following these, syntheses of crown compounds containing diester, dithioester or diamide compounds, carbonyl groups, imine or oxime functions are outlined. The synthesis of acyclic, macrobicyclic, macrotricyclic and macropolycyclic ligands is dealt with, together with that of chiral crown ethers including *meso* compounds, racemic modifications and optically active crown ethers from natural products or from resolved precursors.

Section V reviews briefly the toxicity of crown ethers and some hazards encountered in the processes of their preparations.

194 references up to 1980.
Complementary: none.
Relevant: 27, 2–4; **31**, 21.

27, *2: Crown ethers—complexes and selectivity* by *F. Vögtle and E. Weber*

IV. Crystalline complexes of cyclic and non-cyclic crown ethers (122); **V**. Outlook (144); **VI**. Acknowledgements (144); **VII**. References and notes (144).

Section I reviews the main types of crown ethers and their analogues containing various combinations of O, S and N ring atoms. Open-chain ligands are also discussed, together with the specialist nomenclature of the field.

Section II describes the kinetics and the mechanism of the complexation process and its reversal, involving a wide variety of ligands and metal and other ions (Table 1–7). The experimental methods and the results of the experiments are reviewed in order to understand the stability and selectivity of the crown ether complexes, involving crown ethers proper, and also catapinands (containing two N ring atoms), cryptands (with three-sided enclosed cavities) and podands (open-chain ligands).

Section II.C discusses the thermodynamics of the complexation and presents ΔH°, ΔS°, ΔG° and ΔC_p° values (Tables 8–10) and explains their significance.

Section III deals with complex stabilities and selectivities, the methods for the determination of complex stability and selectivity constants and the factors that influence them. The latter include ligand parameters (binding sites, shape, topology, conformational flexibility or rigidity and substituent effects), type, size and charge of the guest ion, correlation of cation and cavity radii (Table 13), anion interactions and ion-pair effects and, medium parameters (Table 15).

Section IV describes the preparation of crystalline complexes of cyclic and non-cyclic crown ethers with various ions and neutral molecules, the selectivity of crystalline complex formation involving monocyclic crown ethers with alkali, alkaline earth and heavy metal ions and also with neutral molecules. Next, bi- and polycyclic cryptates and open-chain podates and their crystalline structures are discussed.

263 references up to 1979.
Complementary: none.
Relevant: 27, 1, 3, 4; **31**, 21.

27, *3: Organic transformations mediated by macrocyclic multidentate ligands* by *C. L. Liotta*

I. Introduction (157); **II**. Organic reactions mediated by macrocyclic and macrobicyclic multidentate ligands (162); **III**. References (172).

Crown ethers and related multidentate compounds solubilize metal salts in non-polar and dipolar organic solvents (Tables 1–3) with minimal solvation of the anions, which thus become potent nucleophiles (Table 4) and their reaction rates become strongly enhanced (Table 5). Hence crown ethers can be used to catalyse reactions under homogeneous and especially under heterogeneous conditions (Section I).

Section II presents a large number of examples in which organic reactions are mediated and/or catalysed by macrocyclic and macrobicyclic multidentate ligands. Among these are the introduction of fluorine into organic molecules by simple displacement processes (e.g., $CH_3COCl \xrightarrow[\text{18-crown-6}]{\text{KF}} CH_3COF$; 100%), reactions in which carboxylate ions (especially acetate) react as strong nucleophiles (e.g. $PhCH_2Br \rightarrow PhCH_2OAc$; 100%) and reactions of cyanide ions in the presence of crowns and cryptates to give a variety of substitution, elimination and addition processes (e.g. $C_6H_{13}Br \rightarrow C_6H_{13}CN$; 100%).

Other examples include isomerizations, isotope exchanges and fragmentations promoted by metal alkoxides, oxidations by $KMnO_4$ solubilized in benzene by crowns and reactions of potassium superoxide under similar conditions.

78 references up to 1977.
Complementary: none.
Relevant: **27**, 1, 2, 4; **31**, 21.

27, *4: Geometry of the ether, sulphide and hydroxyl groups and structural chemistry of macrocyclic and noncyclic polyether compounds* by *I Goldberg*

Section I reviews briefly the scope and limitations of the three main methods used for the determination of molecular structures, namely electron diffraction (ED), microwave spectroscopy (MW) and X-ray diffraction (XD).

Section II describes the structural parameters obtained by ED and MW for the C—O—C group in ethers (Tables 1 and 2), for the C—S—C group in sulphides (Table 3) and for cyclic compounds such as dioxane, dithiane and thioxane (Table 4) and for various compounds containing C—O—H groups (Tables 5 and 6), and also compares the results for ether, sulphide, hydroxyl and thiol groups (Table 7).

Section III deals with the structural chemistry of the 18-crown-6 system (Figs 1–3 and Table 8), and presents structures of various host–guest complexes of crown ethers (Figs 5–12) and of some inclusion compounds of noncyclic polyethers (Figs 13 and 14).

111 references up to 1979.
Complementary: none.
Relevant: **16**, 2; **17**, 2; **18**, 2; **19**, 2; **20**, 2; **21**, 2; **22**, 2; **23**, 2; **24**, 2; **26**, 2; **28**, 2; **29**, 1; **30**, 1; **31**, 24; **32**, 24; **33**, 12; **34**, 10; **36**, 3.

27, *5: Stereodynamics of alcohols, ethers, thio ethers and related compounds* by *C. H. Bushweller and M. H. Gianni*

Section II deals with rotations about bonds in O- and S-containing compounds, with barriers to rotation in simple systems (Table 1), with barriers about S—S bonds (Table 2) and in detail with *t*-Bu rotation (Tables 3–5) and with rotation of haloalkoxy- and haloacetoxybutanes (Tables 6 and 7), together with inversion processes at O and S.

Section III discusses the stereodynamic behaviour of perfluorotetramethyl Dewar thiophene and its *exo*-*S*-oxide, of monosubstituted cyclohexanes with O or S substituents (Table 8) and of oxacyclohexanes (Tables 9–12), of dihydropyran, of thiacyclohexanes (Tables 13 and 14), of 1,3-dioxacyclohexanes (Tables 15–22) and of 1,3-dithiacyclohexanes (Tables 22–25), including stereoselective reactions of the latter, of other six-membered rings containing O and S, of *s*-tetrathianes (Table 26), and finally of medium-sized (7- or 8-membered) rings (Tables 27 and 28).

179 references up to 1977.
Complementary: none.
Relevant: **19**, 21; **20**, 2; **21**, 1; **22**, 3; **28**, 10; **29**, 2; **31**, 19; **32**, 1; **33**, 2.

27, *6: Chiroptical properties of alcohols, ethers, thio ethers and disulphides* by *G. Gottarelli and B. Samorì*

Section II discusses the OH chromophore and methods for obtaining stereochemical information from CD data and it is concluded that the CD spectrum of OH is not very useful for stereochemical determinations.

Section III deals with the benzoate sector rule (Fig. 3), aromatic chirality methods and the dibenzoate chirality rule and presents various CD spectra of dibenzoates (Figs 4–7).

Section IV describes CD studies on ethers, including ethers of sugars (Figs 8 and 9). Section V deals with thio sugars and episulphides (Figs 10 and 11) and Section VI covers disulphides and presents a quadrant rule for the latter (Fig. 13).

65 references up to 1979.
Complementary: 18, 7.
Relevant: 8, 4; **19**, 6; **20**, 2; **22**, 4; **25**, 3; **26**,3; **28**, 10; **29**, 23; **31**, 1; **36**, 12.

27, *7: The mass spectra of ethers and sulphides* by *C. C. Van de Sande*

Section II discusses the concept of charge localization in MS, the α-cleavage reaction and other general features of the MS of saturated aliphatic ethers and sulphides.

Section III deals with cycloalkyl ethers and sulphides, unsaturated (non-aromatic) compounds such as vinyl and alkyl ethers; cyclic compounds containing O or S in the ring, aromatic ethers and thio ethers and briefly with macrocyclic compounds (crown ethers and S-containing macrocycles).

Section IV discusses the interaction of various functional groups in complex molecules and their combined effects on mass spectra, including migrations, neighbouring-group participations and *ortho*- and *peri*-effects.

Section V deals with the usefulness of MS for the determination of the stereochemistry of the title compounds and Section VI with chemical ionization MS and its use to obtain accurate data for gas-phase proton affinities (Table 2).

296 references up to 1979.
Complementary: 13, 19; **18**, 6.
Relevant: 9, 7; **10**, 5; **16**, 4; **17**, 5; **19**, 5; **22**, 7; **24**, 6; **25**, 4; **26**, 6; **29**, 3; **30**, 22; **31**, 3; **32**, 3; **33**, 4; **36**, 7.

27, *8: The electrochemistry of ethers, hydroxyl groups and their sulphur analogues* by *T. Shono*

Section II describes the cathodic reduction of sulphides in DMF (Table 1), and the effect of methyl groups on the half-wave potentials of methyl-substituted Ph_2S derivatives

(Table 2), together with the effects of the substituents on variously substituted diphenylmethyl phenyl sulphides (Table 3). β-Hydroxy sulphides can be reduced to olefins in excellent yields (Table 4). Next, reductive methylation (Table 6) and reactions with various electrophiles (Table 7) of disulphides are presented, together with anodic reductions of sulphonium salts. Reduction of alcohols requires activation by other functional groups (Table 8), while esters of methanesulphonic acids readily yield the corresponding hydrocarbon (Table 9) and phosphates of phenols give the corresponding arenes [$ArOPO(OEt)_2 \rightarrow ArH$] (Table 10). In some benzylic and allylic ethers, C—O bonds are cleaved (Table 11).

Section III discusses the anodic oxidation of thiols and sulphides, as well as of hydroxyl groups (Tables 12 and 13) and of ethers (Table 14). Isotope effects have been studied in the latter reaction (Table 15). *p*-Methoxybenzyl ethers in the presence of homogeneous electron carriers cleave the C—O bond to yield alcohols and aldehydes in high yields.

56 references up to 1979.
Complementary: **13**, 5.
Relevant: **5**, 2; **8**, 11; **16**, 15; **17**, 14; **19**, 12; **20**, 5; **21**, 5; **22**, 9; **23**, 10; **24**, 17; **25**, 13; **26**, 12; **28**, 7; **29**, 8, 9; **31**, 7; **32**, 6; **34**, 1.

27, *9: Electronic structures and thermochemistry of phenols* by *J. Royer, G. Bertholon, R. Perrin, R. Lamartine and M. Perrin*

I. Electronic structures (352); **II**. Thermochemistry of phenols (360); **III**. References (374).

Section I describes various characteristic physical indices, such as charge densities (Table 1), the influence of methyl groups on charges and reactivity indices (Table 2), dipole moments (Table 3) and conformations. Theoretical studies of the electrophilic substitution on phenols are presented, using a delocalized model and isopotential curves. Next, electronic spectra are discussed (Table 4) and also NMR and MO studies in the pharmacology of phenols and theoretical studies of inter- and intramolecular hydrogen bonds.

Section II deals with thermodynamic properties, heat capacities (Table 5) and physical transformations (melting, vaporization, sublimation and transition enthalpies and entropies). Physical interactions with other substances are treated [complexes or adducts and their enthalpies (Table 6)], followed by chemical transformations [enthalpies of combustion (Table 7), of alkylation (Table 8) and of isothermal chlorination (Table 9)]. In conclusion, relative stabilities (Table 10), thermochemistry and kinetics, resonance energies (Table 11) and reaction orientation are discussed.

133 references up to 1978.
Complementary: none.
Relevant: **16**, 16; **18**, 3; **19**, 3; **20**, 11; **21**, 3; **22**, 6; **23**, 4; **24**, 3; **25**, 2; **26**, 4; **28**, 4; **29**, 24; **30**, 2; **31**, 2; **33**, 3; **36**, 4.

27, *10: Syntheses and uses of isotopically labelled ethers and sulphides* by *M. Zieliński*

I. Syntheses of labelled ethers and sulphides (380); **II**. Tracer and isotope effect studies with ethers (411); **III**. Tracer and isotope effects studies involving sulphides (424); **IV**. Isotope exchange studies with ethers and sulphides (430); **V**. Isotopic studies of complexes with ethers and sulphides (436); **VI**. Isotopic compounds used in cancer studies (437); **VII**. Acknowledgements (438); **VIII**. Bibliography and references.

Section I describes the preparation of various classes of ethers and sulphides labelled with ^{2}H, ^{3}H, ^{14}C, ^{18}O and ^{35}S. A special section deals with ethers and sulphides used in biology, medicine and agriculture (Section I.C).

Section II discusses isotopic studies of the thermal decomposition of ethers in the gas phase and of the mechanism of the Claisen rearrangement. Next, isotopic studies with vinyl ethers, reactions of ethers with organometallic compounds, eliminations, brominations, oxidations and some other reactions are presented.

Section III deals with tracer and isotope studies involving sulphides, such as decompositions, rearrangements, cleavages, eliminations and other reactions, and reactions which result in the production of ethers or sulphides.

Section IV treats ^{2}H, ^{3}H, ^{18}O, ^{35}S and ^{36}Cl exchange studies with the title compounds and the mechanistic implications of these.

Section V discusses the use of isotopes in the study of various complexes formed with ethers and sulphides and the influence of isotopes on the spectra of these complexes and on their equilibria, including the problem of separation of BCl_3 and BF_3 complexes (Containing ^{11}B and ^{10}B) with the title compounds.

Section VI deals very briefly with some of the isotopic title compounds in cancer studies.

445 references up to 1978.
Complementary: **13**, 15; **18**, 18.
Relevant: **5**, 10; **12**, 13; **17**, 12; **19**, 17; **22**, 14; **23**, 16; **24**, 20; **26**, 11; **28**, 8; **29**, 28; **33**, 18; **36**, 12.

27, *11: Gas-phase thermal decompositions of simple alcohols, thiols and sulphides* by *R. L. Failes, J. S. Shapiro and V. R. Stimson*

Section I reviews some techniques and results and tabulates relevant activation energies for various molecular elimination processes and the strengths of the bonds involved (Table 1).

Section II discusses separately the decomposition processes of C_1–C_7 alcohols, including some unsaturated alcohols (Table 2) and compares the rates of elimination from chlorides, acetates and alcohols (Table 3). Reaction mechanisms and kinetics are presented in detail.

Section III deals with decompositions of alcohols (Table 4), ethers (Table 5), carboxylic acids and esters (Table 6), acetals (Table 7), amines and amides (Table 9) and the isomerization of olefins (Table 8), all processes catalysed by acids, especially HCl and HBr and in some cases HI, organic acids, BCl_3 or BBr_3.

Section IV treats the thermal decomposition to C_1–C_5 thiols and of α-toluenethiol and the mechanisms of these processes. Section V presents some results on the decomposition of sulphides and disulphides (Table 10).

108 references up to 1977.
Complementary: **1**, 3; **3**, 4.
Relevant: **16**, 10; **22**, 15; **25**, 15; **32**, 7; **33**, 20.

27, *12: Oxidation and reduction of alcohols and ethers* by *P. Müller*

Section II describes methods of oxidation of alcohols, classifying the material according to the nature of the oxidant. Section II.B presents the mechanism and the effects of the alcohol structure on oxidations involving CrO_3, and *tert*-butyl chromate and chromyl chloride. Section II.C treats similarly processes involving derivatives of Mn ($KMnO_4$, MnO_2) and of Ru (RuO_4, $NaRuO_4$). One-electron oxidants discussed include derivatives of Ce^{IV}, V^{V}, Pb^{IV} and of Ag^{II}. In all these cases the mechanism of the reaction is presented in detail, together with synthetic aspects, scope and application and factors affecting the choice of the oxidant. Section II.E describes the use of dimethyl sulphoxide in the presence of dicyclohexyl carbodiimide and a proton source, and some variations of this method (using acid anhydrides, PCl_5, $COCl_2$, etc.).

Section III deals with the oxidation of ethers under free-radical conditions and by electrolysis. Hydride transfer is achieved also by the use of Ph_3C^+ or diazonium or nitronium ions and by pyrolytic ether cleavage. Other methods of ether oxidation involve CrO_3, RuO_4, $Pb(OAc)_4$, Co^{III} and other reagents.

Section IV treats the reduction of alcohols by catalytic hydrogenation, by dissolving metals (Birch reduction), by Al hydrides ($LiAlH_4$, $HAlCl_2$), silanes or boranes and by reductive alkylation. Indirect procedures include reduction of alcohols by red P in refluxing HI, when the OH group is displaced by I and the latter is reduced by HI($ROH \rightarrow RI \rightarrow RH$), and reductions via sulphonates, isoureas, thiocarbamates and dithiocarbamates.

Section V presents methods of reduction of ethers, including catalytic hydrogenation, dissolving metal reduction, organometallic reagents (Mg and Li compounds) and complex metal hydrides ($LiAlH_4$, BH_3–BF_3, $AlHBu^i_2$, etc.)

397 references up to 1978.
Complementary: 13, 10.
Relevant: 27, 13.

27, *13: Oxidation and reduction of sulphides* by *E. Block*

Section II deals mainly with the oxidation of sulphides to sulphoxides [$RSR \rightarrow RS(O)R$] and less with the conversion of the latter to sulphones. The general methods of oxidation use as oxidants peroxy compounds, $NaIO_4$, hypervalent iodo compounds, boranes and nitrogen oxides. Other methods involve oxidation of sulphides via halosulphonium salts ($R_2S \xrightarrow{X_2} R_2\overset{+}{S}X\,\overset{-}{X} \xrightarrow{H_2O} R_2SO + 2H$), with Me_2SO as oxidant (Cr^{VI}, Au^{III}, Pb^{IV}, Mn^{IV} and Tl^{III} compounds), by organotin derivatives, and by ozone (Table 1) and via photosensitized oxidation by singlet oxygen, one-electron oxidation (Tables 2 and 3), *in vivo* oxidations and using polymer-supported oxidants. The stereochemistry of the oxidation of cyclic sulphides is treated, especially stereoselectivity with different oxidants (Table 4). Asymmetric oxidations by chiral oxidants are treated, together with selective oxidations of dithioethers, methods for the preparation of ^{18}O-labelled sulphoxides, oxidations as proof of the structure of sulphides and oxidations of S in the presence of various other functional groups.

Oxidation of penicillins and the stereoselectivity of this process are treated (Table 5) and also the oxidation of cephalophosporin derivatives. Functional groups other than S can be oxidized selectively. The preparations of perfluoroalkyl sulphoxides, thiophene 1-oxides and thiirane 1-oxides are special cases and are treated separately.

Section III discusses reductions of sulphides to hydrocarbons and thiols or to H_2S. The section is divided according to the nature of the reducing agents: Section III.A deals with Group I and II metals, III.B with Raney Ni and other heterogeneous catalysts, III.C with $LiAlH_4$ and related reagents and III.D with electrochemical and photochemical methods.

Throughout the chapter, reaction conditions and yields are given and the scope and mechanisms of the processes are discussed in detail.

327 references up to 1979.
Complementary: 18, 17.
Relevant: 27, 12.

27, *14: Oxiranes* by *M. Bartók and K. L. Láng*

I. Introduction (610); **II**. Synthesis of oxiranes (610); **III**. Reactions of oxiranes (627); **IV**. References (659).

Section II describes the preparation of oxiranes by oxidation of alkenes with peroxy acids, H_2O_2, hydroperoxides and O_2 and by some other less general methods. Syntheses by 1,3-elimination from 1,2-difunctional compounds:

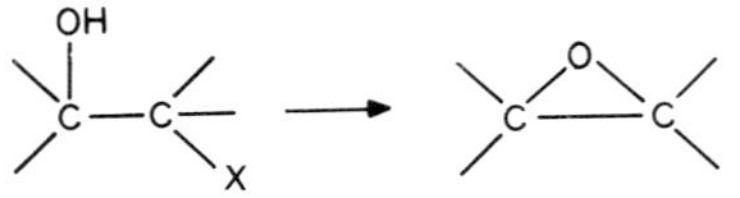

; X = Cl, Br, I, OSO_2R, NR_3^+, N_2^+, OH

and from carbonyl compounds with various nucleophiles, e.g.

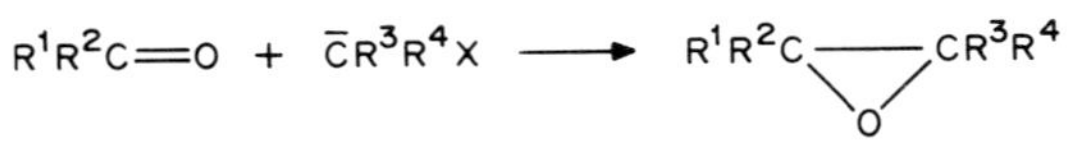

are also treated.

Section III discusses reactions of oxiranes, such as deoxygenation with electrophilic, nucleophilic and other reagents (yielding alkenes), rearrangements catalysed by bases (yielding enols) and by acids (yielding carbonyl compounds), thermal and photochemical rearrangements to a variety of products and rearrangements catalysed by heterogeneous catalysts or by metal complexes. Oxidations by HIO_4 or H_2O_2 give the corresponding dialdehydes with breakage of the C—C bond of the oxirane, but a variety of other products with other oxidants are discussed. Reduction leads to the corresponding alcohols (>C—C< (O) → >CH—COH), using complex metal hydrides (AlH_3, $LiAlH_4$, $LiBH_4$, etc.) or by catalytic hydrogenolysis or by alkali metals. Polymerization (either acid-catalysed or anionic) gives polyethers.

A variety of heterocyclic compounds can be obtained, e.g. by transformation of oxiranes into thiiranes, aziridines or seleniranes or by ring expansion to γ-lactaones, oxolanes or dihydropyrans. Oxiranes can also be transformed to heterocyclics containing two heteroatoms (e.g. oxiranes with carbonyl compounds yield 1,3-dioxolanes), and also to a variety of four-, five- and six-membered heterocycles. Section III.G describes the reactions of oxiranes with Grignard compounds (yielding alcohols), with Mg and Al alkyls, yielding alcohols in both cases:

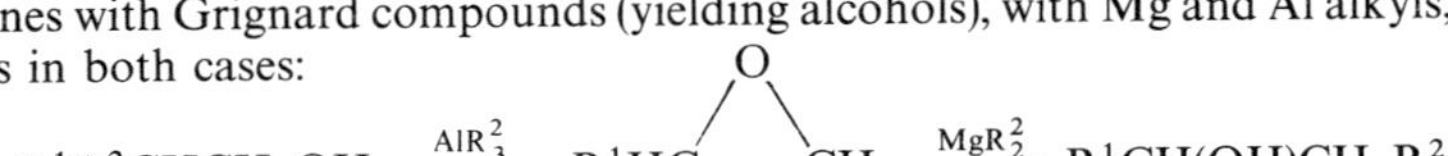

with organolithium compounds or (preferably with $LiCuR_2$: 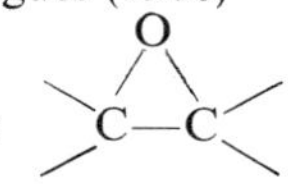 + $LiCur_2$ → 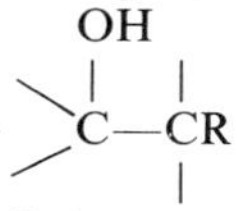 + RCu) and with derivatives of Cd, Zn, Si, etc.

Oxiranes containing double or triple bonds in the molecule give a variety of reactions with organometallic compounds (MeLi, Me_2CuLi, MeMgCl, etc.), yielding mainly 1,4-addition products. Finally, some photochemical and thermal reactions of oxiranes and ring openings with nucleophilic reagents are discussed.

Throughout the chapter, the reaction conditions, scope, yields, stereochemistry and mechanism of the processes are treated in detail.

956 references up to 1978.
Complementary: **3**, 9; **27**, 15.
Relevant: none.

27, *15: Cyclic ethers* by *M. Bartók*

I. Introduction (684); **II**. Synthesis of cyclic ethers (684); **III**. Reactions of cyclic ethers (695); **IV**. References (712).

The chapter surveys cyclic ethers with 4–6 ring atoms, i.e. oxetanes, oxolanes and oxanes.

Section II describes syntheses from monofunctional and difunctional hydrocarbon derivatives, e.g. from 1,3-, 1,4- and 1,5-diols (see **27**, 16) by cyclization. Other methods include ring expansion of oxiranes, reduction of oxacycloalkanones, dihydrofurans or furans, rearrangement of dioxacycloalkanes and 1,2-cycloadditions of olefins and carbonyl compounds to obtain oxetanes and 2,3- and 2,5-cycloadditions of furans to yield condensed polycyclic oxolanes.

Section III deals with the following reactions of cyclic ethers: deoxygenation (e.g. oxetane, yielding cyclopropane and propene); dehydrogenation (e.g. oxolanes, yielding furans); dehydration (e.g. oxolane derivatives to yield butadiene derivatives); rearrangements of oxetanes, oxolanes and oxanes (yielding unsaturated alcohols, ketones and other products); oxidation (e.g. oxolane to γ-butyrolactone); reduction and hydrogenolysis (with complex metal hydrides or catalytically, usually giving open-chain alcohols); polymerization to yield polyethers; formation of heterocyclic compounds either by ring expansion or by transforming the ring O atom to ring S or N and also by both processes together; reactions with organometallic compounds, usually leading to ring opening; and free-radical reactions (such as pyrolysis to olefins and carbonyl compounds, alkylations, ring contractions, insertions of nitrenes and of carbenes and fragmentations). Finally, ring opening with nucleophilic reagents is discussed, including hydrolysis, alcoholysis, reactions with hydrogen halides, with acids and with phosphorus halides and also with some less general reagents.

387 references up to 1978.
Complementary: **3**, 9; **27**, 14, 16.
Relevant: none.

27, *16: Dehydration of diols* by *M. Bartók and A. Molnár*

I. Dehydration of 1,2-diols (722); **II**. Dehydration of 1,3-diols (732); **III**. Dehydration of higher diol homologues (741); **IV**. References (752).

1, 2-Diols eliminate H_2O by pinacol rearrangement, by epoxide formation or by formation of unsaturated compounds, mainly dienes.

Section I.A describes the pinacol rearrangement, its mechanism, stereochemistry, catalysis and intermediates and the accompanying formation of unsaturated compounds. Next, transformations on alumina (yielding either carbonyl compounds or dienes) and dehydrations on metals (Cu, Pd/C, etc. leading mainly to carbonyl derivatives) and in the presence of other catalysts (SiO_2, Na_2HPO_4 or Friedel–Crafts catalysts) are presented, and also intermolecular dehydration leading to cyclic ethers.

Section II discusses the dehydration of 1, 3-diols, which may lead to carbonyl compounds, unsaturated alcohols, dienes, fragmentations or to cyclic ethers, depending on the nature of the diol and the catalyst (Table 1). Reaction conditions, mechanisms and product proportions are given.

Section III deals with higher diols, the dehydration of which may yield cyclic ethers, unsaturated cyclic ethers, unsaturated alcohols and dienes. Again, the proportions of the products depend on the nature of the diol and the catalyst (Table 2). The mechanism of the formation of oxacycloalkanes is discussed in detail, together with the role of the various catalysts and the stereochemistry of the processes. Some less common transformations (formation of unsaturated aldehydes and ketones from unsaturated diols) are also treated.

360 references up to 1979.
Complementary: **13**, 12.
Relevant: **3**, 10.

27, *17: Enol ethers—structure, synthesis and reactions* by *P. Fischer*

Section I reviews the nomenclature (Table 1) and the four main types of reactions (Scheme 1) of enol ethers (vinyl ethers).

Section II describes conformation, relative E/Z stabilities (Table 2), steric interaction energies between two Z substituents (Table 3), PES, Raman, IR and NMR (Table 4) and MS data and summarizes the relationship between conformation and reactivity.

Section III presents methods of preparation by vinylation of alcohols ($HC{\equiv}CH + ROH \rightarrow ROCH{=}CH_2$), by transvinylation ($R^1OCH{=}CH_2 + R^2OH \rightarrow R^1OH + R^2OCH{=}CH_2$), by elimination of alcohols from acetals and by the Horner–Wittig reaction ($Ph_3P{=}CHOR^1 + R^2COR^3 \rightarrow R^2R^3C{=}CHOR^1 + Ph_3PO$).

Section IV discusses electrophilic reactions such as hydrolysis (Table 5), halogenation, reactions with electrophilic O, S, N and P compounds and with carbon electrophiles (carbocations, aldol and other condensations, acylations, etc.). Section V deals with [2 + 2] cycloadditions with tetracyanoethylene (Tables 6–8), giving mechanistic details, and also with ketenes and isocyanates (Table 9). Next, carbene reactions ([1 + 2] cycloadditions), 1, 3-dipolar [2 + 3] cycloadditions (involving aryl, acyl, sulphonyl azides, diazocompounds etc.) and [2 + 4] cycloadditions (Diels–Alder reactions) are described.

Section VI treats the metalation of enol ethers, when both α- and β-vinylic H atoms can be substituted by Na or Li derivatives, e.g. $CH_2{=}CHOEt + t\text{-BuLi} \rightarrow CH_2{=}C(Li)OEt$ and the metal derivatives so obtained can be used for a variety of syntheses.

Section VII deals with the preparation and reactivity of silyl enol ethers, with their reactions with protic electrophilic reagents (HCN, ROH, RSH, HN_3), halogens, sulphenyl or sulphonyl halides, carbon electrophiles, e.g.

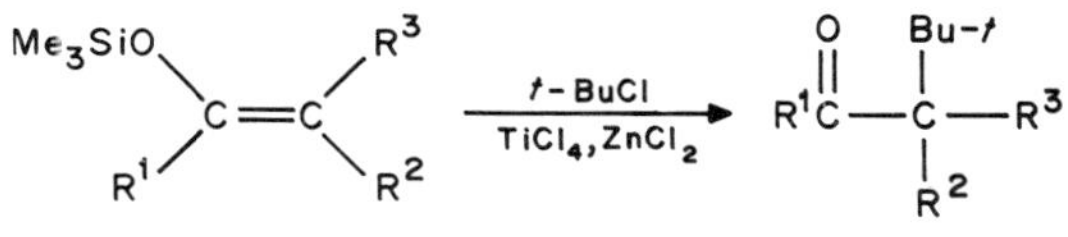

and also with cycloadditions (Section VII.D).

Finally, in Section VIII the physical properties, preparation and reactivity of thioenol ethers are described, including their hydrolysis (Table 10) and TCNE cycloaddition (Table 11 and 12).

409 references up to 1978.
Complementary: none.
Relevant: none.

27, *18: Oxathiacyclanes: preparation, structure and* by *K. Pihlaja and P. Pasanen*

I. Introduction (822); **II**. Four-membered rings (822); **III**. Five-membered rings (823); **IV**. Six-membered rings (837); **V**. Seven-membered and larger rings (850); **VI**. References (853).

Section II deals with four-membered rings containing both O and S, of which only *S*-oxides are known.

Section III treats the preparation, structure and reactions of 2-oxo-1,2-oxathiolanes 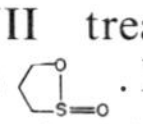. Next, 1,3-oxathiolanes are described, including their preparation and structure, the chemical equilibrations of several epimeric 1,3-dioxolanes (Table 1) and the MS of the latter (Table 2), and also their hydrolysis, protonation, photochemical reactions and reductions. 3-oxo- and 5-oxo-1,3-oxathiolanes are discussed and also 2-oxo-1,3,2-dioxathiolanes.

Section IV describes six-membered rings, such as 2-oxo-1,2-oxathianes, 1,3-oxathianes (Table 3–5), 3-oxo-1,3-oxathianes, 1,4-oxathianes and some oxadithianes and dioxathianes. In each case, whenever available, preparations, structure and reactions are presented.

Section V deals with seven-membered and larger rings and mentions also macrocyclic (9–21-membered) rings.

269 references up to 1977.
Complementary: none.
Relevant: none.

27, *19: Allene oxides and related species* by *P. J. Stang*

I. Introduction (859); **II**. Theoretical calculations (860); **III**. Preparation (862); **IV**. Reactions (866); **V**. Related species (875); **VI**. Acknowledgements (878); **VII**. References (878).

Section II presents theoretical calculations relating to allene oxides, $H_2C{-}C{=}CH_2$ [O bridging], and to two isomeric C_3H_4O molecules, cyclopropanone and oxyallyl (Table 1).

Section III describes the preparation of allene oxides by peracid oxidation of allenes, by exocyclic β-elimination of epoxides:

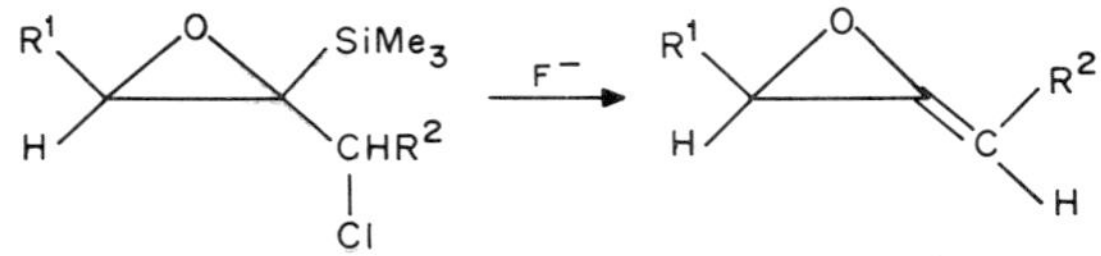

by addition of unsaturated carbenes to carbonyl groups and by some other methods.

Section IV deals with oxidation of allene oxides, their isomerization to cyclopropanones and their reactions with nucleophiles (Table 2).

Section V describes some related species, including oxaspiropentanes, , and allene episulphides, which are the S analogues of allene oxides.

57 references up to 1978.
Complementary: none.
Relevant: none.

27, *20: Advances in the chemistry of acetals, ketals and ortho esters* by *R. G. Bergstrom*

Section II deals with the synthesis of the title compounds. General methods include ketal formation from ketones with diethylene orthocarbonate or with 2-chloroethanol (both methods yielding 1,3-dioxolanes). Ethyl orthoformate reacts with triols to give bicyclic ortho esters. Acid-catalysed reactions of both aldehydes and ketones with methyl orthoformate yield acetals or ketals, respectively [$R_2C{=}O + HC(OMe)_3 \rightarrow R_2C(OMe)_2$]. Thioketals are formed with methylthiotrimethylsilane ($2MeSSiMe_3 + R_2C{=}O \rightarrow R_2C(SMe)_2$). Other, less general methods include formation of α-hydroxyacetals from vinyl ethers and MeOH with peracids, methods involving boron intermediates and electrochemical oxidations.

Section III describes studies relating to the hydrolysis of the title compounds, including the determination of the rate-determining step (RDS), which may involve hemiacetal or oxocarbonium ion intermediates and the study of examples which undergo a change in the RDS during the hydrolysis, as the acidity of the media is varied. In acid-catalysed reactions, evidence for concerted C—O cleavage is presented; structure–reactivity relationships and secondary ^2H isotope effects are discussed. Finally, kinetic solvent isotope effects and salt effects are treated.

109 references up to 1979.
Complementary: **3**, 7; **5**, 13.
Relevant: none.

27, *21: The photochemistry of saturated alcohols, ethers and acetals* by *C. von Sonntag and H.-P. Schuchmann*

sensitized photolysis of alcohols and ethers (917); **VII**. Photolysis of O_2 charge-transfer complexes (918); **VIII**. References (919).

Section II deals with the absorption spectra of the title compounds and presents the molar absorptivities at 185 nm (Table 1). Section III describes the photolysis of primary and secondary alcohols, of *t*-BuOH and of alkoxide ions, giving detailed reaction schemes and mechanisms, and emphasizing the differences between the various classes.

Section IV discusses the photolysis of acyclic ethers (Table 2) and cyclic ethers (Table 3), and Section V treats the photolysis of acetals.

Section VI presents studies on the Hg-sensitized photolysis of alcohols and ethers and Section VII on the photolysis of O_2 charge-transfer complexes.

133 references up to 1979.
Complementary: 3, 8; **13**, 16.
Relevant: 2, 16; **6**, 4; **8**, 12; **9**, 6; **11**, 5; **14**, 8; **15**, 9; **16**, 11; **17**, 9; **18**, 10; **19**, 20; **22**, 10; **23**, 9; **24**, 12; **25**, 11; **27**, 22; **28**, 5; **29**, 5, 6; **32**, 29; **33**, 21.

27, 22: *The photolysis of saturated thiols, sulphides and disulphides* by *C. von Sonntag and H.-P. Schuchmann*

I. Introduction (923); **II**. Photolysis of thiols (924); **III**. Photolysis of sulphides (927); **IV**. Photolysis of dithiaacetals (931); **V**. Photolysis of disulphides (931); **VI**. References (932).

Molar absorptivities of the title compounds near 254 nm are given (Table 1). Section II deals with primary and secondary processes in the photolysis of thiols in the gas phase and also in aqueous solutions.

Section III describes the photolysis of acyclic and cyclic sulphides and Section IV that of disulphides. In all cases detailed mechanistic reaction schemes are presented.

91 references up to 1978.
Complementary: 18, 10.
Relevant: 2, 16; **3**, 8; **6**, 4; **8**, 12; **9**, 6; **11**, 5; **13**, 16; **14**, 8; **15**, 9; **16**, 11; **17**, 9; **19**, 20; **22**, 10; **23**, 9; **24**, 12; **25**, 11; **27**, 21; **28**, 5; **29**, 5, 6; **32**, 29; **33**, 21.

27, 23: *Radiation chemistry of alcohols and ethers* by *C. von Sonntag and H.-P. Schuchmann*

I. Introduction (935); **II**. Neat alcohols in the liquid and solid state (936); **III**. Alcohols in the gas phase (944); **IV**. Neat ethers (945); **V**. Aqueous solutions of alcohols and ethers (947); **VI**. References (961).

Section II deals with energy absorption and primary processes, solvation of the electron, ion–molecule reactions, the formation of hydrogen and the fragmentation of the C—O skeleton. *G*-values of the major products in the radiolysis of liquid alcohols are given (Table 1) and also the H_2, HD and D_2 distribution in the γ-radiolysis of deuterated butanols in the presence of additives (Table 2).

Section III deals with the gas-phase radiolysis of alcohols (Table 3) and Section IV with that of neat ethers.

Section V discusses the primary species in the radiolysis of aqueous solution of alcohols and ethers (Table 4), deoxygenated solutions of saturated alcohols (Tables 5–7), polyhydric alcohols and carbohydrates (Table 8), saturated ethers and acetals and phenols and aromatic ethers. Next, oxygenated solutions of the same substances are treated (Table 9). Throughout the chapter, detailed mechanistic schemes are presented.

336 references up to 1979.
Complementary: 13, 17.
Relevant: 9, 8; **10**, 6; **11**, 6; **15**, 9; **16**, 12; **17**, 10; **18**, 10; **19**, 10; **22**, 11; **25**, 12; **27**, 24; **29**, 7; **31**, 6; **32**, 9; **33**, 21; **36**, 8.

27, *24: Radiation chemistry of thiols, sulphides and disulphides* by *C. von Sonntag and H.-P. Schuchmann*

I. Introduction (971); **II**. Radiolysis in non-aqueous media (972); **III**. Radiolysis in aqueous solution (977); **IV**. References (988).

Section II describes the radiolysis of thiols, disulphides and sulphides in non-aqueous media such as hydrocarbons or cyclohexane, but results in neat liquids and in the gas phase are also presented. Where appropriate, MS, ESR and other studies are also used for the interpretation of the results, and detailed mechanistic schemes are presented.

Section III deals with radiolysis of the title compounds in deoxygenated aqueous solutions. Rate constants of various thiols with some organic and inorganic radicals are given in Table 2 and the reactions of thiols, disulphides and sulphides with hydrated electrons and OH radicals are emphasized, while other radicals are also discussed. Section III.C treats oxygenated solutions (Table 3), which are less well understood than deoxygenated solution. Section III.D presents briefly some biochemical aspects of the radiation chemistry of the title compounds.

183 references up to 1979.
Complementary: 18, 11.
Relevant: 9, 8; **10**, 6; **11**, 6; **13**, 17; **15**, 9; **16**, 12; **17**, 10; **19**, 10; **22**, 11; **25**, 12; **27**, 23; **29**, 7; **31**, 6; **32**, 9; **33**, 21; **36**, 8.

The chemistry of the sulphonium group (1981)

28, *1: General and theoretical aspects* by *M. Simonetta and A. Gavezzotti*

I. Introduction (1); **II**. Quantum chemical description of the sulphonium group (2); **III**. References (14).

Section II presents a simple picture of S valence in the sulphonium group and surveys *ab initio* calculations and the experimentally determined properties of cyclic sulphonium compounds (Table 1), and also some acyclic compounds. Extended Hückel calculations on sulphonium compounds are presented (Table 2) and energy curves and surfaces are given for $^+SH_3$ and $^+SR_3$ groups (Figs 3–9). Table 3 collects data for calculated (EH) properties of $^+SH_3$, $^+SMe_3$ and $^+SPh_3$, including bond angles, barriers, *E*(HOMO), *E*(LUMO), charges on S, H and C atoms and bond overlap populations.

24 references up to 1977.
Complementary: none.
Relevant: **1**, 1; **2**, 1; **3**, 1; **4**, 1; **5**, 1; **6**, 1; **8**, 1; **11**, 1; **12**, 1; **13**, 1; **14**, 1; **15**, 1; **16**, 1; **17**, 1; **18**, 1; **19**, 1; **20**, 1; **22**, 1; **23**, 1; **24**, 1; **25**, 1; **26**, 1; **27**, 1; **31**, 27; **32**, 1; **33**, 1; **36**, 2.

28, *2: The structural chemistry of sulphonium salts* by *E. F. Perozzi and I. C. Paul*

I. Introduction (16); **II**. Structures that have been studied by X-ray methods (21); **III**. Molecular dimensions of sulphonium salts and related compounds (35); **IV**. Intermolecular interactions in sulphonium salts and related compounds (68); **V**. Acknowledgements (74); **VI**. References (76).

The chapter surveys geometric information such as bond lengths and angles, torsion angles, planarity and non-planarity of the groups and geometric aspects of the interaction of sulphonium groups with other molecules or ions. Some compounds related to sulphonium salts (e.g. ylides) are also treated. The methods employed and the factors studied are reviewed in Section I.

Section II describes the structures of sulphonium salts, generally obtained by X-ray analyses (18 salts), sulphonium ylides (4 ylides), two aminosulphonium salts, eleven sulphilimines (i.e. S ylides with N bearing the negative charge) and three Se compounds (selenium salts and Se ylides).

Section III deals with the molecular dimensions of sulphonium ions (Table 1), sulphur–carbon ylides (Table 2), aminosulphonium salts (Table 3), sulphilimines (Tables 4 and 5) and Se compounds (Table 6) and also discusses (Section III.F) chirality, conformations and other structural details of the title compounds.

Section IV treats intermolecular interactions of the title compounds, including electrostatic interactions, intermolecular contacts, crystal packing and polarization effects.

The chapter presents many stereoscopic views of molecules and crystal packings of the title compounds.

59 references up to 1977.
Complementary: 18, 2.
Relevant: 16, 2; **17**, 2; **19**, 2; **20**, 2; **21**, 1; **22**, 2; **23**, 3; **24**, 2; **26**, 2; **27**, 4; **29**, 1; **30**, 1; **31**, 24; **32**, 24, **33**, 12; **34**, 10; **36**, 3.

28, *3: Analysis and determination* by *M. R. F. Ashworth*

I. Introduction (80); **II**. Summary of analytical methods (81); **III**. Chemical methods (83); **IV**. Physical methods (91); **V**. References.

The chapter emphasizes methods relating to sulphonium compounds of biological interest, surface activity and thiazine dyes. Section II gives a brief summary of the methods reviewed.

Section III deals with chemical methods involving ion combination (titrations, precipitations as perchlorate, picrate and other salts and some colorimetric methods). Polarographic and chemical reduction methods can be used for the determination of thiazine dyes (Section III.B).

Degradation of sulphonium salts by acids or alkalis can be used for qualitative and quantitative analysis. Oxidation and halogenation of thiazine dyes (especially methylene blue) can be applied to the titration of the latter.

Section IV deals with spectroscopic methods (which are of very limited use), paper and thin-layer chromatography, ion-exchange, column, liquid and gas–liquid chromatography and electrophoresis.

169 references up to 1977.
Complementary: none.
Relevant: 1, 5; **2**, 8; **3**, 15; **4**, 3; **5**, 17; **8**, 3; **12**, 4; **13**, 6; **14**, 3; **15**, 3; **16**, 3; **17**, 4; **18**, 5; **20**, 3; **22**, 5; **23**, 5; **24**, 5; **25**, 10; **26**, 5; **29**, 21; **30**, 14–22; **35**, 5.

28, *4: Thermochemistry of the sulphonium group* by *R. Shaw*

I. Introduction (101); **II**. Sulphonium salts (102); **III**. Sulphonium ylides (103); **IV**. References (104).

Section II deals with measured heats of formation of ammonium salts (Table 1) and with the heats of formation of trimethylsulphonium salts (Table 2), which have been estimated on the basis of the data in Table 1. The heat of formation of 4-*tert*-butyl-1-thiacyclohexane is estimated by group additivity (Table 3).

Section III discusses sulphonium ylides and the estimation of their heats of formation. Data for the heats of formation of sulphoxides are compared with those of ketones (Table 4).

15 references up to 1977.
Complementary: none.
Relevant: **16**, 16; **18**, 3; **19**, 3; **20**, 11; **21**, 3; **22**, 6; **23**, 4; **24**, 3; **25**, 2; **26**, 4; **27**, 9; **29**, 24; **30**, 2; **31**, 2; **33**, 3; **36**, 4.

28, *5: Photochemistry of sulphonium compounds* by *J. D. Coyle*

Typical photochemical reactions of the title compounds involve homolytic or heterolytic cleavage of the bond between the sulphonium S and one of the adjacent atoms.

Section II describes the conditions, mechanisms and products of the photolysis of sulphonium salts. The processes are complex, e.g. $Ph_3S^+Cl^-$ in CH_3OH produces C_6H_6, PhCl, Ph_2S, Ph_2SO and HCl, whereas $Ph_3S^+I^-$ yields analogous compounds but in very different ratios. Section III deals with sulphonium ylides, the photochemistry of which is strongly influenced by the equilibrium $R_2S^+—\bar{C}R'R'' \rightleftharpoons R_2S + :CR'R''$, although other processes including cleavage of a C—S bond other than the ylidic one and rearrangements may also be important.

Section IV treats sulphimides, which usually fragment to nitrenes and sulphides and can be used as sources of the former. The presence of an S atom in the molecule, even if it is not part of the chromophore, can substantially influence reactions, and sulphonium ylides and similar species have been proposed as photochemical intermediates (Section V).

34 references up to 1979.
Complementary: none.
Relevant: **2**, 16; **3**, 8; **6**, 4; **8**, 12; **9**, 6; **11**, 5; **13**, 16; **14**, 8; **15**, 9; **16**, 11; **17**, 9; **18**, 10; **19**, 20; **22**, 10; **23**, 9; **24**, 12; **25**, 11; **27**, 21, 22; **29**, 5, 6; **32**, 29; **33**, 21.

28, *6: Electronic spectra* by *G. C. Barrett*

The UV absorption of three R_3S^+ salts is shown if Fig. 1 and some CD and ORD studies are quoted (Section II). The nature of the chromophore in R_3S^+ salts is discussed in Section III, including some AMSO–SCF calculations.

The effect of the sulphonium group on the UV spectra is shown in Table 1 and its influence on the behaviour of an adjacent chromophore is discussed (Section IV). Solvent, temperature and the counter ion affect the UV spectra (Section V). UV spectra of sulphonium salts are compared with those of selenonium, telluronium, phosphonium, carbonium and hydroxycarbonium cations (Section VI).

Heteroatom-substituted sulphonium salts (Section VII) are treated briefly.

42 references up to 1979.
Complementary: none.
Relevant: none.

28, *7: Electrochemistry of the sulphonium group* by *J. Grimshaw*

I. Reduction of sulphonium salts (145); **II**. Formation of radical cations and sulphonium salts by oxidation (145); **III**. Reference electrode potentials (154); **IV**. References (154).

Section I deals with the electrochemical reduction of sulphonium salts and presents polarographic half-wave potentials in aqueous solutions (Table 1). Usually, the products are obtained by cleavage of a C—S$^+$ bond, e.g.

$$Me_3S^+ + e \rightarrow Me_2S + {}^{\bullet}CH_3 \xrightarrow[+H^+]{+e} CH_4$$

Section II describes electrochemical oxidations of sulphides with the formation of cation radicals (Tables 2 and 3), which in turn yield a mixture of products, including sulphonic acids, sulphones and sulphonium compounds (the latter on reaction with unreacted sulphide).

Six-membered S heterocyclces are oxidized at a Pt anode to the corresponding radical anion, which is further oxidized to a sulphonium salt (Table 4). The oxidations of thianthrene, phenothiazine and phenoxathiin are described separately.

69 references up to 1976.
Complementary: none.
Relevant: **5**, 2; **8**, 11; **13**, 5; **16**, 15; **17**, 14; **19**, 12; **20**, 5; **21**, 5; **22**, 9; **23**, 10; **24**, 17; **25**, 13; **26**, 12; **27**, 8; **29**, 8, 9; **31**, 7; **32**, 6; **34**, 1.

28, *8: Isotopically labelled sulphonium salts* by *L. F. Blackwell*

I. Introduction (158); **II**. Hydrogen exchange (158); **III**. Sulphonium ylides (162); **IV**. Isotope effects in elimination reactions (170); **V**. Substitution reactions (182); **VI**. References (184).

Section II deals with the acid strengthening effect of positive sulphur and the acidic nature of protons in the α-position to S$^+$, and with structural effects on rates of exchange in these α-protons.

Section III describes the reactions of sulphonium salts and of the corresponding ylides, which are obtained by the action of bases:

$$RCH_2CH_2\overset{+}{S}(R')CH_3 \overset{B^-}{\rightleftharpoons} RCH_2CH_2\overset{+}{S}(R')\bar{C}H_2$$

and which were studied with the aid of isotopes. These include the formation of carbonyl compounds from alkoxysulphonium salts, reaction of the latter with $NaBH_4$, e.g.

$$Me_2\overset{+}{S}OMe + NaBH_4 \rightarrow Me_2S + HCHO$$

their epoxidation and cyclopropanation and reactions with diazomethane and diallyl sulphide. In all these cases, isotopic labelling was used for the elucidation of the mechanism of the reaction. Similarly, eliminations via ylide intermediates (Section III.C) were studied.

Section IV discusses double bond-forming eliminations from compounds containing sulphonium groups [e.g. of the type $ZC_6H_4CH_2CH_2\overset{+}{S}(CH_3)_2$]. Primary, secondary and heavy atom isotope effects and TS structures are reviewed (Section IV.B; Table 1) and the experimental evidence relating to $E2$ TS structures is presented as far as it involves sulphonium compounds (Table 2–4). Section IV.D compares theoretical predictions for the TS for elimination from $ZC_6H_4CH_2CH_2\overset{+}{S}(CH_3)_2$ with the picture obtained by the various studies using isotopes and Hammett ρ values. The Winey–Thorton predictions (Table 1) do not seem to fit the present case.

Finally, substitutions in sulphonium compounds (of both S_N1 and S_N2 types) are treated, describing results obtained mainly with S isotopes, but considering also α-D isotope effects (k^H/k^D values) for sulphonium salts.

71 references up to 1977.
Complementary: none.
Relevant: **5**, 10; **12**, 13; **13**, 15; **17**, 12; **18**, 18; **19**, 17; **22**, 14; **23**, 16; **24**, 20; **26**, 11; **27**, 10; **29**, 28; **33**, 18; **36**, 12.

28, *9: Electronic effects of the sulphonium group* by *J. Shorter*

I. Introduction (188); **II**. The Hammett equation (191); **III**. Sigma values in aromatic systems (197); **IV**. Electronic effects in electrophilic substitution (201); **V**. Electronic effects in nucleophilic substitution (204); **VI**. Stabilization of carbanionic centres by the sulphonium group (207); **VII**. Electronic effects in aliphatic systems (217); **VIII**. The *ortho*-effect of the sulphonium group (220); **IX**. The irrelevance of d-orbital conjugation(?) (222); **X**. References and notes (224).

Section I reviews sulphur bonding, electronic effects and π-bonding and the scope of the chapter.

Section II contains a summary of the features of the Hammett equation and its extensions (Table 1), with special reference to the behaviour of unipolar substituents in LFER relationships. Section III presents sigma values for sulphonium groups obtained from acidity measurements or spectroscopically (by ^1H, ^{19}F and ^{13}C NMR).

Section IV deals with electronic effects in the nitration of sulphonium compounds (Table 3) and in other electrophilic substitutions, and Section V discusses electronic effects in nucleophilic substitutions both in the cases when the sulphonium group is present in the substrate and when it acts as the nucleophile. The stabilizing effect of the title group on carbanionic centres is treated in Section VI, which also presents pK values (Tables 4–7) of various onium salts and ylides. The kinetics of H–D exchanges are dealt with (Section VI.C, Tables 8 and 9), together with the promotion of reactions via carbanionic intermediates (obtained by reactions with bases).

Section VII describes electronic effects of sulphonium groups in aliphatic systems, e.g. in ester hydrolysis (Table 10), and describes a common scale for σ^* with charged and uncharged substituents, first proposed by Palm and co-workers (refs 13–132) for various substituents and extended by others to $R^1R^2S^+$ groups.

Section VIII treats the *ortho*-effect (i.e. the influence of an *ortho*-substituent on the reactivity of a side chain) of R_2S^+ groups, which strongly accelerates the hydrolysis of, e.g., methyl phenylacetate and increases the acidity of phenylacetic acids.

Section IX deals with the as yet unresolved problem of whether it is necessary to invoke a bonding role for the third orbitals of sulphur and quotes papers according to which the π (pd) approach is unnecessary, at least in the relation to bonding.

152 references up to 1978.
Complementary: none.
Relevant: **3**, 3; **4**, 5; **5**, 16; **6**, 8; **7**, 2, 3; **11**, 12; **12**, 5; **13**, 8; **14**, 4; **15**, 5; **16**, 6; **18**, 9; **19**, 8; **23**, 11; **29**, 16; **31**, 8; **36**, 15.

28, *10: Stereochemistry and chiroptical properties of the sulphonium group* by *K. K. Andersen*

Sulphonium salts can be resolved by fractional crystallization of diastereomeric salts formed with various optically active sulphonium ions (Table 1, Section II). Optically active sulphonium salts may also be prepared from enantiomerically pure sulphoxides or from mixtures of sulphides with (–) -menthol and pyridine (yielding chiral *O*-menthylsulphonium salts) (Section III). Absolute configurations are discussed in Section IV, although only few chiral sulphonium ions have been assigned an absolute configuration.

Section V deals with the stereochemistry and configurations of cyclic sulphonium salts containing 3–6 atoms in the ring, with emphasis of six-membered thianium salts (Table 2) and their configurational equilibria. 1,3-Dithianium and thioxanthenium salts are also described.

Section VI treats the stereomutation by various routes (Scheme 6) of trialkyl-, dialkylaryl-, alkyldiaryl- and triarylylsulphonium salts and also of sulphonium ylides.

Section VII discusses processes in which chirality is transferred from S to C, achieving an asymmetric synthesis of the latter. Most of these processes involve sigmatropic rearrangements in which the sulphonium S is converted into a sulphide S.

Section VIII considers the H–D exchange of the acidic protons to the sulphonium S and the stereoselectivity of this process. Section IX presents briefly oxo-, alkoxy- and aminosulphonium salts and Section X reports a study of the ORD and CD of some sulphonium salts.

84 references up to 1978.
Complementary: 28, 17.II.
Relevant: 8, 4; **18**, 7; **19**, 6, 21; **20**, 2; **21**, 1; **22**, 3, 4; **25**, 3; **26**, 3; **27**, 5, 6; **29**, 2, 23; **31**, 1, 19; **32**, 1; **33**, 2; **36**, 17.

28, *11: Synthesis of sulphonium salts* by *P. A. Lowe*

Section II describes the alkylation of saturated, unsaturated, aryl and cyclic sulphides ($R_2S + R'X - R_2\overset{+}{S}R'$); intramolecular alkylation to yield cyclic sulphonium salts and arylation of sulphides.

Section III deals with methods starting from di- and trisulphides ($RSSR + R'I \rightarrow R_2R'\overset{+}{S}I$) and Section IV with those from sulphenyl halides and unsaturated compounds where the sulphonium salt is an intermediate, but can be isolated in some cases:

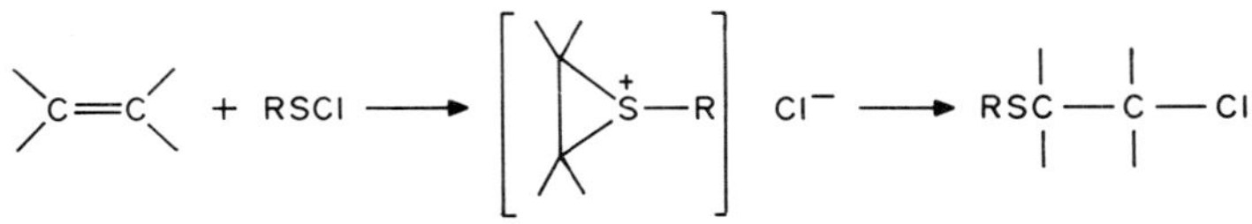

Section V discusses syntheses from protonated sulphoxides, e.g. $Ar_2SO \xrightarrow{H^+} Ar_2\overset{+}{S}OH \xrightarrow{PhOR} Ar_2\overset{+}{S}C_6H_4OR$, from sulphoxides and organometallics, e.g. $Ar_2SO + ArMgBr \rightarrow Ar_2S(OMgBr)Ar \xrightarrow{2HBr} Ar_3\overset{+}{S}\overset{-}{Br} + MgBr_2$, and by alkylation of sulphoxides to yield alkoxy- or arylsulphonium salts.

Section VI considers syntheses of sulphonium salts via *S*-ylides and Section VII the conversion of one sulphonium salt into another by elimination, substitution or rearrangement.

Section VIII deals with the conversion of chloro- and azasulphonium salts into sulphonium salts and Section IX with the formation of bicyclic bissulphonium salts. Section X deals with the methylenesulphonium salts $R\overset{+}{S}{=}CH_2$.

Section XI, XII and XIII discuss the formation of different cyclic sulphonium salts, including isothiophenium, thiopyrylium and dithiolium salts, through a variety of methods.

272 references up to 1978.
Complementary: 28, 7.II.
Relevant: 28, 13, 15, 16.

28, *12: Reactivity of sulphonium salts* by *A. C. Knipe*

Section II.A deals with the unimolecular nucleophilic displacement of sulphonium groups, including solvolysis (Table 1) and racemization. Section II.B considers the stereochemistry and orientation of S_N2 reactions involving sulphonium groups (Table 2), their solvent dependence and effects of ion pairing. The biologically very important methyl transfer is treated separately. Section II.C describes electrophilic and nucleophilic aromatic substitution in the presence of sulphonium groups and Section II.D nucleophilic reactions and the S ion. Next, sulphonium ions as intermediates in substitution at S are discussed.

Section III treats uni- and bimolecular eliminations of sulphonium groups (Table 4), their activation values (Table 5), β-D isotope effects (Tables 6 and 7), solvent effects (Table 8), Hammett correlations (Table 9), Brønsted parameters (Table 10) and product stereochemistry (Table 11), compares rates and alkene proportions in eliminations from sulphonium (Tables 12 and 14) and ammonium (Tables 13 and 15) compounds and tests experimental results vs predictions in the pyrolysis of $RR'R''\overset{+}{S}\overset{-}{O}H$ (Tables 16 and 17). Next, *E1cb* reactions (Table 18), α,β-eliminations (Table 19) and α-eliminations (Table 20) are discussed with mechanistic details.

Section IV deals with polar additions to vinyl and other unsaturated sulphonium salts.

Section V describes the methods of formation of sulphonium ylides from sulphonium salts or from carbenes with sulphides and the stability, structure and reaction of these ylides, including the stereochemistry, mechanism and scope of epoxidation and cyclopropanations by alkylidene transfer, [2, 3]-sigmatropic rearrangements and 1, 2-migrations in sulphur ylides and also some less general reactions of the latter.

508 references up to 1978.
Complementary: **28**, 11, 13, 15, 16.
Relevant: none.

28, *13: Cyclic sulphonium salts* by *D. C. Dittmer and B. H. Patwardhan*

The chapter deals with the synthesis and reactions of cyclic (3–7 and higher membered) sulphonium salts.

Section II describes thiiranium ions, their theoretical treatment, the conditions for their stability, their main methods of preparation and their occurrence as intermediates in reactions such as additions, substitutions and rearrangements. Next, reactions at C and S atoms of the thiiranium ring are discussed and the mechanism, stereochemistry and products are presented. Section II.B treats thiirenium ions, 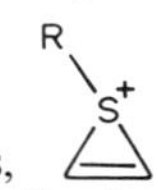 similarly. 1, 2-Dithiiranium ions have only been suggested as intermediates (Section II.C).

Section III presents four-membered rings. Thietanium salts, 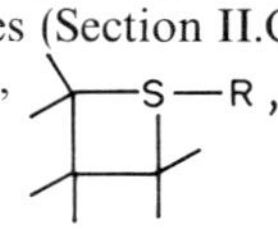 are relatively stable and can be prepared by alkylation of the corresponding thietanes. They appear as intermediates in various reactions of thietanes and during the photolysis of five-membered S-containing cyclic compounds. Their most important reactions involve ring opening, polymerization, *S*-alkylation followed by desulphurization to cyclopropanes and displacements on S with inversion. Section III.B treats briefly thiacyclobutene (thiete) derivatives both as intermediates and as synthones.

Section IV deals with the theoretical treatment, physical properties, stability, synthesis and reactions (both as intermediates and as stable starting materials) of five-membered sulphonium ring compounds. The five-membered ring systems include those with one or two S atoms or with one S atom and one or more other heteroatoms (O, N), together with the analogous unsaturated systems. The sulphonium derivatives are usually easily prepared from the corresponding ring system containing divalent S, by electrophilic attack on the latter with alkyl or allyl or propargyl halides, α-halo ketones and α-halo esters and dimethyl sulphate and similar reagents, e.g.

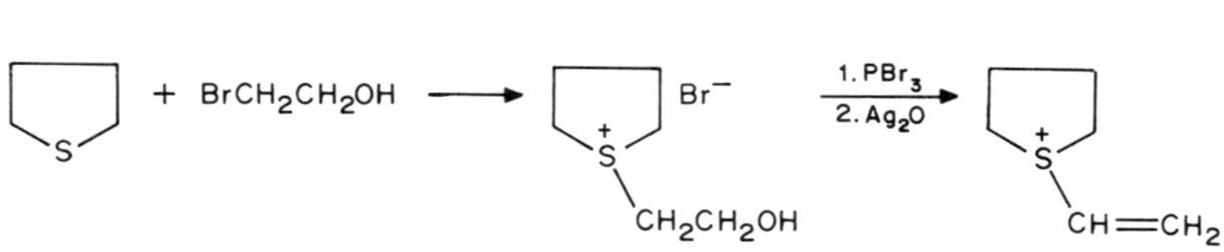

while the preparation of *S*-aryl derivatives is more complicated.

Thiolanium ions and their unsaturated analogues and systems containing additional heteroatoms can be used as synthons in a wide variety of reactions, since they can undergo ring opening, attack on ring-S and ring-C and on the other heteroatoms.

Section V describes thiapyrylium salts, 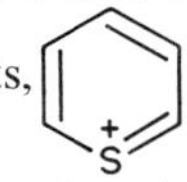 ,their theoretical treatments, syntheses by various methods (e.g. from pyrylium salts with sulphide ions, from 1,5-diketones by cyclization with H_2S, etc.). Among reactions of thiapyrylium salts, nucleophilic attack on either C or S atoms, reactions of side groups, oxidations and one-electron reductions are considered in detail. Next, the structure, synthesis and reaction of non-aromatic six-membered sulphonium compounds and of both aromatic and non-aromatic six-membered sulphonium compounds containing two or more S atoms in the ring are presented.

Section VI discusses more briefly the synthesis and reactions of seven- and higher membered ring sulphonium salts, including macrocyclic compounds.

861 references up to 1978.
Complementary: **28**, 11, 12, 15, 16.
Relevant: none.

28, *14: Organosulphur cation radicals* by *H. J. Shine*

I. Historical developments (524); **II**. Methods of making organosulphur cation radicals (538); **III**. Some newer types of organosulphur cation radicals (549); **IV**. Reactions of organosulphur cation radicals (555); **V**. Acknowledgements (566); **VI**. References (566).

Section I discusses the discovery, identification by ESR and characteristics of phenothiazine (A), thianthrene (B) and phenoxanthiin (C):

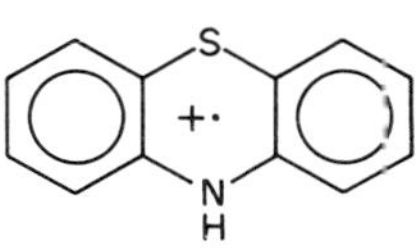
A

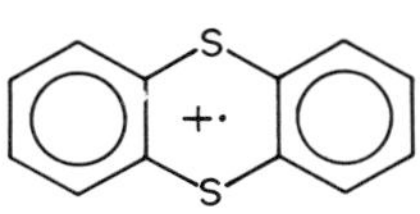
B

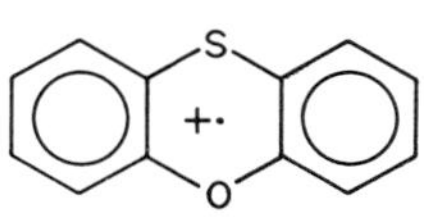
C

and comments generally on the ESR data shown in Tables 1–3. These cation radicals are formed, e.g., from the corresponding sulphoxides with concentrated H_2SO_4, but the mechanism of this reaction is not yet settled, although various proposals have been made (Section I.B).

Section II describes the preparation of the title compounds by the use of H_2SO_4 and other Brønsted acids with persulphuric acid (Table 4), with Lewis acids (Table 5 and 6) and with halogens ($2 \ {>}S{<} + X_2 \rightleftharpoons 2 \ {>}S^{+\bullet} + 2X^-$). They can also be obtained by disproportionation of sulphides with sulphoxides and by anodic oxidation (Tables 7 and 8).

Section III deals with dimer-sulphide, $(Me_2S{-}SMe_2)^{+\bullet}$, and disulphide, $(RSSR)^{+\bullet}$, cation radicals, with tetrathiafulvene (Table 9), 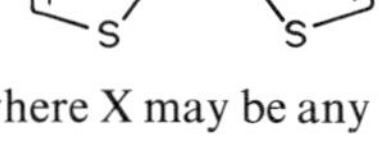 and related cation radicals and also with violenes, $\ddot{X}(CH{=}CH)_n\ddot{X}$, where X may be any organic, particularly heterocyclic, grouping and n may be 0, 1, 2.

Section IV treats the reactions of the title compounds, mainly (owing to their cationic nature) with nucleophiles, e.g. $2 \ {>}S^{+\bullet} + Nu^- \rightarrow {>}S^+{-}Nu + {>}S^\bullet$. The processes presented include reactions with water, ammonia, amines, arenes (which have to be activated by electron-donating substituents, when the title compounds attack through their $S^{+\bullet}$ atom as electrophiles), organometallics (giving *S*-alkylation), ketones (when the S atom displaces a ketonic α-H, alkenes and alkynes (when the two molecules of the title compounds add to the two unsaturated carbons) and finally with inorganic anions (Cl^-, NO_2^-, NO_3^-, which give a variety of products).

161 references up to 1979.
Complementary: none.
Relevant: none.

28, *15: Heterosulphonium salts* by *S. Oae, T. Numata and T. Yoshimura*

Section II.A describes the preparation oxysulphonium (especially alkoxysulphonium) salts, $R\overset{+}{S}(OR'')R'$. The most useful method is alkylation of sulphoxides, e.g. $RSOR' + Me_3O^+ \,^-BF_4 \rightarrow R\overset{+}{S}(OMe)R'BF_4^-$, and also treatment of sulphoxides with alkyl chloroformates, alcoholysis of halosulphonium salts, etc. Dialkoxy- and alkoxyaminosulphonium salts are also treated.

Section II.B deals with unsubstituted azasulphonium salts which are obtained by direct amination of sulphides with hydroxylamines or with chloramine ($Me_2S + NH_2Cl \rightarrow Me_2\overset{+}{S}NH_2Cl^-$) or with a mixture of ammonia with NBS or NCS. *N*-Substituted azasulphonium salts are obtained from sulphides with *N*-haloamides or *N*-haloimides, from sulphonium salts by nucleophilic substitution with amines, e.g. $R_2\overset{+}{S}ClSbCl_6^- + R^1NH_2 \rightarrow R_2S^+NHR'SbCl_6^-$, and by a variety of other similar methods.

Section II.C considers the preparation of thiasulphonium salts (alkylation of disulphides) and Section II.D that of halosulphonium salts (from sulphides and halogens, etc.). Section II.E presents some optically active heterosulphonium salts (Table 1), obtained by alkylation of optically active sulphoxides and related methods.

Section III deals with the reactions of the title compounds. These include nucleophilic substitutions at the sulphonium S atom involving hetero groups containing halogens, O or N, including several examples of neighbouring group participation in racemizations and reductions of sulphoxides in acidic media, also involving the title compounds as intermediates.

Next, nucleophilic substitution on the heteroatom and on the C atom attached to the heteroatom is treated, the latter reaction resulting in transalkylation.

Both the Pummerer and the Sommelet–Hauser reactions involve ylide intermediates and the scope and mechanisms of these two reactions are presented in detail, including isotope, kinetic, product and stereochemical studies.

Section III.E describes the Kornblum oxidation, which usually occurs through unstable alkoxysulphonium salts:

$$RCH_2X + Me_2SO \rightarrow [Me_2\overset{+}{S}OCH_2RX^-] \rightarrow RCHO + Me_2S$$

The reaction can be applied to alkyl halides and tosylates and similar compounds, and has been used extensively for the preparation of aldehydes.

576 references up to 1978.
Complementary: none.
Relevant: None.

28, *16: Synthetic applications of sulphonium salts and sulphonium ylides* by *E. Block*

Section II describes the uses of sulphonium salts for inter- and intramolecular alkylations, e.g. $Me_3S^+OH^- + HOCH_2CH_2SH \rightarrow HOCH_2CH_2SMe$ (Table 1), for the synthesis of alkyl halides by interchange of alkyl groups, e.g. $R_2S + R'Br \rightleftharpoons R_2\overset{+}{S}Br \rightleftharpoons RBr + RSR'$, and for the synthesis of sulphides by ligand interchange (Table 2), e.g. $PhCH_2Br + Me_2S \rightarrow PhCH_2SMe$, when the reaction occurs through a sulphonium salt intermediate.

Section III discusses synthetic methods involving thiocarbonyl ylides, which may lead to syntheses of olefins (Table 3); syntheses of oxiranes from carbonyl compounds and *S*-ylides:

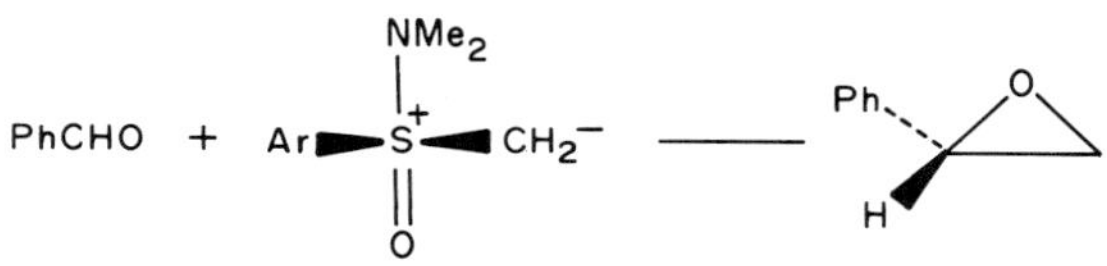

(60%, 20% optical purity)

syntheses of cyclopropanes by Michael addition, alkylations and other displacements, α-eliminations to produce carbenes, α, β-eliminations leading to C=C or C=O compounds (Table 4), fragmentations and [1,2]-, [1,4]-, [1,5]- and [2,3]-sigmatropic rearrangements, the last involving multi-carbon ring expansions (Table 5) and, in the case of azasulphonium ions, leading to substituted anilines and *N*-heterocyclics (Scheme 2).

82 references up to 1979.
Complementary: non.
Relevant: **28**, 12, 13, 15.

28, *17: The biochemistry of sulphonium salts* by *G. A. Maw*

Section II deals with the naturally occurring title compounds: (A) dimethyl-β-propiothetin, (B) *S*-methyl-L-methionine (Table 1), (C)*S*-adenosyl-L-methionine (Tables 2 and 3) (D) *S*-adenosyl-(5′)-3-methylpropylamine.

Section III discusses the biosynthesis of methionine and the biosynthetic pathways to its conversion to the four known natural sulphonium compounds.

Section IV describes the metabolism of sulphonium salts. The processes include with (A) cleavage to thioesters, alkyl transfer (Table 4), oxidation and some other reactions, with (B) cleavage to Me_2S and Me transfer occurs and with (C) a variety of chemical and enzymatic breakdowns occur, including S-, N-, O- and C-methylation and also methylation of polysaccharides, proteins and nucleic acids, decarboxylation and polyamine biosynthesis. Next, the biosynthesis and metabolism of (D) is presented and finally (Section V) the medical and agricultural aspects of the title compounds are described.

442 references up to 1979.
Complementary: **18**, 13.
Relevant: **2**, 7; **3**, 5; **4**, 9; **5**, 18; **7**, 4; **11**, 11; **12**, 12; **13**, 14; **15**, 10; **16**, 13; **17**, 13; **19**, 13; **20**, 6; **22**, 20; **24**, 14; **26**, 13; **31**, 9.

Supplement F: The chemistry of amino, nitroso and nitro compounds and their derivatives (1982)

29, *1: Structural chemistry* by *S. Sorriso*

I. Introduction (1); **II**. The amino group (3); **III**. The nitro group (32); **IV**. The nitroso group (43); **V**. Ring deformations (45); **VI**. Acknowledgements (46); **VII**. References (46).

Section II describes the structure of NH_3 and some of its isotopic species (Table 1), of aliphatic amines (Table 2) and aliphatic ammonium compounds (Table 3), aniline and its nitro- and halo-substituted derivatives (Tables 4 and 5), of some heterocyclic amines (Table 6), *N*-alkyl- and *N*, *N*-dialkylanilines (Table 7) and the $—NH_2$ group in some anilinium compounds (Table 8) and Table 9 gives structural parameters of the $C—NH_2$ group in some aromatic amines.

Section III deals with the structures of aliphatic nitro compounds (Table 10), nitrobenzene and its derivatives (Table 11) and polynitrobenzenes (Table 12).

Section IV discusses the *C*-nitroso group in some monomeric (Table 13) and dimeric compounds. Section V deals with ring deformations caused by substitution of a ring H with an NH_2, NO_2 or NO group and the interpretation of these effects. Throughout the chapter, X-ray, ND, ED and MW data are presented in detail.

243 references up to 1978.
Complementary: 19, 2.III.C.
Relevant: 16, 2; **17**, 2; **18**, 2; **19**, 2; **20**, 2; **21**, 1; **22**, 2; **23**, 3; **24**, 2; **26**, 2; **27**, **4**; **28**, 2; **30**, 1; **31**, 24; **32**, 24; **33**, 12; **34**, 10; **36**, 3.

29, *2: Stereochemistry and conformations* by *M. Raban and J. Greenblatt*

I. Stereochemistry at trivalent nitrogen (53); **II**. Isomerism and topomerism (54); **III**. Effects on torsional and inversional barriers (56); **IV**. Torsional barriers in nitroso and nitro compounds (62); **V**. Torsional and inversional barriers in amino compounds (74); **VI**. References (80).

Section I reviews briefly the principles of the stereochemistry of trivalent N compounds. In these the configurational units (associated with trivalent N) are usually stereolabile and undergo fast stereomutation, as shown by dynamic nuclear magnetic resonance spectroscopy. The concept of enantiotopic and diastereotopic molecules is explained (Section II) and also the concept of topomerization.

Section III discusses steric and inductive effects on torsional and inversional barriers and presents molecular orbital considerations relating to changes in dihedral or torsion angles and to substituent effects and inversion barriers. The results of MO calculations are summarized in Table 1.

Section IV describes torsional barriers in nitroso compounds such as nitrosamines (Table 2), nitrosobenzenes (Table 3), nitric acid and nitrite esters and nitro compounds. Section V deals with torsional and inversional barriers in simple amines, acyclic compounds with N—heteroatom (N, O, P or S) bonds and in saturated *N*-heterocyclics.

106 references up to 1979.
Complementary: none.
Relevant: **19**, 21; **20**, 2; **21**, 1; **22**, 3; **27**, 5; **28**, 10; **31**, 19; **32**, 1; **33**, 2.

29, *3: The chemistry of ionized amino, nitroso and nitro compounds in the gas phase* by *H. Schwarz and K. Levsen*

I. Introduction (85); **II**. Amines (86); **III**. Nitroso compounds (100); **IV**. Nitro compounds (101); **V**. Acknowledgements (122); **VI**. References (122).

Section II deals with neighbouring-group effects in functionalized aliphatic amines and presents detailed reaction schemes for fragmentations via α-bond cleavage of radical cations and other routes (Schemes 1–14). Next, aniline, benzylamine and their derivatives are treated and also the mass spectrometric *ortho* effects.

Section III discusses briefly the MS of nitroso compounds, and Section IV that of nitro compounds; the $RNO_2\urcorner^{+\cdot} \rightleftharpoons RONO\urcorner^{+\cdot}$ isomerization, H-transfer and oxygen migration (Schemes 23–42), chemical ionization of aliphatic nitro compounds (Schemes 43–47) and $RNO_2\urcorner^{-\cdot}$ anions (Schemes 48–53).

Since MS reactions of the simpler title compounds have been very well documented (refs 1a–j), this chapter emphasizes more complex molecules and reactions and especially the MS chemistry of 'naked' ions which contain NH_2, NO or NO_2 groups.

112 references up to 1979.
Complementary: none.
Relevant: **9**, 7; **10**, 5; **13**, 19; **16**, 4; **17**, 5; **18**, 6; **19**, 5; **22**, 7; **24**, 6; **25**, 4; **26**, 6; **27**, 7; **30**, 22; **31**, 3; **32**, 3; **33**, 4; **36**, 7.

29, *4: Rearrangements involving nitroso and nitro compounds* by *D. I. H. Williams*

I. Introduction (127); **II**. The nitramine rearrangement (128); **III**. The rearrangement of *N*-nitrosamines (133); **IV**. Rearrangements involving alkyl nitrites (142); **V**. The rearrangement of nitroolefins (144); **VI**. Aromatic nitro group rearrangements in acid solution (146); **VII**. Rearrangements of *C*-nitro compounds to *C*-nitrites (149); **VIII**. References (151).

Section II deals with the acid-catalysed and thermal nitramine rearrangement, [$C_6H_5N(R)NO_2 \rightarrow o$- + p-$O_2NC_6H_4NHR$] and its scope and mechanism. Section III treats the Fischer–Hepp rearrangement of *N*-nitrosamines [$C_6H_5N(R)NO \rightarrow p$-$ONC_6H_4NHR$], including details of the scope, kinetics and mechanism of the main reaction path, and also routes leading to denitrosation (Tables 1–3). Other rearrangements of nitrosamines (1,2- and 1,3-rearrangements in aliphatic systems, homolytic photochemical cleavage to yield NO, etc.) are also discussed.

Section IV considers thermal, photochemical and acid- and base-catalysed reactions of alkyl nitrites when the *O*-nitroso group may yield a *C*-nitroso product or it may act as a nitrosating agent.

Nitroolefins (Section V) give 1,2-shifts [$RC(NO_2){=}CH_2 \xrightarrow{\text{Base}} RCH{=}CHNO_2$] and also photo-rearrangements, e.g. $PhCH{=}C(NO_2)R \xrightarrow{h\nu} PhC({=}NOH)COR$.

In aromatic nitro compounds acid-catalysed rearrangements occur in which the NO_2 shifts from one ring C atom to an adjacent one (Section VI). Aromatic 1,3- and 1,4-migrations are also known but their mechanism is not clear.

Section VII describes the rearrangement of *C*-nitro compounds to *C*-nitrites, e.g. the photoisomerization $CH_3NO_2 \xrightarrow{h\nu} CH_3ONO$ and similar reactions occuring on aliphatic and aromatic carbons and also with nitroamines and cobalt complexes.

112 references up to 1979.
Complementary: none.
Relevant: **1**, 7; **2**, 15; **3**, 14; **4**, 10; **5**, 8; **8**, 9; **9**, 3; **10**, 3; **11**, 4; **12**, 16; **13**, 13; **14**, 7; **15**, 8; **16**, 17; **17**, 8; **18**, 15; **19**, 18; **20**, 4; **23**, 13; **24**, 10; **26**, 15.

29, *5: The spectroscopy, photophysics and photochemistry of saturated amines* by *A. M. Halpern*

The chapter emphasizes the role of the nitrogen lone-pair electrons in the excited states of saturated amines.

Section II describes NH_3 and alkylamines and presents absorption spectra and summarizes information on cage amines (Table 1). The triplet state of gaseous amines is considered, and also the spectra of amines in the condensed phase.

Section III deals with the photophysics (i.e. the study of the structural and dynamic physical properties which characterize the electrochemically excited state of a molecule) of NH_3, RNH_2, RR′NH, RR′R″N (Table 2), cage amines and diamines in the gas phase and with tertiary amines in non-polar solvents, including fluorescence spectra, quenching reactions (Table 3), energy-transfer reactions, exciplex and excimer fluorescence, excimer formation and temperature dependence. Polar solvents are also considered, and also briefly some photochemical processes (Section IV).

102 references up to 1979.
Complementary: **19**, 20.
Relevant: **2**, 16; **3**, 8; **6**, 4; **8**, 12; **9**, 6; **11**, 5; **13**, 16; **14**, 8; **15**, 9; **16**, 11; **17**, 9; **18**, 10; **22**, 10; **23**, 9; **24**, 12; **25**, 11; **27**, 21, 22; **28**, 5; **29**, 6; **32**, 29; **33**, 21.

29, *6: Photochemistry of nitro and nitroso compounds* by *Y. L. Chow*

VIII. Photochemistry of *N*-nitro and *N*-nitroso compounds (262); **IX**. Photochemistry of *N*-nitrosimines and *N*-nitroimines (280); **X**. References (281).

Section II describes the light-induced reactions of aromatic nitro compounds. It deals with the photoreduction under neutral, acidic and basic conditions and gives detailed mechanistic schemes for all three processes. Next, nucleophilic photosubstitution, photoaddition and photorearrangements are treated, the last including intramolecular redox reactions of *o*-alkylnitrobenzenes, *o*-nitrobenzyl derivatives and nitrobenzenes with *ortho* C=X bonds or with *ortho*-heteroatom substituents, and also the nitro–nitrite rearrangement.

Regarding α, β-unsaturated nitroalkenes (Section III), photoreactions of the olefinic bond, nitro–nitrite rearrangements and intramolecular photoadditions are discussed, and also the possible interfaces of the reaction pathways.

Section IV deals with simple nitroalkanes, with *aci*-nitronates and with *gem*-substituted nitroalkanes.

Section V treats the photochemistry of simple nitrosoalkanes (Table 2), geminally substituted nitrosoalkanes, aromatic nitroso compounds and some related molecules.

Section VI discusses the overall reaction pattern, recombination, disproportionation, α-cleavage, H-abstraction and gas-phase photolysis in the photoreaction of simple alkyl nitrites and also with the photoadditions of the latter, including intramolecular cyclizations and other reactions. The intramolecular exchange of an unactivated H and an X group through the intermediacy of a heteroatom radical (Barton reaction) is treated:

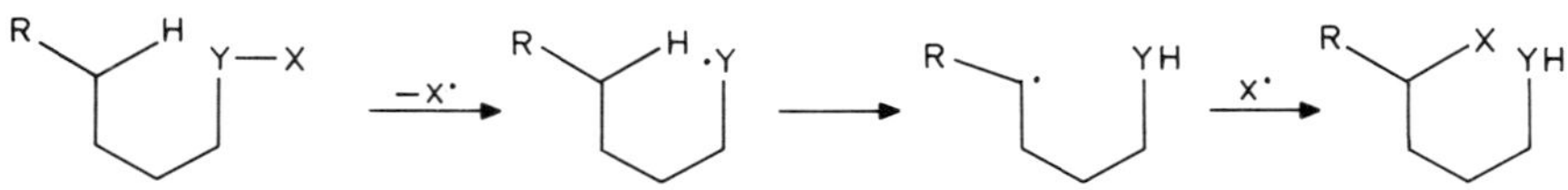

The mechanistic details of this reaction are given, treating intramolecular H-abstraction, reactions of alkyl radicals and their trapping and discussing the interface of the Barton the α-cleavage reactions.

Section VII deals with the photochemistry of nitrate esters, (formation of carbonyl compounds or Barton reactions and also other routes). Section VIII describes the photochemistry of N—NO and N—NO_2 compounds, including primary photoprocesses, photoelimination, photoreduction, photoaddition and photolytic processes of *N*-nitrosamines, and also of nitramines and nitrosamides. Finally, Section IX treats the photolysis of *N*-nitrosimines and *N*-nitroimines.

394 references up to 1979.
Complementary: 6, 4.
Relevant: 2, 16; **3**, 18; **8**, 12; **9**, 6; **11**, 5; **13**, 16; **14**, 8; **15**, 9; **16**, 11; **17**, 9; **18**, 10; **19**, 20; **22**, 10; **23**, 9; **24**, 12; **25**, 11; **27**, 21, 22; **28**, 5; **29**, 5; **32**, 29; **33**, 21.

29, *7: Radiation chemistry of amines, nitro and nitroso compound* by *C. L. Greenstock*

I. Introduction (292); **II**. Radiation chemistry (292); **III**. Radiolysis of amines (300); **IV**. Radiolysis of amino acids (302); **V**. Radiolysis of nitro and nitroso compounds (304); **VI**. Redox reactions (307); **VII**. Miscellaneous (310); **VIII**. Biological applications (312); **IX**. References (316).

Section II reviews general concepts of radiation chemistry, such as energy absorption, early events in spurs and the time scale of the physical and chemical processes that occur, dosimetry, free-radical reactions, radical scavenging, pulse radiolysis and product analysis.

Section III deals with the radiolysis of aliphatic and aromatic amines (Table 1 gives rate constants for the reactions of e_{aq}^-, H· and ·OH) in the presence and absence of oxygen. Section IV describes the radiolysis of amino acids (glycine, Table 2) and Section V that of nitro and nitroso compounds (Table 3) and presents pK_a values for the protonation of nitro radical anions (Table 4).

Section VI deals with redox reactions, electron transfer, self-reactions of some aromatic nitro radical anions (Table 5), one-electron reduction potentials of nitroaromatic compounds (Tables 6 and 7) and the protonation of nitro radical anions.

Section VII considers radiation-induced hydroxylation, carboxylation, deamination and denitration and the competition between radical oxidation vs adduct formation in nitro compounds. Biological applications are treated in Section VIII.

109 references up to 1979.
Complementary: none.
Relevant: **9**, 8; **10**, 6; **11**, 6; **13**, 17; **15**, 9; **16**, 12; **17**, 10; **18**, 11; **19**, 10; **22**, 11; **25**, 12; **27**, 23, 24; **31**, 6; **32**, 9; **33**, 21; **36**, 8.

29, *8: Electrochemistry of nitro compounds* by *A. J. Fry*

Section II deals with the mechanism of the reduction of nitrobenzene in aprotic and protic media, the effect of trapping agents and the choice of experimental conditions (Table 1). Section III describes ring-substituted nitroarenes and Section IV aromatic dinitro compounds which can be reduced by stages and which, in the case of 2, 2-dinitrobiphenyl, can produce tricyclic azoxy, azo or hydrazo compounds.

Section V treats aliphatic nitro compounds which usually yield hydroxylamines as final reduction products, together with oximes:

$$CH_3NO_2 \rightarrow CH_3NO \begin{cases} \nearrow CH_2{=}NOH \\ \searrow CH_3NHOH \end{cases}$$

Section VI considers fragmentations of nitroarene radical anions, indirect electrochemical reductions by producing electrochemically a species capable of chemical reduction of the nitro compounds and reactions of nitroarenes with bases producing (usually coloured) anions.

81 references up to 1979.
Complementary: none.
Relevant: **5**, 2; **8**, 11; **13**, 5; **16**, 15; **17**, 14; **19**, 12; **20**, 5; **21**, 5; **22**, 9; **23**, 10; **24**, 17; **25**, 13; **26**, 12; **27**, 8; **28**, 7; **29**, 9; **31**, 7; **32**, 6; **34**, 1.

29, *9: Electrochemistry of amines* by *W. E. Britton*

Section II gives general considerations regarding the electrochemical oxidation of amines. Section III deals with the aminium radical, which is the first intermediate in the reaction ($R\overset{+}{N}R'H$) and considers its structure and chemical properties.

Section IV presents mechanistic considerations regarding the deprotonation of aminium radicals at either N or C atoms, H-abstraction by aminium radicals and their coupling (either head-to-tail or tail-to-tail).

Section V discusses electrochemical oxidation of triarylamines (to yield benzidines and the corresponding quinonoid dications), of *N*-alkylanilines (giving benzidines and diphenylamines) and of aniline and its derivatives (reactions which are strongly pH dependent and may produce strongly coloured polymeric products, e.g. 'aniline black,' aminophenols, benzoquinones and other products). Section VI treats the various routes occurring with aliphatic amines and Section VII considers oxidation in the presence of nucleophiles, when tertiary amines undergo substitutions both at aromatic positions (when present) and at the α-C of side-chains.

86 references up to 1979.
Complementary: none.
Relevant: **5**, 2; **8**, 11; **13**, 5; **16**, 15; **17**, 14; **19**, 12; **20**, 5; **21**, 5; **22**, 9; **23**, 10; **24**, 17; **25**, 13; **26**, 12; **27**, 8; **28**, 7; **29**, 8; **31**, 7; **32**, 6; **34**, 1.

29, *10: Radical anion reactions of nitro compounds* by *N. Kornblum*

The reaction $p\text{-}NO_2C_6H_4CH_2Cl + Me_2\bar{C}NO_2 \rightarrow p\text{-}NO_2C_6H_4CH_2C(NO_2)Me_2$ is an electron-transfer process in which radical anions and free radicals are intermediates and which is a chain reaction. The scope of these reactions is wide; they proceed under mild conditions and are fairly insensitive to steric hindrance. Sections II–IV give mechanistic details, reaction schemes and yields for the *o*-, *m*- and *p*-nitrobenzyl system and for the *p*-nitrocumyl and other cumylic systems.

Section VII describes the title reactions in purely aliphatic and alicyclic systems, when the displaced atom is the halogen of α-chloro-, α-bromo- or α-iodonitroalkanes or another leaving group in α-nitrosulphones, -ketones, -nitriles, etc. The attacking anion may be derived from secondary (e.g. Me_2CHNO_2) or primary (e.g. $MeCH_2NO_2$) nitro compounds.

Section VIII discusses the reasons for the differences in the reactivity of various nitro compounds to undergo the title reactions. Section IX considers recent studies on the replacement of tertiary aliphatic and alicyclic nitro groups by hydrogen and the replacement of nitro groups by hydrogen from C atoms to which a cyano, ester or keto group is also bonded:

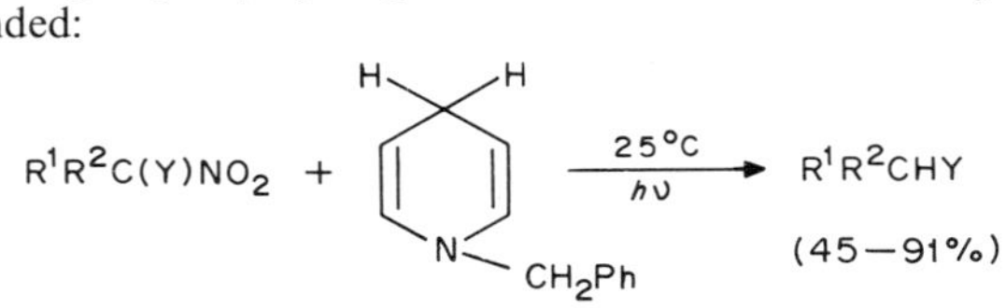

(Y = CN, COOR, COAr). Other systems in which NO_2 groups can be replaced by radical anion chain reactions are also presented.

Section XI treats some related radical anion reactions involing RHgCl reagents and some other mechanistically related processes.

Section XII deals briefly with compounds which undergo electron-transfer substitution, although they do not contain NO_2 groups.

50 references up to 1980.
Complementary: none.
Relevant: none.

29, *11: Direct aminations* by *T. Sheradsky*

I. Introduction (395); **II**. Electrophilic amination (396); **III**. Nucleophilic amination (401); **IV**. Free-radical amination (406); **V**. Miscellaneous aminations (409); **VI**. References (414).

The subject of the chapter is direct displacement of H atoms by amino groups.

Electrophilic amination (Section II) by $ClNH_2$, H_2NOSO_3H and other hydroxylamine derivatives occurs with carboxylic acids and derivatives (Table 1); dialkylamination is facile:

$$\begin{array}{l}\mathrm{PhC(CN)COOEt} \\ \quad | \\ \quad \mathrm{H}\end{array} + \mathrm{Me_2NOSO_2C_6H_2Me_3} \longrightarrow \begin{array}{l}\mathrm{PhC(CN)COOEt} \\ \quad | \\ \quad \mathrm{NMe_2}\end{array}$$

and arenes can be aminated by H_2NOH derivatives in good yields (Table 2).

Nucleophilic amination by alkali metal amides (Chichibabin reaction) of pyridine and related heterocycles is treated in Section III, together with amination by nitriles (Ritter reaction or anodic acetamidation in CH_3CN, Table 3), and by chloramine derivatives and especially by NCl_3 in the presence of $AlCl_3$ (aminating both aromatic and aliphatic systems).

Section IV discusses free radical cuminations involving $ClNR_2$ (Tables 4 and 5), NH_2OH derivatives, NH_3 and H_2NNH_2. Section V deals with nitrene insertion into C—H bonds in aliphatic (Tables 6 and 7) and aromatic (Table 8) compounds and with allylic amination by imido-sulphur and -selenium compounds (Table 9).

97 references up to 1979.
Complementary: 3, 2.
Relevant: none.

29, *12: Pyrolysis of nitrites, nitrates, nitro compounds, nitroso compounds and amines* by *L. Batt*

I. Introduction (417); **II**. Alkyl nitrites (418); **III**. Alkyl nitrates (427); **IV**. Nitroalkanes and nitroarenes (434); **V**. Nitroso compounds (441); **VI**. Amines (447); **VII**. Conclusions (453); **VIII**. Acknowledgements (454); **IX**. References (454).

The decomposition of allyl nitrites is a free-radical chain process which was investigated in great mechanistic detail (Tables 1–6, Section II). Section III treats alkyl nitrates (Tables 7 and 8), polynitrates (Table 9) and peroxynitrates (Table 10). Aliphatic (Tables 11–13) and aromatic (Table 14) nitro compounds and nitrocarbamates (Tables 15) are dealt with in Section IV.

Section V discusses the dimerization of nitroso compounds (Table 16), the conversion of *cis*- to *trans*-dimers (Table 17), Arrhenius parameters for the dissociation of *trans* dimers (Table 18), bond dissociation energies of *C*-nitroso compounds (Table 19) and the decomposition of *N*-nitrosacetanilide (Table 20) and *N*-nitrosoamides (Table 21).

Section VI reports studies on amines (Tables 22, 23 and 25), hydrazines (Table 24) and fluoroamines (Table 26). Little quantitative information is available on the decomposition of these amino derivatives.

213 references up to 1979.
Complementary: 29, 24.
Relevant: 1, 3; **3**, 4; **16**, 10; **22**, 15; **25**, 15; **27**, 11; **32**, 7; **33**, 20.

29, *13: Nitrones and nitronic acid derivatives; their structures and their roles in synthesis* by *E. Breuer*

Section II deals with theoretical calculations relating to the structure of nitrones, $R^1R^2C{=}N\overset{\rightarrow O}{\underset{R^3}{<}}$, their X-ray, dipole moments and acid–base properties, PES, UV (Table 1), IR, NMR (Table 2) and MS and their geometric isomerism (Table 3) and various types of tautomerism. Next, rearrangements of nitrones involving the formation of oxime *O*-ethers, oxaziridines and amides and oxygen migrations and the formation of olefins are treated. Oxidation of nitrones with $Pb(OAc)_4$, $Fe^{(III)}$, periodate, halogens, SeO_2 and O_3 are presented, yielding different products in each case.

Nitrones are reduced by one-electron addition to the corresponding radical anions. Other reductions include deoxygenation to azomethines, reduction to hydroxylamines and amines and the use of nitrones (owing to their ability to act as O-donors) as oxidizing agents.

1, 3-Dipolar cycloadditions yielding a variety of heterocyclic compounds have been studied extensively and are presented in detail in many reaction schemes. Cycloadditions to other 1, 3-dipoles and reactions of *C*-(1-chloroalkyl) nitrones are also considered.

In Section II.B.6 reactions of nitrones with heterocumulenes, such as ketenes, ketene imines, isocyanates and isothiocyanates, carbodiimides, sulphines, *N*-sulphinyl compounds and sulphenes, are dealt with.

Nitrones react with electrophiles. In these reactions nitrones may act either as oxygen or as carbon nucleophiles (the latter type including mainly aldol-type condensations and dimerizations).

Reactions of nitrones with nucleophiles include hydrolysis

$$R^1CH{=}\underset{\displaystyle O^-}{\overset{+}{\underset{|}{N}}}R^2 + H_2O^- \longrightarrow R^1CHO + R^2NHOH$$

and attacks by NH_2OH (to yield oximes), alcohols, thio acids, cyanide ions, carbanions and a variety of other nucleophilic reagents.

Section II.B.9 treats the reactions of nitrones with organometallic compounds containing Li, Mg, Zn, Cu, Ge, Pb, Sn, Hg and transition metals. Next, some reactions of nitrones with free radicals are discussed.

Section III describes the chemistry of nitronic acid derivatives, including the nitroalkane–nitronic acid tautomerism:

$$R_2CHNO_2 \rightleftharpoons R_2C{=}\overset{+}{N}\begin{smallmatrix}O^-\\OH\end{smallmatrix}$$

and the structures of nitronic acid derivatives involving theoretical studies, X-rays, dipole moments, UV and NMR spectra. This is followed by a discussion of the reactions of nitronic acid derivatives, such as oxidation of nitronates to give carbonyl products:

$$R_2C{=}\overset{+}{N}\begin{smallmatrix}O^-\\O^-\end{smallmatrix} + MnO_4^- \longrightarrow R_2CO + MnO_3^- + NO_2^-$$

reduction of nitronates under hydrolytic conditions yields carbonyl compounds probably via the oxime, e.g.

$$RCH{=}\overset{+}{N}\begin{smallmatrix}O^-\\O^-\end{smallmatrix} \xrightarrow{H^+} RCH{=}NOH \xrightarrow{H_2O} RCHO$$

1, 3-dipolar cycloadditions of nitronates leading to heterocycles containing both O and N ring atoms and reactions of nitronates with electrophiles (the nitronate acting either as a O or C nucleophile) and with nucleophiles (e.g. the Nef reaction, Table 4).

439 references up to 1979.
Complementary: 6, 7.
Relevant: none.

29, *14: Nitroxides* by *H. G. Aurich*

I. General and theoretical aspects (566); **II**. Formation of nitroxides (569); **III**. Investigation of nitroxides by physical methods (582); **IV**. Specific properties of nitroxides as studied by ESR (587); **V**. Reactions involving the nitroxide group (594); **VI**. Reactions without direct involvement of the nitroxide group (614); **VII**. Acknowledgements (617); **VIII**. References (617).

Section I reviews the stabilization of nitroxides, in which in $>N-\dot{O} \leftrightarrow >\overset{+\cdot}{N}-\overset{-}{O}$ the unpaired electron is delocalized between the N and O atoms and the various types of nitroxide radicals and the related thionitroxides and iminoxyls.

Section II describes the formation of nitroxides by oxidation of hydroxylamines, amines, nitroso compounds, nitrones, *O*-substituted hydroxylamines and aminyl radicals ($>N^{\cdot} + O_2 \rightarrow [>N-O-O^{\cdot}] \xrightarrow{>N^{\cdot}} 2\,>N\dot{-}O$). Next, the process of 'spin trapping,' i.e. the reaction of a short-lived radical with a nitroso compound or a nitrone or similar compounds yielding a persistent radical (in which the spin is retained), is discussed, together with the properties of spin traps, example of spin trap reactions, the selectivity and kinetics of the reaction and the use of nitroso compounds as spin traps.

Section II.D treats the formation of nitroxides by the reduction of nitro or nitroso compounds.

Section III deals with X-ray analysis and electron diffraction (Table 2), dipole moments (Table 3), ESR (Table 4), NMR and ENDOR spectroscopy and IR and UV (Table 5), PES and MS.

Section IV considers the spin density distribution (Tables 6 and 7), conformation (Table 8) and some dynamic processes (e.g. interchange of conformations, fast exchange of H atoms, spin exchange in diradicals), all studied by ESR.

Section V deals with the main reactions of nitroxides, which are reduction to hydroxylamines and oxidations to nitroso or nitro compounds. It discusses protonation and complex formation, addition of radicals to the nitroxide group, e.g.

$$R'N{=}O \xrightarrow{R^{\bullet}} R'{-}\underset{\underset{O}{|\cdot}}{N}{-}R \longrightarrow R'{-}\underset{\underset{OR}{|}}{N}{-}R$$

dimerization (which is usually not a favoured reaction), disproportionation ($2R\underset{\underset{O}{|\cdot}}{N}H \rightleftharpoons RNHOH + RN{=}O$) (Table 9), reduction (to yield hydroxylamines) (Table 10), and oxidation to yield nitroso compounds, nitrones or oxoammonium compounds (electrochemically) (Table 11). α-Scission of nitroxides is the reverse of spin trapping by nitroso compounds ($R^1\underset{\underset{O}{|\cdot}}{N}R^2 \rightarrow R'N{=}O + \dot{R}^2$), while the oxidation of nitroxides to afford nitrones may be considered as a β-scission.

Section VI presents some reactions in which the nitroxide group is not directly involved, but only other reactive sites in the radical molecule.

229 references up to 1979.
Complementary: none.
Relevant: none.

29, *15: Enamines and ynamines* by *G. Pitacco and E. Valentin*

Enamines exist in tautomeric equilibria in which the imino forms predominate:

unless the amino form is especially stabilized (e.g. by hydrogen bonding).

Section II deals with the preparation of acyclic, cyclic, heterocyclic, functionalized and metallo-enamines and with mechanistic and structural aspects in these syntheses.

Next, basicity, IR (Table 1), UV, NMR and MS are presented. Section II.C discusses the reactions of enamines, including reduction (to amines), oxidation (to carbonyl and amide fragments), protonation (first on N), alkylation (mainly in order to obtain monoalkylated carbonyl compounds, Table 2), alkyl halides, olefins, alkynes, aromatic compounds, carbonyl compounds, carbenes and thiocarbenes. Acylations are described with phosgene, chloroformates, acyl chlorides, ketenes, isocyanates, isothiocyanates and sulphonyl halides. 1,3-Cycloaddition gives usually five-membered heterocycles in a regiospecific manner. In the following, halogenations and reactions with diazonium, azodicarbonyl and sulphur compounds and with cyclopropanes are treated.

Section III describes ynamines ($-C{\equiv}C{-}N{<}$) and their preparation from activated alkynes (e.g. $LiC{\equiv}CNR_2 \xrightarrow{R'X} R'C{\equiv}CNR_2$) from haloethenes (by addition of $LiNR_2$)

and by some other less general methods. The physical properties of ynamines are presented briefly, followed by their reactions such as protonation, hydration, alkylation, reactions with electrophilic olefins and dienes (yielding 1,2- and 1,4-cycloaddition products), activated acetylenes, aromatic systems (e.g. with azulene either sigmatropic rearrangement or valence isomerization occurs), activated heterocycles (yielding cycloadditions as a rule), acylating agents (acid chlorides, phosgene, $SOCl_2$, etc., add to the triple bond to give α-chloroenamine derivatives), carbonyl and azomethine derivatives (1,2-cycloaddition with subsequent ring opening to α,β-unsaturated amides), heterocumulenes (CO_2, isocyanates, ketenes and sulphenes give a variety of cyclizations), 1,3-dipoles (to yield five-membered heterocycles) and halogens and mentions finally some reactions catalysed by coordination and organometallic compounds (oligomerizations, cycloadditions, insertions, etc.).

383 references up to 1979.
Complementary: none.
Relevant: none.

29, *16: Nitro-activated carbon acids* by *E. S. Lewis*

I. Introduction (715); **II**. Equilibrium acidity (715); **III**. Rates of proton transfer from nitroalkanes (723); **IV**. Conclusions (728); **V**. References (728).

Section II discusses equilibrium acidity of Brønsted acids (i.e. substances from which a proton can be removed), including nitroalkanes (Tables 1 and 2), nitroalkenes and nitrophenylmethanes (Table 3), and inductive strengthening of acids by NO_2 groups.

The nitronic acid–nitroalkane tautomerism is treated and the acid strengths of both (Table 4) and also the Lewis acidity of nitro compounds are discussed.

Section III describes the differences between nitroalkanes and other carbon acids, the rates and equilibrium of the ionization of nitro compounds, including the effects of substituents, solvents and bases, and isotope effect studies.

53 references up to 1978.
Complementary: 6, 7.II, **31**, 17.
Relevant: 5, 6; **11**, 3; **13**, 7; **18**, 8; **23**, 6; **24**, 4; **29**, 17.

29, *17: Gas-phase basicity and acidity of amines* by *D. K. Bohme*

I. Introduction (731); **II**. Kinetics of protonation and deprotonation (732); **III**. Proton-transfer equilibria (741); **IV**. Gas-phase basicities of amines (750); **V**. Gas-phase acidities of amines (756); **VI**. Intrinsic effects of molecular structure (758); **VII**. References (761).

Recent techniques which allow direct measurements of the kinetics and equilibria of the protonation and deprotonation of organic molecules in the gas phase include ion cyclotron resonance (ICR) spectroscopy, high-pressure MS, trapped ion MS and selected-ion flow tube and flowing afterglow techniques.

Section II deals with the protonation of NH_3 (Table 1) and amines (Table 2) and with their deprotonation (Table 3) and presents rate constants for all these reactions in the gas phase. Section III describes the preferred direction of proton transfer with various amines, reports orders of basicity and acidity and also equilibrium constants for proton transfer between different amines and presents ΔG_0 values obtained by high-pressure (Fig. 6) and low-pressure (Fig. 7) ICR techniques and by high-pressure MS (Tables 4 and 5). Negative-ion equilibria are summarized in Table 7 and Figs 12 and 13.

Section IV discusses absolute and relative gas-phase acidities (Table 10). In both cases experimental and calculated proton affinities are compared (Tables 9 and 11). Intrinsic effects of molecular structure on the equilibria are considered (Section IV).

63 references up to 1979.
Complementary: 4, 4; **29**, 18.VIII.
Relevant: 2, 9; **3**, 6; **5**, 6; **7**, 6; **8**, 5; **11**, 3; **12**, 3, 10; **13**, 7; **15**, 4; **16**, 5; **17**, 6; **18**, 8; **19**, 7; **20**, 12; **22**, 8; **23**, 6, 7; **24**, 4; **25**, 5, 6; **31**, 17, 21; **32**, 21; **33**, 5.

29, *18: Special properties of di- and poly-amines* by *R. W. Alder and R. B. Sessions*

The chapter emphasizes aspects which depend on the presence of more than one lone pair of electrons in the molecule.

Section II depicts interactions among lone pairs and Section III shows that the title compounds adopt structures which minimize the interactions between lone pairs and brings as examples hydrazines (Fig. 4), 1,2-diamines, 1,8-diaminonaphthalenes (Fig. 5) and macrobicyclic diamines and cryptands (Fig. 6).

Section IV deals with PES (Table 1 gives I_v data of selected diamines) and Section V with reversible electrochemical oxidation (Table 2). Section VI discusses the lifetimes, N hyperfine coupling constants and absorption spectra of diamine radical cations (Table 3) and dications. Gas-phase proton affinities are considered (Table 4, Section VIII) and compared with those of related monoamines.

Section IX treats the solution basicity of diamines and Section X their proton transfer rates, which is often very much slower than the (diffusion-controlled) rates with amines.

Section XI describes metal complexations by the title compounds. This includes chelation by di- and polyamines, 12–16 membered macrocycles with four N atoms which exert a special macrocyclic effect giving enhanced stability to their metal complexes, and also binuclear macrocyclic complexes which can contain more than one complexed metal atom.

The macrobicyclic or cryptate effect is described in Section XI.C, including the selectivity and stability of the complexes. The application of cryptands to ion transport, detoxification, solubilization, anion activation and stabilization is discussed.

Finally, in Section XI.E spherical and cylindrical macrotricyclic compounds are considered (see also **27**, 1–4).

119 references up to 1979.
Complementary: 4, 4, 12.
Relevant: 19, 21.

29, *19: Alkyl nitrate nitrations* by *H. Feuer*

Active methylene groups are nitrated by alkyl nitrates in the presence of bases, e.g. $PhCH_2SO_2Ph + RONO_2 \rightarrow PhCH(NO_2)SO_2Ph$. The reaction is influenced by solvents (Tables 1 and 2) and temperature in the presence of bases, of which *t*-BuOK is very effective. Mono- and dinitration of various compounds and yields are summarized in Table 5. KNH_2 or $NaNH_2$ in liquid NH_3 give successful reactions even with compounds of very low acidity (Tables 6 and 7). The experimental conditions and scope of this reaction system are described for ketones, carboxylic and sulphonate esters (Tables 8 and 9), activated toluenes (Table 10), amides, lactams, alkyl-substituted heterocyclics (Tables 11 and 12), arylmethylene, alkylidene and hetarylmethylene phenylhydrazines (Tables 13 and 14), aldimines (Table 15) and alicyclic ketimines. The use of BuLi and of $(i\text{-Pr})_2NLi$ is recommended in special cases (Section II).

Section III describes some related nitrating agents, i.e. tetranitromethane, fluorotrinitromethane and methyldinitramine, and also some intramolecular alkyl nitrations.

Section IV discusses *N*-nitration of amines with cyanohydrin nitrates, alkyl nitrates and methyldinitramine.

90 references up to 1979.
Complementary: none.
Relevant: 6, 6, 9.

29, *20: Aminals* by *L. Duhamel*

Aminals are *gem*-diamines, $>C(NR_2)_2$, where the aminal function may also be part of a heterocycle.

Section II describes the preparation of aminals by condensation of carbonyl compounds with amines, e.g. $R'CHO + 2HNR_2 \rightarrow R'CH(NR_2)_2$, to yield aldo- or ketoaminals. With 1,2- or 1,3-diamines, cyclic aminals are obtained. Other syntheses involve substitution of dihalo compounds by amines, additions of amines to iminium salts, formation of new C—C bonds and reduction of amidines, $RC(=NH)NH_2 \rightarrow RCH(NH_2)_2$ [$\rightarrow$ RCHO].

Section III treats reactions of aminals, such as the formation of mono- and biammonium salts (with acids of alkyl halides) and of iminium salts, $R_2\overset{+}{N}=CR'R''X^-$ (Table 1). Section IV describes the formation of enamines from aminals, the aminal–enamine equilibrium, the kinetics of the dissociation of aminals into enamines, the stereochemistry of the amine elimination and the formation of enamines from functional and cyclic aminals and aminals which do not undergo this reaction.

Section V discusses reactions of aminals with heteroatom nucleophilic reagents (H_2O, ROH, RSH, RNH_2, $RCONH_2$) and with carbon nucleoreagents (Grignard, diazoalkanes, isonitriles, trihaloacetic acids, C-acids), and with heterocumulenes (forming 1:1 adducts). Reduction of aminals yields amines [$R^1R^2C(NR_2)_2 \xrightarrow{H} R^1R^2CHNR_2 + HNR_2$] and their oxidation gives a variety of cleavage, cyclization and other products.

Section VI considers the use of aminals as potential carbonyl compounds and for the protection of carbonyl groups in other reactions at various sites of the molecule.

286 references up to 1979.
Complementary: none.
Relevant: none.

29, *21: Detection and determination of nitro and nitroso compounds* by *Y. Tapuhi and E. Grushka*

The chapter deals in parallel, in each section, with nitro and nitroso compounds.

Section II describes detection and identification by chemical methods (reduction, various spot tests), IR and Raman spectroscopy (Tables 1–3), electronic spectroscopy and NMR (including ^{1}H, ^{13}C and ^{14}N) and by ^{14}N NQR.

Section III deals with titrimetric methods, Kjeldahl and gasometric methods, electroanalysis (polarographic and coulometric), spectroscopic and gravimetric methods and some indirect methods.

Section IV presents gas, liquid, paper and thin-layer chromatography and paper electrophoresis.

Section V considers some recent methods relating to *N*-nitroso compounds, such as the thermal energy analyser technique.

118 references up to 1978.
Complementary: **6**, 2, 3.
Relevant: **1**, 5; **2**, 8; **3**, 15; **4**, 3; **5**, 17; **8**, 3; **12**, 4; **13**, 6; **14**, 3; **15**, 3; **16**, 3; **17**, 4; **18**, 5; **20**, 3; **22**, 5; **23**, 5; **24**, 5; **25**, 10; **26**, 5; **28**, 3; **30**, 14–22; **36**, 5.

29, *22: Deaminations (carbon—nitrogen bond cleavages)* by *R. J. Baumgarten and V. A. Curtis*

Section I considers the principle of activation, which is essential before (*in vitro*) deaminations can occur, the oxidation states in deaminations and the scope and organisation of the chapter.

Section II deals with the pK_as of the conjugate acids of observed or potential leaving groups in deaminations (Table 1) which involve no change in the oxidation state in the alkyl or aryl moieties. The leaving groups are the following: (A) *N*, *N*-diarylsulphonimide anions from $RN(SO_2R')_2 + Y^- \rightarrow RY + {}^-N(SO_2R')_2$ (+ alkenes) and the scope of the reaction, the effect of acidity, alkene formation and product formation in general (Table 2) and stereochemistry are considered; (B) triflimides, their synthesis and C—N cleavage are treated; (C) various disulphonimide leaving groups; (D) other imides (e.g. saccharin); (E) sulphonamides and (F) carboxamides as leaving groups; (G) 2, 4, 6-triphenylpyridine and related leaving groups (Table 3); (H) pyrrole derivatives; (I) amines and ammonia as leaving groups, e.g. $R_2CHCR'_2\overset{+}{N}Me_3\ \bar{O}H \xrightarrow{\Delta} R_2C{=}CR'_2 + H_2O + NMe_3$ (Hoffmann elimination); and (K) N_2 gas as leaving group on the deamination of aliphatic amines (e.g.

after their nitrosation or after triazene formation with diazonium ions: $RNH_2 + ArN_2^+ \longrightarrow RNHN{=}NAr \xrightarrow{HX} RX + \text{alkenes} + ArNH_2 + N_2$, or from alkane diazotates, etc.); aromatic dediazonization ($ArNH_2 \rightarrow ArN_2^+ \xrightarrow{X^-} ArX + N_2$) also belong to this category. Finally, 2, 4, 6-triphenylpyridine, N_2O gas and CN leaving groups are treated.

Section III discusses reductive deamination. The reactions include as leaving groups sulphonimide anions, succinimide ion, pyridine-derived groups, NH_3, amines, N_2 (in both aliphatic and aromatic cases) and 2, 4, 6-triarylpyridine leaving groups. Replacement of an —NH group by an alkyl group is also a deamination.

Section IV describes oxidative deaminations which are mostly recent procedures. They include oxidations of amines by direct dehydrogenation, e.g. $RCH_2NH_2 \rightarrow RCH{=}NH \rightarrow RCHO + NH_3$, photochemical oxidations of amines to aldehydes and ketones, oxidation of amines to imines ($R_2CHNHR \rightarrow R_2C{=}NR$) with halogens, sulphonyl peroxides and by other methods, oxidative formation of nitriles, carboxylic acids and amides, e.g., $RCH_2NH_2 \rightarrow RCONH_2$, transaminations with aldehydes and ketones and transamination oxidations via imines and oxaziridines, oxidations of primary amines to aldehydes via triazoles, deaminations by formation of *N*-nitroamides and pyrolysis of the latter, conversions to *gem*-dihalides with elimination of N_2 and oxidative degradation of tertiary amines by various methods.

Section V discusses biochemical deaminations (oxidative, reductive and with no change in the oxidation state), considers bioorganic chemistry, compares biochemical and organochemical deaminations and treats enzymatic processes, synthesis and application of chiral methyl carriers, deaminations of amino sugars and 'pseudophysiological' deaminations via pyridinium salts. Environmental studies involving *N*-nitroso and related compounds as mutagens, carcinogens and teratogens are presented.

439 references up to 1979.
Complementary: 4, 8; **29**, 25.
Relevant: 29, 26.

29, *23: Chiroptical properties of amino, nitroso and nitro compounds* by *H. E. Smith*

I. Introduction (1000); **II**. Chiral amino compounds (1001); **III**. Chiral nitroso compounds (1020); **IV**. Chiral nitro compounds (1025); **V**. Acknowledgements (1031); **VI**. References (1031).

Section II describes the sources of chiral amino compounds, application of Brewster's rule regarding atomic and conformational asymmetry, ORD and CD studies with the amino chromophore, the influence of the NH_2 group on other chromophores, chromophoric derivatives (isolated, or formed *in situ*) and the salicylideniminochirality rule.

Section III deals with the preparation of chiral *C*-, *N*- and *O*-nitroso compounds and their ORD and CD. Section IV discusses the preparation of chiral *C*-, *O*- and *N*-nitro compounds and studies describing their ORD and CD.

162 references up to 1979.
Complementary: none.
Relevant: 8, 4; **18**, 7; **19**, 6; **20**, 2; **22**, 4; **25**, 3; **26**, 3; **27**, 6; **28**, 10; **31**, 1; **36**, 4.

29, *24: Thermochemistry of nitro compounds, amines and nitroso compounds* by *L. Batt and G. N. Robinson*

I. Introduction (1036); **II**. Nitro compounds (1038); **III**. Amines (1050); **IV**. *C*-Nitroso compounds (1063); **V**. Alkyl nitrites (1070); **VI**. Alkyl nitrates (1075); **VII**. Conclusions (1077); **VIII**. Acknowledgements (1080); **IX**. References (1080).

The introduction reviews the principles of thermochemistry. Section II deals with the thermochemical properties of nitro derivatives of methane (Table 1) and with other nitroalkanes (Tables 2–4), with the C—N bond dissociation energy of mono- and polynitroalkanes and their halo derivatives (Tables 5–7) and similarly with aromatic nitro compounds, their heats of formation (Table 8) and C—N bond dissociation energy.

Section III discusses amines, their thermochemical properties obtained by group additivity and experimentally (Table 9) and recent thermochemical results for amines (Table 10), kinetic studies on amines and C—N bond dissociation energies by theses (Table 11), heats of formation of amino radicals (Table 12) and C—N bond dissociation energies in amines obtained from kinetic results and compared with those calculated from thermochemistry (Table 13).

Section IV treats *C*-nitroso compounds, calorimetric, EI (Table 14) and kinetic studies and presents values of heats of formation and C—N bond dissociation energies for them (Table 15).

Section V deals with alkyl nitrites, their heats of formation (Table 16) and also with heats of formation of alkoxy radicals (Table 17) and with the RO—NO bond dissociation energies (Table 18).

Section VI considers the thermochemical properties of alkyl nitrates (Table 19), their RO—NO_2 bond dissociation (Table 20) and the activation energies obtained for the latter process in di- and trinitrates (Table 21).

New and amended group values for N-containing compounds are given (Table 22) and C—N bond and RO—N bond dissociation energies in the title compounds (Table 23).

146 references up to 1979.
Complementary: none.
Relevant: 16, 16; **18**, 3; **19**, 3; **20**, 11; **21**, 3; **22**, 6; **23**, 4; **24**, 3; **25**, 2; **26**, 4; **27**, 9; **28**, 4; **30**, 2; **31**, 2; **33**, 3; **36**, 4.

29, *25: Oxidation of amines* by *D. H. Rosenblatt and E. P. Burrow*

I. Introduction (1086); **II**. Chlorine dioxide in aqueous solution (1086); **III**. Halogenating agents (1090); **IV**. Potassium ferricyanide (1098); **V**. Mercuric acetate (1105); **VI**. Lead tetraacetate (1109); **VII**. Other metals (1112); **VIII**. Peroxy species (1119); **IX**. Quinones (1123); **X**. Molecular and atomic oxygen (1126); **XI**. Ozone (1127); **XII**. Electrochemistry (1130); **XIII**. Photochemistry (1134); **XIV**. Radiation chemistry (1140); **XV**. Miscellaneous oxidations (1141); **XVI**. References (1142).

Most of the chapter is organized according to the nature of the oxidants. Thus, Section II deals with ClO_2 oxidation, in which, e.g., Et_3N is converted through a complicated route to $Et_2NH + MeCHO$ and reactions of other amines may produce H-abstraction, oxidative dealkylation, oxidative fragmentation and also other types of reactions.

Halogenating agents (Section III) usually transfer positive halogen to the unshared amine electrons, with formation of haloamines (from RNH_2 and R_2NH) or of *N*-haloammonium ions (from R_3N). The overall reaction is often very complex and the routes depend strongly on the structure of the amine: $ArCH_2NMe_2 + Br_2 \xrightarrow{H_2O} ArCHO + HNMe_2$. Oxidation is often accompanied by (*C*- and *N*-)halogenation.

The $K_3Fe(CN)_6$ oxidation of tertiary amines (Section IV) was studied in detail mechanistically $[RCH_2NR'_2 + Fe(CN)_6^{3-} \rightarrow\rightarrow\rightarrow\rightarrow RCHO + R'_2NH]$.

Section V deals with oxidations by $Hg(OAc)_2$, VI with MnO_2, $KMnO_4$ and $Mn(OAc)_3$ and VII with $Pb(OAc)_4$. The last oxidant converts aliphatic amines and also $PhCH_2NR_2$ into nitriles, while amides, aldehydes, cyclization and fragmentation products are obtained in other cases. Section VIII describes oxidations by Cu^{II}, $Cu_2O + O_2$, Ag^{II}, $AgClO_4$, RuO_4, $PdCl_2$, $AuCl_3$, etc.

Section IX deals with oxidation by hydroperoxides, perbenzoic acids and H_2O_2. The reagent of choice for the oxidation of RCH_2NH_2 and R_2CHNH_2 to the corresponding nitroalkanes is *m*-chloroperbenzoic acid, which also converts substituted piperidines into nitroxyl radicals (Table 2). Peroxydisulphate and diacyl peroxides are also considered.

Quinones (Section X) oxidize aliphatic tertiary amines in benzene to coloured products which are useful in thin-layer chromatography and other analytical methods. Sections XI and XII deal with oxidations by molecular oxygen and by ozone. The latter converts amines to nitro compounds, but much less effectively than peracids. However, the initial amine-ozone adduct may yield amine oxides, hydroxylamines, aminium cations or α-hydroxyamines and their cleavage products.

Section XIII discusses electrochemical oxidations on both inert and on active electrodes, Section XIV deals with anaerobic and aerobic photooxidations, including also those involving chloromethanes, and Section XV treats briefly radiation-induced processes.

317 references up to 1979.
Complementary: none.
Relevant: none.

29, *26:* N*-nitrosamines and* N*-nitrosimines* by *B. C. Challis and J. A. Challis*

Section II describes the formation of *N*-nitrosamines by nitrosation of secondary amines by aqueous HNO_2, including rates and reactivities (Table 1), catalysis and inhibition. Section II.B deals with syntheses by NOCl and N oxides (Table 2) and catalysis by β-substituted alcohols (Table 3) and by other promoters. Section II.C considers the nitrosation of secondary amines by organic nitroso and nitro compounds (*N*-nitrosamines, *N*-nitrosamides, nitrite and thionitrite esters and aliphatic nitro compounds). Section II.D discusses the nitrosation of primary and tertiary amines, quaternary amine salts, amine oxides and tertiary amides and Section II.E deals with catalysed reactions of nitrite ions and with reduction of the NO_3^- ion.

Section III treats the structure, thermochemistry and spectra of *N*-nitrosamines, their acid–base, hydrogen-bonding and complexing properties and their reactions with inorganic acids, electrophilic reagents (*N*-alkylation, acylation, etc.) and organometallic reagents (RMgX and RLi add across the N=O groups). Next, reductions, oxidations and homolysis of the N—NO bond (by thermolysis, trans-nitrosation or photolysis) are described. Section III.J deals with α-substituted *N*-nitrosamines, their α-H exchange reaction:

and also with their preparation via an α-carbanion (Scheme 24) and with their reactions.

Section IV describes the toxicity and carcinogenicity (Table 5), mutogenicity and metabolic pathways of *N*-nitrosamines and their interactions with cellular constituents.

Section V presents the preparation, properties and reactions of *N*-nitrosoimines, which decompose readily even below room temperature to a ketone and N_2 ($R^1R^2C{=}N{-}NO \rightarrow R^1R^2CO + NO_2$).

354 references up to 1979.
Complementary: none.
Relevant: none.

29, *27: The role of Meisenheimer or σ-complexes in nitroarene–base interactions* by *E. Buncel*

I. Introduction and scope (1225); **II.** Meisenheimer or σ-complexes (1226); **III.** Competitive processes involving σ-complexes (1239); **IV**. Acknowledgements (1256); **V**. References (1256).

Section II gives a historical review of Meisenheimer complexes, describes structural features and methods of investigation, H^- ion adducts, spiro complexes and the role of ambident nucleophiles in σ-complex formation [e.g. acetonate complexes and their analogues such as

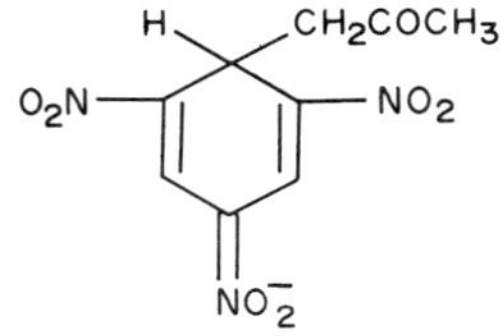

which in basic media undergo abstraction of a γ-H leading to cyclized products, which may be also tricyclic (formed via a second internal nucleophilic addition)]. Other σ-complexes are obtained, e.g., with KF/18-crown-6 in acetonitrile, with sulphites, with organometallic compounds, etc.

Section III discusses nucleophilic displacement *versus* σ-complex formation, formal displacement of an H^- ion involved in nucleophilic substitution, aromatic proton abstraction competing with σ-complex formation and benzylic proton abstraction *versus* σ-complex formation.

212 references up to 1979.
Complementary: **7**, 6.
Relevant: **3**, 6; **4**, 4; **8**, 5; **11**, 3; **12**, 10; **16**, 5; **17**, 6; **19**, 7; **20**, 12; **22**, 8; **23**, 7; **24**, 4; **25**, 5; **31**, 21; **32**, 21; **33**, 5.

29, *28: Uses of isotopically labelled amino, quaternary ammonium and nitro compounds* by *P. J. Smith and K. C. Westaway*

I. Theory of kinetic isotope effects and theoretical approaches to the effect of substituents on transition-state geometry (1262); **II.** Kinetic isotope effects in nucleophilic substitution reactions involving isotopically labelled amines and quarternary ammonium salts (1271); **III**. The use of isotope tracer experiments and isotope effect measurements for the determination of mechanism for an elimination process from a quaternary ammonium salt (1295); **IV**. Use of kinetic isotope effects in the determination of *E*-2 transition-state structure (1300); **V**. The use of oxygen-18, nitrogen-15, carbon-13 and deuterium as tracers in the determination of mass spectral fragmentation pathways (1307); **VI**. References (1310).

Section I reviews the theory of heavy-atom, primary H/D, secondary α- and β-H/D kinetic isotope effects and the effect of substituents on the geometry of transition states.

Section II deals with isotope effects in the Menschutkin reaction ($R_3N + R'X \rightarrow R'\overset{+}{N}R_3X^-$), including secondary H/D (Table 1) and primary nitrogen (Table 2), chlorine leaving group and ^{14}C (Table 3) kinetic isotope effects. Next, kinetic isotope effects in nucleophilic substitutions of quaternary ammonium salts are discussed (Tables 4–7) and substituent effects on the geometry of S_N2 transition states.

Section III deals with elimination reactions from quaternary ammonium salts and considers the use of tracer and kinetic isotope effect studies in the elucidation of the mechanism of these reactions (Tables 8 and 9). Section IV discusses the nature of transition states in reactions involving quaternary ammonium ions and the use of kinetic isotope effects for the determination of the structure of the TS (Tables 10–14).

Section V discusses mass spectral fragmentations of nitroarenes and of monocyanopyridines, and shows how ^2H, ^{13}C, ^{15}N and ^{18}O labelling can be employed in the determination of the MS fragmentation pathways.

88 references up to 1979.
Complementary: none.
Relevant: 5, 10; **12**, 13; **13**, 15; **17**, 12; **18**, 18; **19**, 17; **22**, 14; **23**, 16; **24**, 20; **26**, 11; **27**, 10; **28**, 8; **33**, 18; **36**, 12.

The chemistry of the metal–carbon bond. Volume 1. The structure, preparation, thermochemistry and characterization of organometallic compounds (1982)

30, *1: Structure of organometallic compounds* by *A. D. Redhouse*

I. Introduction (2); **II**. σ-Bonded (monohapto) ligands (3); **III**. Dihapto ligands (13); **IV**. Trihapto ligands (20); **V**. Tetrahapto ligands (24); **VI**. Pentahapto ligands (28); **VII**. Hexa-, hepta- and octahapto ligands (32); **VIII**. References (36).

Section I defines the aims of the chapter and its sources. Section II deals with the structure of complexes which contain anionic σ-donor ligands, such as alkyl, alkenyl, alkynyl, aryl, acyl and ylide complexes, and describes bond lengths and angles. Next, complexes with bridging ligands such as dimeric trimethylaluminium:

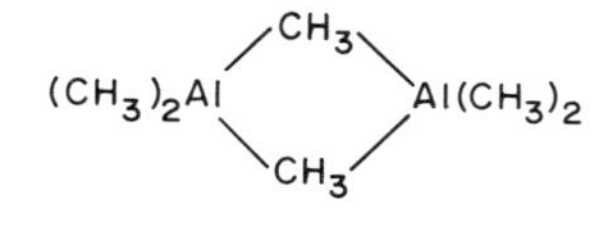

are described, followed by complexes in which the organic ligand can be described as a neutral σ-donor/π-acceptor (such as carbene and carbyne complexes and bridging ligands).

Section III treats transition metal–alkene complexes (η-ethene complexes, alkenes as chelating ligands and also μ-η^2-alkene ligands), complexes with η^2-ligands containing heteroatoms and transition metal–alkyne complexes.

Section IV describes η^3-allyl and cyclic ligands and also η^3-ligands with heteroatoms.

Section V discusses η^4-*cis*-1, 3-diene complexes (e.g. with butadiene and derivatives), cyclic η^4-ligands (with cyclooctatetraene or with cyclobutadiene) and trimethylene–methane complexes.

Section VI considers cyclopentadienyl complexes of the types $M(C_5H_5)X$ and $(C_5H_5)_xML_y$ and similar complexes with cyclohexadienyl and heterocyclic ligands.

Section VII deals with η^6, η^7 and η^8 cyclic ligands (e.g. with benzene, cycloheptatriene and cyclooctatetraene, respectively).

192 references up to 1980.
Complementary: none.
Relevant: 16, 2; **17**, 2; **18**, 2; **19**, 2; **20**, 2; **21**, 1; **22**, 2; **23**, 3; **24**, 2; **26**, 2; **27**, 4; **28**, 2; **29**, 1; **31**, 24; **32**, 24; **33**, 12; **34**, 10; **36**, 3.

30, *2: Thermochemistry of organometallic compounds* by *G. Pilcher and H. A. Skinner*

I. Introduction (44); **II.** Experimental methods (44); **III.** Enthalpies of formation and atomization of organometallic compounds (53); **IV.** Strengths of chemical bonds in organometallic compounds (53); **V.** References (86).

Section II describes experimental methods through which enthalpies of combustion can be obtained, such as static and rotating-bond calorimetry, and deals with the most commonly used reactions in organic thermochemistry, which are hydrolysis, halogenation, redistribution, hydroborination and thermal decomposition. Enthalpies of formation of a very wide variety of organometallic compounds are given in Table 1, including also enthalpies of atomization calculated from data in Table 3.

Section IV treats thermochemical bond strengths, bond dissociation energies and enthalpies (Table 2) and metal—carbon mean bond disruption enthalpies of Me, Et and Ph derivatives (Table 4), metal cyclopentadienyls (Table 5) and metal arenes (Table 6).

Enthalpy and other data for metal cyclopentadienyl complexes are given in Tables 7 and 8. Mononuclear (Table 9) and polynuclear transition metal carbonyls (Table 10) are presented and also their redistribution reactions (Table 11). Next, metal—metal single and multiple bonds are considered (Tables 12–14), followed by discussion of bond energy schemes for organometallic compounds (e.g. the Laidler scheme, Tables 15 and 16), and studies aimed at determining differences in bond dissociation enthalpies rather than absolute values (Table 17).

Finally, bond energy–bond length relationships are discussed (Figs 6 and 7).

200 references up to 1980.
Complementary: none.
Relevant: 16, 16; **18**, 3; **19**, 3; **20**, 11; **21**, 3; **22**, 6; **23**, 4; **24**, 3; **25**, 2; **26**, 4; **27**, 9; **28**, 4; **29**, 24; **31**, 2; **33**, 3; **36**, 4.

30, *3: Synthesis of ylide complexes* by *L. Weber*

I. Introduction (93); **II.** Onium–metallates and metal-substituted onium salts (96); **III.** Metal-substituted ylides (142); **IV.** Metals as anionic centres in ylides (151); **V.** Metal–onium ylides (155); **VI.** Ylide complexes with metal-substituted onium centres (157); **VII.** Reactions of ylides with coordinated ligands (162); **VIII.** Ylide complexes in synthetic chemistry (171); **IX.** References (174).

Section I discusses ylides (i.e. neutral molecules possessing a carbanionic function in an α-position to an onium centre $R^1R^2\overset{-}{C}—\overset{+}{E}R^3_n \leftrightarrow R^1R^2C{=}ER^3_n$; the chapter uses both the ylide and ylene formulations) and reviews the types of complexes which ylides give the metals.

Section II describes the formation of complexes with non-stabilized ylides as terminal monodentate ligands. Direct syntheses include additions of ylides to coordinatively unsaturated metal compounds, ligand substitutions and oxidative additions of halogenated onium salts to metals. Indirect syntheses include additions of donor molecules to carbene complexes, displacement reactions, additions to ylides and alkylations of $[L_nMCH_2SMe]$ or $[L_nMCH_2NMe_2]$.

Non-stabilized double ylides (which contain two carbanionic donor centres) may yield ylide-bridged complexes or chelates. Carbonyl-stabilized ylides may give either metal—C or metal—O coordination as terminal monodentate ligands, or they may act as chelating ligands via the ylidic C and the O atom.

Alkenyl-stabilized ylides (onium allylides) may give η^1 or η^3 coordination (Section II.E). Section II.F considers cyclic ylides as ligands, including η^1, η^3 and η^4 coordination with cyclopentadienylide complexes, 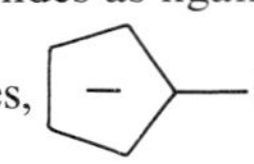 SMe_2, and phosphabenzene and thiabenzene complexes are also treated.

Section III deals with the synthesis of metal-substituted ylides via metallation of simple ylides or by phosphine addition to carbyne complexes.

Section IV describes base-stabilized germylenes and stannylenes and also transition metal complexes with ylides of Ge and Sn.

Section V treats metalonium ylides (containing As, Sb, Bi, Nb, Ta or Ir).

Section VI considers ylide complexes containing three-membered metallocycles or containing bridging phosphinomethyl or thiomethoxymethyl groups.

Section VII describes reactions between free ylides and ligated molecules within the coordination sphere of a metal. These involve organic carbonyl functions in olefinic ligands and CO, carbene, olefinic, organocyanide and halophosphine ligands and also hydrido complexes. Finally, in Section VIII the use of ylide complexes in the Wittig and cyclopropanation reactions and as catalysts is treated.

273 references up to 1980.
Complementary: none.
Relevant: 28, 16.

30, *4: Synthesis of transition metal–carbene complexes* by *H. Fischer*

I. Introduction (182); **II**. Synthesis from non-carbene–metal complex precursors (182); **III**. Synthesis by modification of the carbene ligand (207); **IV**. Synthesis of modification of the metal–ligand framework (217); **V**. Oxidation of carbene complexes (224); **VI**. References (224).

Section II describes the synthesis of metal–carbene complexes by nucleophilic attacks on metal carbonyls and isocyanide and carbyne complexes, by alkylation of acyl complexes, by electrophilic addition to imidoyls, from acetylene or acetylide complexes, from salt-like precursors, from electron-rich olefins, from diazoalkane precursors, from alkylidene complexes and by some other less general methods.

Section III discusses syntheses in which a carbene ligand existing in a complex is modified by nucleophilic or electrophilic attacks or by reactions of the carbene anions in, e.g., acidic alkoxy (alkyl) carbene complexes and again by some less general other methods.

Section IV considers syntheses which involve modification of the metal–ligand framework, e.g. reactions of the metal—carbene bond, ligand substitution reactions and oxidative additions. Oxidation of carbene complexes is treated very briefly in Section V.

310 references up to 1980.
Complementary: none.
Relevant: none.

30, *5: Synthesis of transition metal–carbyne complexes* by *U. Schubert*

Section I deals with syntheses involving carbene complexes and Lewis acids, including the preparation of both neutral and cationic carbyne complexes, and discusses the mechanisms. Next, rearrangements and deprotonation of carbene complexes are treated.

Section II describes methods in which existing carbyne complexes are modified either by carrying out reactions within the carbyne ligand or by substituting it by another carbyne ligand. Finally, in Section III some syntheses from thiocarbonyl and isonitrile complexes are discussed.

56 references up to 1980.
Complementary: none.
Relevant: none.

30, *6: Synthesis of transition metal alkyl and aryl complexes* by *R. J. Puddephatt*

Section II.A deals with the synthesis of the title compounds by alkyl transfer reactions involving alkyl derivatives of Li, Mg, Zn, Hg, Sn and other metals. It considers preparations of methyl and other alkyl derivatives, aryl derivatives and metallacylic compounds. Section II.B describes oxidative additions of alkyl halides, of C—H and of C—C bonds and reactions with metal atoms (obtained by high-temperature vacuum vaporization techniques). Next cyclometallations are presented, followed by syntheses of metallocycles from alkenes and alkynes (Section II.D), of the general type.

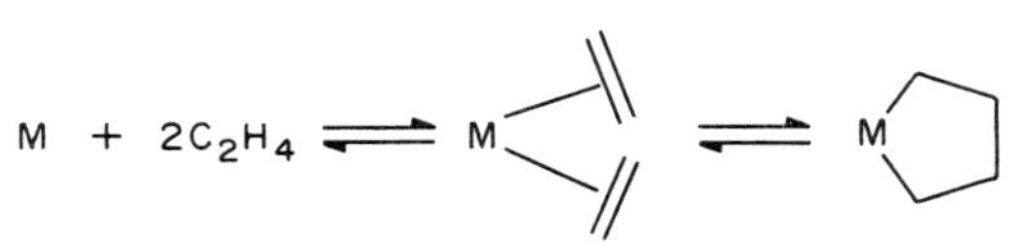

Section II.E discusses insertion reactions of alkenes and alkynes into metal—hydride bonds and insertions involving CO, isocyanides and carbene precursors. Section II.F considers syntheses involving eliminations, which are of limited use only.

Section II.G presents syntheses in which a coordinated ligand is attacked. These may include complexes of alkenes, alkynes, carbenes, η^3-allyl complexes and carbonyl ligands. Finally, syntheses of C-acidic hydrocarbons are treated, e.g. $MR + HC{\equiv}CR^1 \rightarrow RH + MC{\equiv}CR^1$.

289 references up to 1980.
Complementary: none.
Relevant: none.

30, *7: Synthesis of olefin and acetylene complexes of the transition metals* by *G. S. Lewandos*

Section II.A discusses additions to unsaturated complexes. Thus, many transition metal complexes react directly with olefins or acetylenes. Next, methods involving displacement of anions and thermal or photochemical displacement of carbonyl ligands are described. Section II.D deals with displacements of η^2-coordinated ligands such as ethylene, isobutylene, cyclooctene, cycloocta-1,5-diene, norbornadiene and some other π-ligands. Section II.E treats the displacement of tertiary phosphines, nitriles, benzylideneacetone derivatives, solvent ligands, silanes, hydrogen, oxygen and methane.

Section II.F considers ligand rearrangements, involving η^1-allyl, η^1-alkyl and carbene ligands, epoxides (Table 1) and keto–enol tautomers.

Reductive olefination and acetylenation are discussed, including the reduction of metals in positive oxidation states (e.g. metal halides), by metals, metal alkyls or hydrides in the presence of olefins or acetylenes, when a metal–olefin (or acetylene) complex is the product (Section II.G). Finally, a few electrochemical syntheses (Table 2) and some general problems and limitations are discussed.

230 references up to 1979.
Complementary: 1, 6; **9**, 5.
Relevant: 21, 12.

30, *8: Synthesis of η^3-allyl complexes* by *P. Powell*

Section II describes reactions in which allyl groups are transferred from allyl Grignard or other main group derivatives to transition metal halides or by transfer of allyl groups from one transition metal to another.

Section III deals with the reactions of anionic metal carbonyls and of complexes with M—Li, M—Mg and M—Al bonds with allyl halides. Section IV discusses the direct addition of transition metal hydrides to 1,3-dienes (Table 1) yielding allyl derivatives, which can also be obtained by the isopropylmagnesium bromide method and (less frequently) by insertion of 1,3-dienes into metal—C η^1-bonds of alkyl and aryl complexes.

Section V treats protonation, Friedel–Crafts acylation and addition of fluoroolefins to coordinated dienes. Section VI discusses oxidative additions of allyl halides, including reactions of allyl halides with metal carbonyl complexes, direct reactions between allyl halides and metals (e.g. Pd, Ni) and other oxidative additions.

Section VII deals with nucleophilic attack on coordinated 1,3-dienes yielding η^3-allyl complexes, and also with alkoxy carbonylations and related reactions.

Section VIII discusses reactions of metal salts with allyl alcohols, halides and amines, suitable alkenes (which also yield allyl derivatives), 1,3 - and 1,2-dienes, cyclopropenes and cyclopropanes.

Finally, some novel binuclear ('sandwich') complexes, in which allyl groups bridge between two Pd atoms, are presented (Section IX).

260 references up to 1979.
Complementary: none.
Relevant: none.

30, *9: Synthesis of η^4-butadiene and cyclobutadiene complexes* by *G. Marr and B. W. Rockett*

I. η^4-Buta-1, 3-diene complexes (368); **II.** η^4-Trimethylenemethane complexes (388); **III.** η^4-Cyclobutadiene complexes (392); **IV.** η^4-Cyclo-1, 3-diene complexes (412); **V.** References (435).

Section I deals with the preparation of η^4-buta-1, 3-diene complexes from transition metal carbonyls or salts and 1, 3-butadienes, with or without rearrangement or modification of the ligand (Table 1) and from metal carbonyls in reactions involving the cleavage of strained rings (e.g. vinylcyclopropanes, methylenecyclopropanes or thiacyclobutenes), or involving dehydration of allyl alcohols during the reaction. The desired product can also be obtained from butadienes, transition metal atoms and CO and from (η-hydrocarbon)–transition metal complexes such as η^5-dienium–, η^4-trimethylenemethane–, η^3-allyl– and η^2-olefin–transition metal complexes.

Section II describes η^4-trimethylenemethane complexes from allyl halides and $Fe_2(CO)_9$ (Table 2), from methylenecyclopropanes with $Fe_2(CO)_9$ and from η^3-allyl complexes.

Section III treats the preparation of η^4-cyclobutadiene complexes from cyclobutenes and cyclobutanes with metal carbonyls (Table 3), from (η^3-cyclobutenyl)–transition metal complexes, from methylenecyclobutenes and metal carbonyls, by ligand transfer (Table 4), from acetylenes and transition metal carbonyls, from the photolysis product of α-pyrone with transition metal carbonyls, from 1-metallocyclopentadienes and from cyclooctatetraenes and carbonylate anions.

Section IV discusses the syntheses of complexes containing η^4-cyclo-1, 3-dienes. The following ligands are considered: η^4-cyclopentadiene, η^4-cyclopentadienone, η^4-fulvene, η^4-heterocyclopentadienes, η^4-cyclohexa-1, 3-diene (Table 5), η^4-heterocyclohexa-1, 3-diene, η^4-cyclohepta-1, 3-diene and η^4-cycloocta-1, 3-diene.

307 references up to 1979.
Complementary: none.
Relevant: none.

30, *10: Synthesis of complexes of η^5-bonded ligands* by *G. Marr and B. W. Rockett*

I. η^5-Cyclopentadienyl complexes (444); **II.** η^5-Cyclohexadienyl complexes (453); **III.** η^5-Cycloheptadienyl complexes (456); **IV.** η^5-Cyclooctadienyl complexes (457); **V.** η^5-Pentadienyl complexes (457); **VI.** References (459).

Section I describes the preparation of η^5-cyclopentadienyl complexes from metal cyclopentadienides and transition metal compounds (Tables 1–3) and from cyclopentadiene itself, from fulvenes and related compounds and by transfer of a cyclopentadienyl group from one transition metal to another. Other methods involve cycloadditions of alkenes or alkynes with transition metal carbonyls (Section I.F) and electrochemical preparation (Section I.G).

Section II deals with η^5-cyclohexadienyl complexes, which can be obtained by hydride ion abstraction from η^4-diene–transition metal complexes, by nucleophilic attack of H^- or an organolithium reagent on an η^6-arene–transition metal complex or by electrochemical reduction.

Sections III, IV and V treat the synthesis of η^5-cycloheptadienyl, η^5-cyclooctadienyl and η^5-pentadienyl complexes.

113 references up to 1979.
Complementary: none.
Relevant: none.

30, *11: Synthesis of complexes of η^6-, η^7 and η^8-bonded ligands* by *G. Marr and B. W. Rockett*

I. η^6-Arene complexes (463); **II.** η^7-Cycloheptatrienyl complexes (479); **III.** η^7-Cyclooctatrienyl complexes (484); **IV.** Cyclooctatetraenyl complexes (484); **V.** References (485).

Section I describes preparation of η^6-arene complexes. The methods include reduction of a metal salt in the presence of the arene ligand, co-condensation of metal atoms (as vapour) with arenes (Table 1), from metal carbonyls and arenes, when the latter displaces a CO ligand, by cyclic trimerization of acetylenes in the presence of transition metal salts and by displacement of a η^5-cyclopentadienyl ligand by a η^6-arene ligand.

Section I.G, H and I discuss the preparation of η^6-heteroarene (pyridine, thiophene, borazine, etc., behave as six-electron ligands), η^6-cycloheptatriene, η^6-cyclooctatriene and η^6-cyclooctatetraene complexes of transition metals.

Section II discusses syntheses of η^7-cycloheptatrienyl complexes by H-abstraction from η^6-cycloheptatriene complexes (Table 2), by direct reaction of cycloheptatriene with a transition metal derivative or by ring expansion of an η^6-arene ligand.

Sections III and IV treat briefly η^7-cyclooctatrienyl and cyclooctatetraenyl complexes.

135 references up to 1979.
Complementary: 30, 13.
Relevant: none.

30, *12: Synthesis of organolanthanide and organoactinide complexes* by *W. J. Evans*

I. Introduction (490); **II.** General synthetic considerations (491); **III.** Cyclopentadienyl complexes (495); **IV.** Cyclooctatetraenyl complexes (512); **V.** Complexes containing metal–carbon sigma bonds (518); **VI.** Miscellaneous (529); **VII.** References (531).

Section II reviews the special properties of the f elements and their organometallic complexes, the most important synthetic methods and the factors affecting the selection of metals within the lanthanide and actinide groups.

Section III deals with cyclopentadienyl complexes containing trivalent lanthanides {e.g. $[(C_5H_5)_3Ln]$, $[(C_5H_5)_2LnX]$ (X = halide) and $[(C_5H_5)LnX_2]$}, divalent and tetravalent lanthanide complexes, tetravalent $[(C_5H_5)_4An]$ etc., and trivalent $[(C_5H_5)_3An]$, etc., actinide complexes, where An is an actinide metal.

Section IV discusses the complexes of cyclooctatetraene with tri-, di- and tetravalent lanthanides and also with tetra- and trivalent actinides.

Section V treats complexes containing metal—C sigma bonds, including both homoleptic (i.e. containing only one kind of ligand) and heteroleptic (containing mixed ligands) lanthanide and actinide complexes.

Section VI considers some allyl and pyrrolyl complexes and complexes of neutral unsaturated hydrocarbons.

245 references up to 1981.
Complementary: none.
Relevant: none.

30, *13: Metal atoms in organometallic synthesis* by *M. J. McGlinchey*

Section II treats the use of metal atoms for the synthesis of a wide variety of compounds, starting from organic halides, e.g.

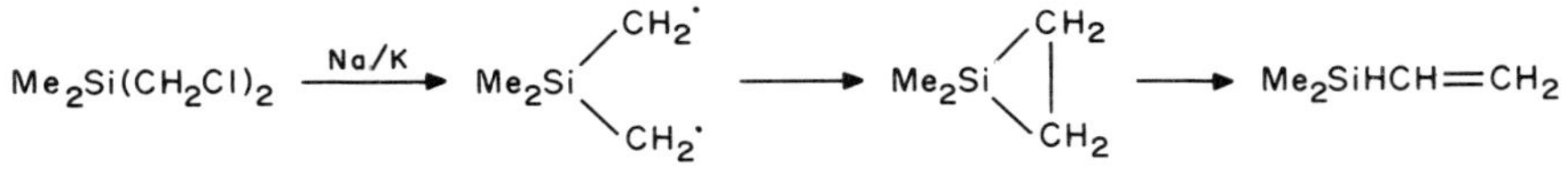

Thus, the use of Na, Li, Mg, Ca, Zn, Ge, Ni, Pd, Pt, etc., is described, also for the direct synthesis of metal complexes.

Section III deals with the reactions of metals with monoalkenes yielding a σ- or π-complexes, e.g. $Ni + C_2H_4 \rightarrow [Ni(C_2H_4)_3]$; with dienes in general the same products are obtained as in the more conventional reactions, although stepwise addition of different ligands is here a route to systems otherwise difficult or impossible to obtain, e.g.

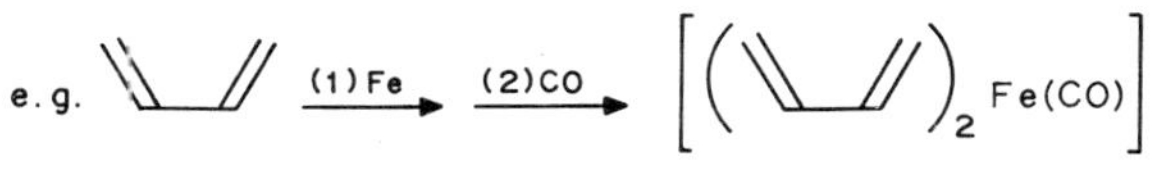

With polyenes, 1:1 addition complexes and sandwich compounds are formed.

Section IV discusses the formation of complexes from arenes and chromium and also with other metals (Mo, W, V, Fe). Section V considers complexes of alkynes with Na, Mg, Al and transition metals and Section VI deals with phosphines as ligands. Section VIII treats some additional ligands obtained by deoxygenation or desulphurization of various organic compounds and a few other reactions.

151 references up to 1980.
Complementary: none.
Relevant: none.

30, *14: Analysis of organometallic compounds: determination of elements and functional groups* by *T. R. Crompton*

Section I and Table 1 list the contents of Chapters 14–19, dealing with each element and technique.

Section II deals with the determination of the various elements and groups in organoaluminium compounds (Tables 2–4); III with the determination of Sb; IV with the determination of As, C, F and S in arsenic compounds (Tables 5–7); V with C and H in Be compounds; VI with Bi; VII with B, C, H, Cl and N in B compounds (Table 8); VIII, IX, X, XI and XII with Ca, Cr, Co, Cu and Ge compounds, respectively.

Section XIII treats Fe (Table 9); XIV Pb, especially in petroleum (Tables 10 and 11); XV–XX with Li, Mg, Mn, Hg, Ni and Pd; XXI with the determination of P, Fe, Si, Ti, As, P, C, H, N, O, halogens and S in P compounds (Table 12); and XXII–XXIX Pt, K, Se, Tl, Sn, Ti, Zn and Zr derivatives.

451 references up to 1979.
Complementary: 30, 15–22.
Relevant: none.

30, 15: Analysis of organometallic compounds: titration procedures by *T. R. Crompton*

The chapter describes titration procedures employed in the analysis of the organometallic compounds listed above. The techniques considered include classical, conductimetric, potentiometric, amperometric, dielectric constant, luminometric, thermometric and high-frequency titration procedures. Typical results are given in seven tables.

204 references up to 1972.
Complementary: 30, 14, 16–22.
Relevant: 1, 5; **2**, 8; **3**, 15; **4**, 3; **5**, 17; **8**, 3; **12**, 4; **13**, 6; **14**, 3; **15**, 3; **16**, 3 **17**, 4; **18**, 5; **20**, 3; **22**, 5; **23**, 5; **24**, 5; **25**, 10; **26**, 5; **28**, 3; **29**, 21; **36**, 5.

30, *16: Analysis of organometallic compounds: spectroscopic methods* by *T. R. Crompton*

Section I describes the visible spectra of organic derivatives of Al (Table 1), As, Bi, B, Co, Cr, Cu, Fe (especially haemoglobin), Pb, Mn, Hg, Ni, P, Se, Ta, Sn (Table 2) and Zn.

Section II deals with UV spectra of B, Li, Mg, Hg and Sn compounds.

247 references up to 1978.
Complementary: 30, 14, 15, 17–22.
Relevant: 1, 5; **2**, 8; **3**, 15; **4**, 3; **5**, 17; **8**, 3; **12**, 4; **13**, 6; **14**, 3; **15**, 3; **16**, 3; **17**, 4; **18**, 5; **20**, 3; **22**, 5; **23**, 5; **24**, 5; **25**, 10; **26**, 5; **28**, 3; **29**, 21; **36**, 5.

30, *17: Analysis of organometallic compounds: polarographic techniques* by *T. R. Crompton*

The chapter describes briefly the various techniques used with different samples in order to permit the polarographic determination of As, B, Fe, Pb, Hg, Mn, Ni, Co, P, Se, Sn and Zn derivatives.

80 references up to 1979.
Complementary: 30, 14–16, 18–22.
Relevant: 1, 5; **2**, 8; **3**, 15; **4**, 3; **5**, 17; **8**, 3; **12**, 4; **13**, 6; **14**, 3; **15**, 3; **16**, 3; **17**, 4; **18**, 5; **20**, 3; **22**, 5; **23**, 5; **24**, 5; **25**, 10; **26**, 5; **28**, 3; **29**, 21; **36**, 5.

30, *18: Analysis of organometallic compounds: gas chromatography* by *T. R. Crompton*

The chapter describes gas chromatographic techniques for the analysis of the organometallic derivatives listed above. In many cases retention times, limits of detection, retention indices and pyrolysis results are tabulated (Tables 1–7) or presented in graphs.

216 references up to 1979.
Complementary: 30, 14–17, 19–22.
Relevant: 1, 5; **2**, 8; **3**, 15; **4**, 3; **5**, 17; **8**, 3; **12**, 4; **13**, 6; **14**, 3; **15**, 3; **16**, 3; **17**, 4; **18**, 5; **20**, 3; **22**, 5; **23**, 5; **24**, 5; **25**, 10; **26**, 5; **28**, 3; **29**, 21; **36**, 5.

30, *19: Analysis of organometallic compounds: other chromatographic techniques* by *T. R. Crompton*

Section I treats the thin-layer chromatography of organic derivatives of Sb, As, Bi, B, Co, Cu, Fe, Pb, Hg, P, Te, Sn and Zn. Section II deals with the paper chromatography of derivatives of As, Bi, Co, Cu, Fe, Hg, Se and Sn. Section III treats the column chromatography of Co, Fe, Pb, Hg, P and Se compounds and Section IV the electrophoresis of Fe, Pb and Ta derivatives.

137 references up to 1977.
Complementary: 30, 14–18, 20–22.
Relevant: 1, 5; **2**, 8; **3**, 15; **4**, 3; **5**, 17; **8**, 3; **12**, 4; **13**, 6; **14**, 3; **15**, 3; **16**, 3; **17**, 4; **18**, 5; **20**, 3; **22**, 5; **23**, 5; **24**, 5; **25**, 10; **26**, 5; **28**, 3; **29**, 21; **36**, 5.

30, *20: Infrared and Raman spectroscopy of organometallic compounds* by *M. J. Taylor*

I. Introduction (776); **II**. The techniques of vibrational spectroscopy (776); **III**. Progress in the interpretation of organometallic spectra (779); **IV**. References (803).

Section II reviews general techniques of IR and Raman spectroscopy. Section III deals with the vibrational spectra of main group organometallic compounds, including spectra of both the organic ligands and Group I, II, III, IV, V and VI organometallics (Tables 1–10). Next, vibrational spectra of organo-transition metal compounds [including alkyl, aryl, ylide, carbene (Table 11), carbyne and other complexes] are treated, followed by spectra of organometallics with donor ligands such as olefin, acetylene, allyl, allene, arene, cyclopentadiene (Table 12), carbonyl, nitrosyl, cyanide and other ligands, and also some coordinated inorganic ligands (Table 13).

434 references up to 1980.
Complementary: 30, 14–19, 21, 22.
Relevant: 31, 4.

30, *21: Multinuclear magnetic resonance methods in the study of organometallic compounds* by *J. A. Davies*

I. Introduction (815); **II**. Magnetic properties of nuclei (824); **III**. Measurements techniques (833); **IV**. Dynamic systems (836); **V**. Paramagnetic systems (839); **VI**. NMR studies of organometallic systems (841); **VII**. High-resolution solid-state NMR studies (896); **VIII**. Conclusions (896); **IX**. Notes added in proof (898); **X**. References (905).

Section I reviews NMR techniques, the nuclear properties of various metals and non-metals (Tables 1–6) and the literature of NMR.

Section II considers relaxation processes (including dipole–dipole, quadrupolar, scalar coupling and spin–rotation interaction and chemical shift anisotropy), nuclear shielding and the chemical shift (conventions, standards, ranges and correlations between chemical shifts and physical properties) and nuclear spin–spin coupling.

Section III describes direct observation methods (continuous sweep, pulsed excitation and others), double resonance techniques and indirect methods of observation (INDOR and triple resonance).

Section IV considers dynamic systems and Section V paramagnetic shifts, spin delocalization, relaxation agents and shift reagents.

Section VI discusses the NMR of organometallic compounds, classified according to the nucleus under study. Thus, in turn, studies are considered relating to H, D and T (Table 7), alkali and alkaline earth metals, transition metals (Tables 8–21), Group IIIB nuclei [boron (Table 22) and Al], C-13, other Group IVB nuclei [Si (Tables 23–26), Sn (Tables 27–29), Pb (Tables 30, 31)], nitrogen and ^{31}P (Tables 32–38) and the other Group VB nuclei, Group VIB nuclei, including ^{17}O (Table 39), ^{33}S, ^{77}Se (Table 40) and ^{125}Te, followed by studies relating to the halogens.

Section VII treats studies on samples in the solid state.

Section IX includes notes added in proof, mainly on publications that appeared in 1980–81, relating to D, T, Li, transition metals (Tables 41 and 42), N, S, Se, Te and the solid state.

610 references up to 1981.
Complementary: 30, 14–20, 22.
Relevant: 6, 2.V; **6**, 3.IV; **9**, 1; **13**, 6.III.B; **16**, 3.V; **17**, 4.II; **20**, 3; **23**, 5.VII; **24**, 5.II.D; **26**, 5.III.C; **31**, 25; **32**, 6.

30, *22: Mass spectrometry of organometallic compounds* by *T. R. Spalding*

I. Introduction: compound identification and fragmentation behaviour (920); **II.** Electron impact studies (925). **III.** Other ionization techniques (942); **IV.** Ion structure and energetics (946); **V.** References (960).

Section I reviews in general compound identification, assignment of spectra and fragmentation behaviour.

Section II describes electron impact studies with organometallic derivatives of the main group metals, including alkyl (Table 1), aryl and halocarbon compounds and derivatives containing M—H, M—N, M—O, M—S, M—halogen and M—M bonds. This is followed by consideration of transition metal complexes, including metal—C σ-bonded compounds, olefin, acetylene, allyl, cyclopentadienyl and arene complexes; η^7-C_7 and η^8-C_8 ligands, organic ligands containing N, P, S, etc., and cluster compounds. Section II.D treats negative ion spectra by electron impact of main group and transition metal compounds.

Section III deals with chemical and field ionization, field desorption and other techniques.

Section IV considers metastable ion studies, photoelectron–photoion coincidence spectroscopy, ion photodissociation, ionization and appearance potential measurements (Table 2), bond strengths (Table 5) and other thermochemical studies.

393 references up to 1980.
Complementary: 30, 14–21.
Relevant: 9, 7; **10**, 5; **13**, 19; **16**, 4; **17**, 5; **18**, 6; **19**, 5; **22**, 7; **23**, 5.VIII; **24**, 6; **25**, 4; **26**, 6; **27**, 7; **29**, 3; **31**, 3; **32**, 3; **33**, 4; **36**, 7.

Supplement C: The chemistry of triple-bonded functional groups (1983)

31, *1: Chiroptical properties of compounds containing triple-bonded functional groups* by *W. Runge*

I. Introduction (2); **II.** Optical rotation of molecules with triple-bonded functional groups (5); **III.** Circular dichroism of molecules with triple-bonded functional groups (25); **IV.** Acknowledgements (45); **V.** References (45).

Section I reviews the role of chiral acetylenes and cyanides in chemistry and spectroscopy, and compares triple-bonded functional groups with other substituents (Tables 1 and 2).

Section II describes the OR of some naturally occurring title compounds and the significance of triple-bonded groups for the theories of OR and also semi-empirical descriptions of the OR of the title compounds using chirality functions. It considers methane derivatives (Tables 4–6), allenes (Tables 7 and 8), 2, 2-spirobiindanes (Tables 9 and 10), and [2, 2]metacyclophanes (Table 11).

Section III deals with the CD of chromophores with triple-bonded sub-units (Tables 12 and 13) and discusses substituent effects of triple-bonded grups on the CD of selected chromophores in [2, 2]cyclophanes, *N*-2, 4-dinitrophenyl and dimedonyl derivatives of 1-alkyl-2-propynylamines (Tables 14 and 15) and carbonyl compounds.

97 references up to 1981.
Complementary: 26, 3.
Relevant: 9, 4; **18**, 7; **19**, 6; **20**, 2; **22**, 4; **25**, 3; **27**, 6; **28**, 10; **29**, 23; **36**, 17.

31, *2: Thermochemistry of the cyano and isocyano groups* by *L. Batt*

I. Introduction (49); **II.** Thermochemistry of cyanides (50); **III.** Thermochemistry of isocyanides (53); **IV.** Group additivity rules (53); **V.** References (55).

The chapter includes in the data presented bond dissociation energies, which are of interest to kineticists.

Section II deals with cyanides and presents thermodynamic data for cyano groups (Table 1), monocyanides (Table 2), polycyanides (Table 3), cyclic nitriles (Table 4) and unsaturated nitriles (Table 5).

Section III discusses the thermochemistry of isocyanides, for which very few data are available, and Section IV treats group additivity rules for the estimation of the thermodynamic properties of compounds for which there are no experimental values.

54 references up to 1981.
Complementary: none.
Relevant: 16, 16; **18**, 3; **19**, 3; **20**, 11; **21**, 3; **22**, 6; **23**, 4; **24**, 3; **25**, 2; **26**, 4; **27**, 9; **28**, 4; **29**, 24; **30**, 2; **33**, 3; **36**, 14.

31, *3: Mass spectra of cyano, isocyano and diazo compounds* by *K.-P. Zeller*

I. Introduction (57); **II**. Cyano compounds (58); **III**. Isocyano compounds (80); **IV**. Diazo compounds (86); **V**. References (104).

The chapter treats —CN and —NC groups and, since almost no material exists on diazonium compounds, the related diazo compounds are included. The MS of C≡C groups was presented in **24**, 6.

Section II deals with aliphatic nitriles (Tables 1 and 2), benzyl cyanides, aromatic nitriles, alkyl cyanoacetates and α, β-unsaturated nitriles.

Section III describes aliphatic (Table 3) and aromatic nitriles and Section IV considers diazoalkanes (Table 4), α-diazocarbonyl compounds (Tables 5 and 6), diazo oxides (Table 7) and some metal-substituted diazo compounds.

In all cases, figures representing the actual mass spectra and also detailed reaction schemes are given.

91 references up to 1979.
Complementary: 24, 6.
Relevant: 9, 7; **10**, 5; **13**, 19; **16**, 4; **17**, 5; **18**, 6; **19**, 5; **22**, 7; **25**, 4; **26**, 6; **27**, 7; **29**, 3; **30**, 22; **32**, 3; **33**, 4; **36**, 7.

31, *4: Infrared spectra of cyano and isocyano groups* by *I. N. Juchnovski and I. G. Binev*

I. Introduction (107); **II**. Nitriles (108); **III**. Isocyanides (128); **IV**. References (131).

The chapter deals with the relationship between IR spectra with the structure of the corresponding compounds.

Section II describes frequencies (Tables 1 and 2) and intensities of saturated aliphatic nitriles and conjugated nitriles, including aromatics (Tables 3–7), with emphasis on the correlations in variously substituted benzonitrilies by using different Hammett-type equations (Tables 5–7). Next, the spectra of anion-radicals (Table 8), carbanions (Table 9) and di- and trianions containing CN groups are considered.

Section III discusses saturated (Table 10) and conjugated isocyanides.

164 references up to 1980.
Complementary: none.
Relevant: 30, 20.

31, *5: Photoelectron spectra of cyano compounds* by *H. Stafast and H. Bock*

I. Introduction and scope (138); **II**. Information from, and assignment of, the photoelectron spectra of cyano compounds (141); **III**. π and σ interactions in monocyano derivatives (149); **IV**. Dicyano compounds and MO parameters (167); **V**. Polycyano compounds (172); **VI**. Concluding remarks (175); **VII**. References (182).

Section I presents a list of cyano compounds investigated by PES (Table 1). Section II discusses the assignment of PES and the information which can be derived from it. Section III deals with π and σ interactions in linear cyano compounds, in C_{3v} molecules (mainly acetonitrile derivatives R_3CCN with R = H, D, F, Cl, CH_3), in molecules of low symmetry, including alkyl cyanides (Table 1), benzonitrile and cyanopyridine derivatives, compounds of type R_2NCN (Table 7) and of the type R_2PCN and R_2AsCN (Table 8) and also CH_3SCN (Table 9).

Section IV discusses PE spectroscopic MO parameters for geminal dicyano compounds (Table 10) and Section V) for polycyano compounds (Table 11).

Section VI considers the characteristics of the PES of cyano compounds and their interpretation and also their analytical application to the optimization of gas-phase reactions.

113 references up to 1981.
Complementary: none.
Relevant: 32, 28; **39**, 5; **40**, 5.

31, *6: Radiation chemistry of triple-bonded molecules* by *Z. B. Alfassi*

I. Introduction (188); **II**. Radiolysis of alkynes (189); **III**. Radiolysis of nitriles (199); **IV**. Radiation chemistry of aqueous solutions of diazonium salts (216); **V**. References (217).

All of the systems —C≡C—, —C≡N and —$\overset{+}{N}$≡N add to the triple bond radicals formed by radiolysis of the compounds or of the solvents.

Section II deals with the radiolysis of acetylene in the gaseous phase, considering the formation of benzene and of polymers such as cuprene. The main ion–molecule sequences are shown in Table 1. Next the radiolysis of acetylene in the liquid and solid phase and in solution is treated and also radiation-induced cross-linking of polyethylene with acetylene.

Section III describes the radiolysis of acetonitrile in the gaseous phase, the initial species formed in the radiolysis of liquid and solid acetonitrile, the final products in liquid acetonitrile, processes occurring with benzonitrile (Table 2), with aqueous solutions of CN^- and of HCN and of aliphatic nitriles (Table 3) and benzonitrile. Radiation-induced production of nitriles from organic radicals and cyanogen is considered.

Section IV treats briefly the radiolysis of aqueous solutions of diazonium salts.

137 references up to 1979.
Complementary: none.
Relevant: 9, 8; **10**, 6; **11**, 6; **13**, 17; **15**, 9; **16**, 12; **17**, 10; **18**, 11; **19**, 10; **22**, 11; **25**, 12; **27**, 23, 24; **29**, 7; **32**, 9; **33**, 21; **36**, 8.

31, *7: Electrochemistry of the cyano group* by *K. Yoshida*

I. Electroreductive formation of stable anion radicals and dianions (222); **II**. Electroreductive elimination (225); **III**. Hydrogenation (236); **IV**. Electrohydrodimerization (240); **V**. Electroreductive nitrile formation (244); **VI**. Miscellaneous reductions (247); **VII**. Electrooxidative nitrile formation (248); **VIII**. Reaction of nitrile solvent with anodically generated cationic species (258); **IX**. Miscellaneous oxidations (260); **X**. References (263).

Section I deals with the formation of stable anion radicals and dianions from aromatic nitriles and from electron acceptors such as tetracyanoethylene and TCNQ. Section II describes various types of electroreductive elimination, such as decyanation, dehalogenation, deacyloxylation, dealkoxylation, denitration and also fission of C—S, N—S and C—P bonds.

Section III treats the hydrogenation of —C≡N and C≡C bonds, benzene and also carbonyl and nitro groups. Section IV considers both self- and cross-coupling of nitriles and the mechanism of the process.

Section V describes electroreductive cyanation and cyanoalkylation and Section VI treats the electroreduction of violenes, hydrazyls and unsaturated compounds containing CN groups.

Section VII discusses anodic cyanation and also anodic formation of CN groups (e.g. from α-amino acids or amines ($RCH_2NH_2 \rightarrow \rightarrow RCH{=}NH \rightarrow RCN$). Section VIII considers the electrochemical reaction ($RCOOH \xrightarrow{MeCN} \rightarrow RNHCOMe$) and other processes occurring with nitrile solvents with anodically generated radicals, cation radicals and cations.

Section IX reviews some miscellaneous electrochemical oxidations of other parts of molecules in the presence of cyano groups which are not themselves oxidized and also some processes promoted by special solvents, reactions of CN-containing ligands in MO complexes, and reactions of isocyanide complexes of transition metals.

262 references up to 1980.
Complementary: none.
Relevant: **5**, 2; **8**, 11; **13**, 5; **16**, 15; **17**, 14; **19**, 12; **20**, 5; **21**, 5; **22**, 9; **23**, 10; **24**, 17; **25**, 13; **26**, 12; **27**, 8; **28**, 7; **29**, 8, 9; **32**, 6; **34**, 1.

31, *8: The directing and activating effects of triply bonded groups* by *M. Charton*

I. Triply bonded groups (270); **II**. Substituent effects in XGY systems (271); **III**. Substituent effects in XY systems (279); **IV**. Directing effects on product formation (292); **V**. Directing effects in chemical equilibria–tautomerism (303); **VI**. Directing effects in stereoisomeric equilibria (312); **VII**. Activation effects (318); **VIII**. Conclusion (319); **IX**. References (320).

Section I defines the contents of the chapter and refers to some uncommon triple-bonded groups (Table 1).

Section II discusses electrical substituent effects and gives the values of various substituent constants of triple-bonded groups (Table 2) and reviews the relationship between the electrical effects of different types of X≡Y groups and the estimation of σ_I, and σ_D for these and also steric effects and polarizability.

Section III treats substituent effects when the substituent is directly bonded to an active site. This includes stabilization of carbenium ions, radicals and carbanions (Tables 4–6). The delocalized electrical effect is considered (Tables 7 and 8).

Section IV reviews directing effects in the product formation of electrophilic aromatic substitution (Table 9), of free-radical addition to olefins (Table 10), in Diels–Alder and 1, 3-dipolar [2 + 3] cycloadditions (Table 11) and in some other reactions. Table 12 shows the application of a correlation equation to the calculation of *r* values for various triple-bonded groups, permitting the prediction of attacks at various positions.

Section V deals with directing effects in various types of 1, 3-prototropy (Table 13), such as allyl–propenyl tautomerism (Table 14), keto–enol tautomerism, acetylene–allene prototropy and other related systems, and also with annular tautomerism in heterocyclic systems (Scheme 2, p. 308). Table 15 again illustrates the use of correlation equations for the prediction of directing effects in the above-mentioned systems.

Section VI discusses directing effects in geometrical isomerism (Table 16) and in conformational equilibria (Table 17). The use of correlation equations in the calculation of the values of directing effects of triple-bonded groups in these systems is demonstrated in Table 18.

Section VII deals briefly with activation effects of triple-bonded groups.

162 references up to 1980.
Complementary: 12, 5; **23**, 11.
Relevant: 3, 3; **4**, 5; **5**, 16; **6**, 8; **7**, 2, 3; **11**, 12; **13**, 8; **14**, 4; **15**, 5; **16**, 6; **18**, 9; **19**, 8; **28**, 9; **29**, 16; **36**, 15.

31, *9: Biological formation and metabolic transformations of compounds containing the cyano group* by *J. P. Ferris*

I. Introduction (326); **II**. Cyanogenic glycosides in plants (326); **III**. Cyanogenic compounds in arthropods (328); **IV**. Cyanogenic lipids (328); **V**. Cyanide formation by moulds and microorganisms (329); **VI**. Cyanide from thiocyanates (332); **VII**. Lathyrogenic cyano compounds (332); **VIII**. Detoxification of cyanide and the metabolism of nitriles (335); **IX**. Speculations on the origins of cyanide biochemistry (337); **X**. Acknowledgements (338); **XI**. References (338).

Section II deals with the occurrence, toxicity, biosynthesis, pharmacology and biological function of cyanogenic glycosides in plants. Cyanogenic compounds are constituents of the defensive secretion of some arthropods (Section III) and cyanogenic lipids occur in several genera of the soapberry family (*Sapindaceae*) (Section IV).

Section V describes the occurrence and biosynthesis of cyanides in moulds in microorganisms including fungi, pseudomonas, chromobacterium and chlorella, and the utilization of HCN for amino acid synthesis. Cyanide is also produced from thiocyanates (Section VI). Some plants may cause the nutritional disease lathyrism, the symptoms of which include skeletal deformation and neurological disorders. The active toxic constituents are β-aminopropionitrile, β-cyanoalanine and some other cyanides. Section VIII discusses the toxicity, detoxification and metabolism of aliphatic, aromatic and multifunctional nitriles. Section IX speculates briefly on the role of cyanides in the original synthesis of various natural products.

82 references up to 1980.
Complementary: 12, 12.
Relevant: 2, 7; **3**, 5; **4**, 9; **5**, 18; **7**, 4; **11**, 11; **13**, 14; **15**, 10; **16**, 13; **17**, 13; **18**, 13; **19**, 13; **20**, 6; **22**, 20; **24**, 14; **26**, 13; **28**, 17.

31, *10: Free-radical reactions involving the C≡C group* by *Y. Amiel*

I. Introduction (342); **II**. General background (342); **III**. Addition of halide-centred free radicals (348); **IV**. Addition of carbon-centred free radicals (353); **V**. Addition of sulphur-centred free radicals (359); **VI**. Addition of oxygen-centred free radicals (369); **VII**. Addition of other Group IVB-centred free radicals (370); **VIII**. Addition of Group V-centred free radicals (372); **IX**. References (377).

The material in the chapter is arranged according to the atom carrying the unpaired electron. Mechanisms, regioselectivity, stereochemistry and relative C≡C versus C=C reactivity are reviewed (Section II).

Section III discusses additions of hydrogen halides and halogenations involving *t*-BuOCl or the halogens. Section IV deals with the addition of CCl_4, $CHCl_3$, $CBrCl_3$, $ClCl_3$, CF_4 and perfluoroiodoalkanes and also with additions of aldehydes, ketones, acids, esters, cyanides, alcohols, ethers, alkanes and arylalkanes. Various amounts of 1:1 and 1:2 adducts, e.g. $MeCHO + HC{\equiv}CH \rightarrow MeCOCH{=}CH_2 \rightarrow MeCOCH_2CH_2COMe$, can be obtained according to the experimental conditions.

Section V describes the addition of thiols, H_2S, disulphides, $(SCN)_2$, sulphenyl halides, elemental sulphur, F_5SCl, RSO_2Cl and hydrogensulphite. Section VI treats the addition of OH, Section VII the addition of silanes, germanes, stannanes and plumbanes and Section VIII considers additions of N_2O_4, NO_2Cl, *N*-chloramines, F_2NNF_2, phosphines, PCl_3 and some other organophosphorus compounds.

208 references up to 1980.
Complementary: none.
Relevant: **24**, 8, 19.

31, *11: Arynes* by *T. L. Gilchrist*

I. Introduction (383); **II.** The structure of benzyne (384); **III.** Methods of generation (385); **IV.** The chemistry of arynes (390); **V.** References (414).

Section II reviews the structure of benzyne and its orbitals. Section III describes the formation of benzyne from 2-halogenophenyl anions, from benzenediazonium salts and by fragmentation of benzo-fused cyclic systems (e.g. 1-aminobenzotriazole) and also considers and tabulates the generation of other arynes (Table 1).

Section IV discusses benzyne reactivity, di- and trimerization, Diels–Alder reactions (Tables 2 and 3), 1,3-dipolar and [2 + 2] cycloadditions (Table 4) and the ene reaction. Next, the selectivities of arynes with different nucleophiles are considered and also nucleophilic addition to unsymmetrical arynes, the fate of the aryl anion intermediate, additions of carbon, nitrogen, oxygen and other nucleophiles and cyclizations involving arynes. Finally, aryne–metal complexes are discussed.

199 references up to 1980.
Complementary: **31**, 12.
Relevant: none.

31, *12: Six-membered didehydroheteroarenes* by *H. C. van der Plas and F. Roeterdink*

I. Introduction (422); **II.** Monocyclic six-membered didehydromonoazaheteroarenes (429); **III.** Bicyclic six-membered didehydroheteroarenes (470); **IV.** Monocyclic six-membered didehydrodiazaheteroarenes (492); **V.** References (507).

Section I discusses reactions in which the title compounds have been suggested as intermediates. These include direct nucleophilic S_N(AE) and S_N(ANROC) reactions, *cine* and *tele* substitutions, halogen migrations and ring transformation reactions.

Section II deals with 3,4-didehydropyridine (**A**) and its *N*-oxide (**B**) and with some of their derivatives, with intermediacy of A in reactions of 3- and 4-halogenopyridines with strong bases, the addition of nucleophiles to A (Table 1), the generation of A by other methods; the generation of derivatives of A, especially in cases when the generating base can also add to the didehydro bond (Table 2), and the generation and reactions of B and its derivatives. Next, 2,4-didehydropyridine (C) and its derivatives

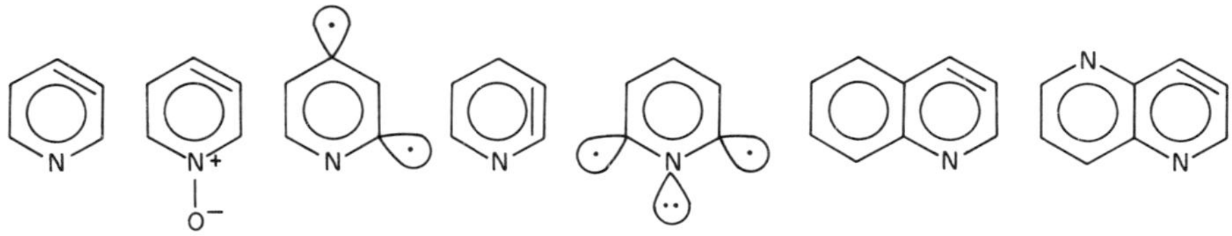

A B C D E F G

are considered, followed by a discussion of the generation, intermediacy and reactions of 2, 3-didehydropyridine (D) and its oxide. Products and yields from the reactions of 2-bromopyridine and its derivatives with strong bases are tabulated (Table 3) and the possible occurrence of 2, 6-didehydropyridine (E) in these reactions is discussed, together with ring closure via polar addition to didehydropyridine intermediates.

Section III treats bicyclic six-membered didehydromonoazaheteroarenes, such as 3, 4-didehydroquinolines (F), their *N*-oxides, and also 3, 4-didehydroisoquinolines and 3, 4-didehydro-1, 5-naphthyridines (G) and didehydro derivatives of 1, 6- and 1, 7- and 1, 8-naphthyridines. Among the oxygen analogues, 3, 4-didehydrocoumarin is mentioned.

Section IV considers the formation as intermediates and the reactions of didehydro compounds obtained from six-membered rings containing two nitrogen atoms, such as pyridazines, pyrimidines and pyrazines.

Through the chapter detailed mechanistic schemes are presented, together with the results of MO calculations relating to the stability of the didehydro intermediates.

184 references up to 1980.
Complementary: 31, 11.
Relevant: none.

31, *13: Oxidation of triple-bonded groups* by *L. I. Simándi*

I. Introduction (514); **II**. Oxidation of alkynes (515); **III**. Oxidation of nitriles (554); **IV**. Oxidation of isocyanides (559); **V**. Oxidation of diazo compounds (563); **VI**. References (566).

Section II deals with the oxidation of alkynes by autoxidation, by catalytic oxidation with O_2, by O_3 and by atomic and singlet oxygen. Next, reactions with peroxy acids, hydroperoxides, peroxides, Fenton's reagent and similar species are described, followed by oxidative coupling of acetylenes and oxidations with HNO_3, *N*-chloro- and *N*-bromosuccinimide, DMSO and halogens. Oxidations by metal ions and salts [$Pb(OAc)_4$, $Tl(NO_3)_3$, Pd^{II}] and by metal oxides (MnO_2, RuO_4, OsO_4, Cr^{IV}, MnO_4^-) (Table 1) are presented.

Section III describes the oxidation of nitriles by Cl_2, O_2, H_2O_2, NaOCl, HNO_3, $Na_2S_2O_8$ and the dehydrocyanation reactions $R_2CHCR_2CN \rightarrow R_2C{=}CR_2 + HCN$ and $R_2CHCN + O \rightarrow R_2CO + HCN$.

Section IV treats the oxidation of isocyanides by O_3, O_2, NO, N_2O, $PhCO_3H$, $(PhCO)_2O_2$, DMSO, some *N*-oxides, $Tl(NO_3)_3$, *t*-BuOCl and Br^+, and also the conversion of isocyanides into isothiocyanates.

Section V considers the oxidation of diazo compounds by O_3, by photooxidation in the presence of various oxygen species, by $CuCl_2$ and with *t*-BuOOH in the presence of $Mo(CO)_6$.

184 references up to 1979.
Complementary: none.
Relevant: none.

31, *14: Reduction of triple-bonded groups* by *R. O. Hutchins and M. G. K. Hutchins*

I. Introduction (572); **II**. Partial reduction of alkynes to alkenes (572); **III**. Reduction of nitriles (586); **IV**. Reduction of other triple-bonded groups (593); **V**. References (595).

Section II describes catalytic hydrogenations with Pd, Ni and other metal catalysts, hydroalumination with $LiAlH_4$ and with $(i\text{-}Bu)_2AlH$, hydroboration and other hydrometalations and lists hydride reagents which in combination with metal derivatives reduce alkynes to alkenes (Table 1). Next, chromium reductions, dissolving metal (Na, Li) reductions and reductions of strained seven- and eight-membered ring alkynes by alcohols alone are treated.

Section III deals with the reduction of nitriles to aldehydes, with metals, metal salts, catalytically and by metal hydrides, especially by $(i\text{-}Bu)_2AlH$ (Table 2) and other borane derivatives. Next, the reactions $RCN \rightarrow RCH_2NH_2$ are dealt with, including catalytic metal–hydrogen and metal hydride reductions, especially by boron reagents (Tables 3–6).

Section IV considers briefly reductions of aromatic diazonium ions (Table 7) and of isocyanides.

323 references up to 1981.
Complementary: **12**, 7.
Relevant: **24**, 7, 8.

31, *15: Dediazoniations of arenediazonium ions and related compounds* by *H. Zollinger*

I. Introduction (604); **II**. Dediazoniations of arenediazonium ions (604); **III**. Comparative discussion concerning the dediazoniations of diazoalkanes and related compounds (651); **IV**. Acknowledgements (661); **V**. References (662).

Dediazoniations are reactions in which an N_2 molecule is one of the products.

Section II discusses the multiplicity of pathways in the title reactions, the S_N1 and aryne mechanisms and the homolytic pathways including thermal reactions in various solvents (Tables 1 and 2) and especially in MeOH and EtOH (Table 3), in alkaline aqueous solutions (Table 4), in highly nucleophilic solvents and in the presence of good nucleophiles (DMSO, pyridine, etc.), in aprotic and relatively apolar solvents (CD_6, C_6D_{12}, C_6D_5Br, $CDCl_3$) and finally under the catalytic influence of metals and metal ions.

Section II.E describes synthetically useful reactions, such as hydroxy-, halogeno-, carboxyanhydro- and hydro-dediazoniations. Section II.F treats the stabilization of arenediazonium salts in the solid state and in the form of crown ether complexes (Table 5); Section II.G discusses photolytic dediazonation (Table 6) and emphasizes the similarities between these and the corresponding thermal processes.

Section III considers the dediazoniation of alkanediazonium ions, which are the intermediates in one of the decomposition pathways of diazoalkanes (the other pathway involving carbenes):

$$\text{products} \leftarrow R_2C\text{:} \leftarrow R_2CN_2 \xrightarrow{H^+} R_2CH\overset{+}{N}{\equiv}N \rightarrow \text{products}$$

the subsection III.A compares aromatic and aliphatic dediazoniations and the differences between their reaction mechanisms.

Section III.B deals with dediazoniations through carbenes, nitrenes and other intermediates.

Throughout the chapter detailed mechanistic schemes are presented for the reactions discussed.

348 references up to 1980.
Complementary: **23**, 8, 12.
Relevant: **31**, 16.

31, *16: Alkenediazonium compounds* by *K. Bott*

Section I reviews the thermal stability of the title compounds and their reactivity toward nucleophilic reagents.

Section II presents syntheses from α-diazocarbonyl compounds from isocyanates, sulphinylamines and sulphonylhydrazones (Tables 1 and 2) under the influence of Lewis acids, e.g.

$$R^1R^2CClCR^3{=}NNHTos \xrightarrow[\text{acid}]{\text{Lewis}} R^1R^2C{=}CR^3N{=}NTos \rightarrow R^1R^2C{=}CR^3\overset{+}{N}_2$$

Section III deals with isolable alkenediazonium salts and with alkenediazonium salts as short-lived intermediates in the nitrosation of vinylamines, in the acidolysis of vinyltriazenes, in the base-promoted decomposition of *N*-nitrosooxazolidones (Table 3) and in the transfer of diazo groups.

Section V describes the vibrational spectra of the title compounds (Table 4) and frequency shifts of the N—N bond when converting an aliphatic diazo group into its conjugated cation (Table 5).

64 references up to 1979.
Complementary: **23**, 2, 14, 15.
Relevant: none.

31, *17: Acidity and proton transfer of cyanocarbon acids* by *F. Hibbert*

Section II deals with the acidity of weak cyanocarbon acids in aqueous solution (Tables 1 and 2), with the activating effect of remote CN groups (Tables 3a and b) and with a comparison of acid-strengthening effects of various groups (Table 4). Next, strong cyanocarbon acids are considered (Table 5), including cyanocyclopentadienes (Table 6), followed by very weak cyanocarbon acids studied by kinetic measurements (Table 7) or by equilibrium acidity measurements in nonaqueous solvents, such as DMSO (Tables 8 and 9). Acidity in the gas phase is treated and results in DMSO are compared.

Section III describes electronic effects in the stabilization of cyanocarbanions (Table 11) and presents the separation of polar and conjugative effects (Table 12) of substituents.

Section IV discusses the rates of proton transfer in aqueous (Table 13) and nonaqueous solvents and also isotope effects on proton transfer (Table 14).

144 references up to 1980.
Complementary: **29**, 16.
Relevant: **2**, 9; **3**, 6; **4**, 4; **5**, 6; **8**, 5; **11**, 3; **12**, 3, 10; **13**, 7; **15**, 4; **16**, 5; **17**, 6; **18**, 8; **19**, 7; **20**, 12; **22**, 8; **23**, 6, 7; **24**, 4; **25**, 5, 6; **29**, 17; **32**, 21; **33**, 5.

31, *18: Recent developments on nitrile oxides, nitrile sulphides and nitrile selenides* by *G. Bianchi, R. Gandolfi and P. Grünanger*

I. Introduction (738); **II**. Structural data on nitrile oxides (738); **III**. Nitrile oxide-forming reactions (742); **IV**. Nitrile oxide–isocyanate rearrangements (748); **V**. Dimerization and polymerization of nitrile oxides (749); **VI**. 1, 3-Dipolar cycloaddition (752); **VII**. Reaction of nitrile oxides with oxygen, nitrogen and carbon nucleophiles (784); **VIII**. Reaction of nitrile oxides with electrophiles (793); **IX**. Nitrile sulphides (793); **X**. Nitrile selenides (797); **XI**. Acknowledgements (798); **XII**. References (798).

Section II presents data relating to the spectroscopy, geometry and electronic structure of nitrile oxides, including vertical ionization potentials (Table 1). Section III treats the preparation of nitrile oxides by dehydrohalogenation of hydroxamoyl halides [ArC(X)=NOH $\xrightarrow{-HX}$ ArCN → O], by dehydration of nitroalkanes, from furazane *N*-oxides and by other methods. Section IV deals with the RC≡N → O → RN=C=O rearrangement and Section V with the di- and polymerization of fulminic acid (HC≡N → O → HON=CHC≡N → O → polymers) and of other nitrile oxides of furazan *N*-oxides under base catalysis.

Section VI considers the mechanism of 1, 3-dipolar cycloadditions including reactions with C=C and C≡C bonds (Table 2), regioselectivity (Tables 3, 4a and 4b), *syn–anti* selectivity, site selectivity and periselectivity, intramolecular cycloadditions, solvent effects, activation parameters and Hammett ρ values. Next, the use of nitrile oxides as synthons for natural products (especially corrinoid and related compounds) and their reactions with C=S, C=N, C=O and C≡N bonds are dealt with.

Section VII presents nucleophilic attacks on nitrile oxides by H_2O (Table 5), RO^-, CH_3COO^-, N_3^-, carbanions, amines (Table 6), hydroxylamines, sulphimines, Cl^-, BH_4^- and AlH_4^-, followed by reactions with electrophilic betaines (Section VIII).

Section IX reviews nitrile sulphides and their preparation and reactions.

Section X deals with the very labile nitrile selenides (RC≡N → Se) and their photolysis.

244 references up to 1980.
Complementary: **12**, 14
Relevant: none.

31, *19: Conformation of cyano and isocyano compounds* by *C. A. Kingsbury*

I. Introduction (805); **II**. (806); **III**. Simple molecules (809); **IV**. Special cases (819); **V**. Complex molecules (821); **VI**. Addenda (825); **VII**. References (830).

Section I and II review the conformational properties of cyano groups, especially when attached to cycloalkanes.

Section III deals with simple molecules such as C_3H_7CH, $NCCH_2CH_2CN$, cyanoalkenes and some cycloheptatriene derivatives.

Section IV discusses mono- and dicarbanions, some free radicals, CT complexes and complexes of cyanocyclohexanes with ICl.

Section V treats relatively complex molecules in which the choice of conformation involves the interplay among a number of factors. These include cyclic and acyclic *tert*-butyl compounds, cyclohexanone derivatives with one CN and several Ph groups, 1,4-dicyanides with bromine atoms in the 2,3-position and also 4-bromo-5-oxo-2,3,5-triphenylpentanenitrile.

Section VI contains additional material relating to each of the other chapters.

130 references up to 1980.
Complementary: none.
Relevant: **19**, 21; **20**, 2; **21**, 1; **22**, 3; **27**, 5; **28**, 10; **29**, 2; **32**, 1; **33**, 2.

31, *20: Recent advances in isocyanide chemistry* by *H. M. Walborsky and M. P. Periasamy*

Sections II, III and IV deal briefly with the toxicity, structure, physical properties and natural occurrence of isocyanides.

Section V describes syntheses by dehydration of formamides ($RNHCHO \rightarrow RN{=}C$), by the Hoffman carbylamine reaction ($CHCl_3 \rightarrow [:CCl_2] \xrightarrow{RNH_2} RN{=}C$), by displacement of halides by cyanide and other methods.

Section VI deals with the rections of isocyanides, especially with cycloadditions to produce 3-, 4- or 5-membered rings, with carbenoid reactions (including $ArN{=}C \rightarrow \rightarrow ArCN$, etc.) and with oxidants (Hg^{II}, Tl^{III} and Pb^{IV}), acetates giving acetoxymetalation products, and Hg salts in the presence of amines yielding carbodiimides ($RN{=}C + R'NH_2 \xrightarrow{HgCl_2} RN{=}C{=}NR'$) or ureas ($RN{=}C + 2R'NH_2 \rightarrow RNHCONHR'$).

Section VII treats the preparation and reactions of α-metalated aldimines (obtained from isocyanide and RM when the former has no α-H atoms: $RN{=}C + R'M \rightarrow RN{=}CR'M$) and also α-metalated isocyanides (obtained when the isocyanide has an α-H atom, by its substitution: $RCH_2N{=}C \xrightarrow{R'M} RC(M)HN{=}C + R'H$). The reactions of the latter are described with alkylating agents, with carbonyl and acyl compounds and with activated olefins. Carbonyl compounds undergo with strongly activated α-metalated isocyanides the formylaminomethylation reaction:

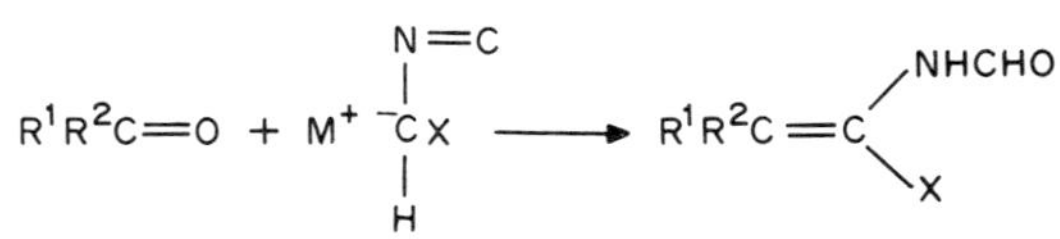

Section VII.C treats the very useful synthetic reactions of tosylmethyl isocyanide [alkylation, ketone formation and especially synthesis of heterocyclics (oxazoles, imidazoles, etc.].

Section VIII discusses the stereochemistry of α-metalated isocyanides and Section IX the reductions of isocyanides (RNC $\rightarrow$ RH) by dissolving metals and by hydrides.

The RNC $\rightarrow$ RCN rearrangement is presented (Section X), followed by a description of metal–isocyanide complexes (Section XI) and polyisocyanides (Section XII).

142 references up to 1980.
Complementary: **12**, 15; **31**, 18.
Relevant: none.

31, *21: Complexation of aryldiazonium ions by polyethers* by *R. A. Bartsch*

I. Introduction (890); **II**. Discovery of the phenomenon (891); **III**. Solid-state complexes of aryldiazonium salts and crown ethers (892); **IV**. Spectral studies of complexes of aryldiazonium salts and crown ethers in solution (896); **V**. Modified reactivity of crown ether-complexed aryldiazonium salts (902); **VI**. Factors which affect the complexation of aryldiazonium salts by polyethers (905); **VII**. Polyethers as phase-transfer catalysts for aryldiazonium salt reactions in solvents of low polarity (911); **VIII**. Conclusions (913); **IX**. Acknowledgement (914); **X**. References (914).

ArN_2^+ compounds complex with crown ethers by insertion of the positively charged rod-like diazonium group into the polar cavity of the macrocycle (Section I and II).

Section III presents the isolation, X-ray diffraction structure, MO calculation, IR spectra (Table 1) and ESCA spectra of the solid title complexes. Section IV deals with the IR (Tables 2–4), UV and visible and NMR (Tables 5 and 6) spectra of the same complexes in solution.

Section IV describes the thermal stabilization of the title compounds in solution and in the solid state, the photochemical stabilization, reduced shock sensitivity, diminished N_α–N_β interchange, deactivation of azo coupling and diminished nucleophilic attack on the aryl rings.

Section VI considers the influence on the complexation of the crown ether (Tables 7–9), of ring substituents, of the anions of the salt (Table 10), of the solvent (Table 11) and of acyclic polyethers (Table 12).

Section VII deals with the effect of polyethers as phase-transfer catalysts in proto- and deuterodediazonization, halodediazonization, aryldediazonization, azocyanide formation, azo coupling and nucleophilic substitution *para* to the N_2^+ group.

Some prospects for the future are presented in Section VIII.

68 references up to 1981.
Complementary: **23**, 7.
Relevant: **27**, 1–4.

31, *22: Poly(diacetylenes) and polyyne polymers containing transition-metal atoms in the main chain* by *W. D. Huntsman*

I. Poly(diacetylenes) (918); **II**. Polyyne polymers containing transition-metal atoms in the main chain (968); **III**. References (974).

Section I describes poly(diacetylenes), monomer crystal packing (Table 1) and molecular structures of poly(diacetylenes) containing also OH, ester, sulphonate ester, urethane, COOH and CONH and other substituent groups (Table 2, a–r). Abbreviations for monomer and polymer names are given in Table 3. Next, the polymerization is treated, giving details on lattices, kinetics, mechanisms (Tables 4 and 5) and on the properties of the products such as structure, bond lengths (Table 6), PES (Table 7), electronic spectra (Table 8), Raman and IR (Tables 9 and 10), electrical properties, defect properties and other physical properties and uses.

Section II discusses polyyne polymers with Pt, Pd, Ni, etc., atoms in the main chain, including their preparation and properties.

256 references up to 1980.
Complementary: none.
Relevant: none.

31, *23: Cyclodimerization of alkynes and reactivity of aluminium halide complexes of cyclobutadienes* by *H. Hogeveen and D. M. Kok*

Section II deals with the cyclodimerization of alkynes to AlX_3 σ-complexes of cyclobutadienes, which in turn may produce 4-, 5- or 6-membered ring compounds. The mechanism of the reaction is described and the cyclodimerization processes promoted by proton acids, organotransition metal complexes and AlX_3 are compared.

Section III treats the chemical reactivity of AlX_3 σ-complexes of cyclobutadienes, in which the AlX_3 group may migrate from one ring position to the other:

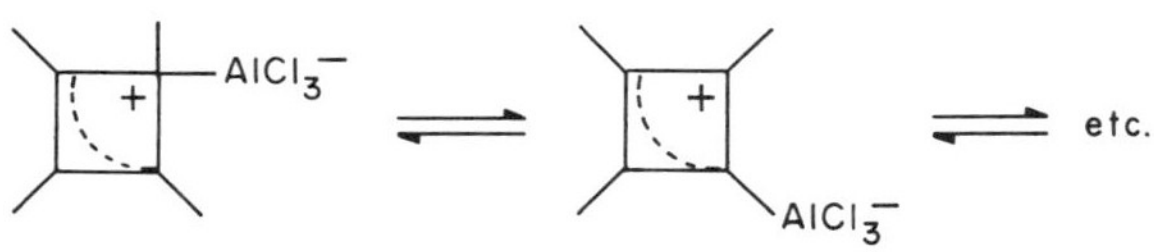

and considers the reaction of these complexes with C≡C bonds (Table 2), C≡N bonds (Table 3), C=C bonds (Table 4), isocyanates (Table 5), carbodiimides, isothiocyanates and sulphinylaniline (Table 6) and some other reagents. Finally, metallocyclopentadienes and transition metal π and AlX_3 σ-complexes of cyclobutadienes are compared.

72 references up to 1981.
Complementary: none.
Relevant: none.

31, *24: Structure of triple-bonded molecules* by *J. B. Moffat*

The chapter reviews experimental and theoretical structural determinations and related spectroscopic measurements, as relevant to the nuclear configuration of the title compounds. Electronic configurations are discussed in **31**, 27.

Section II deals with C_2 and CN^+ species and Section III with triatomic molecules XC≡N (X = H, F, Cl, Bu, I) (Table 1). Section IV treats linear molecules containing both C≡C and CN groups (Tables 2 and 3).

Section V discusses the bond lengths and angles in cyanamides, determined by spectroscopy and by theoretical calculations (Table 4) and Section VI does the same with acetonitriles (Tables 6 and 7), followed by propionitrile (Table 8, Section VIII), byronitrile (Section IX), vinyl cyanides and isocyanides (Table 9, Section X), nitrosyl cyanide, cyanogen azide and isocyanate (Section XI), cyclic molecules such as cyano derivatives of cyclopropane (Section XII) and finally dicyano (Section XIII) and tricyano (Section XIV) compounds.

186 references up to 1980.
Complementary: 24, 2; **31**, 27.
Relevant: 16, 2; **17**, 2; **18**, 2; **19**, 2; **20**, 2; **21**, 1; **22**, 2; **23**, 3; **26**, 2; **27**, 4; **28**, 2; **29**, 1; **30**, 1; **32**, 24; **33**, 12; **34**, 10; **36**, 3.

31, *25: NMR spectra of acetylenes* by *D. G. Morris*

I. ^{1}H chemical shifts (1035); **II.** ^{13}C chemical shifts (1037); **III.** Calculations of ^{13}C chemical shifts and anisotropies (1041); **IV.** Relaxation times (1043); **V.** Coupling constants (1045); **VI.** Deuterium quadrupole coupling (1051); **VII.** Carbon–bromine coupling (1051); **VIII.** Conformational mobility (1052); **IX.** References (1054).

Section I deals with proton chemical shifts in acetylenes (Table 1) and Section II with ^{13}C (Table 2) in simple, substituted and polyacetylenes. INDO, CNDO/2, SCF and other theoretical calculations of ^{13}C chemical shifts and anisotropies are treated in Section III, followed by relaxation times (Section IV). Section V discusses the coupling constants of mono- and polyacetylenes and of heteroatom-substituted acetylenes (Table 3), followed by deuterium quadrupole (Section VI) and carbon–bromine (Section VII) coupling. Conformational mobility in cyclooctyne derivatives, in acetylenes containing substituents with asymmetric carbons and in ditriptycenylethyne and its derivatives is considered in Section VIII.

109 references up to 1980.
Complementary: none.
Relevant: 9, 1; **30**, 21; **32**, 4; **36**, 6.

31, *26: Preparation and synthetic applications of cyano compounds* by *A. J. Fatiadi*

I. Introduction (1064); **II.** Newer methods for synthesis of nitriles: some of their reactions (1065); **III.** Selected synthetic methods and reactions involving cyano cubstrates (1157); **IV.** Selected cyano reagents for organic synthesis (an overview) (1190); **V.** Cyanocarbons and electron acceptors (1126); **VI.** Syntheses of heterocycles via cyano substrates (1243); **VII.** Addenda (1253); **VIII.** (1272); **IX.** References (1272).

Section II treats preparations of nitriles by elimination, e.g. from aldoximes (RCHO $\rightarrow$ RCH=NOH $\xrightarrow{-H_2O}$ RC≡N) with the use of a wide variety of dehydrating agents, from ketoximes by abnormal Beckmann rearrangement, by elimination of H_2O or H_2S

from amides and thioamides and from amines and hydrazones. Next (Section II.D), stereoselective syntheses of unsaturated nitriles are presented (Table 1), involving a variety of condensation and substitution methods. Direct conversion of carbonyl compounds into unsaturated nitriles is treated (Section II.E), followed by syntheses of aminonitriles, enaminonitriles and related compounds. Section II.G deals with the Strecker synthesis of asymmetric amino acids via aminonitriles and Section II.H with the synthesis of saturated nitriles.

Section II.J treats the synthesis of aromatic nitriles by Ni-catalysed cyanation of aromatic halides and by more conventional methods (dehydration of amides, Sandmeyer reaction, etc.). Next, syntheses of heterocyclic nitriles and cyanohydrins are described, followed by the cyanoethylation reaction (Section II.M) via acrylonitrile, treating also related processes and the use of α-metalated nitriles, allylic nitrile anions and other unsaturated and conjugated nitriles, etc.

Section II.N presents cyanomethylation via acetonitrile, including two-carbon elongation involving $^{-}CH_2CN$ anions (e.g. $R^1R^2CO + CH_3CN \rightarrow R^1R^2C{=}CHCN$) or α-substituted acetonitriles, XCH_2CN [e.g. $ArCHO + MeOCH_2CN \rightarrow ArCH{=}C(OMe)CN$].

Section II.O describes the syntheses and reactions of nitriles under phase-transfer conditions and Section II.P the preparation of cyano compounds having O=CCN, $O{=}CCH_2CN$, $O{=}CCH_2CH_2CN$, C=NCN, S=CCN, O=SCN, SC=NCN and O=CSCN functional groups.

The conversion $RCH_2NO_2 \rightarrow RCN$ by the Vilsmeier–Haack and other methods (Section II.O) is followed by photochemical reactions of nitriles (including quenching, photocycloaddition, photosubstitution, photoisomerization and photorearrangement, photocyanation, photochemical benzylation and photolytic formation of —CN groups in special cases).

Section III presents some selected syntheses of cyano compounds, including direct cyanation by $(CN)_2$, aromatizations involving cyanation, Diels–Alder and other reactions with dicyanoacetylene, additions of dicyanocarbene to cyclooctatetraene and various syntheses of cyano sugars.

Section III.B treats selected reactions and transformations of cyano compounds, such as the synthesis of prostaglandins and other cyclizations, decyanation of nitriles, ring expansion by addition of N_3CN to alkenes, transannular cyclization of bicyclic nitriles and the production of aromatic aldehydes, cyanohydrin esters, hydantoins, thiohydantoins, amides, *N*-alkylamides and thioamides—all reactions involving nitriles. Next, the hydrolysis and decarboxylation of nitriles are discussed, and also their transformation into methyl compounds ($ArCN \xrightarrow[Ni/Al_2O_3]{H_2/\Delta} ArMe$). Nitriles as amino-protecting groups and as synthons for many other reactions are also shown briefly.

Section III.C discusses rearrangements of cyano compounds.

Section IV describes the Wittig reaction, hydrocyanation with the Nagata reagent (R_3Al–HCN or Et_2AlCN), the use of Me_3SiCN in cyanosilylation, in the protection of carbonyl groups and in additions and other reactions, the use of solid (inorganic or polymeric) supports in the synthesis of nitriles, cyanoboration, the use of $PdCl_2$–nitrile complexes, dehydrogenation by 2,3-dicyano-5,6-dichloro-1,4-benzoquinone and its use also for benzylic oxidation and hydroxylation and in cycloadditions and the use of $NaBH_3CN$ as a highly selective reducing agent.

Section V treats the versatile use of malononitrile as a synthone in reactions with ketones, oxocarbons, etc., and also the uses of $BrCH(CN)_2$ and of cyanocarbons. Next, it deals with tetracyanoethylene, its reactions with nucleophilic double bonds, with protoporphyrin, with CH acids (leading to heterocycles) and with allylsilane. This is followed by a review of tetracyanoquinodimethane and analogous electron acceptors.

Section VI considers selected syntheses of heterocycles involving nitriles by various cyclization reactions (e.g. with cyanoepoxides, 1-cyanothioformanilide, cyanoimines and cyanamide), by ring enlargements and by cycloadditions involving dicyanomethylides.

Section VII reviews recent results relating to the whole chapter, including reductive decyanation of nitriles by alkali fusion (Table 2) and 34 reactions shown only as equations at the end of the chapter.

1236 references up to 1981.
Complementary: 12, 2, 7, 8; **31**, 28.
Relevant: none.

31, *27: General and theoretical properties of triple-bonded molecules* by *J. B. Moffat*

I. Introduction (1305); **II.** The simple cyanides, CN^-, CN and CN^+ (1306); **III.** Hydrogen cyanide, HCN (1307); **IV.** The nonreactive dimerization of HCN (1309); **V.** Oligomerization of HCN (1310); **VI.** H_2CN^- and H_2CN^+ (1314); **VII.** Cyano and isocyano groups as substituents in carbonium ions, carbanions and radicals (1315); **VIII.** The cyanide–isocyanide isomerization (1320); **IX.** The chemical bond in cyano molecules (1332); **X.** Acknowledgements (1340); **XI.** References (1340).

The chapter is restricted to the treatment of cyano and isocyano groups only. Section II deals with the electronic configurations, energies and calculations of the CN^-, CN and CN^+ species (Tables 1–3). Sections III–V consider the quantum chemistry, structure, dimerization (self-association) and oligomerization (to both linear and cyclic oligomers, including purines) of HCN.

Section VI treats protonated and hydrided HCN (i.e. H_2CN^+ and H_2CN^-) and Section VII the influence of ^-CN and ^-NC groups on ions and radicals, including stabilization energies (Table 4), nuclear configurations (Table 5) and charges (Table 6) of α-cyanocarbonium ions and stabilization energies of α-isocyanocarbonium ions (Table 7).

Section VIII treats the thermal isomerization processes HCN–HNC, MeCN–MeNC and RCN–RNC (R = vinyl, CF_3, Pr, etc.). Data are tabulated for the ionic character of stages in isomerizations (Table 8) and for energies of isomerization and energy barriers (Tables 9 and 10). Section IX presents calculated energy-partitioning bond components (Table 11) of various groups, including triple bonds, partitioned two-centre energies of C≡C bonds (Table 12) and of C≡N bonds (Table 13) in nitriles, bond correlation energies (Table 14) and a listing of electronic, dissociation and molecular correlation energies for 32 simple organic molecules, including several cyanides and acetylenes.

137 references up to 1980.
Complementary: 12, 1; **24**, 1.
Relevant: 1, 1; **2**, 1; **3**, 1; **4**, 1; **5**, 1; **6**, 1; **8**, 1; **11**, 1; **13**, 1; **14**, 1; **15**, 1; **16**, 1; **17**, 1; **18**, 1; **19**, 1; **20**, 1; **22**, 1; **23**, 1; **25**, 1; **26**, 1; **27**, 9; **28**, 1; **32**, 1; **33**, 1; **36**, 2.

31, *28: Recent advances in the synthesis of triple-bonded groups* by *K. Friedrich*

I. Introduction (1346); **II.** Preparation of nitriles by addition of hydrogen cyanide or its derivatives (1346); **III.** Preparation of nitriles by substitution (1351); **IV.** Preparation of nitriles by elimination (1358); **V.** Preparation of nitriles by ring cleavage of heterocycles (1365); **VI.** Preparation of nitriles by conversion of other nitriles (1369); **VII.** Preparation of nitriles by miscellaneous methods (1373); **VIII.** Preparation of acetylenes by elimination reactions (1376); **IX**. Preparation of acetylenes by substitution reactions (1380); **X**. Preparation of diazonium cations (1384); **XI**. References (1385).

Section II deals with the addition of HCN or cyanides to C=C, C≡C, C=O, C=N and C≡N bonds. Section III treats nitrile syntheses by substitutions by cyanide groups of halogens, of groups containing O, S or N, by direct aromatic substitution and by substitution involving C—C bond cleavage. Carbonyl groups can be transformed into cyanoalkyl groups (Section III.B), and ClCN or cyanates also produce nitriles with nucleophiles.

Section IV describes nitrile-forming eliminations such as dehydration of oximes, Beckmann fragmentations, cleavage of hydrazones and by the reaction of carboxylic acids with chlorosulphonyl isocyanate:

$$RCO_2H + ClSO_2NCO \xrightarrow{-CO_2} RCONHSO_2Cl \xrightarrow{DMF} RCN + SO_3 + HCl$$

Section V treats the cleavage of N-containing heterocycles (mainly five- or six-membered) to yield nitriles by thermal or catalytic decomposition.

Section VI considers reactions in which a nitrile is converted into a different one, often making use of the stabilizing effect a CN group exerts on adjacent carbanion or radical centres, in processes involving metalation, condensations, Michael additions, Witting reactions, etc. Section VII treats some other methods not covered in Sections II–VII.

Section VIII describes the synthesis of acetylenes by dehydrohalogenation, dehalogenation and other β-eliminations, elimination of N_2 from hydrazones, ring cleavage of heterocycles and fragmentations.

Section IX covers the syntheses of acetylenes by substitutions in alkali and alkaline earth metal acetylides, through derivatives of Al, Si, Zn, Cu, Pd and B containing C≡C groups, involving substitutions of halogen atoms in a second molecule, or additions to C=O or to C=C groups. The reaction of haloalkynes with trialkylalanes is also mentioned and a few recent reviews regarding the preparation of diazonium cations are quoted (Section X).

281 references up to 1981.
Complementary: 12, 2; **23**, 14; **24**, 18; **31**, 26.
Relevant: 30, 7.

Supplement D: The chemistry of halides, pseudo-halides and azides (1983)

32, *1: Molecular mechanics and conformation* by *A. Y. Meyer*

The chapter discusses mainly aspects of molecular mechanics that relate to halides (Sections I–IV) and especially to their conformations (Sections V–IX), concentrating on atoms and bonds.

Section II deals with molecular-mechanical force-field calculations and their application to the conformations of 1-halopropanes and the *cis–trans* isomerism of 1-halopropanes.

Section III treats intramolecular electrostatics, dipole moments and bond moments of some hydrocarbons and their halogen derivatives (Tables 1–4). It presents the point-dipole approach and local dielectric constants (Table 5). The alternative point-charge approach is considered next, together with the variability in bond moments.

Section IV discusses halide force fields and Section V some fluorine compounds and their energy-component analyses (Table 6).

Section VI considers the phenomenon of non-bonded attraction stabilizing more crowded conformations.

Section VII treats conformational energies and dipole moments (Tables 7 and 8) in 1,2-dihaloethanes, Section VIII in cyclic halides and Section IX discusses solution equilibria (Table 9) and their formulation and computation.

334 references up to 1980.
Complementary: 16, 1, 2.
Relevant: 1, 1; **2**, 1; **3**, 1; **4**, 1; **5**, 1; **6**, 1; **8**, 1; **11**, 1; **12**, 1; **13**, 1; **14**, 1; **15**, 1; **17**, 1; **18**, 1; **19**, 1; **20**, 1; **22**, 1: **23**, 1; **24**, 1; **25**, 1; **26**, 1; **27**, 9; **28**, 1; **31**, 27; **33**, 1; **36**, 2.

32, *2: Diamagnetic behaviour of compounds containing carbon—halogen bonds* by *R. R. Gupta*

Section II describes the experimental methods of Guy and Faraday and the NMR method for measuring diamagnetic susceptibility. Section III treats the calculation methods of Pascal, based on atomic susceptibility (Tables 1–3), the Pascal, Pacault and Hoarau method, which is a modification of the former (Tables 4–6), the Baudet and Tillieu wave-mechanical method (Tables 7–11), the Pascal, Gallais and Labarre method for calculating diamagnetic susceptibility values (Table 13) from bond susceptibilities (Table 12), the Haberditzi method, considering inner shell core electrons, bonding electron increments and π-electrons (Tables 14–17), and a modification of the latter (Table 18).

Section V presents results for alkyl and aryl halides (Tables 19–21), disubstituted benzenes (Table 22), carbonyl compounds (Tables 23–24) and chlorovinyl ketones (Table 25).

78 references up to 1977.
Complementary: none.
Relevant: none.

32, *3: The mass spectra of azides and halides* by *J. M. Miller and T. R. B. Jones*

I. Introduction (75); **II**. The mass spectra of azides (76); **III**. The mass spectra of halides (98); **IV**. References (104).

Section II deals with the MS of aromatic and benzoyl azides (Table 1) and aliphatic azides (Table 2), with MS studies relating to pyrolysis and photolysis, with steroid and alkaloid azides (Table 3) and iodo azides (Table 4), with heterocyclic azides (Table 5) and with sulphonyl and phosphine azides and inorganic systems (Table 6).

Section III treats halides, including environmental applications (e.g. the high sensitivity detection of dioxins), and also halonium ions, halide migrations and the application of the theory of hard/soft acids/bases (HSAB) to MS.

93 references up to 1981.
Complementary: 16, 4.
Relevant: 9, 7; **10**, 5; **13**, 19; **17**, 5; **18**, 6; **19**, 5; **22**, 7; **24**, 6; **25**, 4; **26**, 6; **27**, 7; **29**, 3; **30**, 22; **31**, 3; **33**, 4; **36**, 7.

32, *4: Nuclear quadrupole resonance of carbon-bonded halogens* by *E. A. C. Lucken*

I. The theoretical foundations of nuclear quadrupole coupling (108); **II**. The relationship between electronic structure and nuclear quadrupole coupling constants (116); **III**. The calibration of the method: NQR frequencies in simple systems (120); **IV**. NQR as a tool (131); **V**. References (154).

Section I describes the basics of nuclear quadrupole resonance (NQR) spectroscopy, especially for compounds containing halogens (Table 1), including NQ moments, coupling constants, NQR Zeeman and Stark effects and the effects of temperature and crystalline environment on NQR frequencies, tabulates the vibrational dependence of the coupling constant of some simple organic halides (Table 2) and the related isotope effects (Table 3) and compares gas-phase and solid-state quadrupole coupling constants (Table 4).

Section II discusses attempts to define quantitative relationships between observed NQ coupling constants and derived electronic structures. Section III deals with NQR frequencies and the polarity of the C—halogen bond (Tables 6 and 7), shows ^{13}Cl resonance frequencies for some substituted chloromethanes (Tables 8 and 9) and compares Taft parameters with substituent parameters derived from ^{35}Cl quadrupole resonance frequencies. The NQR frequencies of *o*-, *m*- and *p*-substituted chlorobenzenes are given in Table 11 and their substituent parameters in Table 12. Asymmetry parameters are used as a measure of partial double-bond character of C—halogen bonds (Tables 13–16).

Section IV considers the use of NQR as a tool and shows the attenuation of the substituent effect with chain length in halides (Table 17), and describes also studies with halogenocycloalkanes (Table 18), admantanes, geminally substituted methyl halides (Table 19), halogen derivatives of polycyclic arenes (Table 20) and of heteroarenes (Tables 27 and 28), haloacetylenes (Table 29) and carboranes (Table 30). Donor–acceptor complexes (Tables 31 and 32), sandwich compounds and Van der Waals complexes are also treated.

Section IV.L discusses conformation and configuration in *cis–trans* isomers about double (Table 33) and single (Tables 34 and 35) C—C bonds, *syn–anti* and *endo–exo* isomers (Table 36), hyperconjugation and special effects in α-chloro ethers.

Analytical applications and structure determinations are possible by NQR (Tables 37 and 38) and NQR can be determined also in excited trilet states.

219 references up to 1980.
Complementary: none.
Relevant: none.

32, *5: 1,2-Dehalogenations and related reactions* by *E. Baciocchi*

I. Introduction (162); **II**. Dehalogenations promoted by two-electron reductants (164); **III**. Dehalogenations promoted by one-electron reductants (182); **IV**. Dehalogenations promoted by metals (191); **V**. Electrochemical dehalogenations (195); **VI**. References (197).

The title reactions are β-eliminations involving the loss of two halogens from 1,2-dihaloalkanes and -alkenes.

Section II deals with the use of a wide variety of nucleophiles or two-electron reductants (e.g., Cl^-, Br^-, I^-, RS^-, HS^-, phsophines, arsenides, CN^-, LAH, amines, RO^-, HO^-, etc.): $RCHBrCHBrR' + 3I^- \rightarrow RCH{=}CHR + I_3^- + 2Br^-$.

The mechanisms and stereochemistry of these reactions are described in detail, together with the roles of the halogen, of the organic moiety (Tables 1–3), of the nucleophile (Tables 4 and 5) and of the solvent, and competition with other reactions (substitution, dehydrohalogenation, etc.) is discussed.

Section III treats similarly one-electron reductants, including radical anions, carbanions, metal ions (Tables 6 and 7), radicals and OH^-.

Section IV considers dehalogenations promoted by Zn and other metals. In both Sections III and IV, the mechanisms, stereochemistry, reactivity and structure are presented in detail. Section V describes electrochemical 1,2-dehalogenations.

199 references up to 1980.
Complementary: none.
Relevant: **16**, 9, 10, 15.

32, *6: Electrochemical oxidation, reduction and formation of the C—X bond—direct and indirect processes* by *J. Y. Becker*

I. Introduction (204); **II**. Direct anodic cleavage of the C—X bond (205); **III**. C—X bond formation by direct anodic oxidation of 'X' species (216); **IV**. Indirect cleavage of the C—X bond by electrochemically generated mediators (235); **V**. Direct cathodic reduction of the C—X bond (260); **VI**. Acknowledgements (277); **VII**. References (277).

Section II considers direct anodic oxidative cleavage of C—X bonds, presenting oxidation potentials of alkyl iodides (Table 1), and describing the oxidation of iodoadamantanes (Table 2) and of a variety of C—Br (Tables 3–7), C—Cl and C—F bonds (Table 8). The mechanism of the process is discussed, the results favouring the S_N2 mechanism (Table 9).

Section III describes the formation of C—X bonds by direct anodic oxidation of halogen moieties, considering in turn C—I bonds (Tables 10–12), C—Br bonds (Tables 13 and 14) C—Cl bonds (Table 15), C—F bonds, C—SCN and S—SeCN bonds (Tables 16 and 17), C—N_3 bonds (Tables 18–20) and finally C—OCN bonds.

Section IV treats the cleavage of C—X bonds by indirect pathways through electrochemically generated electron transfer species. The latter include anion radicals, dianions and other mediators (Tables 21–23), cation radicals (Tables 24 and 25) and also inorganic species such as positive iodine (Table 26), positive bromine (Tables 27 and 28), superoxide ion (Tables 29 and 30) and $SO_2^{\bullet -}$. Electrochemical generation of transition metal complexes as electron transfer mediators (Tables 31 and 32) and some less general processes are also considered (templates, modified electrodes).

Section V deals with direct cathodic reduction of C—X bonds in organic halides, including a general description of the reduction pathways and treatment of acyclic mono- and polyhaloalkanes, allyl and benzyl halides, halomethyl arenes, alicyclic , aryl and heteroaryl halides and also organic azides (Tables 33–35).

327 references up to 1982.
Complementary: 16, 15.
Relevant: 5, 2; **8**, 11; **13**, 5; **17**, 14; **19**, 12; **20**, 5; **21**, 5; **22**, 9; **23**, 10; **24**, 7; **25**, 13; **26**, 12; **27**, 8; **28**, 7; **29**, 8, 9; **31**, 7; **34**, 1.

32, *7: Pyrolysis of aryl azides* by *L. K. Dyall*

I. Introduction (288); **II**. Solution pyrolyses involving aryl nitrenes (288); **III**. Solution pyrolyses with neighbouring group assistance (299); **IV**. Pyrolysis with ring opening (310); **V**. Nitrene–carbene rearrangements (313); **VI**. Pyrolysis of sulphonyl azides (316); **VII**. References (318).

Section II discusses the pyrolysis of azides in solution, yielding nitrenes $ArN_3 \xrightarrow{\Delta} Ar\ddot{N}$: (singlet$\rightleftharpoons$Ar$\dot{N}$·(triplet). It describes experimental conditions, the trapping of singlet and triplet aryl nitrenes and their roles in product formation (Tables 1–3), the substitution by nitrenes into adjacent aryl rings, e.g.

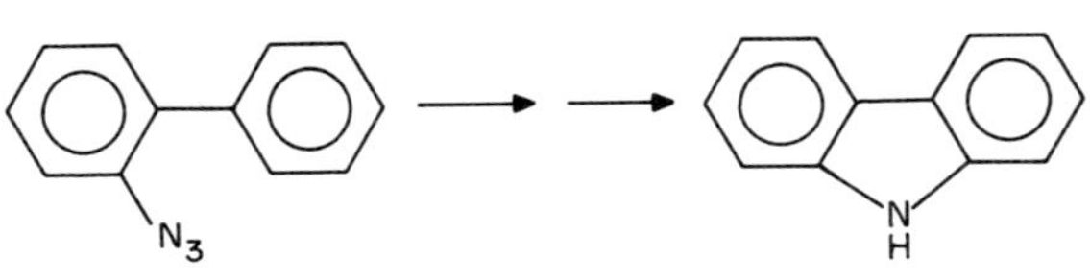

or into adjacent heteroaromatic rings and also reactions via spirodienes and azanorcaradienes. Effects influencing the rates of aryl azide pyrolysis are considered (Table 4).

Section III deals with cyclizations leading to five-membered rings, with their mechanisms, steric (Table 5) and substituent (Table 6) effects and the effects of flanking bromine on the rates of pyrolysis of various *ortho*-substituted aryl azides (Table 7). The magnitudes of neighbouring group effects in azide pyrolysis are treated (Table 8), together with some anomalous neighbouring group effects and examples of possible bridging to incipient aryl nitrenes.

Section IV considers reactions which occur with ring opening in azidoheteroaromatic *N*-oxides, azidoquinones, *ortho*-diazido compounds and azidopyrazoles.

Section V presents nitrene-carbene rearrangements, e.g.

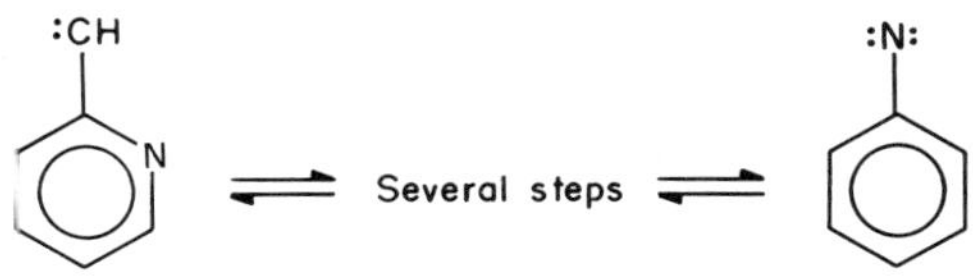

Section VI discusses processes involving sulphonyl azides, such as substitution of the corresponding nitrenes into solvents and cyclizations of sulphonyl nitrenes which in most cases give poor yields.

117 references up to 1980.
Complementary: 14, 5.
Relevant: 1, 3; **3**, 4; **16**, 10; **22**, 15; **25**, 15; **27**, 11; **29**, 12; **33**, 20.

32, *8: Vinyl, aryl and acyl azides* by *H. W. Moore and D. M. Goldish*

The chapter deals with compounds in which the N_3 group is bound to an sp^2 carbon atom.

Section II describes the conversion of triflones and of nitrosooxazolidones to vinyl azides.

Section III treats fragmentations of cyclic vinyl azides to nitriles, with the formation of cyanoketenes, cyano-substituted cyclobutanediones, oxetanones, azetidinones, β-lactones, *N*-heterocyclics and a variety of other examples. Intramolecular cyclizations involving vinyl azide moieties may yield heterocyclics such as indolequinones, indoles, pyridines, oxazoles, isooxazoles, thiazoles and benzoxazines (Section III.B).

Section III.C considers inter- and intramolecular cycloadditions, III.D ring expansions, III.E reactions with electrophiles and III.F the conversion of vinyl azides to aldehydes and ketones.

Preparations of aryl azides are mentioned briefly (Section IV), followed by their photolysis to yield carbazoles, 3*H*-azepines and products of N-migration (Section V.A) and by thermolysis to yield carbazoles, rearrangement products and intramolecular cyclizations (Section V.B).

Section VI deals with the preparation of acyl azides by novel methods avoiding the acyl chloride route, which has many drawbacks.

Section VII describes the thermolysis of acyl azides to isocyanates (Curtius rearrangement) and some related processes involving rapid cyclizations. The photolysis of acyl azides gives both nitrene formation and rearrangement to the isocyanate. Finally, the addition of acyl azides to double bonds is reported.

144 references up to 1980.
Complementary: 14, 9, 10.
Relevant: 14, 2, 5, 6, 7, 8.

32, *9: Recent advances in the radiation chemistry of halocarbons* by *A. Horowitz*

Section II reviews briefly the basic features of the title reaction. Section III deals the reactions of electrons with haloalkanes, including ESR studies and electron scavenging in aqueous and in nonpolar solvents (Tables 1–4), with the identity of cationic species and their charge-transfer reactions (Table 5), with excited states and energy transfer and with CT complexes.

Section IV treats reactions of H atoms in aqueous solutions with haloarenes (Table 6), reactions of OH radicals with various halogen compounds (Table 7), the dehalogenation of halomethanes (Tables 8–10) and haloethanes (Table 11), chlorination reactions and reactions of olefins (Table 12). Finally, Section V considers the overall mechanism of radiolysis in pure compounds (Tables 13–16) and the main products obtained from EtBr, *i*-PrCl, CF_3I and cyclo-C_6F_{12}.

178 references up to 1981.
Complementary: 16, 12.
Relevant: 9, 8; **10**, 6; **11**, 6; **13**, 17; **15**, 9; **17**, 10; **18**, 11; **19**, 10; **22**, 11; **25**, 12; **27**, 23, 24; **29**, 7; **31**, 6; **33**, 21; **36**, 8.

32, *10: Organic chemistry of astatine* by *K. Berei and L. Vasáros*

The fifth halogen, astatine, is the rarest element in the earth's crust and the total amount of its four natural isotopes does not exceed 50 mg. However, isotopes that can be used for chemical studies can be produced in cyclotrons or heavy ion accelerators (Table 1).

Section II describes the preparation of At isotopes, their nuclear properties, decay scheme (Fig. 1) and measurement, and Section III their physical and chemical properties (Tables 2 and 3).

Section IV deals with the compounds of multivalent At (in + 3 or + 5 valency states) and tabulates the preparation and identification of compounds of monovalent At (Table 4). The synthetic methods include homogeneous and heterogeneous halogen exchange, e.g.

$$ICH_2COOH + At^- \rightarrow AtCH_2COOH + I^-$$

$$RI \text{ (vapour)} + At^- \text{ (solid)} \rightarrow RAt + I^-$$

decomposition of diazonium salts (Table 5), astatination via Hg compounds ($ArHgCl \xrightarrow{At^-} ArAt$), electrophilic substitution [including the decomposition of 3-astatotyrosine at various pH values in the absence and presence of H_2O_2 (Table 6) and electrophilic attack by At^+ on H atoms in halobenzenes (Table 7)] reactions of recoil At replacing H in benzene derivatives (Table 8) or halogens in halogenobenzenes (Table 9).

Section V considers the determination of some physicochemical properties by extrapolation from the properties of analogous halogen derivatives (Table 10), C—X dissociation energy values calculated and determined for organic compounds with X = At and X = I (Table 11), the use of gas chromatography to determine some of the features of At compounds [e.g. boiling points (Table 12), retention indices (Figs 8–10), dipole moment, ΔH_v and bond refraction values (Table 13)], the kinetic determination of the C—At dissociation energy by using pyrolytic decomposition with a modified toluene carrier gas method and the determination of dissociation constants in aqueous solutions (Table 14).

Section VI discusses the uses of ^{211}At in nuclear medicine and potential future applications.

116 references up to 1980.
Complementary: none.
Relevant: none.

32, *11: Positive halogen compounds* by *A. Focaud*

I. Introduction (442); **II.** Compounds with halogen—oxygen bonds (443); **III.** Compounds with halogen—nitrogen bonds (452); **IV.** Compounds with halogen—carbon bonds (459); **V.** Compounds with halogen—sulphur bonds (473); **VI.** References (475).

A compound is considered to contain positive halogen when the result of the first step of the reaction is the electron transfer $RX + Nu^- \rightarrow R^- + XNu$, in which the halogen X is transferred with only a sextet of electrons.

Section II deals with the reactions of alkyl hypohalites with sulphides, sulphoxides, amides, amines, imidates, phosphites, phosphines, alkenes, arenes and alkanes and the reactions of chloroperoxy alkanes (e.g. CF_3OOCl) and of halogen derivatives of acids (e.g. MeCOOCl).

Section III treats *N*-haloimides, *N*-haloamides, *N*-sulphonamides, INCO, IN_3, *N*-chlorobenzotriazole, *N*-haloamidines and *N*-haloamines.

Section IV discusses C—X groups (usually in the presence of carbanion-stabilizing groups in the molecule), such as polyhaloalkanes, aryl halides, haloalkynes and a variety of compounds in which electronegative groups are linked to the C atom bearing the reactive halogen (e.g. $ArCHXSO_2Ph$, $BrCH_2COR$) and their reactions with phosphines, phosphites, arene sulphinates, enamines and halide ions.

Section V considers reactions of halosulphonium ions ($R_2\overset{+}{S}Cl$) and related compounds.

246 references up to 1980.
Complementary: none.
Relevant: none.

32, *12: Aspects of the chemistry of halophenols and halodienones* by *J. M. Brittain and P. B. D. de la Mare*

I. Introduction (482); **II.** Properties of the hydroxyl and halogen substituents: acid strengths of substituted phenols (483); **III.** Phenols as bases (489); **IV.** Conversion of phenols to halodienones; the S_E2' reaction (496); **V.** Reactions of halodienones (507); **VI.** Some structural information of cyclohexadienones (521); **VII.** Halogenation of phenols to give halophenols (522); **VIII.** Industrial applications of halophenols and halodienones (532); **IX.** References (542).

Section II reviews the effects of OH and halogen substituents. Table 1 gives substituent constants of various substituted benzoic acids and Table 2 compares the dissociations of halogen-substituted benzoic acids and phenols. Inductive and resonance contributions in ionization of aromatic acids are given in Table 3 and observed and calculated pK_a values of di- and polychlorophenols in Table 4.

Section III deals with general considerations relating to phenols, the protonation of halophenols (Table 5) and protodehalogenation and acid-catalysed rearrangements of halophenols (Tables 6 and 7).

Section IV discusses dienones obtained from phenols by fluorination, chlorination and bromination (Table 8) and intermediates involved in iodination and also halodienones formed by halogenation of phenol derivatives.

Section V treats the conversion of halodienones to phenols and the detailed mechanism of this reaction, in which the *ipso*-substituent (H or other groups) is lost; in addition 1, 3-, 1, 2-migrations of halogen and also the 'quinobenzylic rearrangement' in which the dienone is converted into a (bromomethyl) phenol, can occur:

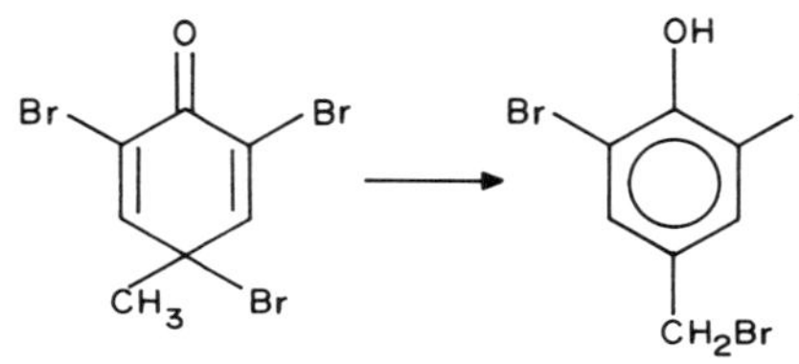

Some polyhalodienones may act as sources of halogen atoms, similarly to NBS.

Halodienones may undergo further addition reactions with excess of halogen; they may also be attacked by nucleophiles, Grignard reagents and also dienophiles (Section V. H–J).

Section VI reviews briefly sources of X-ray, NMR (^{1}H and ^{13}C) and other structural information on cyclohexadienones.

Section VII discusses kinetic forms and structural effects in the chlorination, bromination (Tables 9 and 10) and iodination of phenols and the synthetic aspects of these reactions carried out by a variety of methods (Tables 11 and 12), including stepwise introduction of Br and Cl, halogenations without solvents or catalysts, photodehalogenation, halogenation by rearrangement of dienones, selective *ortho*-halogenation; side-chain halogenation and classical ('indirect') procedures.

Section VIII treats the applications of the title compounds to agriculture, polymers and drugs.

234 references up to 1980.
Complementary: none.
Relevant: 13, 8, 16.

32, *13: α-Halogenated imines* by *N. De Kimpe and R. Verhé*

I. Introduction (549); **II**. Synthesis of a α-halogenated imines (550); **III**. Reactivity of α-halogenated imines (565); **IV**. Properties and applications of α-halogenated imino compounds (589); **V**. References (591).

Section II describes syntheses of the title compounds by condensations of α-halogenated carbonyl compounds with amines ($R^1R^2CXCOR^3 + RNH_2 \rightarrow R^1R^2CXCR{=}NR$), halogenation of imino compounds or of enamines and a variety of other less general methods.

Section III discusses the reactivity of the title compounds, including nucleophilic substitutions (which is not general in these cases as with the corresponding α-halo carbonyl compounds, but strongly structure dependent), C—C bond formations with carbanions, cyanides and organometallic reagents, eliminations of hydrogen halides, nucleophilic additions, rearrangements including Favorskii-type rearrangements and other involving aziridine intermediates, rearrangements of chloroindolienine derivatives and the Beckmann rearrangement of α-bromo oximes. Finally, cycloadditions and some miscellaneous reactions are reviewed.

Section IV treats briefly the properties and applications of the title compounds.

461 references up to 1982.
Complementary: **32**, 19.
Relevant: none.

32, *14: Fluorocarbons* by *B. E. Smart*

I. Introduction (604); **II**. Structure and bonding (604); **III**. Fluorine substituent effects (614); **IV**. Polyfluorocarbon reactions (626); **V**. References (646).

Section II deals with C—X bonds in haloalkanes, tabulating their lengths and dissociation energies (Tables 1–3), and with C—C bonds, the strength of which increase upon halogenation (Table 4). Table 5 demonstrates the dramatic *gem*-difluoro substituent effects in the isomerization of methylenecyclopropanes, and Table 6 presents ethane rotational barriers on progressive fluorination. Similar effects are discussed for fluorinated olefins, acetylenes and benzenes (Tables 7–12).

Section III treats polar and resonance effects (Table 13), steric effects, fluoro carbocations, carbanions (Table 14), radicals and carbenes (Tables 15 and 16) and the 'perfluoroalkyl effect' (i.e. the influence of perfluoroalkyl groups on the stability of strained molecules and of certain functional groups.

Section IV discusses the reactions of polyfluoroalkanes [including nucleophilic displacements (Table 17), and Lewis acid catalysed exchanges of one halogen for another], unsaturated compounds (including nucleophilic, electrophilic and free-radical attacks), perfluorocarbanions, fluorinated cyclobutadienes and cycloadditions of fluoroolefins. Section IV.C deals with nucleophilic (Table 18), electrophilic and photochemical reactions of polyfluoroarenes.

398 references up to 1980.
Complementary: none.
Relevant: **16**, 14; **32**, 22.

32, *15: Xenon halide halogenations* by *M. Zupan*

I. Introduction (657); **II**. Reactions of xenon difluoride with olefins and acetylenes (659); **III**. Reactions of xenon difluoride with aromatic and heteroaromatic molecules (667); **IV**. Reactions of xenon difluoride with various organic molecules (673); **V**. Reactions of other xenon halides (676); **VI**. References (678).

The chapter deals with fluorinations carried out mainly with XeF_2 and also with XeF_4 and XeF_6.

Section II describes the reactions of XeF with olefins and acetylenes, producing mixtures of products, e.g.

$$CH_3CH{=}CH_2 + XeF_2 \rightarrow CH_3CHFCH_3 + CHF_2CH_3 + CHF_2CH_2CH_3 + CH_2FCHFCH_3$$

Norbornene with XeF_2 yields at least seven products (Table 1). The stereochemistry of the addition of halogens to $PhCH{=}CHCH_3$ is treated (Table 2).

Section III discusses reactions of XeF_2 with arenes and with hetarenes and the directive effects of various X groups in PhX (Scheme 3, p. 667). Section IV considers reactions of XeF_2 with various substrates, including adamantane, various diones, thioethers, phosphorus compounds and the preparation of RIF_2 compounds from the corresponding RI molecules.

Section V describes reactions of XeF_4 and XeF_6. The compounds containing more F are much more reactive towards all substrates.

WARNING! XeF_4 and XeF_6 should be handled with great care because of the possible formation of highly explosive xenon oxides.

90 references up to 1980.
Complementary: none.
Relevant: **16**, 7, 8; **21**, 9.

32, *16: The $S_{RN}1$ reaction of organic halides* by *R. K. Norris*

I. Introduction: the $S_{RN}1$ reaction (681); **II**. General scope of the $S_{RN}1$ reaction (682); **III**. Diagnostic and mechanistic features of the $S_{RN}1$ reaction (688); **IV**. Specific halogen effects (694); **V**. References (698).

The $S_{RN}1$ (substitution, radical–nucleophilic, unimolecular) mechanism includes in general the following steps:

$$RX + e \longrightarrow [RX]^{-\cdot} \longrightarrow R^{\cdot} + X^-$$

$$R^{\cdot} + A^- \longrightarrow [RA]^{-\cdot} \xrightarrow{RX} RA + [RX]^{-\cdot}$$

Section II discusses the title reactions in saturated and aromatic systems, the nature of the nucleofuge and the nucleophile and the range of solvents which can be used (DMF, DMSO, HMPA).

Section III considers the initiation process in which a diversity of sources may act as one-electron donors (e.g. solvated electrons, alkali metals dissolving in liquid NH_3), the intermediacy and dissociation of radical anions, the trapping of radicals by anions, chain propagation, termination and competing processes and the inhibition of $S_{RN}1$ reactions.

Section IV deals with relative reactivities of monohaloarenes, ArX: when X changes from F to I, reactivities in $S_{RN}1$ processes increase and the product distributions change. Finally, S_{RN} reactions of dihaloarenes and of benzylidene halides are presented.

141 references up to 1980.
Complementary: **29**, 10.
Relevant: none.

32, *17: Reactions involving solid organic halides* by *E. Hadjoudis*

I. Introduction (703); **II**. Thermal solid-state reactions (704); **III**. Photochemical solid-state reactions (707); **IV**. Gas–solid reactions (709); **V**. Reactions at solid surfaces (714); **VI**. Miscellaneous reactions (715); **VII**. Molecular packing modes (717); **VIII**. Acknowledgement (718); **IX**. References (718).

The unique conformations, limited geometries and possible intermolecular approaches and the molecular packing dominate the chemical behaviour in the solid state.

Section II describes reactions of aromatic diazonium salts, decarboxylations and reactions of biologically important molecules (choline salts, nitrogen mustards, etc.).

Section III deals with photochemical solid-state reactions of anthracene halides (dimerization) and of azobisisobutyronitrile (disproportionation). Section IV discusses the bromination of solid substrates (Table 1) and dehydrohalogenations. Section V treats heterogeneous alkylations and related processes and Section VI solid-state polymerization and Szilard–Chalmers reactions (recoil reactions with rupture of the C—X bond). Section VII considers molecular packing modes of acyl halides which are relevant and important in the solid-state reactivity of RCOX.

58 references up to 1980.
Complementary: 33, 18.VII.
Relevant: 33, 22.

32, *18: Hypervalent halogen compounds* by *G. F. Koser*

The chapter deals with hypervalent organohalogen compounds containing three- or five-valent halogens and 2–5 ligands. Classes and examples of organohalinanes [i.e. organohalogen(IIII) compounds] and of organoperhalinanes [i.e. organohalogen(V) compounds] are shown in Table 2.

Section II discusses the stability and occurrence of organoiodinanes such as ArIO, $ArICl_2$, $ArI(OOCR)_2$, $ArI(SR)_2$ and Ar_3I.

Section III describes the molecular structure of organoiodinanes, tabulating bond distances and angles (Table 3), and presents bonding models for organohalinanes in general and especially for organoiodinanes (Table 4) and also secondary bonding and modes of crystal packing, solution-phase studies and hypervalent bonding in organoiodoso compounds.

Section IV deals with the occurrence of alkyliodinanes and their role as intermediates in various oxidative reactions of iodoalkanes (Table 5). Section V considers cyclic structures and aryliodinane stabilities and the stabilization of various organohalinanes with cyclic structures.

Section VI treats the bonding and structure of organoperiodinanes of four principal structural classes (Structures I–IV, p. 756). It also deals with diaryliodosyl salts, e.g. $[Ar_2I^+{=}O]\cdots X^-$, with bis(acyloxy)iodosylarenes, e.g. $ArI({=}O)(OOCCF_3)_2$, and with tetrafluoro-, tetraacyloxy- and fluoroalkoxyperiodinanes.

Section VII discusses topotactic (solid-state) reactions of organoiodinanes and especially the topochemistry of bis (*o*-iodobenzoyl) peroxide and of some related compounds, and presents generalizations on the topochemistry of bis(aryl) peroxides and of alkoxybenziodoxoles.

Section VIII considers the occurrence and synthesis of halonium ylides:

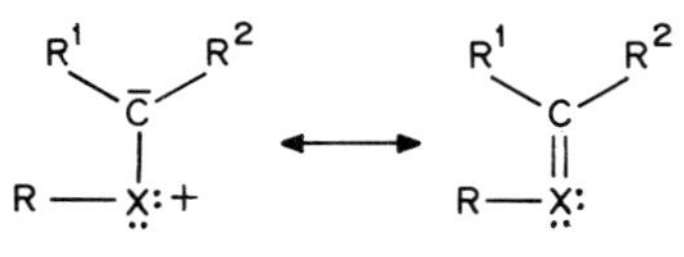

which can be obtained from carbenes, β-dicarbonyl compounds, phenols, cyclopentadienes, monocarbonyl compounds, heterocycles, nitrogen compounds (yielding iodine–nitrogen ylides), from alkanes and from alkenes. All reported halonium ylides of the above types are summarized in Table 8. Section VIII.B discusses the stability, rearrangement and cleavage of these ylides and Section VIII.C their reactions with Brønsted acids, carbon electrophiles and electrophilic halogen species. Finally, transylidation reactions (Table 9), reductive cleavage and molecular structures are presented.

180 references up to 1980.
Complementary: none.
Relevant: none.

32, *19: Synthesis and reactivity of α-halogenated ketones* by *R. Verhé and N. De Kimpe*

I. Introduction (814); **II**. Synthesis of α-halogenated ketones (815); **III**. Reactivity of α-halogenated ketones (850); **IV**. References (919).

Section II.A describes the syntheses of α-fluoroketones from ketones, haloketones (by halogen exchange), diazoketones, carboxylic acid derivatives and α-functionalized epoxides and also the synthesis of α-polyfluoroketones by condensation reactions.

Section II.B deals with the synthesis of α-chloroketones by chlorination of ketones with a variety of chlorinating agents and from alcohols, phenols, diazoketones, alkenes, alkynes, epoxides, carboxylic acid derivatives and aromatic amines.

Section II.C presents similarly the preparations of α-bromoketones from ketones (using Br_2, *N*-bromo compounds and other brominating agents), diazoketones, epoxides and a variety of other substrates.

Section II.D treats briefly syntheses of α-iodoketones and Section II.E discusses the detailed mechanisms of the α-halogenation of ketones.

Section III.A discusses the reactivity of α-haloketones towards nucleophiles and bases, the effect of the CO group (Tables 1 and 2), reactions with inorganic and organic *O*-nucleophiles and with *N*-nucleophiles and bases and inorganic and organic *S*-nucleophiles. Further, reactions with cyanides, carbanions, ylides and enolates are considered, followed by reactions with organometallic reagents (such as derivatives of Mg, Li, B and metal complexes). Reactions with complex metal hydrides and with some phosphorus compounds are also presented.

Section III.B deals with reactions in which α-haloketones act as electrophiles (Friedel–Crafts, etc.) and also with their reactions with alkali metal fluorides, their acid-catalysed rearrangements and with the formation of α-acylcarbenium ions. In the last short subsections the photochemistry, electrochemistry and dehalogenations of α-haloketones is considered.

586 references up to 1981.
Complementary: 32, 13.
Relevant: none.

32, *20: Electrophilic assistance to reactions at a C—X bond* by *D. N. Kevill*

I. Introduction (934); **II**. Relationship between extent of assistance and Lewis acidity (934); **III**. Silver ion assistance to reactions of alkyl halides (939); **IV**. Silver ion assistance to reactions of other organic halides (958); **V**. Mercuric ion assistance to reactions of organic halides (963); **VI**. Electrophilic assistance to reactions of organic halides by non-metallic species (967); **VII**. Electrophilic assistance to reactions of organic cyanides, isocyanides and azides (970); **VIII**. Electrophilic assistance in the gas phase (973); **IX**. Synthetic applications (975); **X**. References (977).

Electrophilic assistance is especially important in nucleophilic substitutions and eliminations involving C—halogen bonds.

Section II considers assistance and Lewis acidity in reactions of alkyl fluorides and in unimolecular solvolyses of RCl and RBr, describes a Brønsted-type relationship for the solvolysis of halides and also electrostatic catalysis by $LiClO_4$ and assistance to bimolecular solvolyses of alkyl halides.

Section III deals with Ag^+ assistance and with the mechanism of reactions of alkyl halides assisted by $AgNO_3$, silver arenesulphonates, $AgClO_4$, $AgNO_2$ and other silver salts. The *cis/trans* ratio of the alkenes produced is considered, together with heterogeneous catalysis in formally homogeneous reactions.

Section IV treats Ag^+-assisted reactions of allyl, vinyl and acyl halides and of halogenated ketones.

Section V discusses Hg^{2+}-assisted reactions of organic halides and Section VI the electrophilic assistance by halogen molecules (including interhalogen compounds), mixtures of HNO_3 with HCl or HBr, peracids, $NO_2^+BF_4^-$ and other stable salts containing nitronium or nitrosonium ions and also salts containing R_2X^+ ions.

Section VII considers the influence of Ag^+ and Hg^{2+} salts, $AlCl_3$, protonic and Lewis acids and other electrophilic catalysts in the reaction of organic cyanides, isocyanides and azides.

Section VIII describes some gas-phase reactions carried out by ion cyclotron resonance spectroscopy and negative ion chemical ionization mass spectrometry with halides in the presence of acids, H_3O^+, NO^+, Li^+ and many other metal cations.

Section IX reviews briefly the synthetic applications of electrophilic assistance to C—X bond reactions.

331 references up to 1980.
Complementary: 16, 7, 9.
Relevant: 21, 4; **32**, 21.

32, *21: Molecular interactions involving organic halides* by *J.-M. Dumas, M. Gomel and M. Guerin*

I. Introduction (986); **II**. Interactions with proton donors (988); **III**. Interactions with electron acceptors (not proton donors) (991); **IV**. Interactions between aromatic halides and electron donors (994); **V**. Hydrogen bonding between aliphatic halides (as proton donors) and proton acceptors (997); **VI**. Interactions between aliphatic halides (not proton donors) and organic bases: 'halogen bonding'? (1001); **VII**. Acknowledgements (1016); **VIII**. References (1016).

Section I describes the scope and methodology of the review. Section II deals with the proton acceptor properties of organic halides, including inter- and intramolecular associations (Tables 1 and 2). Section III treats interactions with electron acceptors such as halides of B and Al, and with I_2 [which associates with both aliphatic (Table 3) and aromatic (Table 4) halides].

Section IV considers interactions between aromatic halides (especially hexafluorobenzene) and electron donors (Table 5) in the solid, liquid and gaseous phase, and discusses the nature of these.

Section V presents interactions between alkyl halides and proton acceptors, when the protons attached to carbon atoms carrying halogens are able to form hydrogen bonds with organic bases (Table 6), but C—X... Y interactions may also be significant (Table 7), in addition to charge transfer from N to Cl atoms (Tables 8 and 9). Data for equilibrium constants, enthalpies and entropies for the interactions of various polyhalomethanes and bases are presented (Tables 10 and 11).

Section VI deals with interactions between saturated and unsaturated halides (not containing, H, so as to avoid interference by potential hydrogen bonding) and organic bases of various kinds [including arenes (Tables 12 and 13), ethers, thioethers and amines (Tables 14–18)]. Solvent effects exerted by aliphatic halides as solvents on various molecular associations are considered [e.g. on the pyrrole–pyridine complex (Table 19)] and the nature of C—X···Y interactions is discussed in detail. The possible influence of 'halogen bonding' on chemical and biological processes is reviewed.

Section VI.E presents a summary of conclusions and possible future developments.

166 references up to 1980.
Complementary: **16**, 5.
Relevant: **2**, 9; **3**, 6; **4**, 4; **5**, 6; **7**, 6; **8**, 5; **11**, 3; **12**, 3, 10; **13**, 7; **15**, 4; **17**, 6; **18**, 8; **19**, 7; **20**, 12; **22**, 8; **23**, 6, 7; **24**, 4; **25**, 5, 6; **29**, 17; **31**, 17, 21; **33**, 5.

32, *22: Formation of carbon—halogen bonds* by *M. Hudlicky and T. Hudlicky*

I. Introductory remarks (1021); **II**. Formation of carbon—fluorine bonds (1022); **III**. Formation of carbon—bromine bonds (1102); **IV**. Formation of carbon—iodine bonds (1142); **V**. Halogenation tables (1162).

The chapter is divided into four main parts according to the formation of C—F, C—Cl, C—Br and C—I bonds, and a fifth part which presents tables correlating halogenating agents with the starting materials and products. Each part has a separate list of references.

Part I (pp. 1022–1066) describes equipment, techniques and fluorinating agents (Table 1) used in the formation of C—F bonds. Part I, Section IV (p. 1022) deals with synthetic methods, including additions of HF, F_2 and halogen fluorides to multiple bonds and to aromatic systems, addition of reagents such as CF_3OF and metal fluorides to multiple bonds, replacement by fluorine of H, halogen or oxygen (the latter including cleavage of epoxides, ethers, esters, replacement of HO and replacement of carbonyl oxygen), replacement of nitrogen by fluorine involving diazo, azido and diazonium groups and cleavage of aziridines and of azirines. Replacement of other elements by F (p. 1058) and some new methods presented in 1981 (p. 1059) are considered.

Part 2 (pp. 1066–1102) reviews the most common chlorinating agents (Table 1) and the addition of HCl, Cl_2 and Cl plus another element to multiple bonds and to aromatic systems (Sections II–V). Section VI deals with the replacement of H by Cl in alkanes, alkenes and alkynes, in functional derivatives at sp^3 carbons, in aromatic systems and their side-chains and in heteroarenes.

Section VII describes the replacement of O by Cl in ethers, epoxides, esters, lactones and sulpho esters and replacement of OH groups by Cl in alcohols, carboxylic and sulphonic acids and finally the replacement of carbonyl oxygen by Cl.

Sections VIII and IX deal with the replacement of S and of N by Cl, Section X with the replacement of C by Cl($C_5F_{11}CO_2Ag \rightarrow C_5F_{11}Cl$) and Section XI with the replacement of other halogens by Cl.

Part 3 (pp. 1102–42) considers (Table 1) the most common brominating agents (Section II) and deals with the addition of HBr, Br_2 and also of Br plus another element to multiple bonds and to aromatic systems. Section VI describes the replacement of H by Br in alkanes, cycloalkanes, alkenes and alkynes. Section VII deals with the replacement of H by Br in functional derivatives at saturated sp^3 carbons and Sections VIII and IX with the same reaction in aromatic systems and their side-chains and in heterocycles.

Section X treats the formation of C—Br bonds by cleavage of ethers, epoxides, esters, lactones and sulpho esters and also the replacement of OH groups by Br in alcohols.

Section XI considers the replacement of N by Br in amides, lactams, nitro compounds, aliphatic diazo compounds and azides, e.g.

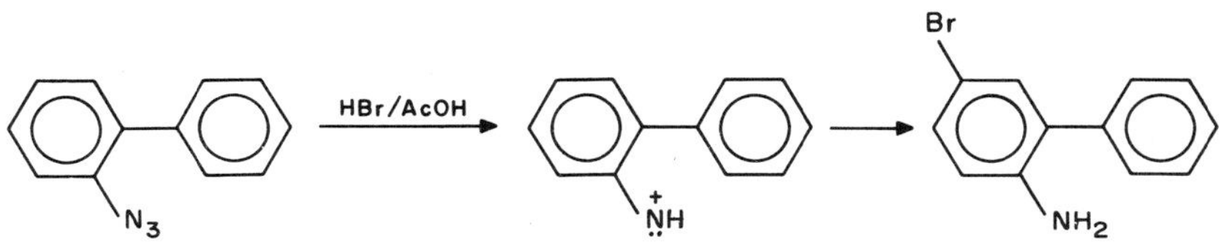

and also in aliphatic amines, amino acids and aromatic amines (Sandmeyer and Gattermann reactions).

Section XII presents the replacement of C by Br (e.g. by the Hunsdiecker reaction) and Section XIII the replacement of B, Si and halogens by Br.

Part 4 (pp. 1142–1161) describes the most common iodinating agents (Table 1) and deals with the addition of HI and of I_2 across multiple bonds (Sections II and IV) and with the addition of I and other groups (e.g. F, Cl, Br, N_3, HO, CH_3O) to double bonds (Section V).

Sections VI, VII and VIII treat the replacement of H by I in aliphatic, aromatic and heteroaromatic systems.

Section IX deals with reactions in which O is replaced by I, such as cleavage of ethers, epoxides and alkyl sulphonates, and replacement of HO in alcohols.

Sections X–XIII consider the replacement by iodine of nitrogen-containing groups, carboxyl groups, halogens and metals.

Part 5 (pp. 1162–1172) attempts to correlate (in tabular form) starting materials, products and halogenating agents, so as to facilitate orientation in the field and to show what reactions are suitable for a specific purpose. The individual reactions are graded with different markings and thus show whether they are in common use, applicable or of limited (rare) applications. The symbols used in the tables are explained (pp. 1162–1163) and methods involving F, Cl, Br and I are presented in separate tables.

282 references to Part 1 up to 1981 (pp. 1059–1066).
266 references to Part 2 up to 1979 (pp. 1096–1102).
302 references to Part 3 up to 1980 (pp. 1136–1142).
125 references to Part 4 up to 1979 (pp. 1158–1161).
28 references (books and reviews) to Part 5 up to 1976 (p. 1172).
Complementary: 32, 6, 15.
Relevant: 16, 7, 8, 14; **32**, 12; 13, 14, 19.

32, *23: Alkene-forming eliminations involving the carbon—halogen bond* by *E. Baciocchi*

The chapter reviews the three main mechanistic pathways of alkene-forming eliminations (*E*2, *E*1cB and *E*1).

Section II describes the structure of the transition state in *E*2 reactions including theories of 'variable *E*2 transition state.' Structural effects are discussed, including theoretical predictions (Table 1), effect of the leaving group, substituent effects at C_β and C_α, effects of the strength, steric requirements, association and solvation of the base (Tables 2 and 3), positional orientation (i.e. the relative proportion of olefins differing in the position of the C=C bond) and geometric (i.e. *cis/trans*) orientation and the effects on these of base strength and association and of the structure of the leaving group and of the alkyl group (Tables 4–10). Section II.D treats the stereochemistry of *E*2 reactions and the effect of the base (Table 11), of the leaving group and of the alkyl structure in both open-chain and cyclic compounds.

Section III discusses the mechanistic variants of stepwise *E*1cB eliminations involving a carbanion intermediate, reactions involving reversibly formed carbanions with halide and other leaving groups and reactions involving irreversibly formed carbanions which occur in the reactions of 2,2-diarylethyl derivatives [e.g. elimination of HCl from $(p\text{-}NO_2C_6H_4)_2CHCHCl_2$ promoted by MeONa], of 1,2-dihalogenoacenaphthenes and 2,3-dihalogeno-2,3-dihydrobenzofurans (Table 12) and of 2-arylsulphonylethyl, indene and other systems.

Section IV reviews the general aspects, stereochemistry (Table 13) and orientation in *E*1 reactions. The effects of the alkyl structure, of the leaving group (Table 14) and of the solvent on the $E1\text{–}S_N1$ competition are discussed.

264 references up to 1982.
Complementary:16, 9; **21**, 4.
Relevant: 1, 2, 3; **13**, 12; **21**, 18; **32**, 5.

32, *24: Structural chemistry of the carbon—halogen and carbon—pseudohalogen bonds* by *M. Kaftory*

The chapter deals with the geometric parameters (bond length, bond angles and torsion angles) which define a specified C—X bond (where X is a halogen atom or a pseudohalogen group).

Sections II–V deal separately with C—F, C—Cl, C—Br and C—I bonds, each section being subdivided into descriptions of saturated, aromatic, olefinic and acetylenic compounds. The scatter of bond lengths is presented in histograms and the data throughout the chapter were critically chosen. Section VI gives conclusions relating to C—X bonds and Table 1 lists the mean values of various C—X bond distances in different types of compounds.

Section VII discusses $C—N_3$ (Table 2), C—OCN and C—SCN bonds (Table 2) and gives some concluding remarks relating to C—pseudohalogen bonds.

130 references up to 1979.
Complementary: 16, 2.
Relevant: 17, 2; **18**, 2; **19**, 2; **20**, 2; **21**, 1; **22**, 2; **23**, 3; **24**, 2; **26**, 2; **27**, 4; **28**, 2; **29**, 1; **30**, 1; **31**, 24; **33**, 12; **34**, 10; **36**, 3.

32, *25: Halonium ions* by *G. F. Koser*

I. Introduction (1266); **II**. Diaryliodonium salts (1267); **III**. Vinyliodonium salts (1316); **IV**. Alkyliodonium salts (1323); **V**. Diarylbromonium and diarylchloronium salts (1324); **VI**. Alkylhalonium ions (1328); **VII**. References (1347).

The chapter deals with R_2X^+-type compounds and also with cyclic halonium ions which have been postulated as intermediates in various reactions.

Section II.A treats the synthesis of Ar_2I^+ salts from iodosoarenes and related compounds with arenes (e.g. $PhIO + PhCH_3 \rightarrow CH_3C_6H_4I^+Ph$) (Table 1); from iodyl sulphate with arenes [$I_2 + KIO_3 \xrightarrow{H_2SO_4} (IO)_2SO_4 \xrightarrow[H_2SO_4]{2ArH} (Ar\overset{+}{I}Ar)HSO_4^-$), from alkali metal iodates or from 'iodoso' intermediates with aromatic substrates, from condensations of iodosoarenes with iodoxyarenes, from aryllithium reagents with (dichloroiodo) arenes [$ArICl_2 + Ar'Li \rightarrow (Ar\overset{+}{I}Ar')Cl^-$] or with *trans*-1-(dichloroiodo)-2-chloroethylene, from [hydroxy(tosyloxy)iodo]arenes with (trimethylsilyl)arenes and through some other pathways.

Section II.B discusses reactions of Ar_2I^+ salts with nucleophiles, including S_N Ar and free radical arylations (Table 2), reductive decompositions by chain mechanisms or in the presence of aliphatic amines in hydroxylic solvents (e.g. $Ph_2I^+ + Et_2NH \rightarrow Ph_2NEt_2$) (Table 3), decomposition in the presence of Cu^+ or Cu^{2+} ions, the formation of benzynes and of tricovalent adducts, and some practical applications of ArI_2I^+ salts (Table 4).

Section III deals with the synthesis of vinyliodium salts by a variety of methods and with their reactions.

Section IV considers briefly alkynyliodonium salts, which are both rare and unstable.

Section V treats the syntheses, structure (Table 5) and reactions of diarylbromonium and -chloronium salts, which are relatively unstable towards nucleophilic cleavage and are therefore usually prepared with anions of low nucleophilicity (BF_4^-, PF_6^-).

Section VI presents classical approaches and a historical perspective regarding alkylhalonium ions, discusses their preparative methodology and isolation (although very often the alkylhalonium ions are *not* isolated from the medium in which they were prepared), e.g. $MeX + MeF–SbF_5 \xrightarrow{SO_2(l)} (Me\overset{+}{X}Me)A^-$, where X = Cl, Br, I and A = Sb_6 or Sb_2F_{11} (Table 6). Some cyclic halonium salts were prepared, especially with three- and five-membered rings (Tables 7 and 8). Section VI.H discusses the reactions of alkylhalonium ions with nucleophiles and Section VI.I their occurrence in the gas phase.

200 references up to 1981.
Complementary: none.
Relevant: none.

32, *26: Carbon—carbon bond formation involving organic halides and transition metal compounds* by *F. Naso and G. Marchese*

I. Introduction (1354); **II**. Reactions of organocopper(I) reagents (1355); **III**. Reactions of organometallic reagents in the presence of transition metal catalysts (1379); **IV**. Self-coupling of aryl or alkenyl halides by means of nickel(0) complexes (1401); **V**. Reactions of halides with alkenes or acetylenes in the presence of palladium catalysts (1403); **VI**. Reactions involving π-allylnickel complexes (1413); **VII**. Reactions with iron carbonyls (1423); **VIII**. Carbonylation reactions (1429); **IX**. Cyanation reactions (1444); **X**. References (1441).

Section II describes C—C bond-forming reactions involving either monoorganic Cu^{I} reagents (i.e. RCu) or cuprates (e.g. R_2CuM, R^1R^2CuM or RYCuM, where M = Li or MgX). Both these types of reagent react with organic halides and with the formation of a new C—C bond, but the bulk of the section is devoted to cuprates, e.g. $C_5H_{11}I + (C_4H_9)_2CuLi \rightarrow C_9H_{20}$. Such 'cross-coupling' reactions are treated with alkyl, alkenyl (Table 1), allenyl, allyl, propargyl, alkynyl, aryl and heteroaryl halides. Section II.I deals with the Ullmann biaryl coupling and related reactions.

Section III deals with the cross-coupling reactions of Grignard reagents in the presence of Ni, Pd or Fe catalysts (Tables 2–4), e.g. $PhCH{=}CHBr + ArMgX \xrightarrow{Ni(acac)} PhCH{=}CHAr$.

Sections III.E–G consider cross-coupling reactions of organolithium, -zinc, -aluminium and -zirconium compounds in the presence of Ni and Pd complexes, e.g. $RZnX + ArX' \xrightarrow{Ni(PPh_3)_4} RAr (R = Ar \text{ or } ArCH_2)$.

Sections III.H–J treat cross-coupling reactions of 1-alkenylboranes, of R_4Sn and of RHgX compounds in the presence of Pd complexes.

Section IV presents the self-coupling reactions of aryl or alkenyl halides ($2RX \rightarrow RR$) in the presence of Ni(0) complexes.

Section V describes procedures in which an olefinic (Table 5) or acetylenic H atom is substituted by the organic residue of a halide, in the presence of a Pd catalyst and a slight excess of base to neutralize the acid formed:

$$RX + \text{H}\!\!>\!C{=}C\!< + R'_3N \xrightarrow{\text{Pd cat.}} \text{R}\!\!>\!C{=}C\!< + R'_3\overset{+}{N}HX^-$$

Section VI deals with the cross-coupling of π-allylnickel complexes with halides, with self-coupling of allyl halides in the presence of $Ni(CO)_4$ in DMF or similar highly polar solvents and with reactions involving π-oxyallylnickel complexes generated from α-haloketones and $Ni(CO)_4$.

Section VII treats cyclocoupling of α,α'-dibromoketones with 1, 3-dienes or aryl olefins in the presence of iron carbonyls and also $Fe(CO)_5$-catalysed reactions of α-haloketones and of *gem*-dihalides, e.g.

$$2Ph_2CCl_2 + 2Fe(CO)_5 \rightarrow Ph_2C{=}CPh_2 + 2FeCl_2 + 10CO$$

Section VIII discusses carbonylation reactions with CO in the presence of Pd and Ni catalysts and aryl, benzyl, vinyl or heteroaromatic halides. These are synthetically useful, since in the presence of methanol, amines or H_2, esters, amides or aldehydes, respectively, are formed. Carbonylation of π-allylnickel complexes with CO accompanied by insertion of olefins or of acetylenes is also presented.

Section VIII.D discusses carbonylations with metal carbonyls only, when the nature of the products depends on the solvent [e.g. ArI with $Ni(CO)_4$ in ROH yields ArCOOR].

Section VIII.E describes carbonylation with metal carbonyl anions (obtained by treatment of metal carbonyls with bases such as RLi compounds or alkoxide ions). Among many other products, e.g. unsaturated ketones can be obtained:

$$MeCONi(CO_3)^-Li^+ + Ph_2C{=}CHBr \rightarrow Ph_2C{=}CHCOMe$$

Section IX deals briefly with cyanation reactions when the X atom of vinyl or aryl halides can be substituted by a CN group by reaction with $K_4[Ni_2(CN)_6]$ or with mixtures of NaCN and $Ni(PPh_3)_3$.

Throughout the chapter the mechanism and steric course of the reactions are described in detail.

360 references up to 1981.
Complementary: **21**, 12; **35**, 1, 2, 3, 6.
Relevant: **1**, 6; **9**, 5; **30**, 4–7.

32, *27: Dihalocyclopropanes* by *P. Weyerstahl*

I. Introduction (1452); **II**. Synthesis of dihalocyclopropanes (1452); **III**. NMR spectra of dihalocyclopropanes (1465); **IV**. Reactions of dihalocyclopropanes (1472); **V**. Acknowledgements (1486); **VI**. References (1486).

The chapter deals with *gem*-dihalocyclopropanes and not with their 1,2-isomers.

Section II describes the synthesis of the title compounds by addition of dihalocarbenes to olefins:

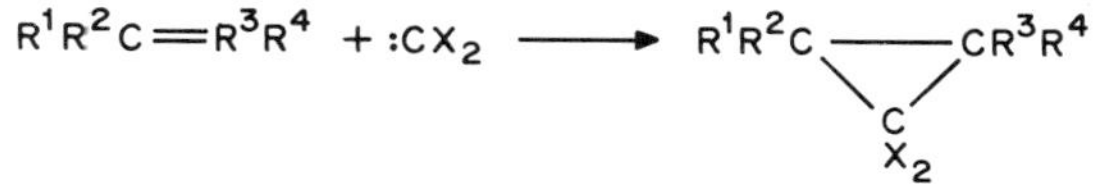

The generation of dihalocarbenes by various methods is discussed, including the use of alkoxides, thermolysis of trihaloacetates, the use of ethylene oxide, phase-transfer catalysis, phenyl(trihalomethyl)mercury as a carbene source ($PhHgCX_3 \rightleftharpoons PhHgX + :CX_2$) and by some other less important methods. Products are listed in Table 1.

Section III treats ^{1}H and ^{13}C NMR spectra and Section IV the reactions of dihalocyclopropanes, including reductive dehalogenation, substitution via Li derivatives, elimination and elimination/addition, carbenoid reactions (to yield, e.g., allenes):

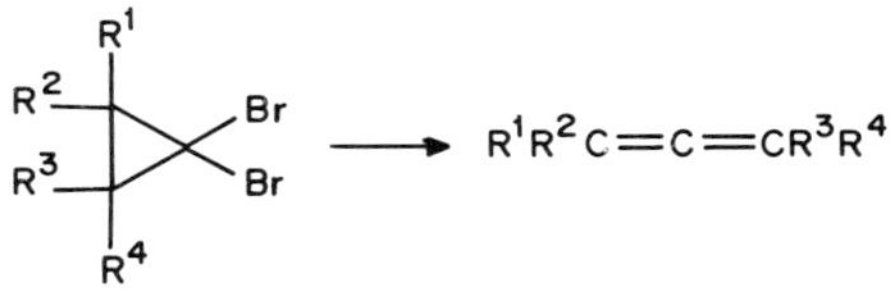

and thermal, H^+-, Ag^+-, Lewis acid- and nucleophile-assisted cyclopropyl–allyl ring-opening reactions.

552 references up to 1981.
Complementary: none.
Relevant: none.

32, *28: Photoelectron spectra of organic halogen compounds* by *K. Wittel and H. Bock*

I. Introductory remarks; photoelectron spectroscopy today and scope of this review (1500); **II**. Information from photoelectron spectra of halogen compounds (1502); **III**. Assignment of PE spectra by comparing equivalent radical cation states of chemically related molecules (1515); **IV**. Small prototype halogen compounds (1522); **V**. Halogenated hydrocarbons and other saturated Group IVB halides (1528); **VI**. Halogen-substituted open-chain and cyclic carbon systems (1537); **VII**. Halogen derivatives of nitrogen, phosphorus and the other Group VB elements (1553); **VIII**. Halogen derivatives of oxygen, sulphur and the other Group VIB elements (1564); **IX**. Electron-deficient halides and organometallic halogen compounds (1574); **X**. Real-time gas analysis in flow systems containing halogen compounds (1585); **XI**. References (1593).

Section I reviews the technique of PE spectroscopy and the scope of the chapter. Section II describes the principles of the measurement, some experimental details, radical cation states and vibrational fine structure, degenerate ionic states (including Jahn–Teller distortions and spin–orbit coupling) and radical cation states and molecular orbitals.

Section III illustrates on four different examples, how PE spectra can be assigned by considering comparisons, correlations and resemblences between the cation states of chemically related molecules.

Section IV deals with hydrogen halides (Table 1), halogens, interhalogens and noble gas fluorides. Section V treats halomethanes including CX_4, HCX_3, H_2CX_2 and H_3CX (Table 2), and also higher allyl halides and halogen derivatives of Si, Ge and Sn (Table 3).

Section VI discusses haloacetylenes, haloethylenes and halogen-substituted C_3 and C_4 olefins; carbonyl and thiocarbonyl halides; halobenzenes (Table 4) and surveys other cyclic systems such as halogenated polycycles, benzyl derivatives and N-, S-, or B-containing heterocycles.

Section VII considers halogen derivatives of Group VB elements (N, P, As, Sb). Inorganic halides and their alkyl derivatives are treated, followed by halogen-substituted nitrogen systems such as the $\cdot NF_2$ radical, difluorodiazene, hexafluorocyclotriphosphazene and nitrosyl and nitryl halides. Section V.C deals with halides of N, P, As and Sb (Table 5) and Section V.D with related halides containing also multiple heteroatom bonds (e.g. F_3NO, F_3PO, Cl_3PO, Cl_3PS).

Section VIII treats halogen compounds of oxygen (F_2O, FOH, Cl_2C, ClO_2, $FClO_3$), saturated sulphur halides (XSX, XSSX, SF_4 and SF_6) and sulphur–halogen derivatives containing multiple N and O bonds ($F_3S{\equiv}N$, $XS{\equiv}N$, $X_2S{=}O$, X_2SO_2, etc.).

Section IX deals with halides of B, Al, Zn, Cd and Hg, halides and oxyhalides of transition metals; low-valent metal complexes with halogen ligands [e.g. $MX(CO)_5$, M = Mn, Re; $M(PF_3)_4$, M = Ni, Pt; $(CH_5H_5)_2MX_2$, M = Ti, Hf, Zn] and ionic halides (especially alkali metal halides). Ionic organic halides (tropylium, pyridinium salts) have not been studied owing to experimental difficulties.

Section X compares the advantages and disadvantages of PE spectroscopic gas analysis with GC, MS and IR methods. Thermal decomposition channels and the generations of short-lived intermediates are treated (Section X.B) and finally the optimization of heterogeneously catalysed gas-phase reactions using PE spectroscopy.

406 references up to 1982.
Complementary: none.
Relevant: **31**, 5; **39**, 5; **40**, 5.

32, *29: Recent advances in the photochemistry of the carbon—halogen bond* by *G. Lodder*

I. Introduction (1605); **II**. Aliphatic, benzylic, homobenzylic and α-keto halides (1606); **III**. Vinyl and acyl halides (1631); **IV**. Aromatic halides (1640); **V**. Heteroatomic halides (1666); **VI**. Acknowledgement (1672); **VII**. References (1672).

Section II discusses the photochemistry of alkyl, allyl, benzyl and homobenzyl halides and of α-haloketones. Section II.F treats the reduction, solvolysis and inter- and intramolecular alkylation reactions of α-halocarboxylic acid derivatives (acids, esters, amides, etc.). The yields, mechanisms and stereochemistry of the processes described are presented in detail.

Section III considers vinyl halides, in which photolysis causes both cleavage of the C—X bond and reactions characteristic of alkenes, involving either homolytic and heterolytic processes. Section III.B deals with acyl halides which yield products arising from acyl and X radicals, e.g. $PhCOBr + Et_2O \xrightarrow{h\nu} PhCHO + CH_3CH_2OCHBrCH_3$.

Section IV treats aromatic halides and their photolysis in H-donor solvents to give reductive dehalogenation (Table 1): $ArX^* \rightarrow X\cdot + Ar\cdot \xrightarrow{RH} ArH$.

Section IV.B discusses intermolecular arylations to produce biaryls and intramolecular photoarylation leading by cyclodehydrohalogenation to the formation of a new aryl—aryl bond, e.g.

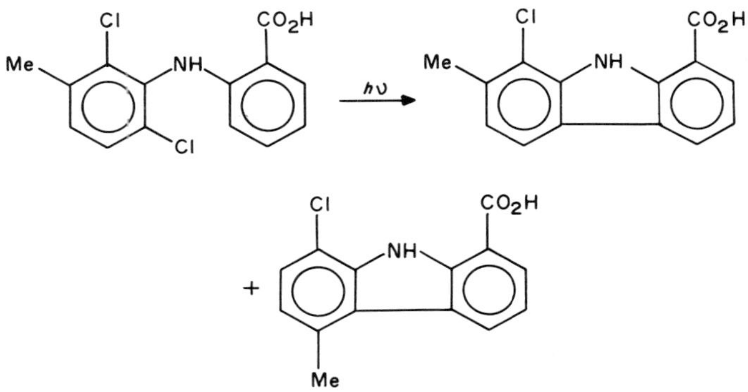

This reaction has been used extensively in the synthesis of alkaloids. Section IV.C considers nucleophilic aromatic substitutions which occur when photolysis is carried out in nucleophilic solvents or in the presence of nucleophiles (Nu), especially easily in the presence of nitro and similar groups on the aromatic ring:

$$ArX^* + Nu^- \rightarrow [ArNuX]^- \rightarrow ArNu + X^-$$

The presence of electron-donating (OCH_3, $—O^-$, NH_2 etc.) groups in the aryl halide has an activating and *ortho*/*para*-directing influence; thus 2- and 4-fluoroanisole give efficient substitution of the F by CN^- in aqueous *t*-BuOH, while the 3-isomer gives replacement of an *o*- or *p*-ring H atom by CN^-.

Section V presents the much less studied photochemistry of heteroaromatic halides. The reactions include reductive dehalogenation, electron donor-sensitized photoreduction, synthesis of aryl-substituted heteroaromatic compounds by arylation, intramolecular heteroarylation and nucleophilic heteroaromatic photosubstitution.

497 references up to 1981.
Complementary: 15, 9; **16**, 11.
Relevant: 2, 16; **3**, 8; **6**, 4; **8**, 12; **9**, 6; **11**, 5; **13**, 16; **14**, 8; **17**, 9; **18**, 10; **19**, 20; **22**, 10; **23**, 9; **24**, 12; **25**, 11; **27**, 21, 22; **28**, 5; **29**, 5, 6; **33**, 21.

The chemistry of peroxides (1983)

33, *1: General and theoretical aspects of the peroxide group* by *D. Cremer*

I. Introduction (2); **II**. Structure (4); **III**. Orbital description (5); **IV**. Properties of XO_2 and X_2O_2 prototypes (21); **V**. Substituent effects (61); **VI**. Abbreviations, symbols, constants and conversion factors (73); **VII**. Acknowledgements (77); **VIII**. References (78).

Section I quotes data (Tables 1–3) which are relevant to the stability of the O—O single bond, to the preference of O atoms to bond to H, C, N or F rather than to another O atom and to the facile scission of O—O bonds and hence the unusual reactivity of peroxo compounds.

Section II deals with the topology of atomic assemblies XO_2 and X_2O_2 (containing the subunit OO, and with the configurational space of XO_2 and X_2O_2.

Section III considers the qualitative valence bond treatment of molecular O_2, of radical, biradical and ionic states of XO_2 and the geometries of X_2O_2. Next, qualitative MO treatment is presented for O_2 (Tables 4 and 5), HO_2 radical, XO_2 including ozone (Table 6), H_2O_2 and X_2O_2.

Section IV discusses the properties of XO_2 and X_2O_2 prototypes, including three-dimensional plots of valence MOs of various peroxides and the energies and geometries of various relevant structures obtained by different methods (Tables 7–10). The conformational subspace of H_2O_2 is treated, followed by total energies, heats of formation and bond dissociation enthalpies, orbital energies, ionization potentials, geometry and vibrational analysis, charge density and one-electron properties and excited states.

Section V deals with general trends in the substituent effects of peroxy compounds XO_2 and peroxides XOOH and XOOX, followed by the treatment of special compounds (peroxy acids, acyl peroxides, polyoxides, ozonides and other cyclic peroxides).

Throughout the chapter many experimental and theoretical data are presented in tabular form (ΔH_f^0, energies and heats of formation, dissociation enthalpies, O—O cleavage energies, ionization potentials, dihedral angles, electronic configurations, vibrational frequencies, etc.) and as figures.

Section VI contains lists of abbreviations, symbols, constants and conversion factors.

330 references up to 1981.
Complementary: none.
Relevant: **1**, 1; **2**, 1; **3**, 1; **4**, 1; **5**, 1; **6**, 1; **8**, 1; **12**, 1; **13**, 1; **14**, 1; **15**, 1; **16**, 1; **17**, 1; **18**, 1; **19**, 1; **20**, 1; **22**, 1; **23**, 1; **24**, 1; **25**, 1; **26**, 1; **27**, 9; **28**, 1; **31**, 27; **32**, 1; **36**, 2.

33, *2: Stereochemical and conformational aspects of peroxy compounds* by *O. Exner*

I. Introduction (85); **II**. Conformation of hydrogen peroxide and general theory (86); **III**. Alkyl peroxides and hydroperoxides (89); **IV**. Acyl peroxides (91); **V**. Peroxy acids (92); **VI**. Peroxy esters (93); **VII**. References (94).

Sections II and III treat the parent compound H_2O_2 and survey dihedral angles (calculated and experimental) in some peroxides and peroxy compounds (Table 1) and values of rotational barriers about the O—O bond (Table 2). New problems and new factors have to be considered with acyl peroxides (Section IV). With peroxy acids (Section V), the hydrogen bond is important in the configurations and dipole moments of the molecules. Section VI deals with the scarce results available on peroxy esters.

110 references up to 1981.
Complementary: none.
Relevant: **19**, 21; **20**, 2; **21**, 1; **22**, 3; **27**, 5; **28**, 10; **29**, 2; **31**, 19; **32**, 1.

33, *3: Thermochemistry of peroxides* by *A. C. Baldwin*

I. Introduction (97); **II**. Measured thermochemical data (98); **III**. Group additivity scheme for thermochemical data (100); **IV**. Conclusions (102); **V**. Acknowledgements (102); **VI**. References (102).

Section II deals with the measured heats of formation of peroxides (Table 1) and hydroperoxides (Table 2) and with bond strengths (Table 3). Entropies and heat capacities are available only for H_2O_2, but can be estimated accurately.

Section III presents the derivation of data by the group additivity method. Thus heats of formation are given for peroxides, hydroperoxides, polyoxides, peroxy and polyoxy radicals and peroxynitrates (Table 4). Entropies and heat capacities can be calculated from the group values listed in Table 5.

37 references up to 1981.
Complementary: none.
Relevant: **16**, 16; **18**, 3; **19**, 3; **20**, 11; **21**, 3; **22**, 6; **23**, 4; **24**, 3; **25**, 2; **26**, 4; **27**, 9; **28**, 4; **29**, 24; **30**, 2; **31**, 2; **36**, 4.

33, *4: Mass spectrometry of organic peroxides* by *H. Schwarz and H.-M. Schiebel*

I. Peresters and peroxylactones (106); **II**. Alkyl hydroperoxides (113); **III**. Acyclic and cyclic peroxides (116); **IV**. Ozonides (123); **V**. Acknowledgements (126); **VI**. References (126).

Most of the data on the subject have been acquired by electron impact mass spectrometry. Detailed fragmentation processes of peresters (including peroxyacetyl nitrate) are shown in Schemes 1–4 and of peroxylactones in Schemes 4–9 (Section I). Section II deals with alkyl hydroperoxides (Schemes 10–12) and III with acyclic (Schemes 13–15) and cyclic peroxides (Schemes 16–20). Schemes 21–23 show the pathways occurring with ozonides (Section IV).

39 references up to 1981.
Complementary: none.
Relevant: **9**, 7; **10**, 5; **13**, 19; **16**, 4; **17**, 5; **18**, 6; **19**, 5; **22**, 7; **24**, 6; **25**, 4; **26**, 6; **27**, 7; **29**, 3; **30**, 22; **31**, 3; **32**, 3; **36**, 7.

33, *5: Acidity, hydrogen bonding and complex formation* by *W. H. Richardson*

I. Introduction (130); **II**. Acidity of peroxy acids and hydroperoxides (130); **III**. Hydrogen bonding (134); **IV**. Complexes (155); **V**. References (157).

Section II deals with the acidity of aliphatic (Table 1) and aromatic (Table 2) peroxy acids and of hydroperoxides (Table 3) and with transmission effects through oxygen. Section III discusses self- and intermolecular association involving peroxy acids (Table 4) and hydroperoxides (Tables 5–7). Data are given for hydrogen bonding between hydroperoxides and arenes, alkenes, alkynes, ethers, alcohols, phenols, carboxylic acids, amines, carbonyls, carboxylic acid derivatives, sulphoxides and other donors (Tables 8–13). A comparison of intermolecular hydrogen-bonded complexes is given in Table 14. Section III.B. 3 treats intramolecular association between hydroperoxide groups on the one hand and arenes, alkynes, ethers, amines, carbonyls and peroxides present in the same molecule on the other (Tables 15–18).

Section IV describes isolable complexes of peroxides, where no σ bonds are formed. Transition metal complexes are omitted. Complexes of peroxy acids with Ph_3P, Ph_3As and pyridine *N*-oxides are treated (Table 18), and also complexes of hydroperoxides with amines (Table 20) and ammonium salts.

97 references up to 1980.
Complementary: none.
Relevant: **2**, 9; **3**, 6; **4**, 4; **5**, 6; **7**, 6; **8**, 5; **11**, 3; **12**, 3, 10; **13**, 7; **15**, 4; **16**, 5; **17**, 6; **18**, 8; **19**, 7; **20**, 12; **22**, 8; **23**, 6, 7; **24**, 4; **25**, 5, 6; **29**, 17; **31**, 17, 21; **32**, 21.

33, *6: Synthesis and uses of alkyl hydroperoxides and dialkyl peroxides* by *R. A. Sheldon*

I. Introduction (162); **II**. Synthesis (163); **III**. General consideration of reaction types and mechanisms (169); **IV**. Metal-catalysed homolysis (173); **V**. Acid-catalysed heterolysis (176); **VI**. Metal-catalysed heterolysis (177); **VII**. Alkyl hydroperoxides as terminal oxidants (188); **VIII**. Reduction of alkyl hydroperoxides to alcohols (193); **IX**. Reactions of α-substituted alkyl hydroperoxides (194); **X**. Concluding remarks (195); **XI**. References (196).

The introduction defines the scope of the chapter and especially the advantages of *tert*-butyl hydroperoxide (TBHP) and of di-*tert*-butyl peroxide (DTBP). Section II describes the synthesis of the title compounds by autoxidation of hydrocarbons (Tables 1 and 2) and of organometallic compounds and also by using H_2O_2 or ozonides. Section III discusses homolytic and heterolytic reactions of the title compounds and their mechanisms.

Section IV describes inter- and intramolecular processes in metal-catalysed homolyses and Section V acid-catalysed heterolyses.

Section VI deals with the use of the title compounds in epoxidation and hydroxyketonization of alkenes, oxidation of compounds containing N, S, OH or C=O groups and alkene ketonization.

Section VII treats allylic oxidation and vicinal dihydroxylation of alkenes and alkynes, alkene epoxidation and alcohol oxidation. Reduction of ROOH to alcohols is described in Section VIII and reactions of α-substituted ROOH in Section IX.

158 references up to 1981.
Complementary: none.
Relevant: none.

33, *7: Singlet oxygen in peroxide chemistry* by *A. A. Frimer*

The chapter discusses the organic chemistry of $^1O_2$ (singlet oxygen) and especially its role in peroxide chemistry. Section II describes the three lowest electronic states of molecular oxygen (Table 1), of which the $^1\Delta$ species is the one relevant to this chapter.

Section III deals with the preparation of $^1O_2$ by photosensitization (Tables 2 and 3), by oxidation of H_2O_2, by decomposition of trialkyl and triaryl phosphite ozonides and endoperoxides and by microwave discharge in a stream of O_2. Section IV treats diagnostic tests and Section V modes of reaction. The latter include singlet oxygen Diels–Alder reactions forming endoperoxides, $^1O_2$ ene reactions forming allylic hydroperoxides and $^1O_2$ reactions forming dioxetanes. In each case the nature of the substrate, the mechanism and reactions of the primary products are described in detail (including rearrangements and other transformations, reductions, solvolyses, homolyses, fragmentations and additions).

227 references up to 1981.
Complementary: none.
Relevant: none.

33, *8: Free-radical reaction mechanisms involving peroxides in solution* by *J. A. Howard*

The chapter describes mechanisms that are involved in the liquid-phase homolytic formation and decomposition of compounds which contain the peroxide function. In Section II rate constants of free-radical reactions involving dialkyl peroxides are given (Tables 1–3) and discussed. The other main classes of compounds are treated in Sections III–XII as shown in the list of contents above. Wherever available, rate constants, Arrhenius parameters and detailed mechanistic schemes of the reactions are presented and catalysis, substituent and steric effects and the nature of the products are discussed.

70 references up to 1980.
Complementary: none.
Relevant: none.

33, *9: Organic sulphur and phosphorus peroxides* by *R. V. Hoffman*

I. Introduction (259); **II**. Sulphonyl peroxides (260); **III**. Organophosphorus peroxides (274); **IV**. References (276).

Section II deals with the synthesis of sulphonyl peroxides (Tables 1 and 2), with their decomposition reactions and also with their reactions with π, μ, and σ donors, including oxidations of amines (Tables 3 and 4), e.g. $3R^1CH_2NHR^2 + (ArSO_2O)_2 \rightarrow R^1CH{=}NR^2 + 2ArSO_3^- R^1CH_2N^+H_2R^2 \xrightarrow{H_2O} R^1CHO + R^2NH_2$. Detailed mechanistic schemes are presented whenever available.

Section III discusses organophosphorus peroxides, among which only bisdiphenylphosphinyl peroxide, $Ph_2\overset{\displaystyle O}{\overset{\|}{P}}—O—O—\overset{\displaystyle O}{\overset{\|}{P}}Ph_2$, has been studied to any extent, including its hydrolysis, decomposition and rearrangement reactions.

65 references up to 1981.
Complementary: none.
Relevant: none.

33, *10: Diacyl peroxides, peroxycarboxylic acids and peroxy esters* by *G. Bouillon, C. Lick and K. Schank*

I. Diacyl peroxides (280): **II**. Peroxycarboxylic acids (287); **III**. Peroxycarboxylates (peroxy esters) (299); **IV**. References (303).

Section I deals with the syntheses of symmetrical, asymmetric and cyclic diacyl peroxides and peroxycarbonates and their stability. Among the reactions of diacyl peroxides with nucleophiles, solvolyses and reactions with carbanions, enamines, aromatic and olefinic hydrocarbons and other reagents are discussed.

Section II treats the syntheses, stability and structure of peroxycarboxylic acids, the mechanism of the epoxidation of olefins (Table 1) and the stereospecificity, stereoselectivity and rate of this reaction. Epoxidations by related peroxycarbon acids (e.g. peroxycarbamic acid, Table 2) are also described. Next, reactions of peroxycarboxylic acids with acetylenes, allenes, alkanes, alcohols, carbonyl compounds, S and N compounds and others are presented.

Section III discusses peroxycarboxylates, their syntheses, stability and reactions, including those with Grignard compounds and phosphines. Substitutions of allylic hydrogens and α-H in ethers and thioethers are also treated.

240 references up to 1981.
Complementary: none.
Relevant: none.

33, *11: Endoperoxides* by *I. Saito and S. S. Nittala*

I. Introduction (312); **II**. Naturally occurring endoperoxides (312); **III**. Synthetic methods for endoperoxides (314); **IV**. Reactions of endoperoxides (342); **V**. References (369).

Section II describes naturally occurring endoperoxides and their biosynthesis. Section III treats the synthesis of the title compounds by nucleophilic displacements and by singlet oxygen reaction with 1,3-dienes (Tables 1 and 2), polycyclic aromatic compounds, vinylarenes (Table 3) and heterocycles (Table 4). Oxidation with triplet oxygen (Section III.C) may lead to cyclic and bicyclic endoperoxides.

Section IV treats the reduction of endoperoxides with diimide, $LiAlH_4$, thiourea, PPh_3 and $P(OPh)_3$. In thermolysis, molecular oxygen may be released (Table 5) or the O—O bond may be cleaved. In photolysis, the reaction mode (O—O or C—O cleavage) depends on the wavelength. Reactions catalysed by bases, acids or metals are presented, followed by the reaction of prostaglandin endoperoxides and their model compounds (Tables 6 and 7). Finally, in Section IV.G the role of endoperoxides in natural product synthesis is reviewed.

213 references up to 1981.
Complementary: none.
Relevant: none.

33, *12: Structural aspects of organic peroxides* by *J. Z. Gougoutas*

Section I presents the background to the chapter, calls attention to some pitfalls and gives a compendium of the geometric parameters of organic peroxide links in a variety of structures (Table 1). Section II deals with acyclic ROOR structures and also with 6-, 5- and 4-membered rings (Table 2). Section III treats cyclic (5-, 6-, 8- and 9-membered) peroxy ketals and acetals (Table 3) and acyclic structures containing the same functions. Section IV presents the structures of alkyl hydroperoxides, hydroperoxy acetals and ketals. Section V reviews peroxy acids and their esters and Section VI treats cyclic and acyclic (Table 4) diacyl peroxides.

Section VII discusses isosteric functional groups (e.g. —CN and —NC or —OCN, —NCO and —NNN) and their similarities. Peroxides (—O—O—) seem to be isosteric with diazo compounds (—N=N—).

141 references up to 1981.
Complementary: none.
Relevant: **16**, 2; **17**, 2; **18**, 2; **19**, 2; **20**, 2; **21**, 1; **22**, 2; **23**, 3; **24**, 2; **26**, 2; **27**, 4; **28**, 2; **29**, 1; **30**, 1; **31**, 24; **32**, 24; **34**, 10; **36**, 3.

33, *13: Polymeric peroxides* by *R. Ceresa*

The chapter deals with true polymers (with molecular weights of several thousands or more) of various peroxidic compounds, especially hydroperoxides.

Section II describes polymers with terminal or pendant hydroperoxy groups, polymers with peroxide groups within the backbone chain, terminal or pendant peroxide groups and polymers with perester groups.

78 references up to 1980.
Complementary: none.
Relevant: none.

33, *14: Organic reactions involving the superoxide anion* by *A. A. Frimer*

The chapter deals with the various modes of reaction of $O_2^{-\cdot}$ (superoxide anion radical). Section II discusses the kinetic and thermodynamic data of deprotonation and H-atom abstraction by $O_2^{-\cdot}$ in reactions involving O—H, S—H, N—H and C—H bonds.

Section III treats nucleophilic reactions between $O_2^{-\cdot}$ and alkyl halides or sulphonates, acid chlorides, anhydrides and esters, diacyl peroxides, aldehydes, ketones, amides, nitriles, S compounds, cations and cation radicals. In all these cases substitution products are formed, which in turn usually undergo further reactions, e.g.

$$RX + O_2^{-\cdot} \longrightarrow RO_2^{\cdot} \xrightarrow{O_2^{-\cdot}} RO_2^{-} \xrightarrow{RX} RO_2R$$

$$RC(=O)OC(=O)R + O_2^{-\cdot} \longrightarrow RC(=O)O^{\cdot} + RC(=O)O^{-} + O_2^{\cdot}$$

$$RC(=O)O^{\cdot} + O_2^{-\cdot} \longrightarrow RC(=O)O^{-} + O_2$$

Section IV describes one-electron reductions by O_2 of conjugated ketones (yielding carboxylic acids), olefins (some of which are totally unreactive but, when CN- or NO_2—substituted, yield acids), aryl systems (again reacting only in the presence of strong electron-withdrawing groups, yielding phenols), peroxides ($ROOH + O_2^{-\cdot} \rightarrow O_2 + RO^{\cdot} + HO^{-}$) and free radicals (yielding peroxide anions, $R^{\cdot} + O_2^{-\cdot} \rightarrow RO_2^{-}$). Finally, the generation of singlet oxygen $^1O_2$ is discussed.

213 references up to 1981.
Complementary: none.
Relevant: none.

33, *15: Transition-metal peroxides—synthesis and use as oxidizing agents* by *H. Mimoun*

The title compounds involve either covalently bound dioxygen resembling O_2^{2-} in the peroxo configuration **I**, or the μ-peroxo configuration comprising bimetallic peroxides **IIa**, hydroperoxides **IIb** or alkyl peroxides **IIc**.

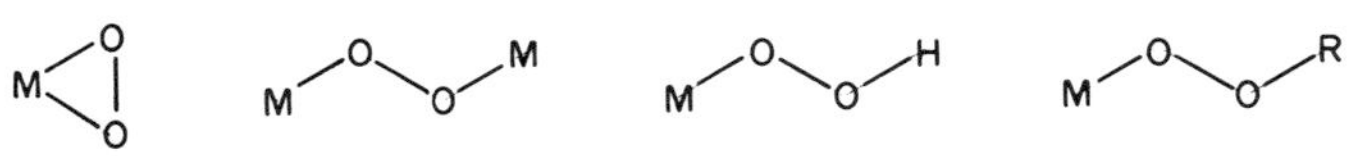

I **IIa** **IIb** **IIc**

Section II deals with the general properties (Table 1) and synthesis of **I**, their general reactivity with electrophiles and nucleophiles and discusses transition metal peroxo complexes.

Section III treats the general properties (Table 2) and syntheses of μ-peroxo complexes and their reactivity. μ-Complexes with Group VIII metals and with early transition metals play an important role as reactive intermediates in numerous metal-catalysed oxidations of hydrocarbons by O_2, ROOH and H_2O_2.

160 references up to 1981.
Complementary: none.
Relevant: none.

33, *16: Organic polyoxides* by *B. Plesničar*

I. Introduction (483); **II.** Hydrogen polyoxides (484); **III.** Dialkyl polyoxides (486); **IV.** Diacyl tetroxides (499); **V.** Alkyl hydrotrioxides (501); **VI.** Acknowledgement (517); **VII.** References (517).

The chapter deals with the formation, identification and decomposition of polyoxides $R^1O_nR^2$, where R^1 and R^2 may be H or other atoms or groups and $n \geqslant 3$.

Section II discusses hydrogen polyoxides HO_nH, their structure, identification and reactions.

Section III treats tertiary tetraoxides (e.g. *t*-BuOOOOBu-*t*); the equilibrium $2R^tOO \rightleftharpoons R^tOOOOR^t$ (Table 1) and the decomposition of some tetraoxides (Table 2). Studies on primary and secondary tetraoxides are reviewed. Section III.B describes trioxides ROOOR, where the R group is *t*-Bu or cumyl (Table 3), CF_3 or SF_5 and also MeOOOBu-*t*, and the structure of dialkyl polyoxides.

Section IV reviews diacyl tetraoxides $R\overset{O}{\overset{\|}{C}}OOOO\overset{O}{\overset{\|}{C}}R$ and their reactions. Section V presents the mechanism of hydrotrioxide formation, the generation, identification and decomposition of hydrotrioxides ROOOH, ether hydrotrioxides obtained by ozonation of ethers (Scheme 6 and Table 4), acetal and aldehyde hydrotrioxides (Tables 5 and 6) and hydrocarbon and alcohol hydrotrioxides. Finally, in SectionV. V.C the structure of hydrotrioxides is discussed.

120 references up to 1981.
Complementary: none.
Relevant: 33, 19.

33, *17: Polar reaction mechanisms involving peroxides in solution* by *B. Plesničar*

I. Introduction (522); **II.** Intermolecular substitution at the peroxide oxygen (522); **III.** Intramolecular nucleophilic rearrangements (551); **IV.** Oxidations involving peroxides and their anions as nucleophilies (566); **V.** Carbonyl-forming eliminations (573); **VI.** References (578).

Section II presents general characteristics, structures and charges (Table 1) of peroxy compounds and describes the epoxidation of olefins (including kinetics, substituent effects, stereochemistry, solvent effects, catalysis, mechanisms and theoretical studies) by peroxy acids and hydroperoxides. Hydroxylation of arenes to phenols and the oxidation of acetylenes, amines, oximes, azo and diazo compounds, sulphides, sulphoxides, thiobenzophenones and organic iodine compounds are described. Section II.H treats the oxidation of amines and of olefins by diacyl peroxides.

Section III discusses the Criegee rearrangement of peroxy esters, the rearrangements of hydroperoxides, dialkyl peroxides and diacyl peroxides and the Baeyer–Villiger oxidation of ketones and aldehydes with peroxy acids.

Section IV treats oxidations with alkaline H_2O_2 and other peroxy anions and the oxidation of imines by peroxy acids. Section V deals with base-catalysed decomposition of dialkyl peroxides and of peroxy esters to form carbonyl compounds.

250 references up to 1981.
Complementary: none.
Relevant: 21, 9; **27**, 14.

33, *18: Preparation and uses of isotopically labelled peroxides* by *S. Oae and K. Fujimori*

Section II describes the preparation of ^{17}O- and ^{18}O-labelled H_2O_2 and of labelled hydroperoxides by autoxidation of ^{14}C-labelled hydrocarbons, by ene reactions of olefins with singlet oxygen, by reaction of H_2O_2 with alkylating agents, by addition of H_2O_2 to double bonds or by perester alcoholysis. Labelled dialkyl peroxides can be prepared by essentially the same methods as hydroperoxides, using either H_2O_2 or ROOH and also cycloadditions of singlet oxygen to olefins. Labelled percids are prepared by autoxidation of aldehydes with O_2, when either of the reactions can be labelled, and also by reaction of H_2O_2 with acylating agents or by reaction of diacyl peroxides with $NaOCH_3$. Section II.E deals with similar methods for the preparation of labelled peresters and Section II.F methods for diacyl peroxides.

Section III discusses the uses of the title compounds in studying the mechanism of the unimolecular homolytic decomposition of peroxides (Table 9), including peresters, diacyl peroxides, *text*-butyl cycloalkaneperformates and polyoxides and also polymerizations initiated by free radicals generated from the unimolecular decomposition of peroxides.

Studies relating to the unimolecular heterolytic decomposition of peresters (Table 10), diacyl peroxides and ozonides and also induced decompositions of the same and similar substrates are described.

217 references up to 1981.
Complementary: none.
Relevant: 5, 10; **12**, 13; **13**, 15; **17**, 12; **18**, 18; **19**, 17; **22**, 14; **23**, 16; **24**, 20; **26**, 11; **27**, 10; **28**, 8; **36**, 12.

33, *19: Ozonation of single bonds* by *E. Keinan and H.T. Varkony*

Ozone usually behaves as an electrophile, which can react with multiple bonds (Section II) and with nucleophilic heteroatoms (Section III), but the chapter emphasizes the ozonations of single bonds in relatively less reactive compounds (Section IV).

Section IV deals with the ozonation of C—H bonds in saturated compounds via 1,3-dipolar insertion, to form an unstable hydrotrioxide and, on decomposition of the latter, the final product. The processes described are ozonation of ethers and acetals (yielding alcohols, aldehydes and $^1O_2$); aldehydes ($\rightarrow$acids + $^1O_2$), hydrosilanes ($\rightarrow$silanols), anthrones ($\rightarrow$anthraquinone), amines (yielding either products of *N*-oxidation, such as nitroxides, or side-chain oxidation) and saturated hydrocarbons ($\rightarrow$alcohols, aldehydes, ketones and acids). Dry ozonation (on silica gel) (Tables 1–3) yields mainly hydroxylic or carbonylic products with a high degree of regioselectivity and stereospecificity. Ozonation of C—C bonds may yield both oxidation and cleavage products (Tables 4 and 5) and ozonations in strongly acidic media (magic acid, etc.) (Tables 6–9) give oxygenated products resulting from C—H or C—C bond cleavage.

72 references up to 1981.
Complementary: none.
Relevant: none.

33, *20: Pyrolysis of peroxides in the gas phase* by *L. Batt and M.T.H. Liu*

The chapter deals with mechanisms of the gas-phase pyrolysis of hydroperoxides (Table 1), dialkyl peroxides, especially MeOOMe (Table 2) and CF_3OOCF_3 (Table 3), peroxynitric acid (Table 4), peroxynitrates (Table 5), ozonidas and dioxetanes. Table 6 presents the correlation of hot-band activity and quantum yield with molecular parameters for alkene + O_2 ($^1\Delta$) reactions.

Detailed reaction schemes are given for the various decomposition reactions discussed.

111 references up to 1981.
Complementary: none.
Relevant: 1, 3; **3**, 4 **16**, 10; **22**, 15; **25**, 15; **27**, 11; **29**, 7; **32**, 7.

33, *21: Photochemistry and radiation chemistry of peroxides* by *Y. Ogata, K. Tomizawa and K. Furuta*

Section I discusses the UV absorption of diacyl peroxides (Table 1), ESR spectra (Table 2) and the photolysis of some peroxidic compounds. Thermolysis and photolysis are compared, involving excited radicals (Table 3) and excited substrates (Table 4), and reactions of peroxide-derived radicals with organic compounds are described.

Section II deals with the photolysis of H_2O_2 (Table 5) and with the photolytic oxidation by H_2O_2 of alcohols and ethers (Tables 6 and 7), carboxylic acids and esters, N compounds and aromatic rings, especially phenols (Table 8).

Section III treats the photolysis of alkyl hydroperoxides (Table 9), ketohydroperoxides, α-azo hydroperoxides and alkenyl and polymeric hydroperoxides. Section IV discusses di-*tert*-butyl, dimethyl, diethyl and bis(trifluoromethyl) peroxides and some cyclic and other peroxides.

Section V deals with aliphatic (Table 10) and aromatic (Table 11) diacyl peroxides and some cyclic and mixed (aliphatic–aromatic) acyl peroxides. Section VI presents photoreactions of aliphatic (Tables 12–14) and aromatic peroxycarboxylic acids with a variety of substrates.

Section VII presents the photolysis of aliphatic and aromatic peroxycarboxylic acid esters (Tables 15–17), including α-acyloxylation by peresters and intramolecular oxidation of unreactive C—H bonds, e.g. α-peroxyacetonitriles (Table 18). Section VIII reviews photoreactions of peroxymono- and peroxydisulphates, P-containing peroxides, sulphoxyperoxides and also Si- and B-containing peroxides.

Section IX reports the radiolysis of H_2O_2, ROOH, ROOR, diacyl peroxides, peroxycarboxylic acids, peroxydisulphate and peroxydiphosphate.

210 references, up to 1981.
Complementary: none.
Relevant: **2**, 16; **3**, 8; **6**, 4; **8**, 12; **9**, 6, 8; **10**, 6; **11**, 5, 6; **13**, 16, 17; **14**, 8; **15**, 9; **16**, 11, 12; **17**, 9, 10; **18**, 10, 11; **19**, 10, 20; **22**, 10, 11; **23**, 9; **24**, 12; **25**, 11, 12; **27**, 21, 22, 23, 24; **28**, 5; **29**, 5, 6, 7; **31**, 6; **32**, 9, 29; **36**, 8.

33, *22: Solid-state reactions of peroxides* by *M. Lazar*

I. Introduction (778); **II**. Isomerization reactions (778); **III**. Evolution of oxygen from endoperoxides of aromatic compounds (781); **IV**. Decomposition of peroxides (783); **V**. Decomposition of low molecular weight peroxides in polymeric media (799); **VI**. Conclusions (803); **VII**. References (804).

Section II describes the topotactic isomerization of 2-iododibenzoyl peroxide derivatives, ^{18}O-scrambling in acyl peroxides and the carboxy inversion (i.e. $PhCMe_2OO\overset{O}{\overset{\|}{C}}Ph \rightarrow PhOCMe_2O\overset{O}{\overset{\|}{C}}Ph$).

Section III deals with the thermolytic release of O_2 from aromatic endoperoxides and with regeneration of the parent aromatic compound. Section IV treats the decomposition of 9, 10-dioxyanthracene and of dibenzoyl peroxide and its derivatives, discussing in both cases kinetics (Tables 1 and 3), topochemistry, mechanisms and products (Table 2). Next, decompositions of cyclic alkylidene peroxides (Table 4), hydroperoxides of peroxy acids (Table 5) and polymer hydroperoxides (Table 6) are discussed.

Section V covers the influence of polymeric media on the decomposition of peroxides (Tables 7 and 8) when the low mobility and relatively great microheterogeneity of the medium influence the reaction.

112 references up to 1981.
Complementary: none.
Relevant: **32**, 17.

33, *23: Organometallic peroxides* by *P.B. Brindley*

I. Introduction (807); **II**. Organometallic peroxides of Group II metals (810); **III**. Organometallic peroxides of Group III metals (812); **IV**. Organometallic peroxides of Group IV metals (818); **V**. Organometallic peroxides of Group V metals (822); **VI**. Peroxy derivatives of Group VI and VIII organometallics (824); **VII**. References (826).

The title compounds are defined as having the metal bonded directly to organic carbon and having dioxygen bonded to the metal, i.e. RMOOH or RMOOR′. Section I reviews general properties and general methods of preparation.

Section II describes derivatives of Be, Mg, Zn, Cd and Hg; Section III those of B (Tables 1 and 2), Al, Ga, In and Tl; Section IV those of Ti, Zr, Hf, Sn and Pb; Section V derivatives of Sb, As and Bi; and Section VI those of Group VI metals (Mo, W, V, Nb, Ta and Te) and of Group VIII metals such as Co, Ir, Rh and Pd.

In each section, detailed reaction schemes are presented for the formation and/or decomposition of the title compounds, whenever available.

87 references up to 1981.
Complementary: none.
Relevant: none

33, *24: Four-membered ring peroxides: 1,2-dioxetanes and α-peroxylactones* by *W. Adam*

I. Introduction (830); **II**. Physical aspects (844); **III**. Synthesis (847); **IV**. 1,2-Dioxetanes as reaction intermediates (860); **V**. Chemiluminescence (880); **VI**. Chemical transformations (909); **VII**. Acknowledgements (912); **VIII**. References (913).

Section I presents the nomenclature, history and literature of the title compounds and tabulates data on 1,2-dioxetanes (Table 1) and α-peroxylactones and 2-imino-1,2-dioxetanes (Table 2).

Section II describes X-ray structural parameters (Table 3) and NMR, IR and UV–visible spectra.

Section III deals with the preparation of 1,2-dioxetanes by dehydrobromination of β-bromo hydroperoxides:

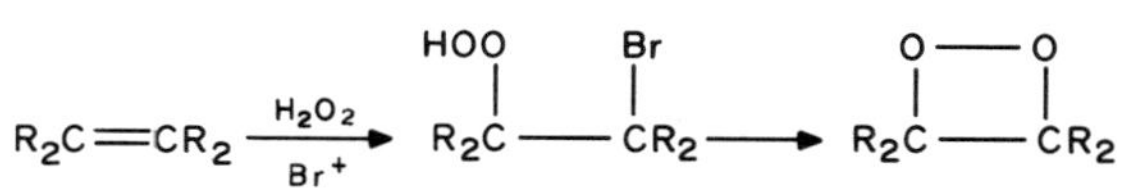

or by singlet oxygenation of electron-rich olefins or by other less general methods. Next the synthesis of α-peroxylactones through α-hydroperoxy acids (Table 4) using various dehydrating agents (Table 5) is described and also the purification and characterization of the title compounds.

Section IV treats the intermediacy of 1,2-dioxetanes in singlet oxygenation of alkenes, cycloalkenes, enols, enol ethers, ketene acetals, enamines and heteroarenes, and in various routes of autoxidations and superoxide cleavages.

Section V discusses the mechanism of light generation in the chemiluminescence of 1,2-dioxetanes. The energy sufficiency of the molecules and the exothermicity of the chemiluminescence process lead to an electronically excited (rather than to a vibrationally excited) product.

The various types of unimolecular decomposition mechanisms are presented and the relevant activation parameters are tabulated (Table 6). Next, catalytic decomposition involving an electron-transfer process is treated, followed by a discussion of excitation parameters (Table 7) and their determination by photophysical and photochemical methods.

Section VI describes those reactions of dioxetanes with nucleophiles and with electrophiles in which no cleavage of the dioxetane C—C bond occurs.

257 references up to 1981.
Complementary: none.
Relevant: none.

The chemistry of the metal–carbon bond. Volume 2. The nature and cleavage of metal–carbon bonds (1985)

34, *1: Electrochemical cleavage of metal—carbon bonds* by *C. J. Pickett*

Section II.B discusses the mechanism of the products and the electrochemical oxidation of Main Group metal—carbon sigma bonds. Cyclic voltammetric oxidation peak potentials correlate linearly with ionization potentials in metal alkyls (Table 1). The organic products may be alkanes, disproportionation or dimerization products, organometallic compounds (by attack on the electrodes) or products obtained by attack on dissolved species, attacks of bases, fragmentation or rearrangement. The data show that the primary irreversible single-electron oxidation $M—R \xrightarrow{-e} M^+ + R^\cdot$ is followed by a second one-electron oxidation, $R^\cdot \xrightarrow{-e} R^+$, and either $R^\cdot$ or R^+ may be the reactive species.

Next the mechanisms and products in reduction processes are described (Tables 2 and 3). Section II.C deals with transition metal alkyls, aryls and acyls, including their reduction (Table 4) and oxidation. Section III treats the electrochemistry of π-bonded hydrocarbon ligands in transition metal complexes with closed shells (containing, e.g., η^5-cyclopentadienyl or η^6-arene ligands).

Section IV considers the oxidation and reduction of transition metal carbonyl, carbene, carbyne and isocyanide complexes.

99 references up to 1982.
Complementary: 30, 17.
Relevant: 5, 2; **8**, 11; **13**, 5; **16**, 15; **17**, 14; **19**, 12; **20**, 5; **21**, 5; **22**, 9; **23**, 10; **24**, 17; **25**, 13; **26**, 12; **27**, 8; **28**, 7; **29**, 8, 9; **31**, 7; **32**, 6.

34, *2: Heterolytic cleavage of Main Group metal—carbon bonds* by *M. H. Abraham and P. L. Grellier*

Section I reviews general aspects including enthalpy changes for the formation of methyl anions from organometallic compounds (Table 1), the different variants of S_E1 and S_E2 substitutions and also mechanisms involving rearrangement, catalysis and electron transfer.

Section II describes the cleavage of Li—C bonds by Si^{IV} compounds, the cleavage of Mg—C bonds by Hg^{II}, B^{III} and Si^{IV} compounds (Table 2) and the cleavage of Zn—C and Hg—C bonds by Hg^{II}, Pd^{III}, Fe^{III}, Tl^{III} and Pb^{IV} compounds (Tables 3–17 list various data for rate constants, activation energies, enthalpies and entropies).

Section II.E deals with cleavage of B—C and Tl—C bonds by Hg^{II} (Tables 18–20) and by Pb^{IV} compounds.

Section II.F treats the cleavage of Group IV.B metal—carbon bonds (Si—C, Ge—C, Sn—C and Pb—C) by Li^{I}, Be^{II}, Fe^{II}, Fe^{III}, Hg^{II}, Ge^{IV}, Sn^{IV}, Cu^{II}, Ag^{I}, K^{I}, Ti^{IV}, V^{V}, Pd^{II}, Pt^{II}, Ir^{III}, B^{III}, Al^{III}, Si^{IV}, Au^{III}, Pb^{IV}, Sb^{III}, Sb^{V} and Bi^{III} compounds. Tables 21–43 present rate constants and activation parameters for many of these reactions.

Section II.G deals with the cleavage of Sb—C bonds by Sb^{III} compounds and Section II.H considers the inversion of primary metal alkyl compounds at the metal—CH centre as shown by NMR studies (Table 44).

Section III discusses the cleavage of non-transition metal—carbon bonds by halogens: $RMX_n + Y_2 \rightarrow RY + YMX_n$.

Sections III.B–E present chlorinolysis reactions of organic Hg^{II}, Tl^{III}, Si^{IV} and Sn^{IV} compounds and Sections III.F–M deal with brominolysis of organic derivatives of Li^{I}, Hg^{II} (Tables 45–48), B^{III} (Tables 49), Tl^{III}, Si^{IV} (Tables 50 and 51), Ge^{IV}, Sn^{IV} (Tables 52–59, including rate and sterochemical data) and Pb^{IV}.

Sections III.N–U treat the iodinolysis of organic Li^{I}, Hg^{II} (Tables 60–66 including rates and activation parameters), B^{III} (Table 67), Si^{IV}, Ge^{IV}, Sn^{IV} (Tables 68–83) and also the cleavage of Sn^{IV} compounds by ICl and IBr (Table 84) and the iodinolysis of Pb^{IV} compounds (Tables 85 and 86). The fluorinolysis of some organotin(IV) compounds is reported and tabular comparisons are given of the results of the Fukuzumi–Kochi analysis (assuming that the thermal ion pair is a good model for the transition state in an electrophilic substitution) with observed data (Tables 87 and 88).

Section IV considers the protodemetallation of organometallic compounds in which a metal-bonded organic group is cleaved and the corresponding RH compound is formed, e.g.

$$H^+ + RHgX \rightarrow RH + HgX^+$$

or

$$R_2Zn + H_2NR^1 \rightarrow RH + RZnHR^1$$

Sections IV.A–D deal with the protodemetallation of organic Li, Mg, Zn, Cd, Hg and B compounds (Tables 89–102) present data for a variety of relevant processes, such as acidolyses with various terminal acetylenes, amines (both primary and secondary) alcohols, inorganic and organic acids and also buffer solutions.

Section IV.E treats organosilicon compounds. In the presence of both Me and other alkyl groups R, the former will be cleaved by H_2SO_4 [$Me_3SiR + H_2SO_4 \rightarrow CH_4 + Me_2Si(HSO_4)R$] (Table 103). Preparative and general kinetic studies of related reactions are described, and also kinetic studies on acidic protodemetallation of $Me_3SiC_6H_4X$ compounds yielding C_6H_5X with acids (Tables 104–106). In the presence of bases, from R_3SiR' compounds R'H is formed preferentially with the ease of removal ($R' = PhC{\equiv}C >$ indenyl, fluorenyl $> Ph_2CH > PhCH_2 >$ aryl groups $> PhCH_2CH_2CH_2 > n\text{-}C_6H_{13} \approx 0$) (Tables 107–114). Section IV.F describes protodemetallation of Ge compounds (Tables 115 and 116) and Section IV.G that of the Sn compounds, including kinetic and stereochemical studies (Tables 117–126) and the proportions of allene and alkyne formed in the reaction of alkenyltin compounds (Table 122). Section IV.H presents acid-catalysed processes of organolead compounds, studied kinetically and mechanistically (Tables 127–129).

591 references up to 1981.
Complementary: none.
Relevant: 34, 3.

34, *3: Homolytic cleavage of metal—carbon bonds: Groups I to V* by *P. J. Barker and J. N. Winter*

Section I presents a historical surveys and defines the scope of the chapter including two main reaction types:

$$R_nM \rightarrow R^{\bullet} + R_{n-1}M$$

$$X^{\bullet} + R_nM \rightarrow R^{\bullet} + XMR_{n-1}$$

Some bond dissociation energies are listed (Tables 1–3) and experimental techniques (ESR, CIDNP) and product studies are briefly reviewed.

Section II deals with thermal and photolytic decompositions and also with radical reactions and reactions with electron acceptors involving Group IA elements, mainly Li.

Section III treats the thermal decomposition of Cu^{I}, Cu^{II}, Ag^{I} and Au^{I} compounds and also photolytic decompositions and radical reactions.

Section IV discusses the thermal and radical reactions of organic derivatives of Ca, Mg and Be, with emphasis on reactions of Grignard reagents.

Section V describes thermal and photolytic decompositions and radical and electron transfer reactions of organic Zn, Hg and Cd derivatives, while Section VI considers the same types of reactions with organic Ga, In, Tl and B (Table 5) and also (Section VI.D) with chemically induced homolytic decompositions in which, e.g., peroxides give radicals and the latter take part in S_H2 reactions at the metal centre.

Section VII similarly treats thermal, photolytic and radical reactions of Si, Ge, Sn, Pb and deals especially with studies of S_H2 reactions at tin, with electron transfer reactions, with the lifetime of $R_4M^{+\bullet}$ moieties appearing in the electron-transfer reactions as an intermediate on electron loss and with the toxicity of metal-alkyl compounds.

Section VII describes thermal and photolytic reactions of organic derivatives of As, Sb, Bi and P. The radical reactions of this group include α-scissions of phosphoranyl radicals, e.g. $Bu^tO^{\bullet} + Me_3P \rightarrow Me_3(Bu^tO)P^{\bullet} \rightarrow Bu^tO(Me)_2P + Me^{\bullet}$, and some radical reactions of Sb and Bi.

276 references up to 1982.
Complementary: none.
Relevant: 34, 2.

34, *4: Insertions into Main Group metal—carbon bonds* by *J. L. Wardell and E. S. Paterson*

alkynes (268); **V**. Additions to carbon—nitrogen multiple bonds (296); **VI**. Miscellaneous insertions (307); **VII**. References (323).

The chapter considers insertions into C—metal bonds of Main Group organometallics involving C—O, C—S, C—C and C—N multiple bonds.

Section II describes reactions with carbonyl groups, of the type

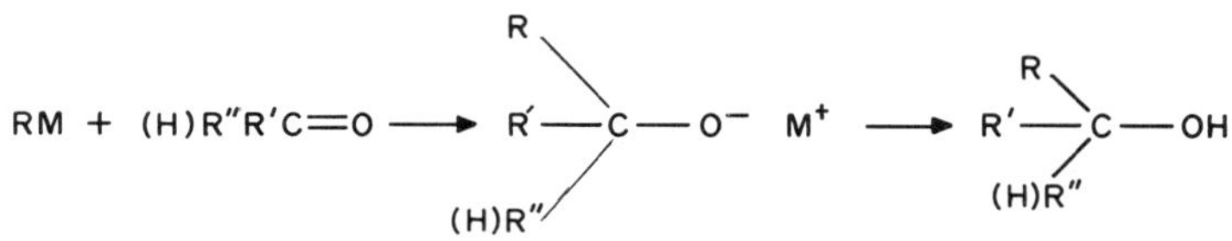

The scope, kinetics and mechanisms of reactions with organic derivatives of Mg, Li and Al are discussed (Tables 1 and 2) and stereochemical considerations are presented in detail (Tables 3 and 4), including 1,2-asymmetric syntheses, 1,3-asymmetric inductions, reactions of alicyclic compounds, asymmetric syntheses using chiral ligands and the stereochemistry of allylic—metal compound additions to aldehydes. Reversible additions, reductions and oxophilic additions (i.e. reactions in which the R of RM is added to the oxygen of carbonyl groups) are also treated.

Section II.B deals with reactions of RM with carboxylic acids, esters, anhydrides, acyl chlorides and amides (Table 5) and Section II.C with insertions of C_3O_2, CO (carbonylation) and CO_2 (carboxylation, Table 6). Section II.D presents additions to β-unsaturated carbonyl compounds, including Michael additions (Tables 7 and 8).

Section III reviews carbophilic additions (to give initial products of nucleophilic attack at C) of C=S bonds, e.g.

$$\mathrm{RM + CS_2 \rightarrow RC(S)SM \xrightarrow[\text{or EY}]{\text{E}} RC(S)SE}$$

(Table 9) and also thiophilic additions, e.g.

$$\mathrm{Ph_2C{=}S + PhLi \rightarrow Ph_2C(SPh)Li \xrightarrow{E^+} Ph_2C(E)SPh}$$

Section IV discusses insertions of alkenes and alkynes into C—Li (Table 10) and C—Mg bonds (including intramolecular additions) and the reverse reactions, intermolecular additions (especially 'assisted additions' of various homopropargylic, allylic, propargylic alcohols and other hydroxy or amino compounds, when the presence of a hydroxy or other donor group enhances the reaction rates of the multiple bond).

Sections IV.E and F deal with insertions into C—Zn and C—B bonds. Section IV.G treats the insertion of alkenes and alkynes into C—Al bonds and catalysed reactions of the same type.

Section IV.H describes insertions of tetracyanoethylene into C—M bonds (with M = Mg, Sn, Pb or Hg) and Section IV.I discusses cycloaddition reactions of allyl- and azaallylmagnesium and -alkali metal compounds with alkenes, alkynes, heterocumulenes and nitriles.

Section V presents additions to nitriles (Table 11) yielding, e.g., imines and/or after hydrolysis carbonyl compounds, $\mathrm{RC{\equiv}N + R'M \rightarrow RR'C{=}NH \rightarrow RR'C{=}O}$, and also to isonitriles, azomethines and nitrogen heteroaromatics.

Section VI reports miscellaneous reactions not classified above. These include reactions with epoxides and related compounds (Table 12), yielding after ring-opening β-substituted ethanols:

$$\underset{\diagdown O \diagup}{H_2C{-}CH_2} + RM \rightarrow RCH_2CH_2OM \xrightarrow{H_2O} RCH_2CH_2OH$$

reactions of organometallic compounds with oxygen under controlled conditions, yielding peroxides (ROOM), alkoxides or aryloxides (ROM), (Table 13), reactions with S, Se and Te (Table 14), which give thiols, selenols or tellurols, insertions of SO_2($RM + SO_2 \rightarrow RSO_2M \xrightarrow{H_2O} RSO_2H$, Table 15), insertions of SO_3 yielding as a rule metal sulphonates (Table 16) and some other, less general reactions such as with >S=O, —N=N— and N=O groups.

556 references up to 1983.
Complementary: **2**, 13; **34**, 5.
Relevant: none.

34, 5: *Insertions into transition metal—carbon bonds* by *J. J. Alexander*

I. Introduction (340); **II**. Co insertion and extrusion, carbonylation and decarbonylation (341); **III**. Insertion of sulphur dioxide and related electrophiles (357); **IV**. Insertion of alkenes and alkynes (365); **V**. Insertion of carbon dioxide, carbon disulphide and isoelectronic species (381); **VI**. Insertion of isocyanides (384); **VII**. Insertion of other unsaturated molecules (391); **VIII**. References (393).

In the title reactions, small molecules X—Y react with transition metal complexes to form products of 1, 1-addition [i.e. MX(Y)C] or 1, 2-addition (i.e. MXYC). The term 'insertion' has no mechanistic significance and refers only to the structure of the product.

Section II describes CO insertions involving Mn and Fe complexes and discusses the intramolecular nature of the reaction (i.e. in these complexes the CO is already coordinated to the metal), alkyl migration vs CO insertion and stereochemical and kinetic studies of both carbonylation (Table 1) and decarbonylation (Table 2). Next, square-planar complexes of Ni, Pd and Pt are treated, followed by studies of the CO insertion mechanism, acceleration of alkyl migration by various catalysts and novel modes of CO insertion.

Section III.A deals with the insertion of SO_2 into metal—alkyl bonds, $[L_nMR] + SO_2 \rightarrow [L_nM(SO_2R)]$, including the scope, mechanism and stereochemistry of the reaction, and with related reactions involving alkenyl and propargyl complexes (Section III.B) and insertions of other species isolectronic with SO_2 (e.g. sulphinyl sulphonamides, SeO_2 and TeO_2), and also of electrophiles such as tetracyanoethylene, SnX_2, GeX_2 and hexafluoroacetone (Section III.C).

Section IV treats the insertion of isolated double bonds and dienes and also of alkynes. In the latter case, the products are alkenyl (vinyl) complexes. The reactions of alkynes with 14-, 16-, and 18-electron complexes are discussed, including insertions into metal—acyl, metal—carbene and metal—μ-alkylidene bonds.

Section V considers the insertions of CO_2, CS_2, COS, RNCO and RNCS into C—metal bonds, although preference is shown for C—N and C—halogen bonds when present. The section deals especially with insertions into Ti, V, Fe, Mn, Ni and Ir complexes.

Section VI presents insertion of isocyanides (RNC) and emphasizes the similarities and differences of this reaction with that of the isoelectronic CO. Square-planar complexes, Cu complexes and also 17- and 18-electron complexes (e.g. of Ti, Mo, Fe and Co) are discussed.

Section VII deals with the insertion reactions involving NO, O_2, S_4 and S-containing molecules (e.g. RNCS), carbenes and N (from IN_3).

280 references up to 1982.
Complementary: 2, 13; **34**, 4.
Relevant: 9, 5; **21**, 12; **30**, 4–7; **32**, 6; **34**, 6–8; **35**, 3, 6.

34, *6: Nucleophilic attack on transition metal organometallic compounds* by *L. S. Hegedus*

Reactions of nucleophiles at the metal centre in transition metal complexes is of major importance for the preparation of various new complexes, while attack at the organic ligands is important in organic synthesis.

Section I.A deals with formation of M—C sigma bonds involving CO, isonitriles, cyanides, η^5-cyclopentadienyl and η^3-allyl, σ-alkyl, aryl and vinyl complexes and emphasizes the wide use of Cu^I complexes (Table 1), metal-catalysed Grignard reactions and transmetallations [e.g. $ArHgX + PdX_2 \rightarrow ArPdX + HgX_2 \rightarrow HgX_2 + ArPdAr \rightarrow ArAr + Pd^0$]. Formation of transition metal hydrides by nucleophilic attack by H^- is considered, and also ligand exchange reactions involving nucleophilic attacks by an external ligand, displacing a coordinated ligand.

Section II.A describes the cleavage of M—C sigma bonds by nucleophilic attack of another metal which displaces the metal present in an alkyl metal complex, including also 'self-exchange' (i.e. replacement of one metal atom of the same metal, but in a different oxidation state: Co^I can attack a complex of Co^{III}). Oxidation of the metal facilitates its removal, and processes utilizing this fact are important in laboratory and industrial syntheses (e.g. the Wacker oxidation of ethylene to acetaldehyde with Pd^{II} catalysts).

Section II.B describes nucleophilic attacks on transition metal CO and isonitrile complexes. The nucleophiles include hydride, carbanions, alkoxides, hydroxide and amines and also alkyl- and aryllithium reagents:

$$NaHB(OR)_3 + [ML_n(CO)] \rightarrow [ML_n(CHO)]$$

$$ArLi + [Ni(CO)_4] \rightarrow [ArC(O)Ni(CO)_3]^- Li^+ \xrightarrow{HCl} ArCOCOAr$$

Section II.C treats nucleophilic attack on metal-complexed olefins. The olefins may be chelating as in complexes of Pd^{II} and Pt^{II} with, e.g., cyclooctadiene (cod) and the nucleophiles may be, e.g., acetate ions, amines or carbanions; for example

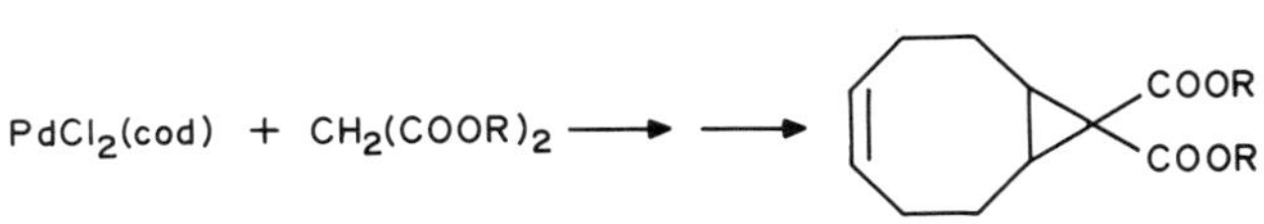

Simple (non-chelating) monolefin complexes also react, although with more difficulty. Even so, several processes reached industrial importance, as in reactions with oxygen nucleophiles leading to aldehydes, vinyl acetates, acetals, glycol ethers, etc., while long-chain olefins are invariably oxidized to ketones. Amines, carbanions, carboxylates, etc., attack olefin – Pd, Pt, Fe and related complexes with the production of a wide variety of new compounds, including aminated compounds, products coupled with carbanions and heterocyclics.

Section II.D considers nucleophilic attacks on π-allyl metal complexes at the π-allyl ligand. Most studies have been carried out with π-allyl palladium halide complexes, according to the general reaction $[PdCl(\eta^3\text{-allyl})]_2$ + excess L + $Nuc^- \rightarrow Nuc—CH_2CH{=}CH_2 + L_nPd$. Again, carbanions, Grignard reagents, amines, carboxylates and other nucleophiles could be used. The stereochemistry and mechanism of the reaction are discussed in detail.

Section II.E describes the Pd-catalysed telomerizations of conjugated dienes. H_2O, alcohols, carboxylates, NH_3, amines, enamines, nitroalkanes and carbanions participate as nucleophiles and the reaction is widely exploited in organic syntheses.

Section II.F deals with nucleophilic attack on cationic π-dienyl complexes, especially complexes of cyclohexadienes with iron carbonyl:

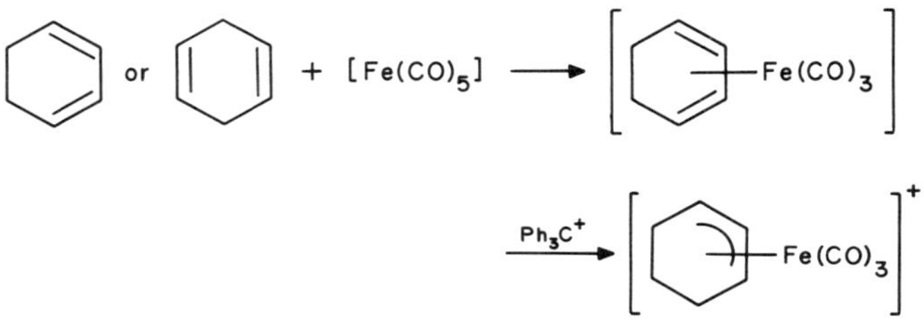

These cationic complexes are generally reactive towards nucleophiles, with substitution into the carbocyclic ring, e.g.

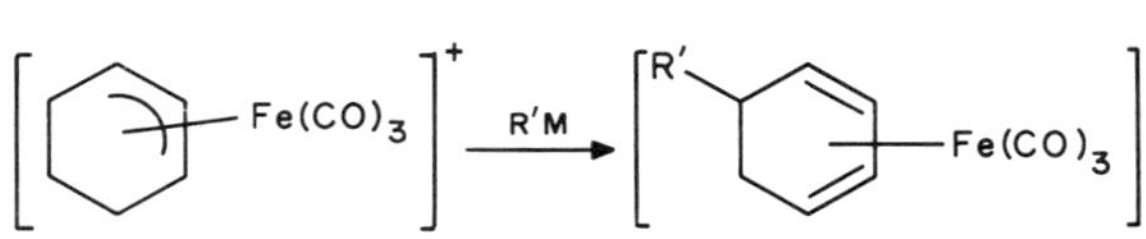

Section II.G discusses nucleophilic attacks on π-arene complexes when both ring and side-chain sites in the aromatic molecule become more reactive towards nucleophilic attack and especially towards carbanions (Table 2).

Section II.H describes more briefly some other reactions of nucleophiles with transition metal complexes.

464 references up to 1982.
Complementary: none.
Relevant: none.

34, *7: Electrophilic attack on transition metal η^1-organometallic compounds* by *M. D. Johnson*

I. Introduction (514); **II**.Ligand modification without substantial change of the carbon—metal bond (516); **III**. Attack of electrophiles at the organic ligand leading to modification of the character of the carbon—metal bond (519); **IV**. Electron transfer reactions (527); **V**. Attack at the metal (532); **VI**. Attack at saturated carbon by the electrophile (541); **VII**. Attack at vinylic carbon (548); **VIII**. Insertion reactions (551); **IX**. References (554).

Section I classifies the reagents and deals with predictions of the sites of attack and with types of reactions.

Section II deals with reactions which do not change the C—metal bond in complexes, such as reversible prototropic processes:

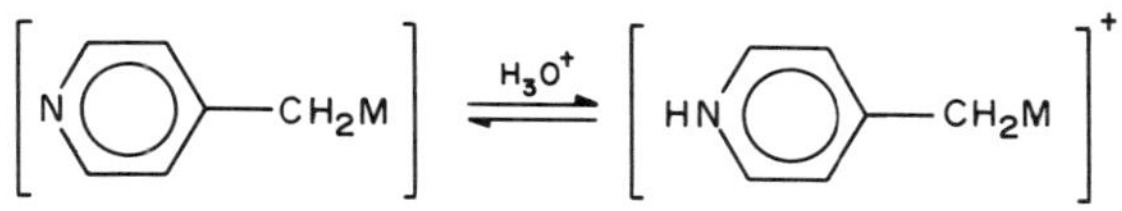

and irreversible processes involving functional groups within the complex (e.g. hydrolysis of esters and amides).

Section III treats electrophilic attacks at the γ-C atom of alkenyl, alkynyl or propadienyl ligands and also with attacks at γ-O, at γ-H, γ-N, β-O, β-N, β-H and β-C atoms of the organic ligands. In these reactions there is a change in the type of the C—metal bond, e.g. a monohapto ligand may change to a dihapto ligand, or an alkyl to a carbene ligand.

Section IV considers one-electron oxidative processes, in which the electrophilic moiety is often (but not always) absent from the reaction products, e.g. $[RCo^{III}(dmgH)L] + Ce^{IV} \rightarrow [RCo^{IV}(dmgH)_2L]^+ + Ce^{III}$, and also reductively induced electrophilic reactions.

Section V discusses reactions of coordinatively unsaturated alkyl and aryl complexes and reactions of alkyl and aryl complexes of coordinatively saturated 18-electron complexes, such as halogenation, acidolysis and cleavage by metal ions in cyclopentadienyliron carbonyl and other complexes.

Section VI considers attacks at saturated carbon atoms by Hg^{II} and Tl^{II} species and by other metal electrophiles, acidolysis and reactions with NO^+ and with NOCl and with Ph_3C^+ cation:

$$Ph_3C^+ + [Ta(Me)_3(Cp)_2] \longrightarrow Ph_3CMe + [Ta(Me)_2(Cp)_2]^+$$

$$\xrightarrow{\text{base}} [Ta(=CH_2)(Me)(Cp)_2]$$

Section VII presents some reactions occurring at vinyl complexes, reactions which are difficult to analyse and the mechanisms of which are uncertain.

Section VIII reviews insertion reactions in which unsaturated electrophiles are inserted between the organic ligand and the metal. The electrophiles include SO_2, SO_3, tetracyanoethylene, hexafluorobut-2-yne and hexafluoroacetone.

210 references up to 1982.
Complementary: none.
Relevant: none.

34, *8: Transition metal—carbon bond cleavage through β-hydrogen elimination* by *R. J. Cross*

I. Introduction (560); **II**. β-Elimination reactions (561); **III**. α- Elimination reactions (599); **IV**. γ-Eliminations (611); **V**. Concluding remarks (618); **VI**. references (619).

Transfer of an H atom from the β-C of a transition metal compound $L_xM\overset{\alpha}{C}H_2\overset{\beta}{C}H_2\overset{\gamma}{C}H_2R$ to the metal or to another substituent yields first an olefin complex in which in turn, by subsequent or concurrent elimination, bond cleavage occurs.

Section II discusses β-eliminations, e.g. *trans*-$[PtCl(Pr)(PEt_3)_2] \rightleftharpoons$ *trans*-

$[PtClH(PEt_3)_2] + CH_2{=}CHMe$ and isotope and other studies relevant to the mechanism of the reaction. For elimination to occur, the presence of a transferable β-substituent (H or D) is essential, in addition to the availability of an extra coordination site on the metal. These factors are considered in cyclopentadienyl, octahedral and square-planar complexes of various transition metals.

Section II.C deals with geometric constraints imposed on the β-H transfer process in metallocyclic compounds and Section II.D presents the results of theoretical calculations.

Section II.E considers examples and applications, including olefin oxidations (e.g. the Wacker process, which involves a β-elimination; Scheme 18, p. 586), substitutions, metal hydride formation and H-transfer, β-elimination from unsaturated groups, isomerization, olefin hydrogenation and hydrosilation.

Section III treats the formally related β-elimination reactions which are less common and more energetic (and hence occur only when no competing β-transfer can proceed):

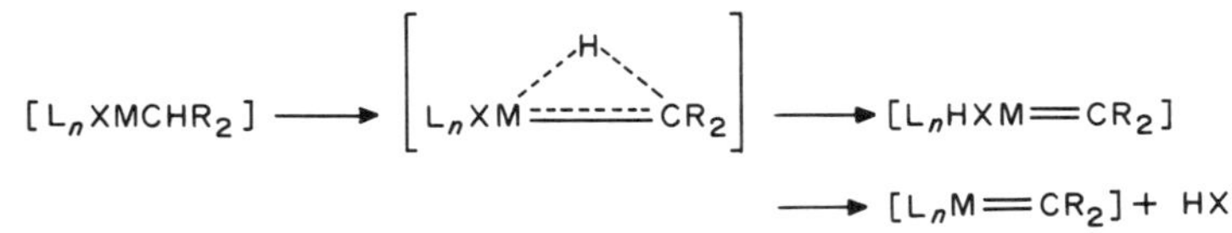

The development of metal—carbon cleavages by α-eliminations is treated, including an orbital description (Fig. 6, p. 601). Section III.B deals with alkylidene and alkylidyne complexes of Nb and Ta, III.C presents the scope of the α-elimination and some theoretical considerations, III.D presents reactions in multinuclear compounds, III.E formyl complexes and III.F cyclic complexes and reactions of generated carbene complexes.

Section IV deals with γ- (and also briefly with δ- and ε-) eliminations, which are again only observable if β-processes cannot occur. The loss of a ligand (presumably to make available a coordination site) is necessary to initiate the process. When the α-atom is a heteroatom such as O, N, S, P or As, a new bond is formed from part of the ligand to the metal and then elimination reactions are usually referred to as metalations. These processes are rare; reactions are intramolecular, since the ligands to be metalated must be coordinated. The steric effects relating to these reactions are discussed.

206 references up to 1981.
Complementary: none.
Relevant: none.

34, *9: Oxidative addition and reductive elimination* by *J. K. Stille*

I. Introduction (626); **II**. Oxidative addition (626); **III**. Reductive elimination (742); **IV**. References (769).

In most organic reactions catalysed by homogeneous transition metal complexes, a C—metal σ-bond is formed and broken. In many of these reactions, the C—metal bond is generated by an oxidative addition and the finished organic product is released from the transition metal by reductive elimination. These reactions are the subjects of the chapter.

Section II deals with reactions in which a metal is oxidized by addition of a substrate XY, with increase of the oxidation state and usually also the coordination number of metal.

Section II.A describes reactions of organic halides and related reactants with Group Ib metals (Cu, Au, Ag) (e.g. $R_fI + Cu \rightarrow R_fCuI$), with oxidative additions to Ni^0 complexes

(Table 1), to Pd^0 complexes (Tables 2 and 3) and to Pt^0 complexes (Table 4), and also to d^8 complexes of Pt^{II} (Table 5) and complexes of Co, Rh^{III} (Table 6), Ir (Tables 7–9) and Fe [dealing especially with reactions of cyclopentadienyldicarbonyl iron, $Fe(\eta^5\text{-}C_5H_5)(CO)_2$(Fp), complexes with allyl halides (Table 11), aryl halides and acid chlorides (Table 12)]. Ru and Os complexes are also described. Among oxidative additions to Group VI.B and VII.B complexes, Mn and Re (Table 14) complexes are presented {including the acylation and subsequent decarbonylation of $[Mn(CO)_5]^-$ (Table 13) and of $[Re(CO)_5]^-$ (Table 14)} and also complexes of Cr, Mo and W. Relevant reactions of early transition metals are scarce, but some processes involving vanadium carbonyls, titanocene complexes and zircocene derivatives are described, in addition to oxidative addition to metal dimers (where the metal is Pd, Pt, Co, Rh or Fe).

Section II.B discusses carbon—carbon insertions, which (e.g. in pericyclic reactions, including valence isomerizations) are key steps in processes involving oxidative addition of a hydrocarbon to a transition metal by breaking a C—C bond to form a metallocycle. The reaction usually occurs with strained small-ring hydrocarbons and is more facile with transition metals of high electron affinity (Table 15). Reactions of strained ring hydrocarbons with $[Fe_2(CO)_9]$ are listed in Table 16, showing both substrates and oxidative addition products.

Section II.C considers carbon—hydrogen insertion occurring through activation of C—H bonds by homogeneous catalysts under mild reaction conditions. Intramolecular reactions are more frequent: cyclometallations occur widely involving cleavage of a C—H bond of a ligand to give a chelate ring containing a C—metal bond. Intermolecular reactions are usually slower by a factor of 10^4–10^6, but are nevertheless observed with Group VIII transition metals (Ni, Pt, Rh, Ir, Ru, Os, etc.) and also with early transition metals (Re, Mo, W, Ti, Zr).

Section III describes processes of reductive elimination, including radical pathways and also α- and β-eliminations (see **34**, 8). Section III.A deals with the elimination of organic halides from transition metal complexes, the conditions for its occurrence and its stereochemistry.

Section III.B treats the 1,1-reductive elimination of organic moieties, which usually only occurs when β-elimination is either impossible or is prevented by some factors.

In complexes of Cu, Ag and Au, whenever possible aryl groups are lost pairwise to yield biaryls, or mixed complexes yield cross-coupling products, e.g.

$$[Cu_6(Ar)_4(R)_2] \rightarrow ArR + (Cu^I(Ar)_3RCu_2^0]$$

$$[Au^{III}XR_2] \rightarrow Au^IX + RR$$

Similar processes occur in complexes of Ni, Pd and Pt, usually yielding coupled products intramolecularly, e.g., *trans*-$[PdR(R')(L)_2] \rightarrow RR' + [PdL_2]$. Mechanisms, kinetic parameters, theoretical calculations, correlation diagrams (Scheme 13, p. 757) and the stereochemistry of the reactions are discussed. Complexes of Rh^{III}, Fe^{II} and of early transition metals are also considered briefly.

829 references up to 1981.
Complementary: 34, 4–8.
Relevant: 9, 5; **21**, 12; **25**, 14; **32**, 26.

34, *10: Structure and bonding of Main Group organometallic compounds* by *J. P. Oliver*

Section II describes organolithium derivatives, many of them occurring as tetramers such as $[LiMe]_4$ and $[LiEt]_4$ and others as hexamers such as $[Li\text{-cyclo-}C_6H_{11}]_6$, $[LiSiMe_3]_6$ and $[LiCH_2SiMe_3]_6$ (Table 1). X-ray structural studies are reported and shown in Figs 2–4. Theoretical studies on organolithium aggregates are discussed and also exchange phenomena involving these aggregates, including changes in aggregation numbers, conformations and rearrangements. Alkyllithium–solvent interactions are treated. Derivatives of Na, K, Rb and Cs seem to form essentially ionic lattices. Some polylithium derivatives are also dealth with briefly (e.g. Li_2CH_2, $Li_2C{=}CH_2$, Li_2-biphenyl).

Section III considers organoberyllium and -magnesium compounds, which have a tendency to form chains in the solid state:

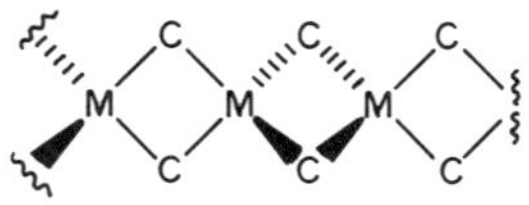

but monomers in gas or in solution.

Section III.B deals with beryllocene and its derivatives, including theoretical and experimental studies.

Section IV discusses organoaluminium derivatives with special regard to multi-centred bonds containing carbon:

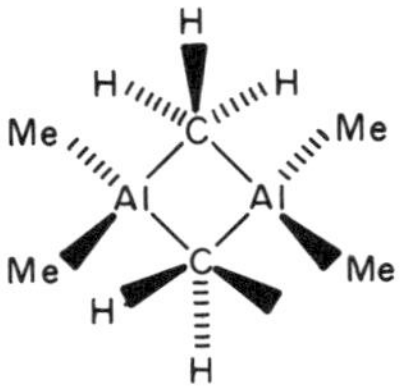

and reviews the various suggestions proposed for the nature of this bonding system. Colligative properties of some organo aluminium compounds are given in Table 2. Interatomic distances and angles for ethynyl-bridged derivatives of Al, Ga, In, Be and Cu are shown in Table 2 and experimental and theoretical studies relating to these compounds are considered. Derivatives of more complex systems [e.g. cyclopentadienyl derivatives of Al, Ga and In (Table 4)] are described.

Section V deals with complexes of Group I metals such as Li_2BeMe_4, $LiBMe_4$ and $[\mu\text{-}Li(tmda)]_2(MgMe_4)]$, and also with Group II–Group III complexes containing both Al and Mg.

Section V.C treats mixed metal systems containing transition metal and Section V.D Group III addition compounds of the type R_3AlNR_3, Me_3GaNMe_3 and Me_3GaPMe_3 (Table 5).

Finally, heteroatom-bridged complexes such as $Me_5Al_2NPh_2$, containing an Al—N—Al bridge, and also compounds containing bridging S- or O-units, are discussed.

272 references up to 1982.
Complementary: 30, 1.
Relevant: none.

The chemistry of the metal–carbon bond. Volume 3. Carbon–carbon bond formation using organometallic compounds (1985)

35, *1: Carbon—carbon bond formation using tin and lead organometallics* by *V. G. Kumar Das and C.-K. Chu*

Section II deals with C—C bond formation via the generation of carbenes by α-elimination from Sn and Pb compounds and also via direct reactions with carbon electrophiles such as aldehydes, ketones or even organic halides, e.g. $Bu_3SnCX_3 + RCHO \rightarrow RCH(CX_3)OSnBu_3$.

Another route involves organo-Sn and -Pb compounds undergoing halogen–lithium and hydrogen–lithium exchange. The lithiated compounds in turn undergo CO additions, e.g.

$$R_3SnCH_2CH_2I + BuLi \rightarrow R_3SnCH_2Li + BuI$$

$$R_3SnCH_2Li + R^1R^2CO \rightarrow R_3SnCH_2C(OLi)R^1R^2 \rightarrow R_3SnCH_2C(OH)R^1R^2$$

$$\xrightarrow{\Delta} R'R^2C{=}CH_2 + R_3SnOH$$

Another route involves transmetallation of Sn or Pb derivatives to yield Li derivatives which in turn react as useful nucleophiles:

$$Bu_3SnCH(OEt)_2 \xrightarrow{BuLi} LiCH(OEt)_2 \xrightarrow{BzBr} BzCHO$$

Section III treats C—C bond formation via β- or γ-substituted derivatives, involving 1,2- and 1,3-eliminations resulting in the formation of double bonds or cyclopropane derivatives, respectively.

Section IV describes reactions of organometal substrates which act as carbon nucleophiles. Thus, vinyl- and alkynyl-metal derivatives may undergo alkynyl transfer (e.g. $Et_3PbC{\equiv}CR + MeCOCl \rightarrow MeCOC{\equiv}CR$), other C—C bond-forming reactions, vinyl transfers and Diels–Alder reactions. Similarly, allyl-metal derivatives may undergo allyl transfer with organic halides, acetates or acid halides and also carbonyl additions and reactions with electrophilic olefins and alkynes. Section IV.C discusses the interaction of aryl- and benzylstannanes and -plumbanes with C-electrophiles (often in the presence of various catalysts): $Ph_3SnMe + BzBr \xrightarrow{BzPdCl(PPh_3)_2} BzPh$.

Section V reviews reactions of tetraalkyltin and -lead compounds and also of derivatives with remote side-chain unsaturation (e.g. R_3Sn-substituted cyclohexanones or cyclohexanols). Section VI presents reactions of *O*-stannyl, *O*-plumbyl, *N*-stannyl and *N*-plumbyl derivatives and organotin hydrides.

Section VII describes some additional processes, namely oxidative addition–reductive elimination, C—C coupling via stannylenes, reactions of $Pb(OAc)_4$ and of aryl lead tricarboxylates, Ziegler-type polymerization, olefin metathesis (see **21**, 10) and the use of organotin and -lead compounds as catalysts.

Throughout the chapter detailed schemes show the mechanism and stereochemistry of the reactions discussed.

322 references up to 1983.
Complementary: none.
Relevant: none.

35, *2: Carbon—carbon bond formation using organometallic compounds of zinc, cadmium and mercury* by *L. Miginiac*

I. Introduction (101); **II**. Reactions of zinc organometallics (101); **III**. Reactions of cadmium organometallics (119); **IV**. Reactions of mercury organometallics (124); **V**. References (132).

Section II treats reactions of organozinc compounds. These include addition reactions of allylic, benzylic, propargylic and allenic, as saturated and other Zn compounds with a variety of unsaturated bonds (C=O, C=N, C=C, C≡C—C≡C, etc.), e.g.

$$R^1CH{=}CHCH_2ZnX + R^2COR^3 \rightarrow CH_2{=}CHCH(R^1)C(OH)R^2R^3$$

In Section II.B substitution reactions of the same Zn derivatives are described with mobile halogens and derivatives with a reactive alkoxy group, followed by addition and insertion reactions of carbene and carbenoid intermediates with C=C, C≡C and CN groups and by description of the Wittig alkene synthesis and the Reformatsky and related reactions.

Section III deals analogously with addition, substitution and carbene reactions of organocadmium compounds.

Section IV describes the following reactions of organomercury compounds: alkene and alkyne additions and the Heck reaction; alkylation and acylation; dimerization; reactions of carbene and carbenoid intermediates (including additions to C=C, C≡C and other multiple bonds and insertions into C—H, B—C, Si—C and Ge—C bonds) and finally Wittig alkene syntheses.

489 references up to 1982.
Complementary: none.
Relevant: none.

35, *3: Carbon—carbon bond formation using η^3-allyl complexes:*

Part 1: *η^3-Allylnickel complexes* by *G. P. Chiusoli and G. Salerno*

Part: *η^3-Allylpalladium complexes* by *J. Tsuji*

Part 3: *Other η^3-allyl transition metal complexes* by *F. Stao*

Part 1, Section II deals with modes of reaction, regiochemistry (Table 1), intermediates and elementary reactivity patterns.

Section III describes C—C coupling of coordinated allyl groups, insertions of double or triple bonds, additions to activated olefins, carbonyl compounds, epoxides and quinones, carbonylations and carboxylations, multiple insertion processes, carbenoid insertions, additions to dienes, HCN addition and reactions with C—Mg bonds.

Part 1: 114 references up to 1984.
Complementary: none.
Relevant: 30, 8.

Part 2, Section II describes the preparation of the title compounds from allylic, olefinic, diolefinic and other compounds.

Section III treats reactions of the title compounds with CO, isocyanides and *C*-nucleophiles (malonates, acetoacetates, β-keto-sulphoxides, -sulphides and -sulphones, enamines, etc.).

Section IV discusses catalytic reactions of conjugated dienes (e.g. in the Pd-catalysed dimerization of butadiene the bis-η^3-allyl complex is involved), oligomerization, dimeriz-

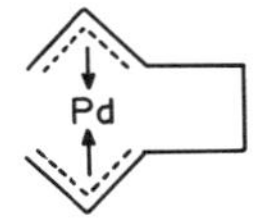

ation with carbonucleophiles, carbonylations, cocyclizations (e.g. two molecules of butadiene with one hetero double bond such as C=N or C=O):

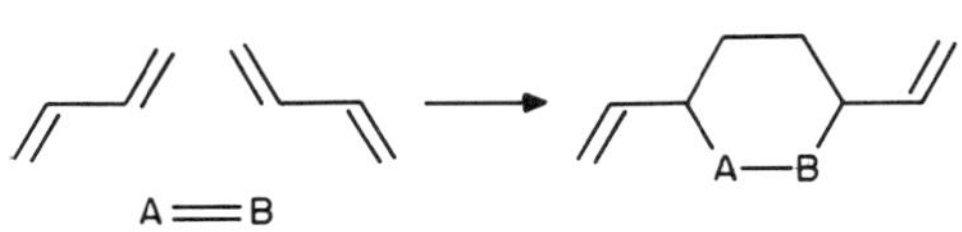

and reactions with CO_2. Next, reactions of allenes and allylic ethers, esters, alcohols, amines and nitroalkanes with carbonucleophiles are reviewed and also reactions of allylic halides (the latter form η^3-allyl complexes with Pd^0 compounds, which then undergo insertion reactions with regeneration of the Pd^0 species, so that the whole process is catalytic).

Part 2: 245 references up to 1982.
Complementary: none.
Relevant: 30, 8.

Part 3 treats some η^3-allyl complexes of Ti and Fe.

Part 3: 12 references up to 1982.
Complementary: none.
Relevant: 30, 8.

35, *4: Olefin oligomerization* by *O.-T. Onsager and J. E. Johansen*

Sections I–III introduce the subject and define its scope and describe the electronic structure of the C=C bond.

Section IV deals with catalytic dimerization of olefins and tabulates possible open-chain C_4, C_5 and C_6 olefins and their abbreviations (Table 1) and also the primary and isomerized products obtained via coordinative mechanisms (Table 2).

Section V treats the anionic dimerization of propene (Table 3) and ethylation of olefins. Section VI presents cationic oligomerization of propene and dimerization of isobutene. Section VII reviews syntheses of organoaluminium compounds using ethylene and propene.

Section VIII discusses the catalysis of the oligomerization of olefins by transition metal complexes (especially of Ni). General mechanistic aspects are reported, including olefin activation via complex formation (Table 4), migratory insertion of olefins, β-H eliminations and reaction rates.

Next, the mechanism of coordinative transition metal catalysis is treated, including the effect of Lewis acids (Tables 5 and 6) and bases (Tables 7–11) on Ni complex catalysis and the kinetics of these reactions. This is followed by descriptions of catalysis by Rh (Table 12), Ti and Ta complexes.

114 references up to 1981.
Complementary: none.
Relevant: 30, 7.

35, *5: Alkyne oligomerization* by *M. J. Winter*

Section II treats linear and cyclodimerization of alkynes. Section II deals with complexes containing two metal atoms ('dimetal complexes'), involving a single alkyne, two linked alkynes or alkyne trimers.

Section IV presents briefly cycloheptatriene complexes of Ni or Pt, obtained via $F_3CC{\equiv}CCF_3$. Production of arenes by cyclotrimerization of three alkynes is discussed in Section V and various types of polymerizations of diynes in Section VI. Cocyclizations of α,ω-diynes with alkynes, catalysed by cyclopentadienylcobalt, $[CoCp(CO)_2]$, is reviewed in Section VII and Section VIII presents complexes of alkyne tetramers with Mo, Re, Pd and Mn. Finally, the formation of cyclooctatetraene and other tetramers by cyclotetramerization of alkynes is treated.

Detailed mechanistic schemes are presented for the reactions discussed, whenever available (Schemes 1–17).

116 references up to 1982.
Complementary: none.
Relevant: 30, 7.

35, *6: Transition metal carbonyls in organic synthesis* by *J. A. Davies and R. J. Shaver*

The chapter describes the use of metal carbonyls in stoichiometric and catalytic organic syntheses.

Section II deals with activation of simple alkanes and with valence isomerization and carbonylation of strained cycloalkanes. Section III treats protection, carbonylation, cyclooligomerization, nucleophilic and electrophilic attacks on coordinated olefins and Diels–Alder reactions of the same and with cyclopropane formation.

Section IV describes the protection, storage, carbonylation and cyclooligomerization of alkynes (Table 1) and Section V carbonylation and coupling of alkyl aryl and allyl halides and decarbonylation of acyl halides. Section VI reviews the carbonylation and hydrosilylation of alcohols and Section VII the decarbonylation, reductive amination (Table 2), reductive alkylation and arylation of ketones and ketenes.

Section VIII discusses reactions of aldehydes, including carbonylation, decarbonylation, reductive amination, *N*-acylamino acid synthesis, reductive alkylation and arylation. Section IX treats processes involving epoxides and ethers, including rearrangement, reduction, carbonylation and polymerization under hydrosilylation conditions.

Section X presents briefly the decarbonylation and hydrosilylation of carboxylic acids and Section XI the decarbonylation of acid anhydrides.

Section XII deals with the carbonylation of esters and lactones and the decarboxylation of pyrones, Section XIII the carbonylation of dialkyl acetals and Section XIV the carbonylation of ortho-esters.

Section XV treats the deoxygenation of sulphoxides, Section XVI the dehydrosulphuration and desulphuration of >C=S and —N=S compounds (Table 3) and

Section XVII the carbonylation of $>C{=}N$, $-N{=}N$ and $-C{\equiv}N$ compounds (Table 4).

Section XVIII describes reactions of alkyllithium and Grignard reagents and organomercury(II) halides and Section XIX the thermolysis and carbene cleavage of transition metal carbene complexes.

186 references up to 1982.
Complementary: none.
Relevant: **1**, 6; **9**, 5; **21**, 12; **25**, 14; **30**, 3; **32**, 26.

35, *7: Olefin and alcohol and carbonylation* by *G. K. Anderson and J. A. Davies*

The chapter discusses CO insertion (mainly transition metal catalysed) into olefins and alcohols.

Section II and III describe general aspects of the production of acid derivatives (Table 1) and the mechanisms involving Co, Ni and Pd systems in olefin carbonylation.

Section IV treats the carbonylation (and, in the presence of H_2O, the hydrocarboxylation) of monoenes (Table 2), dienes, cumulenes and unsaturated amines, amides, alcohols and halides.

Section VI presents the $ROH + CO \xrightarrow{[M]} RCOOH$ reaction and especially the carbonylation of methanol by Co, Rh and Ir catalysts, and also some other carbonylations with other metal catalysts, usually leading to products containing more than one CO or RO unit. Section VII reviews some recent advances in the same reaction.

155 references up to 1982.
Complementary: **21**, 12.
Relevant: **1**, 6; **9**, 5; **30**, 7.

35, *8: Olefin hydroformylation* by *J. A. Davies*

Olefin hydroformylation (oxo reaction) converts C_n olefins into C_{n+1} aldehydes in the presence of Co or Rh catalysts:

$$RCH{=}CH_2 + CO + H_2 \rightarrow RCH_2CH_2CHO + RCH(CHO)CH_3$$

Sections II and III deal with historical and industrial (Table 1) aspects of the reaction.

Section IV.A treats the mechanistic aspects of Co-catalyses systems involving non-modified catalysts {i.e. Co metal, salts or $Co_2(CO)_8$, all of which are converted into $[CoH(CO)_4]$ under the reaction conditions (Table 2)}. Olefin complexation, metal alkyl and metal acyl formation, cluster and by-product formation and free-radical reactions are discussed. The catalyst stability, selectivity, hydroformylation and hydrogenation activities with modified Co catalysts (i.e. containing phosphines, PR_3) are reviewed.

Section IV.B discusses similarly Rh-catalysed systems. For unmodified catalysts the formation of $[RhH(CO)_4]$, olefin complexation, metal alkyl formation, metal acyl formation and cleavage and cluster formation are treated. For PR_3-modified catalysts the reaction mechanisms, including associative and dissociative cycles (Scheme 12) are presented.

112 references up to 1982.
Complementary: 21, 12.
Relevant: 1, 6; **9**, 5; **30**, 7.

35, *9: The Fischer–Tropsch synthesis* by *G. Henrici-Olivé and S. Olivé*

The Fischer–Tropsch (FT) synthesis may be formulated as $n(CO + 2H_2) \xrightarrow{\text{cat.}}$ $-(CH_2)_n- + nH_2O$, where n is usually not a very high number and oxygenated by-products are also formed. Sections II and III describe the history and present status of the title reaction (Tables 1 and 2).

Section IV deals with primary products, with secondary reactions of the α-olefins formed and with various products which can be obtained using modified forms of the FT reactions (e.g. polymethylenes, oxygenated compounds and amines in the presence of NH_3). Section V treats theoretical Schulz–Flory molecular weight distributions (Table 3) and the actual experimental distributions. Kinetics and thermodynamics are discussed in Section VI. Section VII reviews suggested mechanisms, the carbide theory and details and support for the CO insertion mechanism (Scheme 1), including evidence for the primary reactions, the structure and role of various metal complexes and the role of alkali metal promoters. Section VIII discusses product selectivity, the consequences of the Schulz–Flory distribution and deviations from it.

143 references up to 1983.
Complementary: none.
Relevant: none.

35, *10: Olefin carbonylation* by *D. M. Fenton and E. L. Moorehead*

Reaction of olefins with CO with an Ni carbonyl catalyst in the presence of alcohols may give (in what is defined as a non-redox carbonylation) saturated esters:

$$CH_2{=}CH_2 + CO + ROH \rightarrow CH_3CH_2CO_2R$$

If there is also substrate oxidation, the reaction is called oxidative carbonylation:

$$CH_2{=}CH_2 + CO + ROH + (O) \rightarrow CH_2{=}CHCO_2R + RO_2CCH_2CH_2CO_2R + H_2O$$

Section II discusses the first reaction (Tables 2 and 3) and its most important variants, and Section III the second reaction. Carbonylation intermediates (Section IV) and mechanistic aspects of non-redox carbonylation (Section V) and oxidative carbonylation (Section VI) are presented.

69 references up to 1982.
Complementary: 21, 12.
Relevant: 1, 6; **9**, 5; **30**, 7.

The chemistry of organic selenium and tellurium compounds. Volume 1 (1986)

36, *1: Organic derivatives of sulphur, selenium and tellurium—an overview* by *K. A. Jensen and A. Kajaer*

I. Introduction (2); **II**. Nomenclature (3); **III**. History (5); **IV**. Analogues of alcohols and ethers (5); **V**. Onium salts and ylides (7); **VI**. Insertion compounds (8); **VII**. Analogues of sulphoxides, sulphones and related compounds (9); **VIII**. Analogues of carbonyl compounds (10); **IX**. Oxo acids of sulphur, selenium and tellurium (13); **X**. Halogen compounds (15); **XI**. Chalcogen derivatives of Group V elements (16); **XII**. Heterocyclic compounds (17); **XIII**. References (17).

Sections I–III present some general characteristics and the nomenclature of the title compounds (Table 3) and a brief summary of their history. Section IV compares the chemical and physical properties of alcohols, selenols and tellurols and of sulphides, selenides and tellurides. Section V deals with trioganylsulphonium, -selenium and -tellurium ions and with the ylides obtained by their deprotonation. Section VI treats the insertion of S, Se and Te into chains, rings or clusters and Section VII treats sulphoxides and sulphones and their Se and Te analogues. Section VIII discusses S, Se and Te analogues of aldehydes, ketones, carboxylic and carbonic acids. Section IX describes oxo acids of S, Se and Te in valency states six, four and two and Section X mono-, di- and trihalides. Finally, Section XI deals with N, P and As derivatives of the title compounds and Section XII with heterocycles containing S, Se or Te.

105 references up to 1983.
Complementary: none.
Relevant: 3, 13.

36, *2: General and theoretical aspects of organic compounds containing selenium or tellurium* by *R. A. Poirier and I. G. Csizmadia*

I. Introduction (21); **II**. Theoretical background (23); **III**. Results (33); **IV**. Discussion (45); **V**. Conclusions (57); **VI**. Acknowledgements (60); **VII**. References (60).

Sections I and II review elementary concepts and compare first ionization potentials (Table 1) and total electronic energies of O, S, Se and Te (Table 2), atomic orbital basis sets (Tables 3–7) and *ab inito* computational methods.

Section III presents previous results of calculations on organic Se and Te compounds

(Table 8) and more recent results of total molecular energies, using different basis sets (Tables 9–11) and the corresponding geometries (Tables 12–14). Tables 15–19 summarize proton affinities, energies of hydrogenation, first ionization potentials (and also experimental ionization potentials) and conformational stabilities. These four energy-related properties are further discussed in Section IV.A, and non-energy-related properties such as charge distributions (Tables 20–22) and dipole moments (Tables 23 and 24) in Section IV.B, which also shows experimental geometries (Table 25) and the deviation of calculated geometric parameters from the experimental values (Table 26).

94 references up to 1983.
Complementary: none.
Relevant: **1**, 1; **2**, 1; **3**, 1; **4**, 1; **5**, 1; **6**, 1; **8**, 1; **11**, 1; **12**, 1; **13**, 1; **14**, 1; **15**, 1; **16**, 1; **17**, 1; **18**, 1; **19**, 1; **20**, 1; **22**, 1; **23**, 1; **24**, 1; **25**, 1; **26**, 1; **27**, 9; **28**, 1; **31**, 27; **32**, 1; **33**, 1.

36, *3: Structural chemistry of organic compounds containing selenium or tellurium* by *I. Hargittai and B. Rozsondai*

Section II deals with geometric data relating to one-coordinated Se and Te, in most cases referring also to the corresponding O and S analogues (Tables 1–6).

Section II treats acyclic selenides and tellurides (Tables 7–10), acyclic diselenides and ditellurides (Tables 11–12), O-, S-, Se- and Te-containing heterocycles (Tables 13–19), including alicyclic and aromatic rings, and organometallic complexes in which a chalcogen atom is two-coordinated and directly bonded to C and a metal atom (Table 20).

Section IV describes three-coordinated (Tables 21–23) and Section V four-coordinated (Tables 22–28) Se and Te compounds. Section VI discusses five-coordinated Se (the only one being CF_3SeCl_3) and also five- and higher-coordinated Te compounds.

The chapter contains many structural drawing and projections, including stereoscopic views.

333 references up to 1983.
Complementary: none.
Relevant: **16**, 2; **17**, 2; **18**, 2; **19**, 2; **20**, 2; **21**, 1; **22**, 2; **23**, 3; **24**, 2; **26**, 2; **27**, 4; **28**, 2; **29**, 1; **30**, 1; **31**, 24; **32**, 24; **33**, 12; **34**, 10.

36, *4: Thermochemistry of selenium and tellurium compounds* by *L. Batt*

The review is confined to Se compounds, since no publication on Te compounds was found at the time of the writing the chapter.

Section II deals with the heats of formation (Tables 1 and 2) and bond dissociation energies (Table 3) of organoselenium compounds.

16 references up to 1983.
Complementary: none.
Relevant: **16**, 16; **18**, 3; **19**, 3; **20**, 11; **21**, 3; **22**, 6; **23**, 4; **24**, 3; **25**, 2; **26**, 4; **27**, 9; **28**, 4; **29**, 24; **30**, 2; **31**, 2; **33**, 3.

36, *5: Detection and determination of organic selenium and tellurium compounds* by *K. J. Irgolic and D. Chakraborti*

I. Introduction (162); **II**. Determination of Se/Te inorganic compounds (163); **III**. Identification and determination of Se/Te compounds (165); **IV**. Polarographic and other electrochemical methods (183); **V**. References (185).

Section II reviews qualitative tests for Se and Te and their determination in organic compounds. Section III deals with the identification and determination (mainly by chromatographic methods) of selenols, tellurols, selenenyl compounds (RSeX), trihalides ($RSeX_3$ and $RTeX_3$), seleninic ($RSeO_2H$) and selenonic ($RSeO_3H$) acids, diselenides (R_2Se_2) (Table 1), selenides (R_2Se) (Table 2), tellurides (R_2Te) (Table 3), dihalides (R_2SeX_2 and R_2TeX_2) (Table 4), selenonium and telluronium salts ($Ar_3Se^+X^-$ and $Ar_3Te^+X^-$) (Table 5), selenamino acids and related compounds (Table 6), selenoxides (R_2SeO), selenourea, selenophene, tellurophene and related compounds, piazselenols, Se and Te diethyldithiocarbamates [$(Et_2NCSS)_2Se$ and $(Et_2NCSS)_2Te$] and trialkylsilyl selenides and tellurides.

Section IV treats polarographic and other electrochemical methods for the (mainly qualitative) determination of Se and Te compounds.

181 references up to 1983.
Complementary: none.
Relevant: 1, 5; **2**, 8; **3**, 15; **4**, 3; **5**, 17; **8**, 3; **12**, 4; **13**, 6; **14**, 3; **15**, 3; **16**, 3; **17**, 4; **18**, 5; **20**, 3; **22**, 5; **23**, 5; **24**, 5; **25**, 10. **26**, 5; **28**, 3; **29**, 21; **30**, 14–19.

36, *6: Nuclear magnetic resonance and electron spin resonance studies of organic selenium and tellurium compounds* by *N. P. Luthra and J. D. Odom*

I. Introduction (190); **II**. NMR studies (191); **III**. ESR studies (234); **IV**. Acknowledgements (238); **V**. References (238).

NMR properties of Group VI nuclei are given in Table 1. Section II deals with NMR of organic Se compounds, including relaxation times (Tables 2) and chemical shifts (Table 3), followed by a review of the various classes of Se compounds such as RSeH (Table 4), R_2Se (Table 5), Se-benzylselenoalkanoic acids (Table 6), ArRSe (Table 7), Ar_2Se (Table 8), $(RSe)_2$ (Table 9), $(ArSe)_2$ (Table 10), selenenyl sulphides (Table 11), seleno esters (Table 12), C=Se compounds (Table 13), RSeCN and ArSeCN (Table 14), selenenyl halides (Table 15), Se heterocycles (Tables 16–20), selenium–oxygen compounds (Table 21) and some others (Table 22). Isotope effects on ^{77}Se nuclear shielding are discussed and also coupling constants.

Next, the NMR of Te compounds is reviewed, including relaxation times (Table 23), solvent shifts (Table 24), R_2Te (Table 25), ArRTe (Table 26), Ar_2Te (Table 27), ditellurides (Tables 18–30) Te trihalides (Table 31), Te dihalides (Tables 32) and Te halides (Table 33), Te heterocycles (Tables 34–36) and some others (Tables 37 and 38).

Section III treats the ESR of Se compounds (Table 39).

148 references up to 1984.
Complementary: none.
Relevant: 6, 2.V; **6**, 3.IV; **9**, 1; **13**, 6.III.B; **16**, 3.V; **17**, 4.II; **20**, 3; **23**, 5.VII; **24**, 5.II.D; **26**, 5.III.C; **30**, 2[illegible]; **31**, 25.

36, *7: Mass spectrometry of organic selenium and tellurium compounds* by *G. D. Sturgeon and M. L. Gross*

I. Introduction (244); **II**. Mass spectra (249); **III**. Summary (282); **IV**. Acknowledgement (283); **V**. References (283).

Section I compares relevant characteristics of O, S, Se and Te compounds and reviews ionization energies and the role of charge localization in fragmentation (Table 1).

Section II deals with the mass spectra of R_2Se and R_2Te, ArRSe and ArRTe, Ar_2Se and Ar_2Te, cyclic selenides and tellurides, diselenides and ditellurides, other selenides, dialkyl, cyclic, alkyl aryl and diaryl selenoxides, rearrangements to selenoxides, dialkyl (R_2SeO_2) and diaryl (Ar_2SeO_2) selenones, Se and Te dihalides, Te dicarboxylates, seleninic acids and selenocyanates, carbonate and urea analogues, heteroaromatic Se and Te compounds (Table 2), fulvalene and pentalene analogues and complex heterocycles and organometallics of Se and Te involving also transition metals.

92 referneces up to 1983.
Complementary: none.
Relevant: 9, 7; **10**, 5; **13**, 19; **16**, 4; **17**, 5; **18**, 6; **19**, 5; **22**, 7; **24**, 6; **25**, 4; **26**, 6; **27**, 7; **29**, 3; **30**, 22; **31**, 3; **32**, 3; **33**, 4.

36, *8: Radiation chemistry of organic selenium and tellurium compounds* by *R. Badiello*

I. Introduction (287); **II**. Selenium compounds (287); **III**. Tellurium compounds (302); **IV**. Concluding remarks (303); **V**. References (304).

Section II discusses the radioprotective effects of Se-containing compounds on various biological systems (Table 1) and the radiation chemistry of Se compounds, including transient spectra (Tables 2 and 3) and rate constants (Table 4) of selenourea reactions (Scheme 1), radiolytic products from selenocystine (Table 5) and selenomethionine (Tables 6 and 7), rate constants of reactions of e_{aq}^- and OH with selenoamino acids (Table 8), radical yields by X-ray irradiation of S- and Se-amino acids (Table 9), phosphorescence (Table 10) and fluorescence studies and also interactions between $(CNSe)_2^{-\bullet}$ and $SeO_3^{-\bullet}$ on the one hand and various amino acids on the other (which is relevant to the inactivation of some enzymes by these radical anions).

Section III reviews the radiation chemistry of Te compounds.

75 references up to 1980.
Complementary: none.
Relevant: 9, 8; **10**, 6; **11**, 6; **13**, 17; **15**, 19; **16**, 12; **17**, 10; **18**, 11; **19**, 10; **22**, 11; **25**, 12; **27**, 23, 24; **39**, 7; **31**, 6; **32**, 9; **33**, 21.

36, *9: Selenium-stabilized carbenium ions and free radicals* by *L. Hevesi*

I. Introduction (307); **II**. Selenium-stabilized carbenium ions (308); **III**. α-Seleno free radicals (321); **IV**. References (328).

Section II deals with cyclic selenocarbenium ions, including 1, 2-diselenolylium ions (Table 1), 1, 3-dichalcogenolylium ions (NMR data: Tables 2 and 3), 1, 3-diselenolan-2-ylium and 1, 3-thiaselenolon-2-yliun ions and some other cationic systems. Next, aliphatic selenocarbenium ions are treated and examples of the seleno-Pummerer rearrangement are tabulated, including conditions, key (cationic) intermediates and products (Table 4). Bis(seleno)carbonium ions and selenoallyl cations are also discussed.

Section III treats ESR investigations of Se-containing free radicals (Scheme 9 and Tables 5 and 6).

73 references up to 1984.
Complementary: none.
Relevant: none.

36, *10: Selenium- and tellurium-containing organic polymers* by *Y. Okamoto*

The general routes to produce Se-containing polymers are

$$X(CH_2)_nX + Na_2Se \longrightarrow \text{⟮}(CH_2)_nSe\text{⟯}_m$$

$$X(CH_2)_nX + Na_2Se_2{}^- \longrightarrow [(CH_2)_nSe_2]_m$$

$$NCSeASeCN \longrightarrow HOSeASeOH \longrightarrow \text{⟮}SeASe\text{⟯}_n$$

and ring-opening polymerization, e.g.

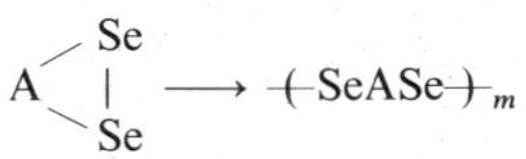

(where A is a divalent alkylene or arylene radical).

Section II deals with polyselenoalkylenes, $(C_nH_{2n}Se)_n$, O- and Se-containing copolymers, $(CH_2O(CH_2Se)_2)_n$, poly(seleno-*p*-xylene), poly(seleno-1,4-phenylene) ('PPSe'), $(C_6H_4Se)_n$ (Table 1), poly(seleno-4, 4-biphenylene), $(C_6H_4C_6H_4Se)_n$, poly(seleno-2, 5-thienylene) and also with poly(carbondiselenide), $[C({=}Se)Se]_n$.

Section III treats poly(diselenoalkylene)s, $(C_nH_{2n}Se_2)_n$. Section IV reviews briefly polymers of the type $(ASe_x)_n$ and Section V presents the less well known organotellurium polymers, e.g. $[Te(Ph_2)OOCACOOTe(Ph)_2]_n$.

38 references up to 1983.
Complementary: none.
Relevant: none.

36, *11: Organometallic compounds with selenium and tellurium atoms bonded to Main Group elements of Groups IIIa, IVa and Va* by *R. A. Zingaro*

Section II reviews compounds containing the bonds B—Se [NMR of $(RB)_2Se_3$, Table 1], B—Te, Al—Se, Al—Te, Ga—Se and Ga—Te.

Section III deals with compounds containing Si—Se or Si—Te bonds, and also Ge—Se, Ge—Te, Sn—Se, Sn—Te and Pb—Se bonds. Table 2 gives NMR parameters of compounds in which Se is bonded to Si, Ge or Sn, Table 3 those of some chloro(methyl)seleno derivatives of Si and Ge and Table 4 results of studies of reactions of $Me_3SiSeMe$ with various halides containing Si, Ge or Sn.

Section IV treats compounds with P—Se, P—Te, As—Se and As—Te bonds and tabulates reactions between Me_3SnH and complexes which contain bonds of the above-mentioned types in ligands coordinated to $Cr(CO)_5$ or $Mo(CO)_5$ (Table 5).

73 references up to 1983.
Complementary: none.
Relevant: none.

36, *12: Synthesis and uses of isotopically labelled selenium and tellurium compounds* by *K. Fujimori and S. Oae*

Section II deals with the preparation of labelled Se or Te derivatives, including inorganic compounds, acyclic and cyclic selenols, diselenides, selenides, tellurides and seleno amino acids.

Section III describes physicochemical uses of Se/Te-labelled compounds for IR, Raman, microwave and NMR studies. Next, mechanistic studies of organic Se/Te compounds, including isotope exchange and photochemical reactions are discussed, followed by biochemical uses including the study of the metabolism of Se/Te compounds and of selenoproteins, Se-containing t-RNA and Se analogues of non-haem iron proteins.

The medicinal uses of compounds containing Se and Te include scintiscanning employing a variety of Se- and Te-labelled amino acid, fatty acid and other natural product analogues. Similar compounds have also been used in immunological and other studies.

160 references up to 1983.
Complementary: none.
Relevant: **5**, 10; **12**, 13; **13**, 15; **17**, 12; **18**, 18; **19**, 17; **22**, 14; **23**, 16; **24**, 20; **26**, 11; **27**, 10; **28**, 8; **29**, 28; **33**, 18.

36, *13: Selenium and tellurium heterocycles* by *M. Renson*

The chapter deals with heterocycles containing divalent Se or Te. Tetravelent Se/Te derivatives are described in **36**, 14.

Section II treats three-membered rings containing one Se/Te atom. Section III describes four-membered rings having one Se/Te atom, or O and Se, S and Se, two Se or two Te atoms or three heteroatoms.

Section IV deals with the synthesis, aromaticity and reactions of selenophenes and tellurophenes, with benzoselenophenes and dibenzoselenophenes and with selenophenes condensed with heterocycles containing O, S or N atoms. Next, tellurophenes condensed to homo- and heterocyclic rings are presented, followed by five-membered rings with two, three or four heteroatoms (Se/Te and also O, S and N as heteroatoms).

Section V discusses six-membered Se/Te rings. Systems containing one Se or Te atom in a single ring and condensed ring systems with additional benzo-, naphtho- or heterocyclic rings are reviewed. Next, monocyclic systems with two heteroatoms (either two Se or Te atoms or one of these together with O or S) are described, followed by treatment of benzo- and dibenzo-condensed six-membered Se/Te compounds. Six-membered heterocycles containing either Se/Te and also N are reviewed next, followed by compounds with three heteroatoms (one or more Se/Te and also O, S or N atoms) and with four heteroatoms.

Section VI deals similarly with seven-membered rings containing one Se/Te atom (benzo- and dibenzoselenepins), two heteroatoms with or without N and three heteroatoms (Te and two oxygens or Se and Two nitrogens).

Sections VII–IX describe briefly eight-membered rings, macrocycles, spiroheterocycles and bridged Se/Te heterocycles.

1075 references up to 1983.
Complementary: none.
Relevant: none.

36, *14: Tetra- and higher-valent (hypervalent) derivatives of selenium and tellurium* by *J. Bergman, L. Engman and J. Sidén*

Section I reviews the nomenclature and structure of 4- and 6-valent derivatives of Se/Te.

Section II deals with general aspects and preparation of $ArZX_3$, Ar_2ZX_2 and Ar_3ZX (where Z = Se or Te and X = halogen), with reaction of ZX_4 and RZX_3 with carbonyl compounds, e.g.

$$PhCOMe + ZCl_4 \rightarrow PhCOCH_2ZCl_3 \xrightarrow{PhCOMe} (PhCOCH_2)_2ZCl_2$$

with reactions of ZX_4 and RZX_3 with alkenes and acetylenes and the regio- and sterochemistry of the adducts produced.

Section III treats the preparation and properties of tetraorganyl Se and Te compounds and Section IV the organic derivatives of tetravalent Se and Te which contain also Z—O, Z—S or Z—N bonds, such as R_2TeO, $Se(OR)_4$, $Te(SR)_4$, $Se(SR)_4$ and $ArSO_2N{=}Se{=}NSO_2Ar$, etc.

Section V deals with tetrahaloorganyl tellurates which are readily obtained: $ArTeX_3 + Y^+X^- \rightarrow Y^+ + (ArTeX_4)^-$. Related aliphatic tellurates are known, but have not been studied systematically.

Section VI discusses the general aspects of hexavalent Se/Te derivatives and their preparation, structure and reactions. These include selenones (R_2SeO_2) and tellurones (R_2TeO_2), which are the analogues of sulphones (R_2SO_2). Selenones have been studied extensively and are used in organic syntheses. The existence of aliphatic tellurones has been questioned, but the aromatic ones [e.g. Ar_2TeO_2 and $PhTe(R)O_2$] have been studied.

Section VII reviews some 4- and 6-valent Se/Te compounds which contain Se—Se or Te—Te bonds.

290 references up to 1984.
Complementary: none.
Relevant: none.

36, *15: Directing and activating effects involving selenium and tellurium* by *F. Fringuelli and A. Taticchi*

Section II discusses the various types of substituent constants for Se/Te-containing groups, including constants for groups with bivalent chalcogen atoms (Table 1) and tetravalent Se and Te atoms (Table 2), constants for heterocyclic groups containing chalcogen atoms (Table 3) and 'replacement constants' σ_{Ar} or σ_a in which the Se or Te atom of the heterocycle is considered as a substituent *replacing* a —CH=CH— moiety in the benzene ring (Table 4). 'Special' substituent constants are used when the concept of replacement is extended to other aromatic rings besides benzene (Table 5), and hydrophobic substituent constants (Table 6) are for groups in which the chalcogen atom is bound to an alkyl (Me, Et, Pr) or to a $—CH_2COOH$ group.

Section III deals with the reaction constant ρ, with electronic transmission through heterocyclic systems (Table 7) and through Se or Te atoms (Tables 8–10), with substituent effects in electrophilic and nucleophilic aromatic reactions (Tables 10 and 11) and with NMR studies of substituent effects (Tables 13 and 14).

Section IV treats electrophilic aromatic substitutions in five-membered O, S, Se and Te heterocycles (Tables 15 and 16), isomer distributions in their benzo derivatives (Table 17) and in other fused ring systems (Tables 18–20), nucleophilic aromatic substitutions and formation of Meisenheimer complexes (Table 21), Diels–Alder and other addition reactions, side-chain reactions of heteroaryl derivatives (Tables 22 and 23), protophilic reactions, metalations and finally equilibrium reactions such as ionizations of heterocyclic OH, COOH and NH_2 derivatives (Table 24) and pK_{BH^+} values of selenachromone, selenaxanthone and their O and S analogues.

Section V describes seleninic acids (pK values in Table 26), selenols and thiols (Table 27), onium salts and their electrophilic substitutions (Tables 28–30), selenonium and telluronium ylides, selenides and tellurides and the acidity of their α-CH groups (Table 31), α:γ ratios in the alkylation of allyl selenide anions (Table 32), cyanohydrin equilibria of benzaldehydes substituted by —SeMe, —SMe and —OMe groups (Table 33) and finally effects on various reactions of selenoxides and selenones.

335 references up to 1983.
Complementary: **3**, 3; **13**, 8; **18**, 9; **28**, 9.
Relevant: **4**, 5; **5**, 16; **6**, 8; **7**, 2, 3; **11**, 12; **12**, 5; **14**, 4; **15**, 5; **16**, 6; **19**, 8; **23**, 11; **29**, 16; **31**, 8.

36, *16: Functional groups containing selenium and tellurium in various oxidation states* by *N. Sonoda and A. Ogawa*

Section II reviews the preparation of selenols and selenolates (Table 1), their reactions with alkylating agents and their use in reductions, ring openings and additions.

Section III deals with selenenic acids (RSeOH) and their anhydrides, including their preparation and reactions, and with selenenyl halides (RSeHal). The reactions of the latter are described with carbonyl compounds (leading to the corresponding α,β-unsaturated carbonyl derivatives (Table 2) and unsaturated compounds (alkenes, alkynes, allenes), followed by cyclofunctionalization (i.e. electrophilic addition of a PhSe group to an olefin possessing nucleophilic moieties such as OH and COOH, followed by ring closure to yield stereospecifically cyclized products) (Table 3). In Section III.C the synthesis and reactions of selenenic esters (RSeOR′) and selenocarboxylates (RSeOCOR′), selenenamides ($RSeNR'_2$) and selenocyacanides (RSeCN) are presented.

Section IV treats the synthesis of seleninic acids [RSe(=O)OH] and anhydrides (RSe(=O)—OSe(=O)R) and their uses in oxidation of alcohols, α-hydroxylation of ketones, oxidation of benzylic hydrocarbons, epoxidation of olefins, dehydrogenation, oxidation of N-, S- and P-containing compounds and deprotection of ketones and aldehydes from their thioacetals.

Section V presents the synthesis and reactions of selenonic acids ($RSeO_3H$).

Section VI discusses tellurols (RTeH), tellurenyl compounds (RTeHal, ArTeCN, etc.), tellurinyl compounds [$RTeX_3$(X = Hal, OH, etc.), R_2TeX_2, R_3TeX, RTe(O)X] and telluronyl compounds [RTe(=O)Ar], including methods of preparation and reactions.

492 references up to 1983.
Complementary: none.
Relevant: none.

36, *17: Stereochemistry and chiroptical properties of organic selenium and tellurium compounds* by *G. Snatzke*

Section II discusses selenonium and telluronium compounds, selenoxides, diselenides and ditellurides, in which the chalcogen atom is part of the chirality element.

Section III describes selenides, tellurides, selenoketones (R^1R^2C=Se), biopolymers and selenophosphinic acid derivatives, in which the chalcogen is not part of the chirality element.

ORD and CD, Cotton effects and the geometry of the molecule are treated.

38 references up to 1984.
Complementary: none.
Relevant: **8**, 4; **18**, 7; **19**, 6; **20**, 2; **22**, 4; **25**, 3; **26**, 3; **27**, 6; **28**, 10; **29**, 23; **31**, 1.

36, *18: Ligand properties of organic selenium and tellurium compounds* by *H. J. Gysling*

Section I reviews general aspects of complexes containing Se or Te ligands and the notations for ligand bonding modes (Table 1) and lists the abbreviations used in the chapter.

Section II.A deals with neutral monodentate Se ligands, including selenourea (Table 2), substituted (Table 3) and cyclic (Table 4) selenourea derivatives and Ni^{II} complexes of polymethylene bis(phenylselenoureas) (Table 5), complexes of triorganophosphine selenides ($Se{=}PR_3$, Table 6) and selenium heterocycles (Tables 7–13), complexes of RSeSeR and related ligands (Table 14) and $Se(EMe_3)_2$ where E = Ge, Sn or Pb, transition metal complexes of selenocarbonyl derivatives (Table 15) and diorganoselenoxides (R_2SeO) (Table 16), Se-functionalized carbene complexes (Table 17) and miscellaneous monodentate ligands (Table 18).

Section II.B describes anionic monodentate ligands including selenocyanates and isoselenocyanates (Tables 19–25), selenols as terminal (Table 26) and as bridging (Table 27) ligands, SeH^- ligands, organoseleninates ($RSeO_2^-$) (Tables 28 and 29), dialkylselenocarbamates [SeC(O)NR_2] (Table 30) and some other less important ligands.

Section II.C treats neutral bidentate ligands such as (Se, C) donors, including CSe_2, CSSe and CH_2Se, (Se, N) and (Se, P) donors (Table 31) and also (Se, O), (Se, S) and (Se, Se) donors (Table 32).

Section II.D discusses anionic bidentate ligands, including $(Se, N)^-$ and $(Se, P)^-$ donors (Table 33) and $(Se, O)^-$, $(Se, S)^-$ and $(Se, Se)^-$ donors (Fig. 10 and Tables 32–42), presenting metal complexes with selenocarbamates (Table 35), selenoxanthates, -thiophosphates, -thiophosphinates, -thiocarbamates, etc. (Tables 36–38), dialkyldiselenocarbamates (Tables 39 and 40), 1, 1-diseleno ligands (Table 41) and 1, 2- and 1, 3-diseleno-type ligands (Table 42).

Section II.E deals with tridentate and Section II.F with tetradentate (Table 43) ligands. Section II.G describes complexes incorporating $Se_n (n = 1, 2, 4, 5$; Fig. 11) ligands μ_2-Se, μ^2-Se_2, μ^2-Se_2, $\mu_2\eta^2$-Se_2, μ_2-Se_4 and η^2-Se_5 ligands and also transition metal cluster compounds containing μ_3-Se ligands (Table 44).

Section III tabulates complexes with monodentate Te ligands (Table 45). Section III.A describes neutral monodentate ligands such as dialkyl and diaryl tellurides, diorganoditellurides and some others. Section III.B deals with anionic monodentate Te ligands, mainly tellurols, and their complexes (Tables 46 and 47), Section II.C with bidentate ligands of telluroformaldehyde and of telluroketene (Table 48) and Section III.D with complexes incorporating $Te_n (n = 1, 2)$, including dimers and cluster compounds (Table 49), and also ^{125}Te NMR data for metal complexes containing Te ligands (Table 50).

Throughout the chapter detailed molecular structures of complexes are presented, together with reaction schemes for their syntheses and chemical behaviour.

675 references up to 1985.
Complementary: none.
Relevant: **1**, 6; **9**, 5; **19**, 9; **30**, 4–7.

Author index

Subject index